Handheld XRF for Art and Archaeology

Studies in Archaeological Sciences 3

The series Studies in Archaeological Sciences presents state-of-the-art methodological, technical or material science contributions to Archaeological Sciences. The series aims to reconstruct the integrated story of human and material culture through time and testifies to the necessity of inter- and multidisciplinary research in cultural heritage studies.

Handheld XRF for Art and Archaeology

Edited by
Aaron N. Shugar
Jennifer L. Mass

Leuven University Press

ISBN hardcover 978 90 5867 907 9
ISBN paperback 978 90 5867 934 5

D / 2012 / 1869 / 54
NUR: 682/950

Lay-out: Friedemann BVBA (Hasselt)
Cover: Jurgen Leemans

Acknowledgements

The editors would like to give their profound thanks to Dr. Bruce Kaiser for helpful discussions, for his encouragement, and for his generous support of cultural heritage science (including the Anaheim meeting to discuss initial contributions to this volume). We are also grateful to the authors of each of the following chapters for their perseverance and dedication to this endeavor.

A.N.S. would like to thank my co-editor Jennifer Mass for being a pleasure to work with. She made editing the volume all the more enjoyable and had been, and always will be a good friend. I would like to give a special thanks to my wife Natalie. She has been patient, understanding, forgiving, and tolerant of me during this entire process. Even better, she makes every day worth waking up to. I would also like to thank my son Ethan, who, although he does not realize it yet, had to live with limited access to me (and me from him) for quite some time.

J.L.M. would like to thank my co-editor Aaron Shugar for his vision in initiating this timely project, for his invitation to collaborate, for his patience every time I left the country in the middle of the project, and above all for his valued friendship. I would also like to thank Winterthur Museum Director of Conservation Lois Price and Winterthur/ University of Delaware Program Director Debra Hess Norris for their encouragement and support of my participation in this project. Winterthur Scientific Research and Analysis Laboratory scientists Catherine Matsen, W. Christian Petersen, and Joseph Weber are greatly appreciated for their patience during my extended office-bound bouts of editing. Finally, I would like to thank my husband Adam for his extensive X-ray physics expertise, the many weekends that he graciously lost me to this project, and for his unwavering support and encouragement.

Table of Contents

List of Illustrations (with partial captions)

List of Tables (with partial captions)

Chapter 1

Introduction

Aaron N. Shugar and Jennifer L. Mass

Introduction

Over the last decade the advancement of X-ray generation and detection technology has allowed the instrumentation of X-ray fluorescence (XRF) to evolve from laboratory-based standalone units to highly portable and lightweight handheld devices (Piorek 1998; Potts, 2008). These instruments [termed handheld XRF or portable XRF (PXRF) spectrometers] have given researchers in the fields of art conservation and archaeology the opportunity to study many of the materials encountered in their work with greater accessibility and flexibility than ever before. As a result, handheld XRF represents a great potential advance for these fields. Handheld XRF spectrometers allow, for example, for the nondestructive and *in situ* major and minor elemental analysis of objects found during excavations, the soils surrounding these objects, monumental sculptures, and the materials employed in the construction and finishes of standing buildings. They also assist in the preservation of material culture by allowing the spectrometer to travel to the object requiring analysis rather than having to incur the risks of having delicate, unstable, or friable objects travel to the spectrometer, the cost of insuring the object for travel, and the potential loss of access for museum viewers if the object needs to be taken off display. The low relative cost of these spectrometers (compared to the laboratory-based instruments) has prompted many museums, academic institutions, and cultural centers to invest in handheld XRF for general elemental analysis/materials characterization purposes. It is also notable that the low cost of these instruments has meant that, for the first time, cultural institutions without scientific staff have in large numbers started to pursue collections-based materials research using XRF. Speakman *et al*. (2011) conservatively estimate over 200 handheld spectrometers are currently being used in such situations. We estimate that the total worldwide use of handheld XRF in cultural heritage applications approaches 1500 units or more. The broad dissemination of XRF technology into cultural heritage institutions is an exciting advance for the understanding and preservation of art and archaeological collections. It has the potential for allowing one to rapidly address questions of condition, authenticity, technology of manufacture, and (somewhat less commonly) provenance. However, while handheld XRF instruments can greatly increase access to the technology and simplify data collection, the proper selection of analysis conditions and interpretation of the data generated still require a detailed understanding of the principles of X-ray spectrometry that govern the technique. For example, when X-ray photons interact with an object five different phenomena occur simultaneously: incident beam absorption, elastic scattering (Rayleigh), inelastic scattering (Compton), diffraction (Bragg), and X-ray fluorescence. All of these phenomena except absorption will be detected during an XRF experiment, and it is critical that they can be identified by the researcher. Too often these instruments are sold and used as rapid "point and shoot" solutions for materials analysis problems in cultural heritage with little training on X-ray physics and spectrometry, and the identification of instrument artifacts. In addition, there is insufficient discussion on XRF being a surface analysis technique only (the penetration depth of the X-rays varies depending on the object's density and the tube voltage), and that cultural heritage objects almost invariably have complex, layered surfaces with heterogeneous microstructures. As a result, they often generate deceptive qualitative and erroneous quantitative data. This is especially a concern given the broad range of materials encountered in art conservation and archaeology, and their widely varying conditions/states of alteration and degradation. Artists' materials start out as highly complex composite materials, often inorganic-organic composites, and once they have been subjected to hundreds or thousands of years of alteration they are often characterized by features such as high levels of surface porosity and/or layered microstructures that make them particularly unsuitable candidates for a nondestructive surface analysis technique such as XRF. This is particularly true when quantitative data about the bulk composition of the object is desired.

The total number of publications that have used handheld XRF as their main analytical technique is growing so rapidly as to be nearly exponential. There have been several volumes and articles published in the last few years dealing specifically with X-ray fluorescence analysis of art objects and archaeological materials (e.g. Mantler and Schreiner 2000; Schreiner *et al.* 2004; Potts and West 2008; Shackley 2011).

A quick look at the Art and Archaeology Technological Abstracts (AATA) database (The Getty Conservation Institute 2012) indicates that prior to 2005 (see Table 1.1) there were no publications related to art or archaeology which had handheld XRF in their abstracts (Note that AATA cannot be taken as a comprehensive list of all the technical publications that deal with art and archaeological material, but it is a good source for observing general trends in the fields). In the five years following, there were six publications using handheld XRF and 38 using portable XRF (conservation laboratory-based spectrometers on wheels that can, if necessary, be moved into nearby galleries). There has also been a steady proliferation of papers, sessions, and workshops at professional conferences in art conservation and archaeology focused on handheld XRF applications (see The Getty Conservation Institute 2006; Sirois 2007; Podsiki 2009; Society for Historical Archaeology 2012 for examples of workshops focusing on the use of handheld XRF in the fields of art and archaeology). Although the data is not in yet for the years 2010-2012, there is already an indication that the number of publications will continue to increase rapidly.

Date	Portable XRF	Handheld XRF	Total XRF
2010-	14	2	54
2005-2009	35	5	218
2000-2004	20		151
1995-1999	7		50
1990-1994	1		64
1985-1989	2		73
1980-1984	0		28
1975-1979	3		26
1970-1974	2		5

Table. 1.1
Number of publications including the terms "Portable XRF", "Handheld XRF", and "XRF" in the AATA database. Numbers for 2010 are skewed as they only compromise a partial year of data (accessed October 2011).

Discussions concerning the proper implementation and limitations of these relatively new handheld XRF instruments have been largely missing from the literature to date. Topics that need to be addressed include the selection of appropriate analysis conditions, the identification of spectral artifacts, the selection of proper calibration standards to ensure the best possible quantitative results based on the materials investigated, and the field conditions in which the instrument can be used to obtain reliable data. Additional topics that should be discussed include deciding when some surface preparation (such as polishing) is required prior to analysis to obtain information about the bulk material, designing appropriate experimental setups to limit background interferences and high X-ray scatter, and attempting to quantify the errors introduced by working with the non-ideal analysis conditions imposed by nondestructive analysis (such as curved geometries, or small objects).

The impetus for putting together this volume on handheld X-ray fluorescence (XRF) is the recent proliferation of these spectrometers in cultural heritage institutions, where they are often incorporated into collections-based research despite the fact that operators often have minimal training in the appropriate applications of the technique and the interpretation of the resulting data. What has become apparent when reading the current literature that incorporates handheld XRF data on cultural materials, is that there is often a clear misunderstanding of what researchers expected to obtain from their analysis, and what was possible to achieve. This problem has become even more relevant, and more pressing, with the recent development, miniaturization, and accessibility of analytical instrumentation (not only XRF spectrometers but Raman and FTIR spectrometers as well). In addition, the push in industry to create single "push button" instruments to provide rapid raw quantitative data using manufacturer calibrations (often in instruments originally designed for use sorting modern scrap metals) has aided in the perception that users can collect accurate quantitative data from their analytes without any development of a customized calibration. Also lacking are sufficient training in the fundamentals of instrumental analysis, quantitative analysis, spectroscopy, or X-ray physics. In our collective goal of maintaining the quality and usefulness of the data collected and reported on cultural heritage materials (which of course directly impacts preservation decisions made on these materials), data that will amenable to and accessible for future mining and interpretation, this should be seen as a major step backward for the field.

In spite of these issues, handheld XRF has the potential to be a great advance for the art conservation and archaeology communities, opening the door to analysis for many museums, academic departments, curators, conservators, and archaeologists. The ability to rapidly, non-invasively, and non-destructively characterize cultural artifacts is truly exciting, as is the potential to research artifacts that have been previously inaccessible due to their fragility or locations. This technology unlocks a world of opportunities for a better understanding and exploration of artists' materials, allows for the development of new theories of trade interactions, the establishment of alloy compositions of immovable statuary and architectural metalwork, and the study of artifacts *in situ* that could never before be investigated without sampling (decreasing the sampling of cultural heritage objects is a major current initiative in cultural heritage preservation). In addition, the ability to collect data both non-invasively and non-destructively means that, unlike analytical techniques which require removal of samples for analysis, often restricting the number of samples, or location from which samples can be taken, an unlimited number of analysis spots can be studied and these analysis spots can include prominent features on the object. For conservators, it can help address questions regarding prior intervations and can guide them towards better and safer treatment protocols. For archaeologists, handheld XRF can be taken into the field and help direct excavation in real time. All of this new research will result in publications leading to the expansion and improvement of the knowledge base in both fields. It is the goal of this volume to ensure that this new research and its interpretation is achieved using rigorous analytical methodologies and a sound background in the theory and practical application of X-ray fluorescence.

This volume focuses specifically on the applications, possibilities, and limitations of handheld XRF in the fields of art conservation and archaeology. It is important to note that all handheld XRF instruments work under the same principles of X-ray physics despite the instrument manufacturer. While there are differences in how the spectrometers are assembled and the software provided, they all deliver similar data and have similar issues associated with obtaining high quality results. Therefore, the discussions concerning how to obtain useful results from specific XRF spectrometers in the following chapters can, and should be applied to any manufacturer's instrument.

The papers presented here focus on the experimental methodologies, protocols, and limitations of handheld XRF analysis in dealing with the complexity of the individual materials encountered in conservation and archaeological research, including glasses, ceramics, metals, paintings, manuscripts, photographs, paper, soils and on site application for archaeological materials. They are not intended to be isolated studies providing the results of analysis for a particular class of material. They are instead intended to be guides that present and address current problems in the material being examined, the issues confronted when characterizing this material using handheld XRF, and the methodologies used to overcome these issues (as well as a discussion of the problems that simply cannot be resolved using this technique alone). In many of the chapters, the authors encounter similar issues even though they are studying different materials (such as the inherent difficulties of quantification of layered or corroded surface structures, which applies to historic glasses, manuscripts, paintings, ceramics, and silver) but these issues are resolved uniquely in relationship to the specific material being investigated. In this way, each chapter can be looked at as a suggested guideline for best practices for any researcher considering investigating these materials with handheld XRF. While detailed case studies are presented, the main intention of this volume is to provide standards of best practice for the XRF examination of the materials encountered in cultural heritage research, and also to highlight the limitations of this surface analysis technique as it is currently applied, so that misinterpretation and over-interpretation of data can be avoided.

In chapter 2 Smith develops a methodology for quantifying the elemental compositions of Renaissance bronzes. He provides details concerning the proper methods of data acquisition and the potential difficulties that one can encounter if not properly prepared with appropriate calibration standards. Stulik and Kaplan address the challenges associated with investigating photographs in chapter 3. Their long-term research in this area has resulted in the development of specific thin film standards to aid in the proper identification of photographs using quantitative XRF, including identifying the imaging materials and the different toners used to modify and enhance images. The complex problems associated with investigation of paintings using XRF and handheld XRF are dealt with by McGlinchey in chapter 4. These layered structures offer one of the more difficult tasks for spectral interpretation in cultural heritage research. Recommendations concerning the limitations of XRF for these works and appropriate expectations for one's results are discussed. High Z and low Z analysis protocols are also provided. In chapter 5, Trentelman, Patterson and Turner focus on the delicate nature of investigating illuminated manuscripts with X-rays. The discussion ranges from creating interdisciplinary working relationships to building custom supports to ensure artifact safety during analysis. The inherent difficulties of elemental characterization of works of art on paper are also discussed (including obtaining data from both sides of the page simultaneously), and recommendations are made to help researchers glean as much information as possible from the data collected. Barrett, Shannon, Wade and Lang take on the complicated task of creating a calibration allowing for the quantitative characterization of the chemistry of historical papers in chapter 6. This chapter discusses the physics behind the process and offers the development of a methodology to obtain useful data from objects of variable thickness. The issues related to establishing limits of detection are discussed and the manufacture of a custom support is presented. Although the study of metal artifacts is in many ways an ideal fit for handheld XRF analysis, a number of complications can arise, including altered/depleted surface microstructures and surface oxidation, both of which can create substantial problems for quantification and bulk composition determination. These issues are discussed by Mass and Matsen in chapter 7 in dealing with the quantitative analysis of historic silver hollow wares. Bezur and Casadio take on the challenging task of analyzing the complex layered structures of porcelains in chapter 8. They carefully describe the physics behind establishing X-ray depth penetrations and extracting the relevant information from the analysis to describe the differences found between Du Paquier porcelains and Böttger porcelains. Chapter 9 presents the challenging issues related to the identification of heavy metal pesticides in ethnographic collections. Shugar and Sirois review these issues and present the current expectations for the use of XRF to identify and quantify these persistent poisons that make repatriation and conservation of these collections so challenging. Recommendations are made regarding the proper calibration of instrumentation and the limitations of the technique. In chapter 10 Donais and George present the beneficial use of XRF as a field instrument on archaeological sites. A detailed methodology is presented to ensure proper recording of location-specific data and *in situ* interpretation of results. The advantages of having real-time data to aid archaeologists in answering questions on site, which might help better direct excavation, are also discussed. Neff, Voorhies and Paredes Umaña present the use of handheld XRF in the analysis of soils from archaeological sites in chapter 11. Post-collection data processing is a focal point of this chapter including a discussion of how to extract the most useful information possible using statistics and principle component analysis. The analysis of obsidian and the calibration of handheld XRF to provide useful quantitative data for this important class of archaeological artifacts are discussed by Ferguson in chapter 12. Taking the compositional data of known samples collected using NAA and XRF, Ferguson provides a thorough overview of the issues related to calibration of handheld XRF, and discusses the potential problems one encounters when dealing with small samples and unknown variation in artifact composition. In chapter 13 Aimers, Farthing and Shugar review the creation of a calibration for the analysis of archaeological ceramics and include a comparison to laboratory-based WD-XRF results. The process of preparing appropriate reference materials is detailed along with an explanation of its importance for obtaining meaningful and reproducible results using handheld XRF. Finally, in chapter 14, Kaiser and Shugar discuss the issues involved in obtaining useful XRF data from glass artifacts and offer suggestions to mitigate or eliminate these issues. In addition, they probe the limits of the technique's potential in the investigation of historical and archaeological glass. Calibration issues are discussed as are the alkali-depleted surfaces of these glasses and the manipulation of the instrumental setup to achieve higher count rates for low Z elements.

Application of commercial X-Ray fluorescence spectrometers to the study of objects of art

An understanding of the basic physics behind XRF is necessary for anyone intending to interpret XRF spectra and to use XRF for the study of cultural heritage objects, and will make the chapters that follow all the more useful. Such a discussion is beyond the scope of the current volume, but it has been well reviewed in the scientific literature (see, for example, Jenkins and Vries 1973; Jenkins, Gould, and Gedcke, 1995; Jenkins 1999). We will instead provide an overview the application of XRF for the study of cultural materials, the rise of handheld XRF spectrometers, an overview of the X-ray physics necessary to understand the spectral artifacts encountered in XRF data, and the proper instrument settings to reduce these artifacts.

Although this volume deals with the use of handheld XRF in the fields of art and archaeology, handheld XRF has been used extensively in other fields of study. There is a long history of its use in geology for the characterization of rocks, ores and soils as well as the identification of precious metals and trace elemental analysis (Shackley 2011). Handheld XRF has recently seen rapid development as a result of the recycling industry, where it is commonly used for alloy identification and waste processing. In the last several years handheld XRF has also established itself as a go – to – tool for environmental remediation, being used for the identification of lead in paint and measuring heavy metal contents in soil (EPA method 6200; EPA 2007) for example. Although the potential benefits of using handheld XRF in art and archaeology are clearly apparent, our fields are in reality included in a group of users (including forensics and industrial hygiene), that are considered "novel users" by the manufacturers. As with the other analytical instrumentation employed in our fields, we do not comprise major sales revenue and so the instruments are not specifically designed for our purposes. Our work does, however, provide very useful insights into the limitations of these instruments for manufacturers. The materials we study and the problems we encounter typically have little resemblance to the materials that are industry's focus. Not only are our materials constructed differently than modern materials, but hundreds or thousands of years of use or deposition along with chemical and physical aging over that time have given them surface structures completely unlike their modern analogues. Since modern industries (such as alloy sorting – a billion dollar business) do not have to confront the broad range of alloy compositions encountered in cultural heritage materials, manufacturers have dedicated their calibrations and instrumental parameters to support the characterization of modern materials only. *This alone is a major source of difficulty for our fields in attempting to rapidly acquire useful quantitative data from these instruments – our alloys are typically outside of the range of the calibrations provided with the instruments, typically rendering these calibrations useless for our purposes.* An additional factor that sets us apart from industrial use is that for cultural heritage objects it is often impossible to do any sample preparation – our work typically has to be performed in a non-destructive and non-invasive manner regardless of the objects' surface morphology or condition. This is in contrast to the amount of sample preparation typically required for reproducible XRF analysis (such as, for example, borate bead preparation for silicates).

Analytical questions posed by cultural heritage researchers

The typical questions that cultural heritage researchers have when attempting to obtain data from handheld XRF spectrometers fall into two categories: those that can be addressed through qualitative analysis, and those that can be addressed through quantitative analysis. In some cases the questions are straightforward including:

- What is this object prepared from/what material is on the surface of this object?
- Is a particular element present in this object?
- What alloy is this metal (e.g. bronze or brass)?
- Is this darkened soil found during an excavation contaminated with heavy metals?

These questions can often be answered fairly rapidly and comprise ideal uses of handheld XRF for qualitative analysis (assuming the minimum detection limits of the instrument are known). In fact, this is often the case during an archaeological excavation where a quick qualitative or comparative analysis can help direct further excavations and ultimately save time and money (for example by noting that the soil three meters north of a trench has demonstrably less phosphorus than the soil one meter away from this trench; see Chapter 10 by Donais and George for examples of on-site XRF use). But very often, the questions asked by cultural heritage professionals require far more detailed investigations and more rigorous experimental methodologies, such as:

- What are all the pigments and fillers present in this painting?
- How do the compositions of the four different components of this artifact differ?
- What are the trace element concentrations of this ceramic?
- What is the provenance of this piece of glass, ceramic, obsidian, or colorant?

- Is this material safe to handle given that is has likely been exposed to inorganic pesticides containing lead, mercury, barium, and/or arsenic?
- Why did this alloy fail?
- What is causing this pigment to alter or change color?

These questions offer up an array of problems for the museum scientist or researcher to take on when confronted with artifacts that cannot be sampled, or sometimes altered in any fashion (cleaned or polished). A brief overview some of these issues follows (and they are all discussed in detail in various chapters of this volume). One of the most important things to note is that XRF provides elemental data only, but that artists' pigments are often composed of compounds that are made up from the same elements (for example red lead Pb_3O_4 and litharge PbO), and so XRF cannot discriminate between these compounds. XRF cannot offer the stoichiometry for these compounds or identify any low atomic number counterions (distinguishing, for example between a copper carbonate and a copper acetate). This is a result of handheld XRF not being able to identify elements lighter than Al (in most cases; see chapter 14 by Kaiser and Shugar for details on how to overcome this issue and obtain data for elements as low as Na) leaving the carbon- and oxygen-containing components out of the data set altogether (see chapters 4 by McGlinchey, and 5 by Trentelman, Patterson and Turner for discussions of how this impacts the analysis of pigments).

The question of addressing the composition of an artifact might seem straightforward enough but one must investigate deeper and ask several further questions to ensure the results obtained will be adequate. The most critical question is: is the surface composition of the object likely to represent the bulk? Recall that XRF is a surface analysis technique only (the emission depth of the fluorescent X-rays is provided by:

$$I/I_0=e^{[-\mu\rho x]} \quad (1)$$

Where I_0 is the initial intensity, I is the attenuated intensity, μ is the X-ray mass attenuation coefficient, and ρ is the density of the material. Note that the mass attenuation coefficient appears in some texts as mu/rho and other texts as mu. Both forms are present in this volume, and as long as the appropriate definition for the thickness component of the equation is used (x = thickness or x = mass thickness) the two equations will provide identical results. To interpret the XRF data collected and choose the appropriate analysis conditions one must first consider: What is the object likely to be composed of? Is it largely organic such as a basket or a painted wooden mask? If this is the case, determination of the composition might be difficult based on the limitations of the elements one can identify with XRF (for elements lighter than potassium the fluorescent X-rays are typically sufficiently absorbed and scattered by the air that they are greatly attenuated before they reach the detector). If the object is metallic (a best case scenario) is there a possibility of plating or surface enrichment? A nickel-plated brass will be challenging to distinguish from German silver – a copper, zinc, and nickel alloy – using standard XRF protocols. It is also important to consider what level of determination you are you expecting to achieve. Major and minor elemental composition is not too difficult to obtain, but trace elements pose more of an analytical challenge (see chapters 2 by Smith, and 7 by Mass and Matsen).

Provenance determination might well prove to be one of the most difficult quantitative challenges for handheld XRF of art and archaeological materials. To do this one needs to have samples and data from the sources of the raw material used to make the object (such as obsidian or clay) that can then be compared to data from the unknown. In addition, the data from this source material ideally should be collected on the specific XRF instrument being used to analyze the unknown for a custom calibration to be created (some success of cross-instrument calibration has been reported for Si-PIN diode detectors but this has not yet been demonstrated with SDD detectors). What is even more challenging is ensuring that the concentrations of the elements of interest are represented in a greater range in the source material and references material than what you would expect to see in your unknown samples (see Chapters 12 by Ferguson and 13 by Aimers, Farthing and Shugar for examples).

The problem of determining safe handling protocols for contaminated cultural heritage objects is a difficult question often asked of museum scientists and conservators. Although we can help determine what elements are present (above our minimum detection limits) in a material, specific statements concerning handling safety must be made by medical or toxicological specialists only (see Chapter 9 by Shugar and Sirois for a detailed discussion). Several companies dedicated to the analysis of lead and cadmium in children's toys have tried to address this problem and have encountered legal issues based on inaccurate information they reported (see Zhu Zhu pets for example – Grant 2012). While full quantification is a challenge given the absence of appropriate reference standards in such cases, the elements in question can at least be identified and broader statements of handling guidelines can be recommended.

Ultimately, the archaeological or conservation scientist will likely ask more detailed questions of their collaborators to more fully frame the problem to be addressed and in many cases, will guide the collaborator toward alternative techniques of analysis, or strongly suggest a combination to analytical techniques to most fully answer materials analysis questions about cultural heritage objects. This approach is frequently necessary as demonstrated in Chapters 4, 5 and 8.

Analytical challenges of cultural heritage materials

Since we often are unable to sample works of art and archaeological artifacts, which would be required for us to prepare samples for XRF analysis in a traditional manner (such that they have flat, polished, and homogeneous surfaces), we are left with the *in situ* analysis of the complete object – an often highly difficult situation for generating high quality and reproducible analysis results. There are several key issues that can have an effect on the resulting data derived from this type of analysis.

The basic requirements for an XRF sample are that it be of homogenous material that is infinitely thick to X-rays (at the analysis conditions) and of uniform particle size. These requirements are rarely met for cultural heritage materials. Many works of art, such as photographs and slivers of obsidian, are too thin to meet the infinite thickness requirement and so quantitative analysis of these materials is possible only with thin film standards (which is an acceptable calibration technique). In addition, works of art are typically heterogeneous, often on many length scales. Consider, for example, a glazed ceramic that is not only a layered structure consisting of a body, and interaction zone, and a glaze, but also may contain intermingled vitreous and crystalline regions within the body. No set of calibration standards is 100% appropriate to reproduce this type of inhomogeneity. The surface preparation required to make an object homogenous, such as the removal of a corrosion layer on a metal, a depleted surface on an alloy, or an alkali-depleted layer on a glass may not be permissible depending upon the condition of the object and the policy of the institution. Finally, ideal and reproducible measurement geometries are not achievable for most cultural heritage objects, an inherent limitation of analysis without sampling. The problem related to these issues and the specifics of the materials we encounter are important to understand and can best be presented in discussing the feature common to almost all objects we encounter - layered structures.

Layered microstructures

In almost all cases of material culture objects, one has to contend with layered surface structures caused by the object's technology of manufacture and/

or alteration over time. This can be in the form of surface contamination such as debris from excavation and/or attached concretions, or from layered paint structures, corroded surfaces, and artifacts with leached components (i.e. glass and copper-depleted silver-copper alloys). In all of these cases, we run into the same issue of the data under-reporting the composition of the bulk since the instrument identifies the composition of the surface layers with greater sensitivity than the bulk (assuming the incident X-ray beam is not fully absorbed before it reaches the bulk region, and that the fluorescent X-rays emitted from the bulk are not fully absorbed before they reach the detector). This may be best exemplified in the change in compositional data obtained from even a lightly corroded metal coupon (Ehrhardt and Kaiser 2011). The effect is so dramatic that even a thin organic coating or oxidation layer can have an effect on the overall compositional results (see Smith, Chapter 2 and Mass and Matsen, Chapter 7). The thickness and penetration of corrosion films on metals can introduce additional elements into the analysis that are not part of the base metal as well (Constantinides, Adriaens and Adams 2002; Neff *et al.* 2004) and even a minute amount of contamination can alter quantitative results as seen in the analysis of silver coins by Notis *et al.* (2006).

In addition to the potential of surface accretions developing on an artifact, the possibility of elemental leaching can occur. This is most relevant in glass artifacts, and is observed in particular in archaeological glasses that have undergone corrosion in humid/damp environments. The depletion of alkali cations (i.e. Na, K) as well as glass stabilizers (i.e. Ca) can extend well below the level of visual perception of alteration (manifest by an iridescent surface or an opaque surface crust), creating the potential for one to think that a cleaned glass is an accurate representation of the bulk composition of the glass when it in fact still has a remaining alkali-depleted surface (Weber *et al.* 2002). In addition, glass recovered from a buried environment will not only leach and corrode but that corrosion layer can take in elements from the surrounding soil and any heavy metal elements in the local groundwater (diagenesis, Frame *et al.* 2008). These examples only touch the surface of issues related to layered surfaces and a quick glance at any object being investigated will confirm some level of layering exists. The demand of handheld XRF to provide quantitative data is hampered by this factor alone. Remember, that accurate quantification is dependent on homogenous samples that are infinitely thick and contain similar particles sizes, and so the errors reported for cultural heritage object analyses will be necessarily larger than for modern materials.

The standard methods used for calibration of instrumentation typically falls under four main categories which will be discussed below.

Traditional methods offered to achieve quantitative analysis: 'values', 'numbers', or 'concentrations', and their relevance to cultural heritage materials

There are four methods used by XRF manufactures to obtain quantification of data. These are; 1. Fundamental parameters, 2. Compton normalization, 3. Empirical calibration, and 4. A combination of one of the first two with the third. These are described in brief below.

Fundamental parameters (FP) uses algorithms to solve for non-linear equations that ultimately describe the dependence of the intensity of X-ray lines from each element in a material on its concentration (Thomsen 2007). Calibration procedures typically rely on the measurements of pure elements in conjunction with the known physics behind the spectrometer's response to those calibrated elements. The calculations are included for each element and the parameters considered include the fluorescence intensity, emission wavelengths, absorption edges and absorption coefficients. In addition, the instrumentation settings are taken into consideration and the end result is a normalized quantification that adds to 100%. As a result, as long as the material to be analyzed is comprised of elements that can be measured by the XRF (greater than Ca in standard modes) and enough data has been collected to develop trace element peaks, the data can be reasonably accurate. But if the sample contains organic materials, or is comprised of elements below Ca, those elements will not be recognized, and the data will still be normalized, altering the true compositional quantification.

Compton normalization provides quantitative data using a different calibration process. It relies on an algorithm that is developed from standard compounds (i.e. Fe_2O_3) that are calibrated against the Compton peak (inelastic scatter of tube X-rays from the sample). The intensity and shape of the Compton peak changes with the different densities and compositions of different materials. For that reason it can be quite powerful in helping to reduce matrix effects that can result when analyzing materials of varying composition. The Compton peak acts as an internal standard for this quantification process which is typically used when investigating materials that are primary composed of light elements (i.e. glass, ethnographic artifacts, some ceramics). This method can provide data for low concentrations (ppm range). It is developed from standards calibrated against the Compton peak, or incoherent backscatter radiation observed with materials composed primarily of light elements (EPA 1998:43).

For an empirical calibration, the researcher chooses a number of reference materials that closely match the chemistries of their unknown samples. The concentrations of the elements present in the standards must cover the range (and ideally exceed the range) of the potential concentrations of the elements in the unknown. This can be difficult, and requires that a rough knowledge of the composition of your unknown, whether that comes from performing alternative analysis (e.g. ICP, NAA, WD-XRF), reviewing published data of similar samples, or qualitatively assessing your sample to determine the potential compositions. The effectiveness of this calibration technique is that it uses closely matched standards that resemble the density and composition of your unknown, allowing for a good linear correlation. The number of reference standards used can vary but it is recommended to use as many as possible to cover the elements of interest. The known samples are plotted to create a curve for peak area vs. concentration. The unknown can then be analyzed and the peak area for each element measured and plotted on the curve to provide a concentration (see Ferguson, Chapter 12 and Aimers, Farthing and Shugar, Chapter 13 for examples).

In the fourth version, one can take either an established FP or Compton normalization and add in specific reference standards that better represent the unknown samples in question.

All of these calibration methods have their appropriate uses, but are all based on those fundamental requirements of samples. That is why the traditional method used for sample preparation for XRF analysis is to create powdered pressed pellets, or fused glass beads. This methodology has been in place since the late 1950's (Instituut 1957) and continues to be the standard practice for academic and industrial XRF users today. To make matters more difficult for the quantification of heterogeneous materials, the standards used for these calibration programs are all flat and highly polished. The surfaces of the unknowns that are being quantified need to match the state of these standards directly to have any correlative relationship. Based on this, few the objects we analyze in the cultural heritage community can be reliably quantified. Three choices can be made here, the first is to accept qualitative data, the second is to accept and report a high degree of error with your results, or third, one can attempt to make reference materials that match their unknown's heterogeneity as well as composition (see Chapter 9 by Shugar and Sirois).

Evaluation of spectra

Although we have not directly addressed the basics of X-ray fluorescence, we (and other authors in this volume) have offered references to seminal literature on the subject and suggest anyone embarking on XRF analysis of cultural materials study them prior to beginning experimental work. However, it is important to recognize that in addition to the elemental peaks that are visible in the XRF spectrum there are often additional peaks present that can directly affect ones' ability to accurately interpret an XRF spectrum which are necessary to discuss here. These include elastic and inelastic scatter, escape peaks, Bragg diffraction peaks and sum peaks. In all cases for objects with elements heavier than aluminum, the X-rays produced from the tube (Rh, W, Ag for example) will interact with the sample to produce characteristic X-rays. In addition to this, some of the bombarding X-rays strike the object without producing any fluorescence. These X-rays are deflected by the object and lose no energy in the process - this is termed Rayleigh scatter ($E_I = E_O$). These X-rays are collected by the detector and appear in the spectrum as the characteristic peaks related to the element that makes up the X-ray tube's anode (i.e. Mo tube: $K\alpha_1$=17.479 keV, $K\alpha_2$=17.374 keV, $K\beta_1$=19.608 keV, $L\alpha_1$=2.293 keV, $L\beta_1$=2.395 keV, $L\beta_2$=2.518 keV, $L\gamma_2$=2.623 keV). These peaks have no relationship with your object's chemical composition and are strictly the result of elastic scattering of the X-ray beam. In direct relationship with this process is inelastic scattering in which the bombarding X-rays give up part of their energies to the electrons in the object being analyzed, but not enough to cause fluorescence. This loss of energy is termed Compton scatter and will appear in the spectrum as broad peaks (broader than fluorescence peaks) with slightly less energy than the Rayleigh scatter ($E_I > E_O$). The Compton peak can provide you with an enormous amount of information about the makeup of the sample. The density of a material has a direct effect on the Compton scatter. Consider a beam of photons interacting with a dense material. The photons cannot penetrate deeply into the material and the scattering is limited to a reduced volume of area. The result is a small Compton peak since there is less chance of the interacting beam having the volume interaction to lose energy. Now think of a less dense material, like softwood. The beam penetration is has a greater volume area allowing the bombarding photons to interact with more material, resulting in a higher probability for inelastic scattering. The result will be a much larger, potentially broader Compton peak. By visually assessing your spectrum you can make general statements about your material types by looking at this phenomenon.

Additional peaks unrelated to fluorescence can also create confusion when inspecting a spectrum and attempting to perform peak assignments. Sum peaks (also known as double photon counting events) are peaks that created when two characteristic fluorescent X-rays arrive at the detector at the same time, and so are experienced by the detector as a single X-ray at twice the photon energy (see Figure 1.1). They are added and counted as a single event ($E_{measured} = E_1 + E_2$). Very often the resulting peaks, such as with copper, will not cause significant confusion in the interpretation of the spectra (since 2×Cu Kα = 16.08 keV which does not interfere with any other element's X-ray line), but in some cases, such as with lead, the sum peaks overlap the Kα peaks for cadmium and tin which can be quite misleading (see Chapter 4 for an example of this issue). In these situations, sum peaks can be recognized because they occur in groups (for example, for the lead L lines the observed peaks will be 2×Pb Lα, 2×Pb Lβ, and Pb Lα+Pb Lβ). The solution for resolving this issue is to reduce the X-ray flux hitting the detector. This can be accomplished with filters or by reducing the X-ray tube current. Silicon escape peaks are an additional spectral feature one might observe. These escape peaks are the result of the characteristic X-rays from the sample hitting the detector and the silicon present (remember that the detectors used are Si-PIN diodes and silicon drift detectors) fluoresces, reducing the characteristic X-rays by the energy of the silicon Kα X-ray (Si energy = 1.74 keV, therefore $E_{measured} = E_1 - 1.74$ keV). The effect does not have a solution and escape peaks are always present but they are usually weak and only appear with strongly emitting elements. A classic mistake made by novice XRF spectral interpreters is the misidentification of the Si escape peak for Cu when looking at a Cu alloy. Copper's Kα energy is ~8.04. The Si escape peak is then 8.05 – 1.74 = 6.31. The Kα energy for Fe is 6.41. Without careful examination of the peak location the Si escape peak for Cu is often misidentified as an Fe Kα peak (see Figure 1.1).

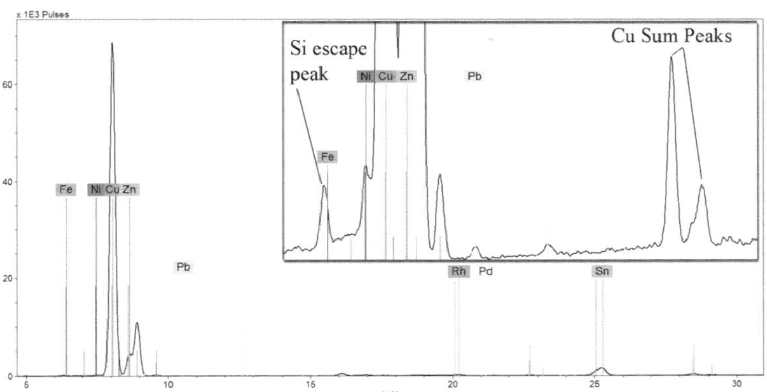

Figure 1.1:
Spectrum of copper standard 89b showing both sum peaks and the Si escape peak for copper. This particular Si escape peak often gets mis-interpreted as an iron peak (TRACER III Rh tube with an 279 μm Al, 25.4 μm Ti filter, run at 40 kV and 3.15 μA).

Finally, for the analysis of crystalline materials such as metals, minerals, gemstones, and pigments, it is possible that Bragg, or diffraction peaks could be present in one's spectrum. These peaks can be identified by changing the orientation of the object being analyzed, if they are truly a result of diffraction, then the peaks will shift position or disappear entirely. Bragg peaks can also be recognized because they will not have the narrow, Lorentzian peak shape expected for fluorescence peaks (see Figure 1.2).

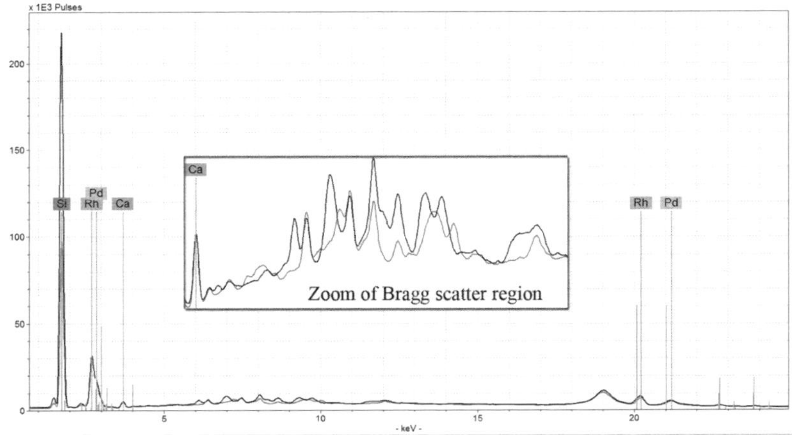

Figure 1.2:
Two spectra of a quartz sample showing the effect of coherent Bragg scattering. The orientation of the crystal is different for each scan causing the Bragg peaks to shift slightly with no change in the positioning of the Si and Ca Kα peaks (TRACER III-SD Rh tube no filter, run at 40 kV and 9.8 μA).

Conclusions

We have seen that the proper application of XRF to the study of cultural materials requires a thorough understanding of the interaction phenomena between the X-ray beam and the object, the surface structure of the object to be studied, and an understanding of the X-ray emission depths for the elements in question in the material being studied. We have also noted the inherent limitations of carrying out XRF non-destructively (lack of ideal and reproducible geometries, for example), as well as the limitations of manufacturer-generated calibrations supplied with the instrument for quantitative analysis. The authors of the following chapters have demonstrated how these principles and limitations function when applied to the study of a broad range of cultural materials, and we hope that they will serve as a best practices guide for future cultural heritage researchers.

To conclude, we hope that this volume will be of use to art conservators, archaeologists, and students of art conservation and archaeology, as well as museum scientists embarking on handheld XRF analysis of cultural heritage materials. While we have highlighted many potential limitations of this technique for the study of such materials in this volume, we feel that handheld XRF holds great potential for increasing the range of cultural heritage materials available for study. With a careful understanding of its limitations, handheld XRF can continue to be of great use to the researchers involved in the understanding and preservation of our global material culture.

Acknowledgements

The authors are grateful to Bruce Kaiser for providing the figure presenting Bragg scattering in this chapter. We are grateful to Natalie, Adam, Ethan, Hershey, Rufus, and Laszlo for their patience with us during the assembling of this volume.

References

Constantinides, I., A. Adriaens, and F. Adams. 2002. "Surface characterization of artificial corrosion layers on copper alloy reference materials." *Applied Surface Science* Vol. 189 (1-2):90-101.

Ehrhardt, K., and B. Kaiser. 2011. "TRACER III V+ and ARTAX Compositional Analysis of the Copper-base Metal Industry at the New Lenox Site, Will County, Illinois." *FIRST Newsletter* Vol. March 2011 (08).

Grant, D. 2012. *Zhu Zhu Pets unsafe? Hardly, says manufacturer, as consumer group backtracks*. The Christian Science Monitor, December 7 2009 [cited January 2012]. Available from http://www.csmonitor.com/Business/new-economy/2009/1207/zhu-zhu-pets-unsafe-hardly-says-manufacturer-as-consumer-group-backtracks.

Instituut, Rijksuniversiteit te Utrecht. Geologisch. 1957. *Geologica Ultraiectina*: L. van der Velder.

Jenkins, R., R.W. Gould, and D. Gedcke. 1995. *Quantitative X-ray spectrometry*: M. Dekker.

Jenkins, R., and J.L. Vries. 1973. *Practical X-ray spectrometry*: Springer-Verlag.

Jenkins, Ron. 1999. *X-ray Fluorescence Spectrometry*. 2 ed, *Chemical analysis*. New York: John Wiley & Sons, Inc.

Mantler, M., and M. Schreiner. 2000. "X-ray fluorescence spectrometry in art and archaeology." *X-Ray Spectrometry* Vol. 29 (1):3-17.

Neff, Delphine, Solenn Reguer, Ludovic Bellot-Gurlet, Philippe Dillmann, and Régis Bertholon. 2004. "Structural characterization of corrosion products on archaeological iron: an integrated analytical approach to establish corrosion forms." *Journal of Raman Spectroscopy* Vol. 35 (8-9):739-745.

Notis, M., A. Shugar, D. Herman, and D. Ariel. 2007. "Chemical composition of the Isfiya and Qumran coin hoards." In *Archaeological Chemistry: Analytical Techniques and Archaeological Interpretation*, edited by M. Glascock, R.J. Speakman and R.S. Popelka-Filcoff, 258-274. The American Chemical Society 968.

Podsiki, Cheryl. 2009. "XRF Seminar, The Field Museum, Chicago, June 2008." *ICOM-CC Ethnographic Conservation Newsletter* (30):15-17.

Potts, P. J. 2008. "Introduction, Analytical Instrumentation and Application Overview." In *Portable x-ray fluorescence spectrometry: capabilities for in situ analysis*, edited by P. J. Potts and Margaret West, 1-12. Cambridge, UK: Royal Society of Chemistry.

Potts, P. J., and Margaret West. 2008. *Portable x-ray fluorescence spectrometry: capabilities for in situ analysis*. Cambridge, UK: Royal Society of Chemistry.

Schreiner, M., B. Fruhmann, D. Jembrih-Simburger, and R. Linke. 2004. "X-rays in art and archaeology: An overview." *Powder Diffraction* Vol. 19 (1):3-11.

Shackley, M. Steven. 2011. "An Introduction to X-Ray Fluorescence (XRF) Analysis in Archaeology." In *X-Ray Fluorescence Spectrometry (XRF) in Geoarchaeology*, edited by M. Steven Shackley, 7-44. New York: Springer.

Sirois, P. Jane. 2007. "XRF Pesticide Workshop, Arizona State Museum, October 23, 2006." *ICOM-CC Ethnographic Conservation Newsletter* (28):29-30.

Society for Historical Archaeology. 2012. *Portable X-Ray Fluorescence in Archaeology: Fundamental Principles, Applications and Concerns* 2012 [cited January 2012]. Available from https://www.conftool.com/sha2012/index.php?page=browseSessions&form_session=120&CTSID_SHA2012=hZPXPq,K9b6rNlY7Ip5LHMLgsp1.

Speakman, Robert J., Nicole C. Little, Darrell Creel, Myles R. Miller, and Javier G. Iñañez. 2011. "Sourcing ceramics with portable XRF spectrometers? A comparison with INAA using Mimbres pottery from the American Southwest." *Journal of Archaeological Science* Vol. 38:3483-3496.

The Getty Conservation Institute. 2012. *X-ray Fluorescence Workshop*. The Getty 2006 [cited Januaray 2012]. Available from http://getty.edu/conservation/publications_resources/newsletters/21_3/gcinews1.html.

The Getty Conservation Institute. 2012. *AATA Online: Abstracts of International Conservation Literature*. The Getty 2012 [cited January 2012]. Available from http://aata.getty.edu/NPS/.

Thomsen, V. 2007. "Basic Fundamental Parameters in X-Ray Fluorescence." *Spectroscopy* Vol. 22 (5):46-50.

Weber, G., D. Strivay, L. Martinot, and H. P. Garnir. 2002. "Use of PIXE–PIGE under variable incident angle for ancient glass corrosion measurements." *Nuclear Instruments and Methods in Physics Research Section B: Beam Interactions with Materials and Atoms* Vol. 189 (1-4):350-357.

Chapter 2

Handheld X-ray fluorescence analysis of Renaissance bronzes: Practical approaches to quantification and acquisition

Dylan Smith

Introduction

Handheld X-ray fluorescence spectrometry (XRF) has become a highly desirable technique for the investigation of works of art that makes it possible to perform quantitative elemental analysis regardless of location. Until now XRF analysis has been limited to the collections of institutions with scientific resources, temporary loans to those institutions, or, less commonly, objects transported for testing with the associated costs. One of the earliest applications of XRF analysis to works of art was on a Renaissance 'bronze' (Jestaz 1972) and subsequent studies have demonstrated the value of this technique for the study of this important form of sculpture. ('Bronze' is used here in its conventional sense to describe any copper alloy work of art.) However, because of the limitations of access, many masterpieces remain unstudied and the understanding of Renaissance practices is fragmentary at best; the alloys of few Renaissance artists have been comprehensively studied (Bewer 2001, Bassett *et al.* 2008, Smith and Sturman 2011). Indeed, the use of copper alloys in the Renaissance (and many later periods) remains less well studied than those in the Bronze Age and Classical Antiquity (Craddock 2009). Handheld XRF analysis has the potential not only to fill the gaps in knowledge about specific objects and artists, but also to expand the current understanding of Renaissance bronzes across geographic regions and historical periods.

Since 2006, the Objects Conservation Department of the National Gallery of Art (NGA) has been developing methods for the application of handheld XRF to the investigation of Renaissance bronzes, utilizing a Bruker Tracer III-V. Although this type of XRF instrument was designed for analysis of modern industrial metals, the factory methods for quantification can be customized for the investigation of historic alloys. The working methods for acquiring spectra with the handheld XRF have also been adapted to take into account the particular demands of analyzing art, including unusual working conditions, unpredictable surface characteristics, and an often incomplete understanding of construction. Although the methodology discussed was developed using the Bruker Tracer III-V instrument, all of the recommendations can be applied to other handheld XRF instruments currently available. Throughout the process of development, practical experience analyzing a wide variety of alloys and objects has motivated improvements in data acquisition and quantification, which will both be considered in this paper.

XRF analysis of renaissance bronzes

Many types of inorganic historic objects have been studied using XRF and copper alloys are well suited to elemental analysis using this method (Glinsman 2004). As a surface technique, XRF analysis will only obtain results representative of the bulk alloy when the surface is consistent with the underlying metal. Most Renaissance bronzes, particularly small bronzes, have characteristics that are appropriate for this type of analysis, having remained in protected indoor environments, unaltered by weathering or burial. Furthermore, Renaissance bronzes are unlikely to have original chemical patinations, although recipes are mentioned in period documents (Stone 2011). At present only one artist is known to have consistently used this technique, the late 15th-early 16th c. Mantuan artist, Pier Jacopo Alari Bonacolsi, known as 'Antico' (Bormand 2008, Stone 2011, Smith and Sturman 2011).

Renaissance artists typically used drying oils mixed with various natural resins, to impart color, unify appearance, and disguise casting flaws on their works in bronze (Stone 1990, Pittard *et al.* 2007, Stone 2010). After application these varnishes were sometimes 'stoved', heated in an oven, to more rapidly produce a hardened and durable coating. Some of these 'applied patinas' are pigmented, such as an opaque type colored with carbon black commonly used in Venice and Padua. Other achieve a color only through the preparation of the oil, such as the translucent red, brown, and tawny yellow patinas found on Florentine bronzes, notably those from the 16th century workshop of Giovanni Bologna (Stone 2010). In addition to these original organic coatings, waxes and other materials have often been applied at various times over the course of time. (For the effect of these varnishes on analysis, see 'Spot Selection'.)

Many of the artists producing Renaissance bronzes were highly skilled and the alloys can be very consistent throughout, particularly in smaller works. However, various types of inhomogeneity may be present that must be considered during analysis in order to obtain a representative result. Variations on a microscopic scale (under 200 microns) occur as a result of distinct phases present within the metallographic structure, dependent upon the particular alloy as well the method of manufacture (Scott 1991; Fig. 2.1). Use of an XRF instrument with a sufficiently large spot size (in this study approximately 4.5 mm^2) will tend to average out such variations and produce a representative assessment. The presence of tin at very high levels, above 14%, can cause phase segregation and selective surface enrichment, producing a result that does not represent the bulk alloy (Scott 1991). However, during the Renaissance such alloys were used primarily in specialized classes of objects including bells as well as mortars and pestles (Motture 2001).

SEM-EDS backscatter image of cast bronze simulating a Renaissance alloy

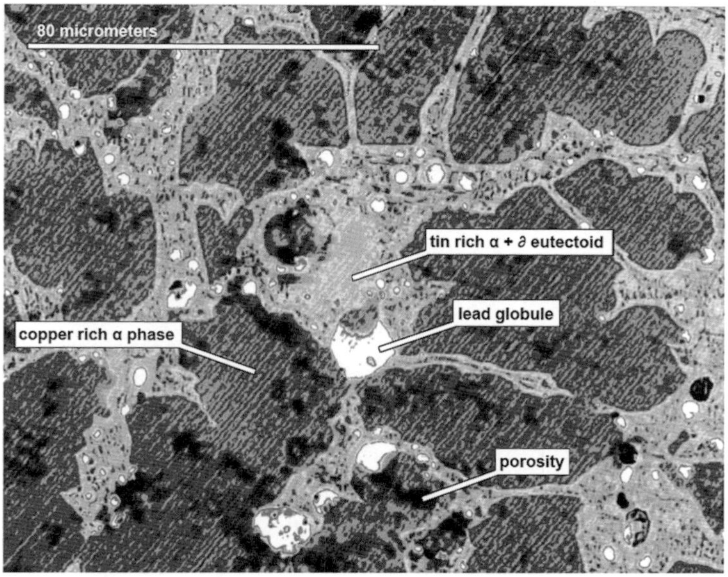

83% copper, 14% tin and 4% lead (polished and etched with 10% ammonium persulfate)

Figure 2.1:
SEM-EDS backscatter image of cast bronze simulating a Renaissance alloy (sample and image prepared by L. Glinsman).

Non-uniformity may still be noted in alloys with a significant content of lead, which is essentially insoluble in copper. At low levels, lead will tend to form finely dispersed globules along grain boundaries, however, with increasing contents and slower cooling rates, larger globules of lead can form (Scott 1991; Fig. 2.1). Starting at around 8 wt. % (weight percent) lead, significant variation within a single object has been noted for Renaissance bronzes, which is likely due not only to the presence of large globules, but also the smearing of lead caused during tooling of the surface after casting. In one highly leaded Renaissance medal, testing with the handheld XRF reported as much as 20 wt. % on the obverse (reported as 1957.14.656, Table 2.1), while a spot tested on the reverse was closer to 10 wt. %. Using a Bruker Artax XRF spectrometer, an area was mapped on the highly leaded side of this medal, revealing significant variations across the span of a centimeter (Fig. 2.2). Lead can also vary widely within large sculptures due to separation by gravity during the casting process (Craddock 2009). Because XRF allows spots to be non-destructively tested, such large scale variations can be correctly identified and a representative result obtained. This is a significant advantage over methods that require sampling, where the result from a single analysis may or may not properly represent the alloy of an entire object.

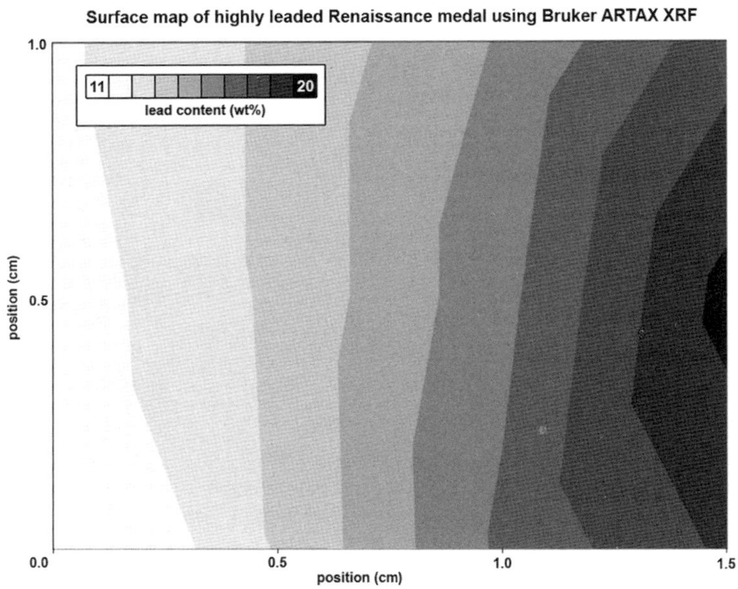

Figure 2.2:
Map of lead on surface of Renaissance medal (analysis by L. Glinsman).

Although qualitative analysis using XRF can identify distinct parts within a sculpture or the similarity of two or more objects, quantification of spectra is necessary to resolve more complex questions about Renaissance bronzes. Used in conjunction with statistical analysis, quantitative elemental data gathered from large groups of objects may be able to associate alloys with a particular artist, region, or time period (Glinsman 1997, Smith 2011). To date, variations noted in Renaissance objects tend to emphasize deliberate alloy selection based on function, workshop, or region, rather than accidental inclusion of trace characteristics that might be linked to ore sources. This is consistent with the extensive trade of raw materials and the established techniques of refining methods in Europe at that time.

NGA ACCN	Inst	Cu	Zn	Sn	Pb	Ni	Fe	As	Ag	Sb	Total
significant arsenic											
1957.14.737	Kev	81,2	8,9	5,5	2,3	0,1	0,2	1,0	0,1	0,5	99,8
	Cus	80,5	10,5	5,6	1,8	0,2	0,4	1,0	0,2	0,5	100,6
	Fac	80,1	10,2	5,0	2,5	0,3	0,4	1,1	0,2	0,4	100,0
1957.14.689	Kev	86,3	6,9	2,4	0,8	0,9	0,2	2,2	0,2	0,1	100,0
	Cus	87,7	6,5	2,1	0,8	0,8	0,5	1,0	0,2	0,1	99,7
	Fac	87,2	6,4	1,9	1,4	0,8	0,5	1,5	0,2	0,1	100,0
significant arsenic and antimony											
1957.14.717	Kev	79,4	6,5	4,7	4,2	0,4	0,3	2,0	0,2	2,0	99,7
	Cus	79,1	6,1	4,7	4,6	0,5	0,7	1,4	0,2	1,9	99,3
	Fac	73,6	6,1	3,9	8,3	0,6	0,6	1,9	0,2	1,5	96,7
1942.9.167	Kev	79,9	0,6	7,5	5,9	0,3	0,3	1,2	0,1	4,0	99,8
	Cus	76,2	0,0	8,3	6,5	0,5	0,1	1,3	0,3	4,0	97,2
	Fac	67,9	0,1	5,6	11,6	0,6	0,1	1,3	0,2	2,8	90,3
higher tin with significant arsenic and antimony											
1957.14.653	Kev	75,2	0,5	14,6	3,5	0,6	0,2	0,9	0,2	4,0	99,8
	Cus	71,6	0,0	18,1	2,9	0,7	0,2	1,6	0,4	3,6	99,0
	Fac	65,2	0,1	11,6	4,4	0,8	0,2	1,6	0,3	2,1	86,5
high lead with significant arsenic and/or antimony											
1957.14.650	Kev	74,8	0,3	6,2	12,9	0,4	0,2	0,9	0,2	4,8	100,7
	Cus	68,7	0,0	6,2	15,9	0,5	0,1	1,4	0,3	4,3	97,4
	Fac	60,0	0,1	4,1	33,3	0,5	0,2	2,6	0,2	3,1	103,9
1957.14.726	Kev	77,3	9,1	3,3	9,1	0,2	0,3	0,0	0,1	0,4	99,8
	Cus	75,3	9,2	3,5	11,6	0,2	0,8	0,3	0,2	0,4	101,5
	Fac	73,2	9,4	2,8	20,1	0,1	0,7	2,5	0,2	0,3	109,2
1957.14.656	Kev	76,2	0,4	6,3	14,1	0,2	0,3	0,0	0,2	2,2	99,8
	Cus	67,2	0,0	8,3	20,1	0,2	0,2	0,5	0,4	2,2	99,1
	Fac	64,1	0,0	6,3	37,3	0,1	0,2	2,6	0,2	1,6	112,4

[shaded] = significant difference in calculated value

Table 2.1:
Surrogate calculated values for Kevex and Tracer with current custom and factory calibrations.

Quantification is also essential for successful meta-studies, where alloy data must be assembled from many different instruments and institutions. If each instrument can independently achieve a sufficient level of accuracy across the range being studied, then, in theory, data can be compiled successfully for study. Recent inter-laboratory studies performed to investigate this potential have yielded positive results when certain parameters are taken into consideration (Heginbotham *et al.* 2011).

Customizing quantification for the NGA Tracer III-V

Renaissance artists used a great variety of copper alloys including a broad range copper, zinc, tin and lead contents, as well as a many other elements, such as arsenic and antimony, sometimes at significant levels. The empirical method for quantification provided by the manufacturer has proven insufficient to accurately analyze these historic alloys, partly due to its primary design for analysis of modern alloys. However, the software provided with the instrument allows a custom calibration to be created to accommodate different elemental ranges and improve accuracy. Each custom calibration may include changes to the set of reference standards used, the inter-element corrections applied, the elements of interest selected, and the spectral lines measured. Acquisition run conditions, including use of a vacuum, filters, and instrument distance are also specific to each calibration. Acquisition time does not need to be selected within the calibration settings and is determined instead by the practical realities of gathering spectra (discussed below under 'Working Methods').

Run conditions

The NGA handheld XRF has a rhodium tube and a silicon-pin diode detector with a resolution of 220 eV. When the NGA obtained this instrument in 2006, a rhodium anode was selected for its effectiveness at exciting a wide range of elements, particularly those of interest in copper alloys. Handheld and lab-based instruments are available with a range of targets, including rhenium, tungsten and silver.

The NGA spectrometer has continued to be run at the setting conditions recommended for analysis of copper alloys: 40 kV anode voltage, 1.8 μA anode current, with a sandwiched tube filter of 25.4 micrometers titanium and 304.8 micrometers aluminum. This filtering selection allows X-rays from 4 keV to 40 keV to excite the sample, filters out the rhodium L-lines, and flattens the background in the spectral region for many of the elements of interest in copper alloys. Run at these settings, the raw and valid counts per second (cps), as well as deadtime, vary significantly depending upon the composition of the object being analyzed. For a pure copper standard, the raw cps are approximately 13000, valid cps are 6800, and the dead time, 48%. For a bronze standard with 11% tin, the raw cps drop to approximately 8400, valid cps to 5500, and dead time, 35%. The software provided with the NGA spectrometer compensates for this response range, as well as minor fluctuations in run conditions, by normalizing each peak measurement to the total measured counts. Ideally, deadtime should fall within the range of 20-35%

to ensure that the detector is properly saturated but not overloaded with signal. Higher deadtimes may cause shifts in peak location but if these are checked prior to acquisition, and no shift is present, a higher deadtime percentage is acceptable.

Modification of the run conditions has not been considered until recently because of the limitations imposed by the inclusion of factory-run standards in the custom calibration. Although the factory standards can contribute to the range and breadth of a calibration, they cannot be retested at new run conditions, therefore any newly added standards must conform to the existing conditions. For this reason, it is desirable to establish a fully independent set of standards that allows unhindered modification of run conditions. Changes to voltage and amperage, as well as the use of additional filtering options, could enhance detection of certain elements and improve quantification on particular classes of alloys.

Instrument to surface distance

Run conditions in the calibration must also be modified in order to incorporate changes in analysis geometry. Although handheld XRF spectrometers are designed for contact analysis, the instrument should not be used this way for handheld analysis of works of art because of the risk to the surface. Alternatively, a tripod can be used to carefully position the instrument until it just contacts this surface, however, this presents certain practical difficulties (see 'Working Methods' below). To prevent direct contact of an object with the metal head during handheld analysis, the front and sides have been covered with a 1.5 millimeter layer of inert closed cell polyethylene foam. Although thinner materials are available, it was felt that a certain amount of cushioning is desirable to minimize the risk of marking the surface.

To better understand the effect of distance on quantification, tests were performed on a copper alloy standard (MBH 33XGM8) adjusting the distance to the instrument in half millimeter steps from contact to two millimeters (Fig. 2.3). The standard used was selected as a generic Renaissance alloy including all of the typical major and minor elements (wt. % = Cu: 83.81, Zn: 5.45, Sn: 4.13, Pb: 6.11, Fe: 0.10, Ni: 0.15, As: 0.02, Ag: 0.10, Sb: 0.01). The alloy content was calculated at each distance from a series of five spectra collected for 60 seconds using a custom calibration where the standards were analyzed with the foam. The average value was compared to the quantified result for the reference spectra for this standard included in the calibration. For the major elements, the measured differences converge toward zero within the typical working distance including the foam (1.5 mm, down to 1 mm compressed), and diverge outside of this range;

the minor elements show little change in actual value (although the percent change can be significant at trace levels). By including the small distance introduced by the foam in the analysis of the standards, the custom calibration now reflects the actual working distance used for objects and obtains a more accurate result. The results of these tests also suggest a certain tolerance for slight variations from the optimum distance that are unavoidable in actual practice.

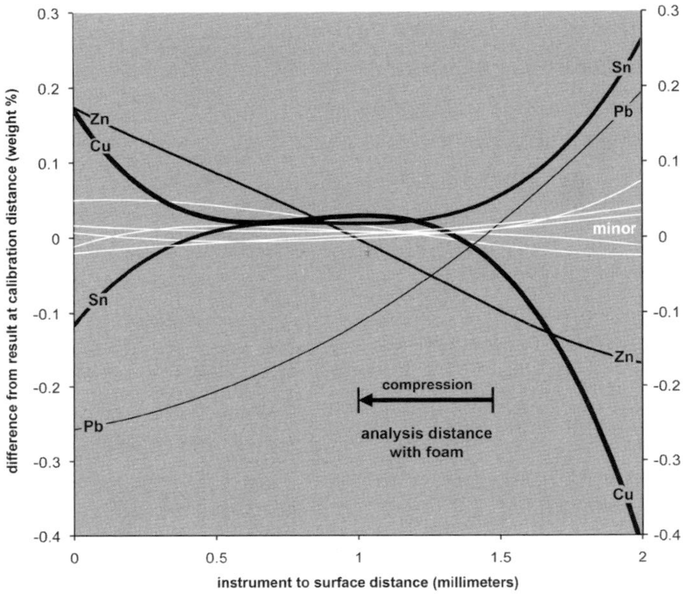

Figure 2.3:
Effects of distance on Tracer quantification.

Initial modification of the set of standards

The set of copper alloy reference standards chosen is the most critical component of a calibration. These standards are used within the spectrometer software to establish the necessary points of reference for the regression curves used to calculate values for unknowns. These calculations are particularly complex due to matrix effects, where the characteristic radiation produced by each element in the alloy may be absorbed or enhanced by interaction with the other elements present. As a result of these effects, is critical that the calibration include standards that are close in composition to the unknowns to be analyzed. The range and variety of these standards define the types of alloys that can be accurately quantified with a given calibration.

Each standard must be analyzed under the same settings and conditions that will be used to analyze unknowns. However, acquisition times for the standards should be significantly longer than those for typical analyses; the NGA custom calibration uses 180 seconds. Subtle differences between individual instruments are introduced into the calibration with the analysis of the standards; these differences make each calibration instrument-specific. As a result, the custom calibration cannot be modified by the addition of standards if an instrument is repaired or replaced. Changes caused by instrumental drift over the lifetime of the X-ray tube may also be sufficient to require re-calibration. This effect has not been noted in a comparison of results for four standards analyzed in 2006 and in 2011. A spot on each standard was analyzed five times for 30 seconds, however, the same spot was not necessarily analyzed in 2005 and 2011, although in theory the standard should be uniform. (Runs of 30 seconds were the practice recommended when the instrument was first acquired.) The difference of average value from the actual and standard deviation were calculated; the results are nearly identical despite the six year time span (Fig. 2.4). Analyzing the same representative standard at the beginning of each analysis session after warming up the tube is recommended to more clearly document tube performance over time.

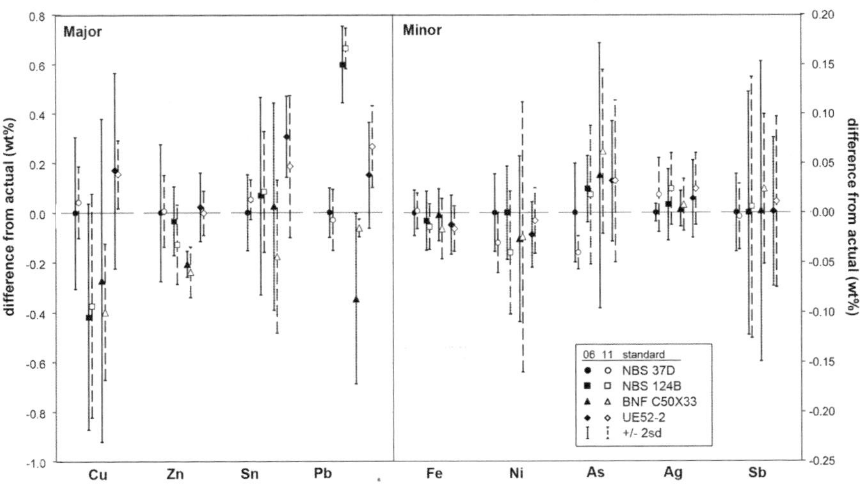

Figure 2.4:
Instrument drift comparison.

The set of copper standards included in the NGA custom calibration has been significantly altered and expanded to obtain more accurate quantification across a broad range of historic alloys. The original calibration provided by the manufacturer in 2006 included thirty-three factory-run standards with a number of gaps in the elemental ranges (Fig. 2.5). Tin bronze alloys, common in Renaissance objects, were not well represented, nor were highly leaded alloys. The zinc contents in the standards were clustered at low and high ends of the range with only a single moderate value. The addition of thirteen certified and reference copper alloy standards from the collection at the NGA has extended and filled in these elemental ranges. Seven factory standards were eliminated, notably those including significant amounts of light elements. Alloys containing these elements would not be expected in Renaissance works and quantifying them accurately would require different run conditions and a distinct calibration.

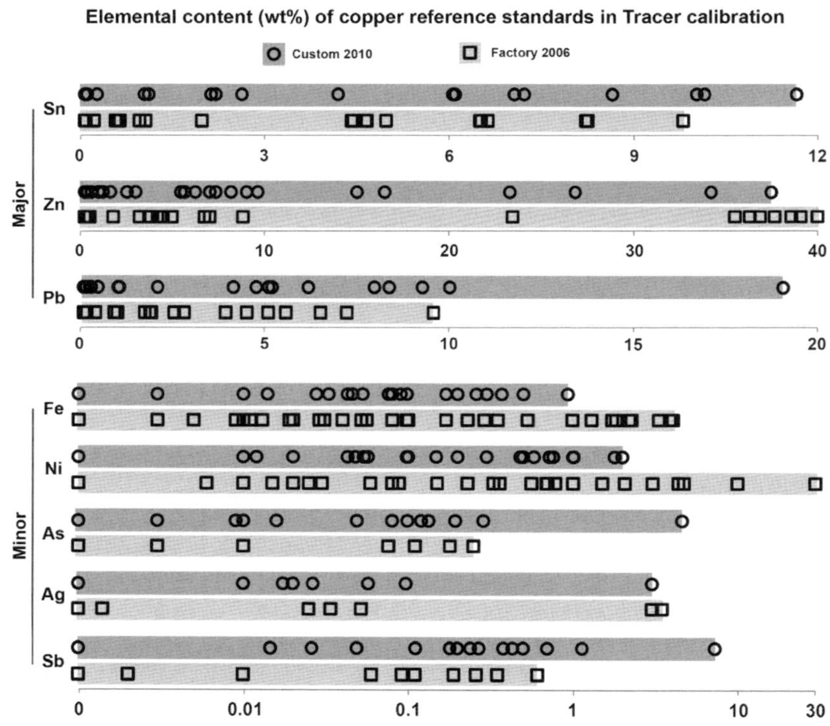

Figure 2.5:
Comparison of current custom and factory calibrations' elemental ranges.

Expansion for arsenical, antimonial, and high lead alloys

The custom calibration has been developed in response to ongoing research on Renaissance objects and the need to assess an ever increasing variety of alloy types. The elemental range of the earliest custom calibrations were based upon alloys typical of 16th c. Florentine works, particular those produced by the workshop of the artist Giovanni Bologna, who had been the focus of several prior studies at the NGA (Sturman 2001, Sturman 2004, Sturman and Glinsman 2004). This custom calibration was found to perform well on similar alloys during analysis of a large group of bronzes by Giovanni Bologna carried out in 2006 (Smith 2011). However, the unintended bias of this calibration for tin bronzes with low trace element contents would ultimately prove inadequate for the full diversity of Renaissance alloys.

One of the first indications of the need for further development of the custom calibration occurred in late 2006 during testing of two reliefs by Vittore Gambello, known as Camelio, at the Ca' d'Oro Museum, Venice. Quantification of spectra from these works produced low overall elemental totals, low reported values for copper, and large negative values. Such errors can be produced by a number of factors. Incorrect geometry can introduce errors, however, this is usually minimized by careful observation of procedure as well as a certain degree of spatial tolerance included in the design of the NGA spectrometer (see 'Distance to Surface' above, and 'Hand-held Analysis' below). Errors have been found to most often result from the surface condition of the tested spot or from a calibration with insufficient range; identifying the actual cause requires confidence in the calibration being applied. In the case of the Ca' d'Oro objects, the alloys contained high levels of lead (in excess of 10 weight percent (wt. %) and significant levels of antimony (2-4 wt. %) and arsenic (above 1 wt. %), elemental contents well beyond the limits of the calibration in use at that time. Subsequent research has identified similar alloys in many Renaissance sculptures, particularly those of late 16th c. Venice; ongoing research on these objects has motivated further development of the custom calibration (Sturman and Smith 2008). It should be noted that the numerical errors produced may not be clearly associated with the element that is out of range; for the Ca' d'Oro objects, the largest errors were in the lead and arsenic values although elevated antimony content appeared to be the primary cause.

How far the value for an element must be out of range to cause a calibration to fail depends on the excitation characteristics of the element. In general, the higher atomic weight elements would be expected have a more significant impact across the entire spectrum. To better understand this effect for the three major alloying elements, three limited calibrations were created with maximum values of approximately 2 wt. % of zinc, lead, and tin respectively. The full standard set was then quantified and the calculated values compared to the actual values. The limited zinc calibration introduced a systematic linear error, increasing every zinc value by approximately 20%. For the limited lead calibration, a linear error was also introduced, but reporting lead values reduced by approximately 12%, much less of an effect than was expected. This result suggests that inhomogeneity, rather than simply elevated lead content, is responsible for much of the error observed in actual objects. For tin, calculated values were dramatically depressed even just beyond the range of the calibration and this effect was non-linear (Fig. 2.6). The complex nature of this effect makes the inclusion of the correct range of tin standards particularly critical in obtaining a meaningful result.

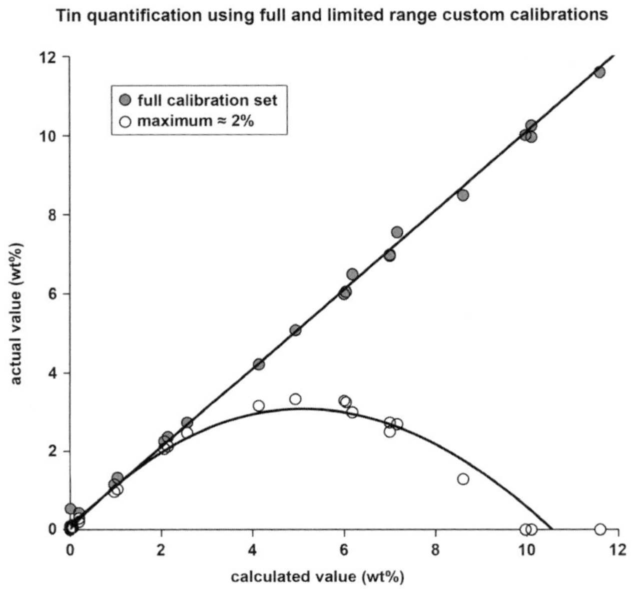

Figure 2.6:
Comparison of tin quantification for full and limited range custom calibrations.

Quantification can be improved by altering the calibration and adding standards in the appropriate compositional range. However, for the types of arsenical and antimonial alloys encountered, no copper alloy standards were available in the NGA collection, nor were they readily commercially available, a problem noted by other authors (Motture 2001, Motture 2003). An interim solution for the custom calibration was inclusion of a group of secondary standards, a group of eight Renaissance medals from the NGA collections. These medals were selected for their arsenical, antimonial and leaded alloys and had the advantage of being actual objects with similar characteristics to the unknown objects to be tested. The drawback of this approach was that the 'actual' values were themselves of limited accuracy; these results had been obtained with the National Gallery's Kevex 0750A XRF spectrometer and quantified with EXACT fundamental parameters software using standards (Glinsman 2007). Even with this limitation, emulating the performance of the well-tested Kevex would mark a significant improvement in the results obtained using the handheld instrument without the secondary standards.

Each medal was analyzed with the handheld spectrometer in the same spots previously tested with the Kevex. These spots were also mapped using a Bruker Artax XRF spectrometer to map surface characteristics and identify medals with pronounced alloy segregation (see Fig. 2.2). At that time, nine more standards were also added to the custom calibration reference set, reducing the size of remaining gaps in elemental content. The total number of reference spectra in the custom calibration became fifty-six: twenty-six factory standards, twenty-two added certified and reference standards, and eight secondary standards. The new calibration produced meaningful results on the Venetian-type alloys, and proved useful in the interpretation of objects for several research projects, although doubts remained about the accuracy in the value ranges covered only by the secondary standards.

Further modification of the set of standards

In 2007, a commercially available standard was located that contained 4.6 wt. % arsenic, part of a set of five standards produced specifically for the study of art objects (European Commission 2001). This arsenical standard provided a new means to test the accuracy of the custom calibration and allowed all but one of the secondary standards to be eliminated; the remaining medal continued to serve as a reference point for unknowns with high levels of lead and antimony. The new calibration allowed prior results from the Venetian investigation to be quantified more accurately for inclusion in published research on the *Sacristy Door* by Jacopo Sansovino in San Marco (Sturman and Smith 2008).

In 2010, the final secondary standard was replaced by a standard with a high antimony content, 7.3% (NBS 54C). Although this standard is a tin-based 'bearing metal', testing suggests that the spectrometer software is able to extrapolate from this extreme point of reference, possibly because the alloy contains some copper. However, inclusion of another bearing metal standard (NBS 53C) failed to compensate for high lead values; it is likely the influence of lead on the spectra is too pronounced for the software to compensate. Ultimately, a commercially available standard with a lead content of 19 wt. % was purchased (IARM 184A) and introduced into the calibration, greatly extending the range of leaded alloys that can theoretically be quantified. However, it must be kept in mind that this high concentration of lead is certain to be much more evenly distributed in this standard than it would be in an actual historical object.

Four more standards (MBH 33XGM4, 33XGM7, 33XGM8, 33XRB1) were acquired by the NGA in 2010. These new standards were also gunmetals--mixed alloys-- found in many Renaissance objects, which were under-represented in the calibration. More importantly, these standards contained trace levels of silver (0.02 wt. % - 0.1 wt. %), levels that had previously been represented in the calibration only by factory standards. The assessment of silver at very low levels is particularly important because its absence can serve as an indicator of alloys that were produced after the introduction of electrolytic refining in the 19th century (Barbour and Glinsman 1993).

In 2012, eleven additional standards (MBH 31X7835-8, 31X7835-9, 31XTB5, 32XLB10, 32XLB14, 32XLB17, 32XSN5, 32XSN6, 33XGM20, 36XCAS3, 36XCAS4) were added to the custom calibration. These standards were developed to advance inter-institutional consistency for XRF analysis and more accurately reflect the diversity of elemental content present in historic alloys (Heginbotham *et al.* 2011). Inclusion of these standards extended the range for antimony (MBH 32XLB17, 4.2 wt. %) and the tin-based standard containing antimony (NBS 54C) was eliminated. These standards also improved the assessment of arsenic with the addition of two standards (36XCAS3, 2.95 wt. % and 36XCAS4, 3.70 wt. %) that raised doubts about the accuracy of the European Commission arsenical standard, which was removed from the calibration. The addition of these standards also significantly expanded the set of trace elements that could be considered, including reference values for chromium, cobalt, bismuth and cadmium.

With the addition of these new standards, the set of reference alloys in the custom calibration was sufficiently broad to eliminate the factory-run standards altogether, allowing for flexibility in the run conditions used for quantitative testing. The acquisition of new standards has also allowed for greater selectivity among the NGA standards, eliminating those that have a less standard preparation (such as compacted granules, rather than cast).

Inter-element corrections

After the set of standards, the inter-element corrections included in the NGA handheld software play the largest role in determining the accuracy of a particular calibration. These corrections must compensate for the complex X-ray physics occurring during excitation, notably matrix effects caused by inter-element excitation and self absorption. A number of significant peak overlaps are also present in copper alloys, including copper Kβ and zinc Kα lines, copper Kα and nickel Kβ lines, lead Lα and arsenic Kα lines, as well as the escape peak of copper with iron Kα line. The computations used for these corrections are not made available to the user in the software for the Bruker handheld instruments; their effect can only be observed in the quantified results. Although these settings are specific to the software used with the NGA spectrometer, the methods used to refine the settings could be applied to other XRF instrumentation.

To determine if a particular correction should be applied, it is useful to observe the effect on a standard with the highest content of the influencing element. For example, when deciding if a correction for lead should be applied to the calculation of tin, the standard with the highest lead value will most clearly demonstrate if the calculated tin values are being shifted correctly toward the actual values. If the impact of an influencing element is small even at the maximum values, the setting should probably not be applied. Excessive inter-element corrections will force the calibration to produce accurate values for the reference standards, but may also result in unpredictable and inconsistent results when applied to unknowns (Shugar 2010).

Usually a reference standard that cannot be fit well for a particular element should not be omitted from the inter-element corrections for that element, even if the spectrometer software provides this option. The inability to fit a particular standard most often indicates a need to change the inter-element settings, however, it may also result from a problem with the spectrum or actual values for that standard included in the calibration. During the development of the custom calibration, a number of errors have been identified among the factory-run reference standards provided by the manufacturer. The trace levels of silver in four of the factory standards (Cu 7131 - Cu 7134) were over-reported by a factor of ten in the original calibration. The trace level of lead in another factory standard (CuB10) also appears to have been over-reported by a factor of ten.

Elements of interest and selecting spectral lines

The elements of interest in the NGA custom calibration have been limited to those considered most relevant to the study of Renaissance copper alloys (major elements: copper, zinc, tin, lead; minor/trace elements: iron, nickel, arsenic, silver, and antimony). For particular investigations or for other historical periods, other trace elements could prove important. Whatever elements are selected must be represented in the reference standards by an appropriate range and diversity of values.

The software for the NGA handheld XRF also allows the spectral lines measured to be selected for each element of interest. This is particularly useful for quantifying elements, such as arsenic, where a significant overlap is present for the Kα peaks with lead Lα lines. Although the software performs a certain amount of modeling of the lead Lα peak based on the Lβ, correctly measuring low levels of arsenic using the Kα lines remains difficult in the presence of significant amounts of lead. In the custom calibration, assessment of arsenic has been shifted to the Kβ peak in an attempt to reduce the impact of this overlap. Advances in the calibration appear to have minimized the difference between measurements made on the two peaks; although the Kβ peak is generally more reliable for low arsenic levels in the presence of lead, the small peak is more sensitive to changes in the overall spectral background. An alternative custom calibration using the Kα peak has been maintained for spectra where overlaps occur with the Kβ peak, such as surfaces that contain traces of gold or mercury from fire gilding. Assessment of other spectral lines, such as the zinc Kβ peak, has been tested, however, none have been found to improve accuracy.

The spectrometer software permits the measurement of multiple lines for a particular element, although for consistency, only one line should be included in the inter-element corrections. Quantification of additional lines can be used to assess the interference of coatings and other surface phenomena. By comparing the calculated values for the tin Kα to the Lα peak or the lead Lβ to the Mα peak, for instance, the preferential absorption of the lower energy line can be identified and measured (discussed below in 'Spot Selection').

Normalizing and reporting

If the analyzed alloy is in the range of the calibration and the spot tested has the proper conditions, the quantified total should be close to 100 wt. %. Experience with the NGA spectrometer has suggested a reasonable tolerance is ±2 wt.%, and, given this narrow range, normalization has little effect on reported values. The copper value is most sensitive to small variations in test conditions, therefore, the practice has been adopted to normalize the total to 100% using only the copper (equivalent to calculating the copper as the remainder). Tests on the calibration set of standards indicate this approach slightly improves the fit of the calculated copper values to the actual content.

Using this quantification method, the calculated element content may be reported as a negative number. In practice, these values are set to zero provided the value is small. Larger negative values indicate a more significant problem, possibly with the sample surface or the range of the calibration, which will require investigation to identify and resolve the issue.

Testing custom calibrations

The factory software for the NGA spectrometer provides a rather narrow understanding of the effect of changes in the calibration settings on the accuracy of a particular calibration. For each standard in the reference set, this software reports the actual value, calculated value, and the difference between these values. An average of these differences across all of the standards is also given which serves as the primary guide to improving accuracy. However, the average difference can be misleading; correction of a single large error might significantly reduce this value, but introduce a small systematic error for every other standard (Shugar 2009). Close inspection of the differences reported for individual standards may identify such a mistake, however, better methods of review are possible. These alternative approaches can be used by manually quantifying and re-quantifying the standards as a calibration is changed, however, these methods are greatly enhanced if batches of spectra can be readily processed and exported.

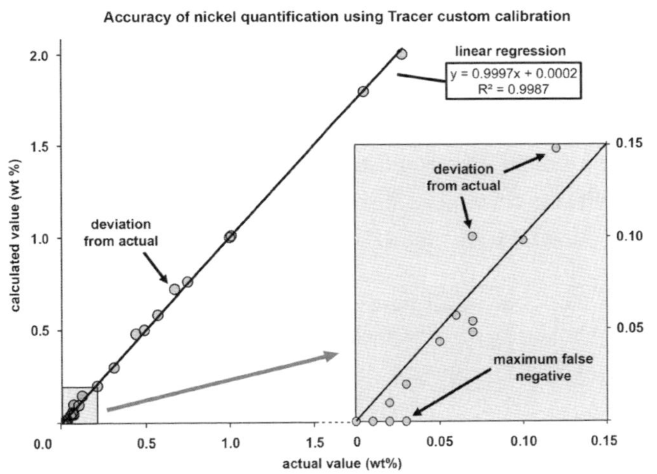

Figure 2.7:
Elemental actual/calculated fit for nickel using custom calibration.

The relationship between the calculated and actual values for the calibration reference standards is much more easily understood graphically (Fig. 2.7). Using a best fit linear regression plot, poorly fit individual standards, as well as more significant systematic errors, are easily noted (see 'Comparison of factory and custom calibrations' below). Inspection of the lowest values reveals the concentrations at which false positives and negatives are reported, providing one indication of the limits of detection (LOD), the minimum content of an element that can be identified (but not necessarily quantified) with certainty. The exported data also allows other measures of accuracy to be calculated. At present two indicators of accuracy are used, an average of the relative errors for each element (the difference between the calculated and actual values as a percentage of the actual) and the linear regression fit of calculated to actual values. Measurements of relative error will tends to overemphasize small errors in trace values; to produce a more representative measure of accuracy, standards with values below the LOD are not considered.

Limits of detection and quantification

Testing of the NGA handheld spectrometer was carried out to more accurately estimate the LOD and the limit of quantification (LOQ) for all of the elements of interest (Thomsen 2003, Shugar 2009). Eight standards from the calibration set were selected that included a range of low values for the nine elements being assessed. Each standard was analyzed ten times for 60 seconds to determine the standard deviation for each reported elemental value, which was then used to calculate a relative standard deviation (RSD), the standard deviation as a percent of the actual value. The RSD was plotted against the actual value for each standard and a best fit power regression was determined (Fig. 2.8). The regression equation was used to predict the corresponding elemental content at RSD = 10% and RSD = 33%, the IUPAC standards for the LOD and LOQ respectively (Fig. 2.8, Table 2.2).

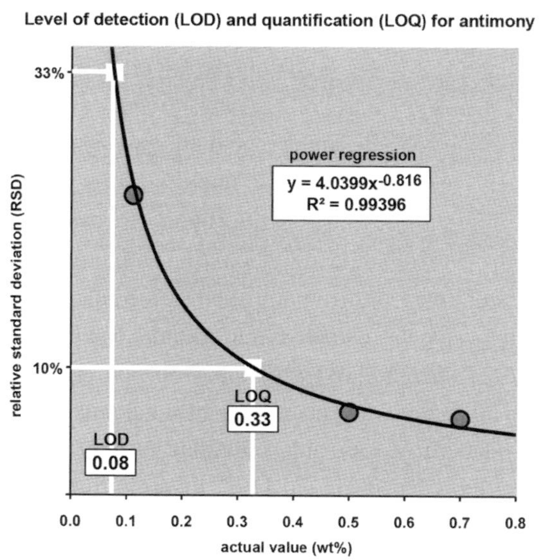

Figure 2.8:
Calculation of antimony LOD and LOQ using RSD power regression method.

An alternate estimate of precision based on spectrum background variation was calculated with a series of ten 60 second tests on a pure copper standard; three standard deviations provides an estimate of the LOD and ten standard deviations the LOQ (Table 2.2). For the NGA instrument using the custom calibration, the LOD was under 0.1 wt. % for all of the elements; for most elements the LOQ values were around 0.2 wt. %, with slightly higher values for antimony (0.33 wt. %), arsenic (0.38 wt. %), and tin (0.45 wt. %). Most of the LOD and LOQ calculated from the copper background are similar, suggesting this test, which requires only a single standard, is also effective at estimating instrument precision. However, significant discrepancies are noted, such as in the LOQ for tin and arsenic. If the necessary standards are available, the RSD estimates, based on a range of alloys, should provide a better indication of the performance of the spectrometer and calibration on historic alloys.

		Zn	Sn	Pb	Fe	Ni	As	Ag	Sb
Tracer III-V	Actual/RSD regression								
	LOD (RSD = 33%)	0,05	0,10	0,03	0,03	0,04	0,07	0,03	0,08
	LOQ (RSD = 10%)	0,22	0,45	0,20	0,11	0,16	0,38	0,22	0,33
	Pure copper background								
	LOD (3sd)	0,05	0,06	0,03	0,03	0,05	0,05	0,03	0,08
	LOQ (10sd)	0,17	0,20	0,10	0,10	0,16	0,15	0,10	0,27
Tracer III-SD	Actual/RSD regression								
	LOD (RSD = 33%)	0,02	0,02	0,01	0,01	0,01	0,03	0,01	0,02
	LOQ (RSD = 10%)	0,07	0,15	0,04	0,04	0,05	0,11	0,06	0,09
	Pure copper background								
	LOD (3sd)	0,02	0,03	0,00	0,01	0,02	0,02	0,01	0,02
	LOQ (10sd)	0,07	0,09	0,00	0,03	0,06	0,07	0,04	0,05

Table 2.2:
Calculated LOD and LOQ using RSD power regression method and pure copper analysis.

The levels of precision indicated by these tests for the NGA spectrometer using a custom calibration are adequate for meaningful quantitative analysis of Renaissance copper alloy sculpture. Because of the limitations of historical refining processes, even the most refined alloys, such as those used in the Florentine workshop of Giovanni Bologna, contain variable traces of all of the elements of interest (0.1 – 0.4 wt. %) (Smith 2011). The presence of these low levels in every alloy, considered with the variation typically found within an individual object, means that elemental content ranges below the LOQ for the NGA handheld instrument are unlikely to play a significant role in interpreting the relationships between sculptures. However, the LOD for a particular spectrometer must be considered carefully when trying to determine with certainty if trace elements are present or absent.

Comparison of factory and custom calibrations

The NGA custom calibration demonstrates significant improvements in accuracy over the original factory calibration in assessing the range of alloys included in the current standard set (Table 2.3). For all elements, the custom calibration reduced the average difference from the actual values and improved the linear regression fit of the calculated values to the actual. The custom calibration performance was significantly improved at higher levels of lead; the factory calibration reported large differences from the actual for lead concentrations above 9 wt. %. Differences between the calibrations were more pronounced in the minor elements, nickel, arsenic, and antimony; the factory calibration demonstrated systematic errors where the calculated content deviated consistently across the range of standards (Fig. 2.9). The custom calibration demonstrates none of these defects and quantifies minor elements more accurately.

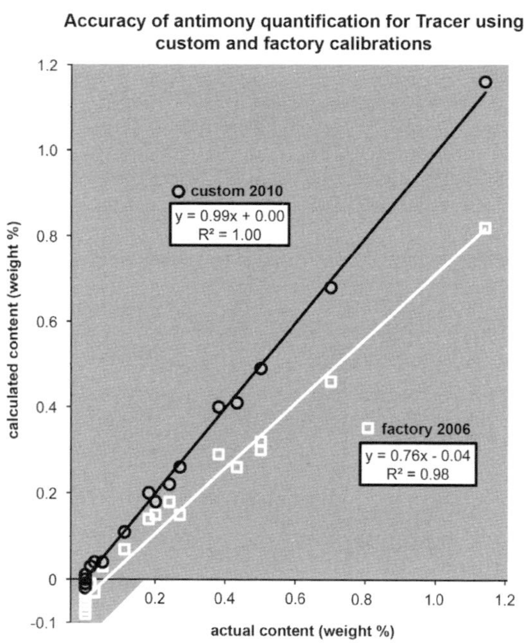

Figure 2.9:
Antimony fit of current custom and factory calibrations.

standards			difference from actual			linear regression fit		
	avg wt%	cal	avg	%	max	RSQ	slope	int
Cu	83,08	Cus	0,37	0	1,27	1,00	1,00	0,23
		Fac	1,29	2	2,66	0,97	0,94	4,81
Zn	9,50	Cus	0,12	7	0,47	1,00	1,00	-0,00
		Fac	0,30	22	0,43	1,00	0,98	0,01
Sn	4,83	Cus	0,05	7	0,14	1,00	1,00	0,00
		Fac	0,43	20	0,04	1,00	0,95	-0,14
Pb	4,68	Cus	0,19	14	0,63	1,00	1,00	0,01
		Fac	1,52	51	13,70	0,96	1,49	-0,61
Fe	0,23	Cus	0,01	8	0,05	0,99	0,99	0,00
		Fac	0,02	13	0,06	0,98	0,89	0,03
Ni	0,61	Cus	0,02	7	0,03	1,00	1,00	0,00
		Fac	0,07	18	0,34	0,99	1,14	-0,03
As	0,71	Cus	0,03	17	0,10	1,00	1,00	0,00
		Fac	0,30	65	2,25	0,87	1,30	0,19
Ag	1.06*	Cus	0,003	5	0,02	1,00	1,00	0,00
		Fac	0,045	24	0,09	1,00	1,03	-0,00
Sb	0,42	Cus	0,02	5	0,02	1,00	0,99	0,00
		Fac	0,14	32	-0,02	0,98	0,76	-0,04

* trace values, except one standard with 3.03%

Table 2.3:
Current calibration vs. factory fit measurements.

Comparison to other instrumentation

Throughout the development of the custom calibration, results for various Renaissance objects analyzed with the Bruker Tracer III-V have been compared to those obtained on other instrumentation at the NGA, particularly the Kevex 0750A XRF spectrometer (Glinsman 2004). Quantification of the group of Renaissance medals previously used as secondary standards, but no longer in the calibration, provides an indication of the ability of the current custom calibration to generate meaningful values for these complex alloys (Table 2.1). These results are comparable in accuracy to those reported by Kevex, which is not the case for the original factory calibration.

A more detailed comparison of the Kevex and the handheld spectrometer was included in a recent study examining quantitative results on thirteen alloys from a group of Renaissance statuettes tested at various times with the two instruments (Smith 2011). This study begins to address the alloy variations found within an individual object relative to the differences in content reported by different instruments. For the major elements, the variation measured between the handheld XRF and Kevex (inter-instrument) was less than that reported between analyzed spots on a single object (intra-object), suggesting that these data comparisons between the instruments could be made without further corrections (Table 2.4). Intra-instrument differences did exceed the intra-object variation for some of the copper and tin values, which appears to result from the normalization of the Kevex results. The inter-instrument differences were slightly greater than the intra-object differences for the minor elements. For nickel, a 0.1 wt. % artifact had previously been identified in the Kevex, which was approximately the difference reported by the handheld spectrometer. The cause of the difference in antimony values for the two instruments was less clear, nor was it apparent which was reporting more accurately. However, the antimony values in the tested objects were within the range of standards included in the handheld calibration and values for those standards were calculated correctly. With some reservations and corrections, the outcome of this testing indicated that data from the two instruments could be meaningfully integrated for the purposes of the study. More importantly, this study also suggested that the level of accuracy for the handheld spectrometer using an empirical method was approaching that obtained by the well-established Kevex using fundamental parameters.

	comparative values (wt%)			**intra-object variation (IOV)**				**inter-instrument difference (IID)**				**linear regression fit**		
	avg	**min**	**max**	**Tracer**		**Kevex**		**avg**	**%**	**max**	**# of IID > IOV**	**RSQ**	**slope**	**int**
				2sd	**%**	**2sd**	**%**							
Cu	86,73	80,63	91,81	0,86	1	1,27	1	0,75	1	2,34	4	0,98	0,80	16,60
Zn	2,17	0,11	8,53	0,30	14	0,25	11	0,16	8	0,29	0	1,00	0,95	0,19
Pb	3,97	0,98	10,03	0,83	21	2,06	52	0,22	6	0,53	0	0,99	0,99	0,02
Sn	6,08	2,45	9,47	0,68	11	0,63	10	0,52	9	1,38	3	0,92	0,96	0,61
Fe	0,27	0,09	0,55	0,11	41	0,13	50	0,14	54	0,38	9	0,04	0,18	0,20
Ni	0,16	0,10	0,25	0,06	39	0,10	62	0,11	72	0,20	10	0,57	0,69	0,17
As	0,20	0,14	0,34	0,09	47	n/a	n/a	n/a	n/a	n/a	n/a	n/a	n/a	n/a
Ag	0,09	0,03	0,19	0,04	43	0,01	12	0,03	29	0,06	4	0,70	0,60	0,02
Sb	0,34	0,16	0,69	0,08	23	0,13	40	0,18	55	0,29	9	0,83	1,18	0,12

Table 2.4:
Comparison of quantitative analysis of Renaissance bronzes using Tracer and Kevex.

The NGA spectrometer has also been included in two systematic investigations of quantification relative to other analytical instrumentation. In 2006, a workshop held at the Getty included comparative testing of four unknown copper alloys by thirteen institutions using nineteen different instruments, including handheld and lab-based XRF, as well as scanning electron microscopy-energy dispersive spectroscopy and -wavelength dispersive spectroscopy (Getty Conservation Institute 2006). Surprisingly, the results obtained with the NGA spectrometer using an early custom calibration were found to be the most accurate for a set of copper alloy standards based on the sum of absolute errors. The NGA quantified results had a sum of absolute errors of 4.9% for the four samples tested, compared to an average score of 21% among the 19 institutions participating in the workshop. Although the unknowns were standards and less complex than historical alloys, it appears significant that the most pronounced differences between the NGA spectrometer and other instruments were reported for the standard with the most diverse elemental content (Cu_WSB7). It should be noted that three of the standards tested were present in the NGA calibration set, however, the NGA instrument outperformed two other Tracer III-V spectrometers that also had these standards available, again emphasizing the importance of the custom calibration.

Another recent study of instrumental accuracy compared analyses from nineteen instruments at different institutions on copper alloy standards as well as actual historical material (Heginbotham *et al.* 2011). Like the prior informal study, the results demonstrated that the NGA spectrometer could obtain levels of accuracy similar to those of more established instrumentation, and that identical Tracer III-V instruments produced very different results depending on calibration. This study also indicated that further effort is required by the various institutions to obtain consistent alloy results across different types of analytical instrumentation. In order for data to be combined in future meta-studies, greater collaboration will be required to develop more consistent acquisition procedures and more accurate methods for quantification (National Gallery of Art 2007, Heginbotham *et al.* 2011).

Discussion

Future improvement of the NGA custom empirical calibration remains possible through introduction of additional standards with a greater range and variety of antimony, arsenic, and lead contents. Discussion is ongoing among a number of institutions about the production of additional standards specifically for the study of art (National Gallery 2007, Heginbotham *et al.* 2011). Partitioning the current general calibration into two or more specialized calibrations for specific alloy ranges—brasses, bronzes, or leaded alloys—would be expected to improve accuracy. However, preliminary testing of calibrations focused on elevated ranges of zinc, tin, and lead, produced similar results to the general calibration when tested on standards. Tests on a recently obtained standard with higher tin (MBH 32XSN5, 16.05 wt.%) suggested that this content—5% more than the previous maximum—is beyond the range that can be quantified accurately within a single broad calibration. The quantification method supplied by the manufacturer could also be replaced entirely by third party software that would provide either an empirical method with greater transparency or a method based on fundamental parameters (Heginbotham *et al.* 2011). Whatever quantification method is applied must remain practical for application outside of a lab environment; alternatively, two methods could be used, one for collection in the field and one for final processing.

An important recent development in handheld XRF analysis is the introduction of silicon drift detectors (SDD) which have an improved resolution of 145 ev. A Bruker Tracer III-SD was acquired by the NGA in 2010 and testing has indicated significant improvements over the PIN diode instrument. A simple comparison of count rates on a copper alloy standard (MBH 33XGM8) found that raw cps increased by nearly a factor of six (10800 to 64700), valid cps by nearly a factor of nine (6300 to 55500), and deadtime was significantly reduced (48% to 15%). With the greater capacity and resolution of the SDD detector, the level of precision at 60 seconds is approximately three times that of the PIN diode instrument (see Table 2.2). However, for many of the Renaissance alloys analyzed, the improvement in accuracy provided by the SDD detector over the PIN diode may be negligible because of the variation present in the objects themselves. Improvement would be expected for the detection and quantification of trace elements with greater confidence, which can be a critical factor in establishing ore processing methods and the associated date of production (Barbour and Glinsman 1993).

Working methods

In parallel to the development of the custom calibration, significant changes have been made to the working methods used to acquire spectra with the NGA handheld spectrometers. These changes have been prompted primarily by practical experience during analysis of objects in the field. A summary of the current NGA working method is given at the end of this paper.

Handheld analysis

Early tests on sculpture at the NGA demonstrated that working handheld with the spectrometer is preferable to use of a tripod. Handheld analysis provides the necessary flexibility in examining the complex surfaces of many objects, particularly when they cannot be moved, because of size or installation. Using the spectrometer handheld also significantly reduces the set up time required for each acquisition by eliminating the need to repeatedly reposition the tripod or, less desirably, the object itself. Economizing the time for each acquisition can greatly reduce the total time necessary for a project, particularly when analyzing a large group of sculptures or a single complex object. When performing analysis on site, working efficiently is critical, because access to the works of art is usually very limited. Time for analysis and examination is always likely to run short, because of the unpredictable nature of examining objects outside of a controlled environment.

In order to work handheld, the Tracer must be used in contact with the surface of the work of art to maintain position. The surface condition and structural stability of each object and each spot on that object should be carefully considered prior to analysis. Handheld analysis will not always be appropriate. When working at another institution, the requirement of surface contact should be clearly explained in the planning stages as well as just prior to the start of analysis. An advantage of working in contact is that the distance to the tested surface may be more reproducible. Using a tripod, this distance has to be re-adjusted for each run, without the advantage of focusing lasers present on many lab-based instruments.

In practical application, when working hand-held, it is difficult to maintain a fixed position for much longer than 60 seconds. For this reason, test locations should always be selected to minimize the physical effort necessary to support the instrument. (Note that this is 60 seconds of actual time, as reported in S1PXRF; live time, computed using the raw and valid count rates, is typically about 2/3 of the actual depending on the alloy). As with any XRF instrument, it is important to establish the correct orientation to the surface for analysis. Tests were carried out to evaluate the effect of instrument-to-surface angle on quantification. Three 60 second acquisitions were run on the same location on a copper alloy standard (MBH 33XGM8) with the surface oriented at 90° (optimum), 95°, 100°, and 105° (Fig. 2.10). Even at 9° off axis, little change is noted in the quantification of the major elements (although greater variation in composition has been noted by Mass and Matsen for silver alloy quantification (this volume). At 12° off axis, an effect is clearly noted which is roughly equivalent to introducing about 0.5 mm of distance, however, the effect is not identical (see Fig. 2.3). These tests also demonstrate that the markedly different effects of geometry on the counts measured and the quantified value for a particular element, as well as the variations noted for different elements.

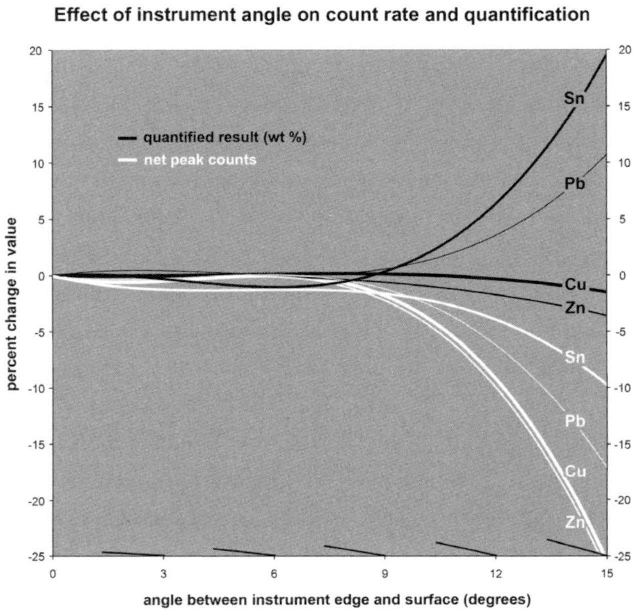

Figure 2.10:
Effect of instrument angle on quantification.

Time, precision, limits of detection and quantification

Reducing the acquisition time reduces the precision of XRF analysis and increases the probable error in alloy measurements, therefore, longer acquisition times are always desirable. Decreasing the number of counts collected for each peak being measured increases the impact of random fluctuations present in the spectral background. This effect is more pronounced for smaller peaks associated with lower elemental contents or limited excitation, and may be further exacerbated by the presence of peak overlaps.

To assess the impact of acquisition time on precision for the NGA spectrometer, the same spot on a copper alloy standard (MBH 33XGM8) was analyzed ten times for four different durations (15, 30, 60 and 120 seconds). The resulting spectra were then quantified using the current custom calibration. As expected, the standard deviation for each elemental value tends to decrease as time increases because of the exponential relationship between background and peak variability (Table 2.5). However, for most elements, little difference is observed for values obtained at 60 and 120 seconds (twice the standard deviation provides an approximate 95% confidence interval). The small variations observed at 60 seconds (0.1 – 0.3 wt. % for the major elements, 0.02 – 0.04 wt. % for the minor elements) will tend to be overwhelmed by the alloy variation found between spots in a typical Renaissance sculpture (Smith 2011). The high percentages of variation present for all acquisition times for arsenic and antimony indicate that the values present in the tested standard are well below the LOD.

	C33XGM8		15 secs		30 secs		60 secs		120 secs	
	actual wt%		2sd	%	2sd	%	2sd	%	2sd	%
Cu	83,81	+/-	0,44	0,5	0,37	0,4	0,25	0,3	0,21	0,2
Zn	5,45		0,25	5	0,13	2	0,11	2	0,10	2
Sn	4,13		0,33	8	0,26	6	0,17	4	0,12	3
Pb	6,11		0,27	4	0,18	3	0,15	2	0,10	2
Fe	0,10	+/-	0,03	31	0,03	28	0,02	20	0,02	23
Ni	0,15		0,04	28	0,03	22	0,04	24	0,02	15
As	0,02		0,04	278	0,02	136	0,02	122	0,03	212
Ag	0,10		0,05	51	0,04	39	0,04	38	0,04	38
Sb	0,01		0,07	492	0,06	406	0,03	233	0,04	252

Table 2.5:
Effect of acquisition time on elemental precision of Tracer.

Spot selection

In practice, spots are typically selected on Renaissance bronzes where the original organic coatings are thin or absent to minimize the potential impact on XRF analysis of the underlying metal. However, it is not always possible to select uncoated areas. The question has been raised recently whether spots lacking the coating are necessarily the most representative of the original alloy; the coating has usually been removed by wear, which can also affect the alloy content of the surface. An assessment of the influence of organic films on quantification was done using Mylar® film ($(C_{10}H_8O_4)_n$; 36 micron thickness) (Wilson *et al.* 1999). The density of this film (approx. 1.38 g/cm^3) is similar, but slightly denser than a pure drying oil coating, which is around 0.91 g/cm^3 (Feller *et al.* 1985). Renaissance bronze coatings may include pigments, however, the most common--carbon black--would not significantly alter the density or response of the film (Stone 1990, Pittard *et al.* 2007, Stone 2010). The color of many translucent patinas has also been shown to be due to the preparation of the oil itself rather than added colorants.

A series of three 60 second acquisitions were collected on the same spot on a copper alloy standard (MBH 33XGM8) placing layers of Mylar® over the surface (from 1 up to 5 layers). The count rates and quantified values were determined for each of the major elements (Fig. 2.11). The predicted attenuation for each element was also calculated using the unattentuated count rate and the attenuation coefficients for Mylar® (Hubbell and Selzter 1996). For copper and zinc, the predicted attenuation conforms closely to the measured count rate, however, in the quantified values, little change is observed due to normalization to the total count rate in the Bruker software. For tin and lead, although a slight attenuation is predicted, in the measured counts, there was in fact a slight increase in counts, which appears to be at least partly due to an overall shift in the background level. This increase is further amplified in the quantified values for tin and lead by the normalization process, as the counts measured now represent a greater portion of the total. Overall, the non-uniform attenuation for elemental peaks at different energies, combined with the normalization of peak measurements, results in significant errors in the quantified values with increasing film thicknesses.

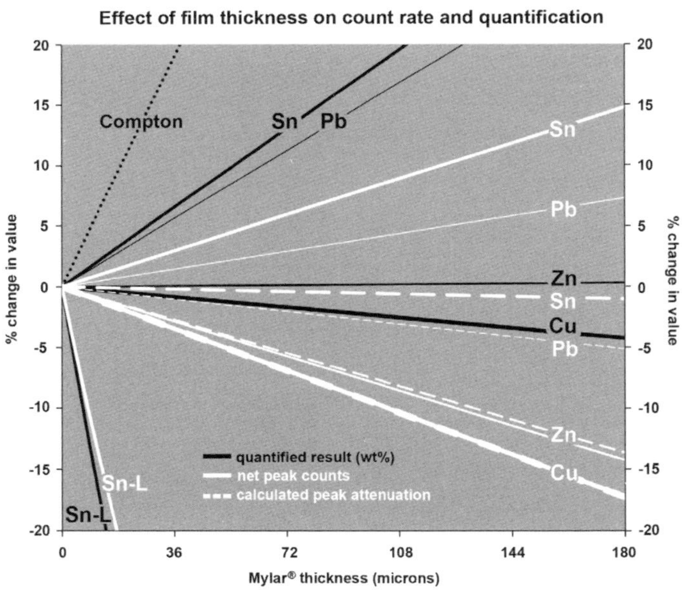

Figure 2.11:
Effect of film thickness on count rate and quantification.

Fortunately, these phenomena do not appear to play a significant role in XRF analysis of Renaissance bronzes. For layers of 100 microns or less, the effect on the count rate and quantified value is minor and would be difficult to separate from variations due to instrumental error and small alloy variations within the analyzed object. The presence of a thinner surface coating may be more readily recognized through comparative measurements of the Compton peak (18.5 – 19.7 keV) or the tin Lα line (Fig. 2.11). Reviewing spectra taken from a variety of Renaissance bronzes, the attenuation of the more sensitive Compton peak by the film was only clearly noted on a few objects with unusually thick coatings. Tests performed on a simulated unpigmented Renaissance organic patina suggests the thickness of these films is less than 20 microns (sample prepared by Richard Stone, see Stone 2010).

Acquiring spectra in the field

During analysis of works of art outside of the lab environment, it is particularly critical to obtain high quality spectra, because performing additional analyses at a later date may be impossible. Without reliable spectra, quantitative results cannot be obtained and even qualitative interpretation may prove difficult. Working on site, it is important to emulate lab practices, including careful selection of test areas with the minimum texture and curvature that fill the field of the beam, as well as maintaining consistent test geometry. The additional challenges of working in the field require close monitoring of the spectra during acquisition and immediate review of the quantified results.

The most important change in the working method for the NGA spectrometer is the consistent use of a laptop and S1PXRF commercial software for the acquisition of spectra. In the generic field method presented by the manufacturer, spectra are acquired with the software PMI on a PDA attached to the spectrometer and only transferred and fully processed on a computer at the end of a session. This technique has been used successfully with the NGA spectrometer, notably to examine works by the Florentine artist Giovanni Bologna while on exhibit at the Kunsthistorisches Museum, Vienna in 2006 (Smith 2011). However, these analyses were performed under nearly ideal conditions, in a closed gallery with good lighting, on sculptures that were well-studied, in good condition, and easily accessible.

Experience during subsequent projects has indicated the importance of collecting spectra with the added capabilities available using a laptop. At the end of 2006, a team organized by the NGA performed a preliminary survey of bronze sculpture in Venice, including Jacopo Sansovino's *Sacristy Door* in the Church of San Marco (Sturman and Smith 2008). Data was collected on the PDA and reviewed after the study session was complete. Many of the spectra collected on the lowest registers of the door were found to have depressed count rates and produced inconsistent quantified results. This was partly due to the presence of alloys beyond the calibration range at that time, which was later extended and the data re-quantified. However, difficulties in certain spectra persisted, likely due to the interference of dust and debris present on the surface during acquisition. Using PMI on the PDA provided insufficient information during the acquisition process to recognize and potentially remedy such problems.

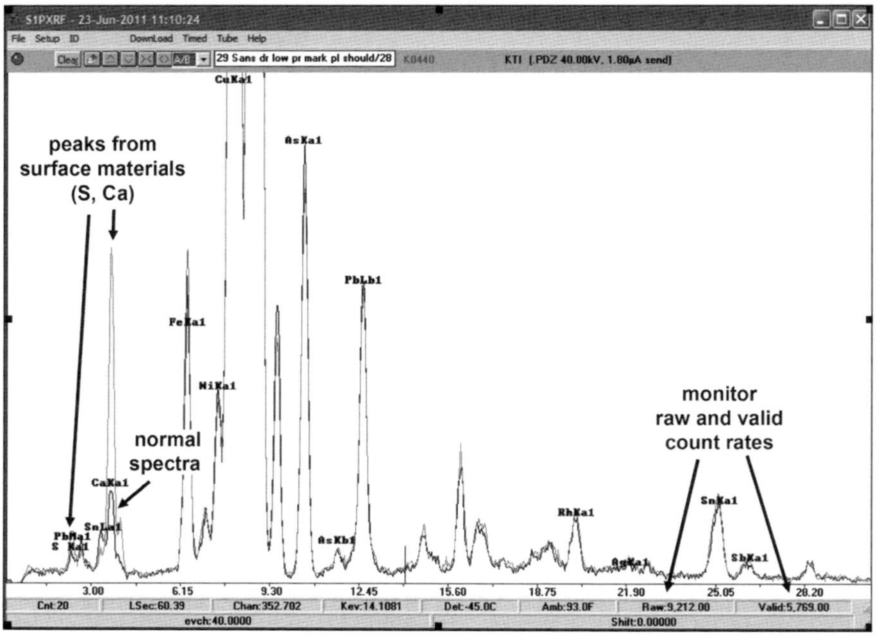

Figure 2.12:
Examination of spectra using S1PXRF.

Using a laptop for acquisition, each spectrum can be examined immediately for unexpected peaks contributed by surface materials that might interfere with quantification (Fig. 2.12). Using the small image available on the PDA, this is extremely difficult. In S1PXRF, the raw and valid count rates can also be monitored to insure the quantified results fall into expected ranges and are generally consistent from spot to spot. For works of art outside of one's own collection, any sort of cleaning prior to analysis may not be appropriate or permitted. Direct monitoring of the spectra using a laptop makes it possible to identify surface contamination and select the optimum locations for analysis. Monitoring count rates can also help identify areas with the least curvature and texture. On surfaces that prove particularly difficult to analyze, a series of short untimed assays can be run, slightly shifting the instrument each time, until the best spot for a timed analysis is identified. Even when applying these strategies, additional analyses should always be made with the expectation that certain spectra may have to be excluded from the final results due to problematic surface conditions.

Reviewing and assembling data

Working with a laptop, each quantified result can be copied from the spectrometer software and collected in a spreadsheet. Developing methods to more rapidly export and view data allows each quantified result to be easily inspected for indications of non-optimal spectra, such as low elemental totals and large negative numbers. Assembling the alloy data during acquisition will also help to reveal patterns within an object or a group of objects being analyzed. This instant feedback allows XRF analysis with the Tracer to be more fully integrated into the process of examination and proved invaluable during research in the Church of Saint Anthony, Padua, on the monumental *Paschal Candlestick* by Andrea Briosco, known as Riccio. Immediate review of the alloy results assisted in the identification of joins between separately cast sections that were not readily visible (Sturman *et al.* 2009). A series of spots were analyzed across an area of interest until a significant shift in the alloy revealed the location of the join.

Working in teams

Another significant improvement in the working method for the NGA handheld spectrometer is the division of the task of acquisition between two people. Working as a team, one person can operate the XRF and focus on inspection of the surface, spot selection, and holding the instrument steady. The second person operates the laptop, monitors count rates during acquisition, inspects and organizes the quantified results, and documents spot locations. Although one person can perform analyses, the object and laptop must be set up in close proximity, so that both can be easily observed during acquisition. For the investigation of the massive *Paschal Candlestick*, much of the work was done on scaffolding with team members often working several meters apart by using long USB extension cables. Team members can also periodically trade tasks to reduce the physical strain of supporting the XRF during difficult and extensive research sessions. In Padua, the team approach was taken a step further—an additional team with a second instrument examined other bronzes in the church while the first team worked on the *Candlestick*. Prior to the trip, the second instrument had been prepared with the latest custom calibration to insure that results conformed on the two spectrometers.

Safety

Although the spectrometer is designed to be used safely hand-held and operates at very low energies, certain precautions should be observed in addition to those recommended by the manufacturer. Works should be examined in a location where access can be controlled by the operator and areas clearly designated as safe and unsafe. The operator should pay close attention to the location of people in the vicinity of the analysis area and be ready to interrupt analysis if necessary. For the operator, the most likely form of exposure is the accidental placement of a hand near the front of the instrument during analysis. For this reason, at the NGA, ring dosimeters are worn on the non-trigger hand, rather than the standard badges. During analysis of copper alloys, the object itself is of sufficient density to absorb the incident X-ray beam and the close proximity of the instrument blocks the reflected beam. However, operators should be aware that the presence of space between the instrument and object, including that introduced by the thin layer of foam, diverges from standard practice and creates a potential path for scattered radiation to escape. The risk of exposure is significantly increased when the spot being tested is smaller than the X-ray beam. Extra care should be taken in these situations to minimize the risk of exposure and ideally the instrument directed toward an adequate barrier. It is recommended to test the potential level of exposure in the laboratory using a Geiger counter while operating the instrument under conditions that reflect the range of operating conditions in the field, including those that are less than optimal.

Conclusion

The ultimate test of the validity of any analytical method applied to works of art is whether it can provide meaningful insight into the objects being studied. In this sense, handheld XRF has proven its usefulness by contributing vital information toward the study of Renaissance bronzes. Alloy analysis provided has help to confirm authenticity and help establish attribution (Cristanetti *et al.* 2007), link material use and historical documentation (Sturman and Smith 2008), reveal construction methods (Sturman *et al.* 2008), and indicate authorship through changes in workshop practices (Smith 2011). Neither the factory quantification method provided for the spectrometer, nor the standard field approach using a PDA, has proven adequate for the accurate assessment of historic copper alloys. Successful application of the handheld spectrometer to the study of Renaissance bronzes has been achieved through creation of a custom empirical calibration and improvement of the working methods, refined through the experience of analyzing objects in the field. Successful analysis must recognize the inherent limitations of surface analysis, but also consider the actual level of accuracy necessary to meaningfully characterize historic materials. Ultimately, no study of Renaissance bronzes can draw conclusions based solely on alloy analysis and must be considered alongside other forms of technical investigation, notably X-radiography, as well as traditional art historical methods. Although a great variety of copper alloys have been observed in Renaissance objects, the range of alloys that can be cast is inherently limited. Identical alloys may occur in two completely different contexts, and the interpretation of the significance of that alloy will be dependent upon context.

Handheld XRF has the potential to transform the study of Renaissance bronzes. By taking the technology to the art, it is possible to select and analyze the most securely dated and attributed works to represent the production of a particular artist, rather than relying solely on those objects that happen to be in institutions with analytical resources. By increasing the number and variety of objects analyzed through application of portable technology, a coherent picture of the full 'alloy landscape' for Renaissance bronzes can be assembled, connecting the islands of data previously gathered by lab-based instrumentation. However, even with handheld instrumentation, analyzing every work of interest with the same instrument to insure consistency in the results is impractical. Obtaining a broad understanding of Renaissance bronzes will also require the promotion of inter-instrument accuracy to allow the successful creation of alloy meta-studies that combine the technical resources of many institutions.

Acknowledgments

I would like to thank Jennifer Mass and Aaron Shugar for their invitation to contribute my experiences to this publication.

Many current and former colleagues at the National Gallery have contributed to this research and collaborated on the projects described in this paper, including Shelley Sturman, Nicholas Penny, Lisha Glinsman, Karen Serres, Simona Cristanetti, Alison Luchs, Debra Pincus, and Kathryn Morales. All of the members of the Department of Object Conservation at the National Gallery have supported my work, as well as the Department of Scientific Research, and the Department of Sculpture and Decorative Arts. Bruce Kaiser and Bruker AXS have offered crucial technical advice throughout my investigations. Robert H. Smith and the Smith Family Foundation have provided continuing support of my research at the National Gallery.

Much of my research has been performed carried out at other institutions with the generous cooperation of many colleagues, including: the Ufficio Tecnico of the Basilica of San Marco; Victoria Avery, Fitzwilliam Museum; the Veneranda Arca del Santo and the friars of the Basilica of Saint Anthony, Padua; Davide Banzato, Musei Civici di Padova; Claudia Kryza-Gersch, Kunsthistorisches Museum; Denise Allen, The Frick Collection; James Draper and Richard Stone, The Metropolitan Museum of Art; Philippe Malgouyres, Musée du Louvre; Johann Kräftner, Liechtenstein Museum; Manfred Leithe-Jasper; Patrizia Piergiovanni, Galleria Colonna; Adriana Augusti, Ca' d'Oro; and Beatrice Paolozzi Strozzi, Museo del Bargello.

A very special thanks to my wife, Patricia Favero, whose patience and support made my research and this paper possible.

Summary of current NGA working method

Prior to research

- Estimate the types of alloys that may be present in the objects to be studied based upon available research.
- Develop a custom calibration appropriate to the alloys expected.
- Use the number of objects and knowledge of their construction to estimate the number of analyses and the time required to perform them. Be as generous as possible within the practical limits of working at the particular site.

During research

- Warm up instrument to insure settings are correct and tube has stabilized.
- Analyze a representative standard prior to each session to record tube characteristics.
- Work hand-held to maximize efficiency and access to all parts of the study objects.
- Use an acquisition time of 60 seconds for spectra to be quantified; shorter times can be used for qualitative results.

- Work in teams whenever possible. If working alone, arrange object and laptop so that both can be easily observed during testing.
- Examine each object carefully prior to analysis to determine optimum spots for testing: clean, uncoated, and representative.
- Be certain to position the instrument so it can be held stably and comfortably for the duration of the analysis.
- Use a ring dosimeter on the hand not holding the instrument and establish safe and unsafe areas for personnel prior to operation.
- Monitor count rates during acquisition. Note typical ranges for the object and confirm consistency from spot to spot.
- Inspect spectra and quantify data while working. Compile data while working and review for variations due to problems or areas that require additional analyses.
- Process and inspect results as much as possible while on site to allow for re-testing and/or additional testing.

AFTER RESEARCH

- Evaluate errors that occurred during testing to determine whether the cause is surface conditions of the object or the quantification of the custom calibration.
- Re-evaluate effectiveness of custom calibration. Alter and expand calibration as necessary.

References

Barbour, D., and L. D. Glinsman. 1993. "An Investigation of Renaissance Practices as a Means for Identifying Forgeries." In *Conservation Research, Studies in the History of Art*, 14-29. Washington: National Gallery of Art.

Bassett, J., P. Fogelman, D. A. Scott, and R. C. Schmidtling. 2008. *The craftsman revealed: Adriaen de Vries, sculptor in bronze*. Los Angeles: Getty Conservation Institute.

Bewer, F. 2001. "The Sculpture of Adriaen de Vries: A Technical Study." In *Small bronzes in the Renaissance*, edited by Debra Pincus, 159-193. Washington DC: National Gallery of Art.

Bormand, M. 2008. "Catalog entry for Antoninus Pius and Faustina." In *Bonacolsi l'Antico: uno scultore nella Mantova di Andrea Mantegna e di Isabella d'Este*, edited by F. Trevisani and D. Gasparatto, 266-269. Milano: Elacta.

Bruker Corporation. 2005. KTI Emperical Spreadsheet Manual.

———. 2006. Tracer III-V User's Manual.

Craddock, P. T. 2009. *Scientific Investigation of Copies, Fakes and Forgeries*. 1st ed. Oxford: Butterworth-Heinemann.

Cristanetti, S., A. Luchs, N. Penny, K. Serres, D. Smith, and S. Sturman. 2007. Recent acquisitions made to the Robert H. Smith Collection of Renaissance bronzes. *Burlington Magazine*.

European Commission, Joint Research Centre, Institute for Reference Materials and Measurements. 2001. The certification of arsenic, lead, tin and zinc (mass fractions) in five copper alloys, BCR-691. Report EUR 19778/1 EN.

Feller, R. L., N. Stolow, and E. H. Jones. 1985. *On Picture Varnishes and their Solvents*. Rev. and enl. ed. Washington: National Gallery of Art.

Getty Conservation Institute. 2006. XRF Workshop – Part III : Quantitative XRF Analysis using handheld instrumentation. July 6-7.

Glinsman, L. 1997. "Renaissance portrait medals by Matteo de'Pasti: a study of their casting materials." In *Conservation research 1996/1997*, edited by National Gallery of Art (U.S.), 93-108. Studies in the history of art, Hanover, N.H.: University Press of New England.

———. 2004. *The Application of X-ray Fluorescence Spectrometry to the Study of Museum Objects*. Ph.D., University of Amsterdam.

Glinsman, L. and L. Hayek. 2007. "A Multivariate Analysis of Italian Renaissance Portrait Medals: An Expanded Nomenclature for Defining Alloy Composition." In *Renaissance Medals*, edited by J. Graham Pollard, Eleonora Luciano and Maria Pollard, appendix LVI-LXII. Washington: National Gallery of Art.

Hanlan, J., N. Stolow, J. Grant, and R. Tolmie. 1970. "Application of Non-Dispersive X-ray Fluorescence Analysis to the Study of Works of Art." *Bulletin of the American Group, IIC* Vol. 10 (2):25-40.

Heginbotham, A., A. Bezur, M. Bouchard, J. M. Davis, K. Eremin, J. H. Frantz, L. Glinsman, L. Hayek, D. Hook, V. Kantarelou, A. G. Karydas, L. Lee, J. Mass, C. Matsen, B. McCarthy, M. McGath, A. Shugar, Sirois J, D. Smith, and R. J. Speakman. 2011. "An Evaluation of Inter-Laboratory Reproducibility for Quantitative XRF of Historic Copper Alloys." In *Metal 2010: International Conference on Metal Conservation*, edited by P. Mardikian, C. Chemello, C. Watters and P. Hull, 244-255. Clemson, South Carolina: Clemson University.

Hook, D. 1988. "Scientific Analysis of the Copper-Based Medals." In *A Catalogue of the French Medals in the British Museum*, edited by M. Jones, 305-312. Vol. 2. London: British Museum.

Hubbell, J. H., and S. M. Seltzer. 1989, 1990, 1996. Tables of X-Ray Mass Attenuation Coefficients and Mass Energy-Absorption Coefficients from 1 keV to 20 MeV for Elements Z = 1 to 92 and 48 Additional Substances of Dosimetric Interest. Ionizing Radiation Division, Physics Laboratory, National Institute of Standards and Technology (NIST).

Jestaz, B. 1972. "Une statuette de bronze: le Saint Christophe de Severo da Ravenna." *La Revue du Louvre* Vol. 22:67-78.

McGlinchey, C., and K. Trentelman. 2008. "Portable XRF for the examination of artists' materials and cultural heritage." In *Conservation Science 2007: papers from the Conference held in Milan, Italy, 10-11 May 2007*, edited by Joyce Townsend, Lucia Toniolo and Francesca Cappitelli, 195-196. London: Archetype.

Motture, P. 2003. "The Production of Firedogs in Renaissance Venice." In *Large bronzes in the Renaissance*, edited by Peta Motture, 277-307. Washington D.C.: National Gallery of Art.

Motture, P., and G. Martin. 2001. *Bells & mortars and related utensils: catalogue of Italian bronzes in the Victoria and Albert Museum*. London: V&A Publications.

National Gallery of Art. 2007. The Application of X-ray Fluorescence Analysis to the Study of Renaissance Bronzes: A Roundtable Discussion. National Gallery of Art, Washington, D.C., May 14-15, 2007.

Pittard, V., S. Stanek, M. Griesser, H. Hanzer, and C. Kryza-Gersch. 2008. "Comprehensive investigation of the 'organic patina' on Renaissance and Baroque indoor bronze sculptures…." In *Conservation Science 2007: papers from the Conference held in Milan, Italy, 10-11 May 2007*, edited by Joyce Townsend, Lucia Toniolo and Francesca Cappitelli, 49-55. London: Archetype.

Rousseau, R. 2001. "Detection Limit and Estimate of Uncertainty of Analytical XRF Results." *The Rigaku Journal* Vol. 18 (2):33-47.

Scott, D. A. 1991. *Metallography and microstructure of ancient and historic metals*. Los Angeles: Getty Conservation Institute.

Shugar, A. 2009. "Peaking Your Interest: an introductory explanation of how to interpret XRF data." *Western Association for Art Conservation Newsletter* Vol. 31 (3):8-10.

———. 2010. "Variation in data reported from handheld XRF analysis resulting from minor alterations to machine calibration" ISA archaeometry symposium, Tampa FL.

Smith, D. 2011. "The Application of Alloy Analysis to Questions of Attribution: Giovanni Francesco Susini and the Workshop of Giambologna." In *Metal 2010: International Conference on Metal Conservation*, edited by P. Mardikian, C. Chemello, C. Watters and P. Hull, 189-197. Clemson, South Carolina: Clemson University.

Smith, D., and S. Sturman. 2011. "The art and innovation of Antico's bronzes: a technical investigation." In *Antico: the golden age of Renaissance bronzes*, edited by Eleonora Luciano, D. Allen, Claudia Krysa-Gersch and Stephen J. Campbell, 157-177. Washington D.C.: National Gallery of Art.

Stone, R. 1981. "Antico and the Development of Bronze Casting in Italy at the End of the Quattrocento." *Metropolitan Museum Journal* Vol. 16:87-116.

———. 2010. "Organic patinas on Renaissance bronzes." *Metropolitan Museum Journal* Vol. 45:107-124.

———. 2011. "Understanding Antico's patinas." In *Antico: the golden age of Renaissance bronzes*, edited by Eleonora Luciano, D. Allen, Claudia Krysa-Gersch and Stephen J. Campbell, 178-182. Washington D.C.: National Gallery of Art.

Stone, R., R. White, and N. Indictor. 1990. "Surface composition of some Italian renaissance bronzes." In *ICOM-CC Preprints of the 9th Triennial Meeting, Deresden, German Democratic Republic, 26-31 August 1990*, edited by Roy Vontobel, 26-31. Los Angeles: ICOM Committee for Conservation.

Sturman, S. 2001. "A Group of Giambologna Female Nudes: Analysis and Manufacture." In *Small bronzes in the Renaissance*, edited by Debra Pincus, 121-141. Washington DC: National Gallery of Art.

———. 2004. "Interpreting the Bronzes: A Technical View." In *Art of the Renaissance bronze, 1500-1650: The Robert H. Smith Collection*, edited by Anthony Radcliffe and Nicholas Penny, 297-310. London: Philip Wilson Publishers.

Sturman, S., S. Cristanetti, D. Pincus, K. Serres, and D. Smith. 2009. "Beautiful in form and execution': the design and construction of Andrea Riccio's Paschal candlestick." *Burlington Magazine* Vol. CLI (1279):666-672.

Sturman, S., and L. Glinsman. 2004. "Technical Reports" and "Explanation of XRF Table." In *European bronzes from the Quentin collection: an exhibition at the Frick Collection*, edited by Manfred Leithe-Jasper and Patricia Wengraf, 302-323. New York: M.T. Train/Scala.

Sturman, S., and D. Smith. 2008. "La porta della sacrestia di San Marco di Jacopo Sansovino: resulti preliminari di un'indagine." In *L'Industria Artistica del Bronzo del Rinascimento a Venezia e Nell'Italia Settentrionale*, edited by M. Ceriana and V. Avery, 447-458. Venice: Foundazione Giorgio Cini.

Thomsen, V., D. Schatzlein, and D. Mercuro. 2003. "Limits of Detection in Spectroscopy." *Spectroscopy* Vol. 18 (12):112-114.

Wilson, A., D. Turner, and A. Robbins. 1999. "Evaluation of Some Sample Support Films for Microsample X-Ray Analysis." In *JCPDS-International Centre for Diffraction Data*, 301-307.

Chapter 3

Application of a handheld XRF spectrometer in research and identification of photographs[1]

Dusan C. Stulik and Art Kaplan

Introduction

The era of classical or chemical photography began in the early 19th century with key camera based experiments by Thomas Wedgwood (Eder 1978) and Joseph Nicephore Niepce (Marignier 1999), and with the introduction of the first "practical" photographic process by Daguerre in 1839 (Daguerre 1839). Since then more than 150 different photographic processes and process variants have been invented, tested and introduced into photographic practice (Woodbury 1898, Jones 1911, Nadeau 1989, Cartier-Bresson 2008, Lavedrine 2009). Some photographic processes such as albumen, tintype, silver gelatin or chromogenic C-print processes were used for many decades. Millions of photographic images have been created using these processes and can be found in a variety of public and private photographic collections. A whole range of other photographic processes such as matte collodion, platinotype, or autochrome were introduced and used for a limited period of time or for specialized applications and later superseded by other more advanced or more convenient photographic process. Other photographic processes such as heliography, chrysotype, eburneum or 3M Electrocolor™ processes had only limited practical applications and only rare examples of these processes can be found in museum collections, archives of inventors or deposits of patent submissions.

All of these "chemical" photographic processes have been recently almost completely replaced by digital photography and digital printing. Millions of photographs of the chemical photography era are now kept in museums, archives, libraries and private collections, often without adequate care or protection and rather poor registration collection documentation. The correct identification of photographs is a mandatory prerequisite for the development of an adequate strategy for the long term preservation, storage, exhibition and conservation treatment of any individual photograph or collection of photographs. Problems of misrepresentation and authentication of photographs is also a growing concern for curators, collectors and art dealers due to the constantly increasing monetary value of some vintage photographs and often confusing issues related to later prints.

A number of different working methods have been developed and introduced to aid in the identification of photographs (Coe and Haworth-Booth 1983, Knodt and Pollmeier 1999, Reilly 1986). The majority of these methods are based on the visual examination of photographs or in combination with a detailed inspection under a loupe, low power microscope or a stereomicroscope. It is our estimate that these methods of photograph identification are able to solve about 80% of identification cases. The remaining 20% of cases require the use of instrumental and analytical tools to identify the material composition of unknown photographs (Stulik 2008: 436-451). X-ray fluorescence spectroscopy (XRF) is the most powerful analytical method (almost ideal) for the identification of the elemental composition of photographs and photographic materials. A great advantage of XRF analysis, in comparison with microscopic investigation, is its high level of objectivity. In the majority of analytical cases XRF clearly identifies the presence or absence of key elements needed for the identification of the photographic process.

XRF is able to detect and theoretically quantify almost all elements of the periodic table excluding hydrogen and helium. The practical analytical range of detection for most handheld XRF instruments is from atomic number Z=11 (sodium) to Z=92 (uranium). XRF analysis is also nondestructive / noncontact and it can be applied to the analysis of solids, powders, liquids and thin film material for major, minor and often trace concentrations of chemical elements. Recent instrumental advances in miniaturization and portability of XRF instrumentation (Potts and West 2008) make it a very versatile and almost universal tool for both laboratory and "field" elemental analysis of art objects.

The portability of XRF instrumentation allows a change in strategy of the analysis of art objects. No longer is it required that art be brought to a conservation science laboratory. Handheld XRF spectrometers allow for an *in situ* investigation or conducting the analytical work directly inside of museum galleries or art storage areas without any dangerous and often expensive transport of art objects to a conservation laboratory which can require extensive paperwork and insurance costs. Such a development is highly appreciated by many museum directors, curators and collection managers and it often brings many new opportunities for more advanced and extensive analytical research of valuable and previously inaccessible photographic collections.

A great deal of information and experience has been gathered within the art conservation community during the last decades on the application of XRF to the study of paintings, metals, glass, stone and illuminated manuscripts (Mantler and Schreiner 2000: 3-17). All of this accumulated knowledge has been very helpful in developing an analytical strategy for the use of XRF in the study of photographs and photographic materials.

So far most analytical applications of XRF analysis in the study of photographs have focused on the use of qualitative XRF analysis and its use in identification of photographic processes (McCabe and Glinsman 1996: 13-40, Stulik *et al.* 2002: 658-663, Brown *et al.* 2005: 57-72, Stulik 2005: 556-564). Our analytical research, which began in 2000, also showed that research in photographs can greatly benefit from quantitative XRF analysis. The quantitative methodology for the analysis of photographs was developed, laboratory tested and applied to problem solving when dealing with issues related to provenancing and authentication of photographs (Stulik and Khanjian 2002: 188-194, Brown 2002, Messier 2005: 123-130, Kaplan and Stulik 2008, Stulik and Kaplan 2008). XRF analysis of photographs has also been responsible for several new discoveries related to photographic processes and changes and corrections of the history of photography (Stulik *et al.* 2002: 658-663, Wright and Stulik 2007: 28-32).

At the same time XRF analysis of photographs has a large number of specific needs, restrictions and limitations that have to be considered when developing an analytical strategy for both qualitative and quantitative analysis. The specific needs, restrictions and knowledge of analytical limitations of modern handheld XRF instrumentation when analyzing photographs will be the main focus of this chapter. Several advanced working methodologies developed for research in photographs will be also introduced and explained using illustrative examples and practical case studies.

Internal physical and chemical structure of photographs

Photographic materials (negative or positive) can have either a simple or complex multilayer structure. Based on their internal structure, photographs, (which form the core of both museum and private photographic collections), can be divided into four major categories for purposes of XRF analysis. This division can account for more than 95 percent of collection holdings (Fig. 3.1).

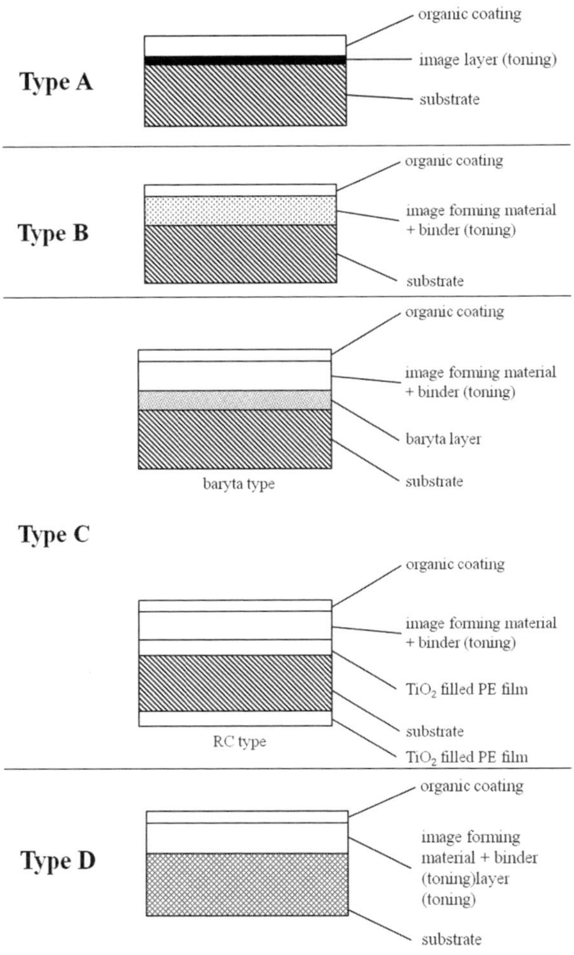

Figure 3.1:
Schematic depiction of the internal structure of different types of photographic images.

Type A are photographs with a very thin image layer without organic binder deposited directly on top of, usually, a paper base but sometimes on other materials such as textiles. The image forming layer is deposited directly on the top layer of paper fibers or on a thin layer of sealant which prevents deeper penetration of the image forming chemistry into the paper base during processing. The image forming material can be unprotected or it can be coated with a layer of organic coating or varnish. Some early paper negatives might also represent the Type A photographic material. Many paper negatives were made transparent by varnishing and waxing but the fact that silver particles are present directly on fibers of the paper substrate still puts paper negatives to the Type A category of photographic material.

From an XRF analysis point of view we can see the X-ray spectrum of Type A photographic material as a spectrum formed by the spectral superposition of the X-ray spectrum of the image layer (which can be toned), with the X-ray spectrum of the paper substrate. Salt prints, platinotypes, cyanotypes, kallitypes and many others belong to the Type A category of photographic images.

Type B are photographs with imaging material distributed in a layer of organic material or binder. The primary example of such material are 19th century albumen prints that consist of a paper substrate coated with albumen layer that contains toned or untoned particles of silver. Some special types of gelatin silver photographs are also of this type. XRF spectra of the Type B photographs are very similar to spectra of Type A material and differentiation between these two categories of photographic material needs, in most cases, to be done with the aid of microscopic investigation and by analysis of organic components using FTIR or Raman spectrometry. The final interpretation of microscopic and analytical results might be rather complicated. The organic material can be present as a part of an image layer, surface sizing or as a protective coating. In some cases no final interpretation can be done properly without a (micro-destructive) cross-section analysis. When concentrations of organic components are rather low, even the cross-section analysis might not provide conclusive results.

Type C photographs have a much thicker image layer (5-20 μm) containing an organic binder. The emulsion of Type C material has the image-forming metal dispersed usually within a layer of collodion or gelatin emulsion forming material. Another important feature of Type C photographic material is that these materials very often (but not always) contain a baryta or clay layer between 6-50 μm thick. The baryta layer is a layer of, usually, barium sulfate alone or in combination with some clay-type fillers that protect the emulsion layer from the impurities of the paper base and increase the brightness of photographic image. The presence of this layer is very important for XRF analysis because the layer is formed of simple or complex inorganic components mixed with an organic binder. As such, the baryta or clay layer makes the resulting X-ray spectrum of such a material more complicated. The X-ray spectrum of Type C photographic material is formed by the spectral superposition of X-ray signals from the substrate of the photograph (baryta or clay layer) and the image forming material, which can be toned or untoned. The most important examples of the Type C photographic material are modern silver gelatin and matt collodion printing papers. Some of these photographic materials may also contain some matting agents to decrease the glossiness of the emulsion layer. The most important matting agents are starch particles which do not provide a problem for XRF analysis, but some photographic paper manufacturing companies also used some inorganic matting agents. These matting agents include barium sulfate, clay or silicate based materials, which may provide another level of complication when interpreting XRF spectra of these materials.

During the second half of the 20th century a number of photographic manufacturing companies introduced so-called resin coated (RC) papers which have a paper substrate sealed from both sides by a thin polyethylene (PE) film. The main advantage of such photographic paper was faster washing due to the presence of the PE film which does not allow the processing liquids to enter the paper substrate. To provide similar optical properties as baryta coated photographic papers the majority of these RC papers had the emulsion facing side made with a titanium dioxide loaded PE film. The presence of titanium is clearly visible in XRF spectra of both B&W and color RC photographic papers.

Extreme cases of the Type C materials are some modern photographic materials that have an even more complicated internal physical and chemical structure. For example some instant photographic materials (Polaroid, Instant Kodak photographs, etc.) can have up to twenty-one individual layers containing a large number of organic and inorganic components both in high and trace concentrations (Rogers 2007).

Type D photographic material comprises a broad range of early photographic images (daguerreotypes, ambrotypes, tintypes, glass negatives and positives as well as special photographic processes (photoceramics) and photographs printed or transferred on different permanent supports (metal, glass, fabric, plastic etc.). The X-ray spectrum of such a material is formed by spectral superposition of analytical signals of all of the layers of the photographic image which contains metal or inorganic components.

Type A, B, C and D photographic materials represent all major classes of photographic material but a distinction between individual types of material can be more gradual. For example in a pigment print, particles of pigment, either inorganic, organic or organic on an inorganic carrier, are bound together and to the paper substrate using an organic binder (gum Arabic for example).

To make the analytical situation even more complicated, the presence of materials added to modify the physical or chemical properties of photographic material (fillers or pigments in the paper substrate), impurities due to raw materials (baryta layer composition, metal impurities from paper making process), impurities due to contamination of photographic chemicals (traces of iridium in sensitizing solution for platinotypes) or contamination deposited during the processing of photographic material (metal impurities from ground water used for washing photographs) can further complicate the interpretation of the XRF results.

Elemental composition of photographs

Figure 3.2 shows the periodic table of chemical elements with color highlighted elements that we have encountered during analysis of thousands of photographs from our study collection of photographs and a number of collections of photographs containing a large selection of different historical photographic processes of the chemical photography era.

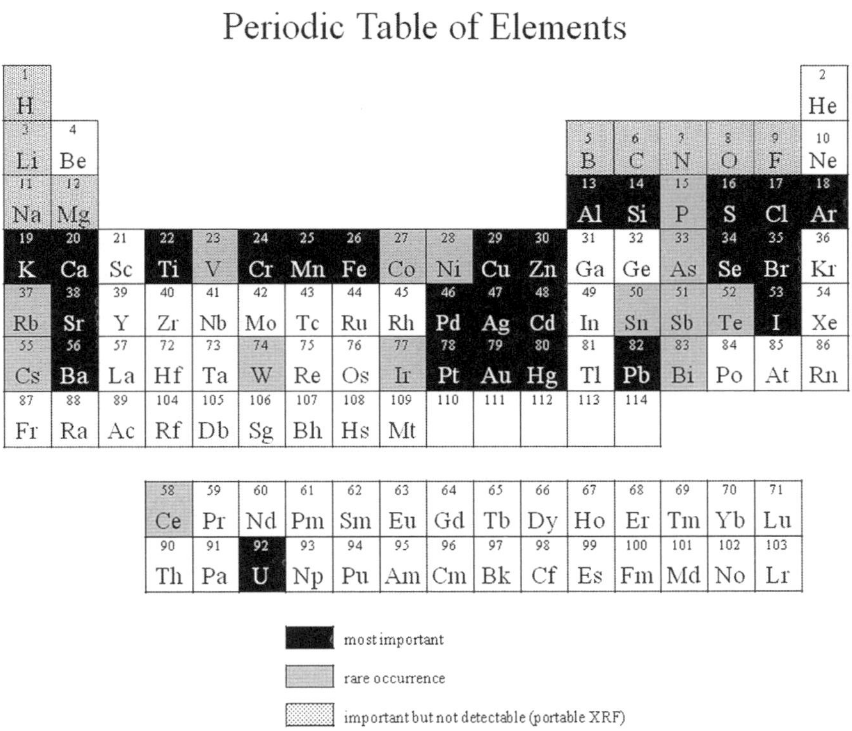

Figure 3.2:
Periodic table showing the relative importance of each element in XRF analysis of photographs.

Chemical elements highlighted represent the most important chemical elements detected in photographs. Some of these elements are primary image formers and toners. These elements can be present both in their elemental form (silver, gold, platinum, palladium, iron etc.) or in the form of more or less defined chemical compounds (reddish iron oxide, Fe_2O_3, of some pigment or bromoil prints, cyano complex compounds of cyanotype blue). Chemical compounds can also be found in the baryta layer (barium and strontium sulfates together with clay minerals).

A paper substrate may also contain inorganic fillers (calcium and magnesium bicarbonate buffers), mineral coloring matter and paper making process incorporated impurities. In some cases the analytical signal of an element in the XRF spectrum of an analyzed photograph may have two or more sources. This is very important to consider when interpreting the results of XRF analysis. An example of this analytical situation can be the analysis of a platinotype photograph printed on an off-white paper. Off-white papers are often prepared by adding some iron based pigments (ochres, umbers) to the paper mass during its manufacturing. The resulting XRF peak of iron represents the spectral superposition of iron from both the paper substrate and from iron residue from the incomplete clearing and washing of the print during darkroom processing. A complete separation of contributions of the different iron signals contributing to the final intensity of the iron signal is complicated by the absorption of the iron radiation by other components of the photograph. Another important example can be the detection of sulfur in the XRF spectra of brown, sulfur toned, gelatin silver photographs. There is a very high level of probability that some of the detected sulfur signal originated from silver sulfide produced by various methods of sulfur toning but that the majority of the sulfur signal comes from barium sulfate $BaSO_4$ which is typical for the majority of existing 19th and 20th century gelatin silver POP (printing-out paper) and DOP (developing-out paper) photographs.

Table 3.1 shows the most important sources of detected elements when analyzing photographs. The table also indicates the relative (high/low) concentrations of elements and whether their occurrence is common or rare.

Table 3.1 deals more with elements detected in paper and polymer based photographs and photographic materials (photographs and photographic negatives). It does not deal with metal (daguerreotype), glass (ambrotype, glass negative) or ceramic or enamel based photographic materials. The XRF spectra of these materials usually exhibit a high concentration of all chemical elements of the substrate in combination with a weaker analytical signal from the image forming material. The analytical signal of some chemical elements that are present in glass and ceramics (Na, Si, Al, K, Ca etc.) might be weak even at high concentrations due to the low sensitivity of XRF spectrometers for low atomic weight elements.

XRF instruments for research in photographs

Early handheld XRF instruments, developed during the last decades of the 20th century, for both bulk and thin film analysis of different materials used one or several sealed radioactive sources (such as Am-241, Cd-109, Fe-55 etc.) as the

Chemical Elements Detected	Common Sources of High Intensity X-ray Signal	Common Sources of Low Intensity X-ray Signal	Notes
Al		fillers, colorant	low sensitivity
Si		fillers	low sensitivity
P		fillers, phosphoproteins	low sensitivity
S		Toning, $BaSO_4$, $SrSO_4$ from baryta layer	low sensitivity
Cl	PVC (polymer)	silver halide	low sensitivity
Ar		atmosphere (air has ~1% Ar)	
K		fillers, unprocessed chemicals	
Ca		fillers, buffers	
Ti	RC polyethylene	optical brighteners, fillers, colorant (for hand-colored images)	
V		toner (rare)	
Cr	matte board colorant	gelatin hardener, chromium based processes, colorant (for hand-colored images)	
Mn	matte board colorant – add other layers Cr found in	colorant, reducer	
Fe	matte board colorant, tintype substrate (steel)	impurity, iron based processes, colorant	
Co		1840s smalt particles in paper, colorant (for hand-colored images)	
Ni		toner (rare)	
Cu	daguerreotype substrate	toner, colorant (for hand-colored images)	
Zn	paper filler, whitening compound	Paper, fillers, colorant (for hand-colored images)	
As		glass support, glass negatives	
Se		toner	
Br	silver halide	silver halide	unprocessed material
Rb		impurity	
Sr	*$SrSO_4$ ($BaSO_4$ impurity)	$SrSO_4$ ($BaSO_4$ impurity)	*rare at high levels
Pd	palladiotype	toner (rare)	
Ag	silver based processes	toner (rare)	
Cd		soluble halide, colorant	
Sn	glass (rare)	toner (rare)	
Sb		Pigment (rare)	
Te		toner (rare)	
I	silver halide	silver halide	unprocessed material
Cs		modern “toner” (rare)	
Ba	$BaSO_4$	$BaSO_4$, filler	
W		toner (rare)	
Pt	platinotype	toner	
Au	chrysotype	toner	
Hg		toner, bleacher, colorant (for hand-colored images)	
Pb	matte board, paper	toner, colorant (for hand-colored images)	
Bi		colorant (for hand-colored images)	
Ce		toner, bleacher	
U	uranotype, wothlytype	toner	

Table 3.1:
Main sources of analytical signal encountered during XRF analysis of photographs. Several elements can occur in either low or high concentrations as noted.

source of primary X-ray radiation. Combined often with rather low (250-350 eV) spectral resolution, solid state detectors, powerful spectral de-convolution and qualitative analysis software, these types of instruments were responsible for the very early application of quantitative XRF analysis in the investigation of photographs.

Later developments in handheld XRF spectrometry led to the replacement of radioactive sources by miniature X-ray tubes. New generations of X-ray detectors also provided much higher (~150 eV) spectral resolution. These types of handheld XRF instruments are now widely used in art conservation for non-destructive and non-contact elemental analysis of art objects.

A number of different handheld XRF spectrometers are available today and can be used successfully in art conservation (Table 3.2).

Instrument	Manufacturer	kV range	Current (μA)	Anode Material*	Detector
TRACeR III-V / III-SD	Bruker	40	15	Rhodium or Silver	Si PIN / SDD
S1 Turbo sorter	Bruker	40	20	Silver	Si PIN
S1 Turbo SD	Bruker	40	20	Silver	SDD
S1 Turbo SD (LE)	Bruker	40	20	Silver	SDD
S1 Turbo SDR	Bruker	40	20	Silver	SDD
Delta	Olympus Innov-X	40	200	Silver, Gold, Rhodium or Tantalum	Si PIN / SDD
Omega	Olympus Innov-X	40	100	Silver	SDD
X-MET 5000	Oxford Instruments	45	50	Rhodium	Si PIN
X-MET 5100	Oxford Instruments	45	50	Rhodium	SDD
Niton XL2 800	Thermo Scientific	45	100	Silver	Si PIN
Niton XL2 Gold	Thermo Scientific	45	100	Silver	SDD
Niton XL3t Gold Plus	Thermo Scientific	50	200	Silver	SDD

*Many companies offer custom modification and options for application specific anode materials

Table 3.2:
A partial list of handheld XRF spectrometers on the market today.

A number of art conservation laboratories and research groups already use handheld XRF spectrometers in their work but none of the available spectrometers answer all of the needs of art conservation research and an even smaller segment of the available instruments can answer the many specific needs and requirements of research in photographs. Because the analysis of photographs is a fast developing field of art conservation research today we are providing at least some general guidelines that may help researchers and laboratories interested in the analytical study of photographs in selecting and specifying instruments better suited to the different analytical tasks required to analyze photographs and photographic material.

Several instrumental factors are of primary importance when selecting a handheld XRF spectrometer for the analysis of photographs:

Anode material

Many handheld XRF instruments available today use silver or gold anodes. The spectral peaks of anode elements usually show in a recorded XRF spectrum of analyzed material. This is a problem when analyzing photographs, which in most cases contain silver as the imaging material and in many instances gold that was used for toning of many 19th century and some 20th century photographs. When selecting an XRF spectrometer for the analysis of photographs it is much better to select a spectrometer with an anode made of material that does not interfere or interferes minimally with analytical tasks. Some available instruments use Re or Rh anodes that represent a much better choice when analyzing photographs. Some XRF companies also provide their customers with a choice of several X-ray tubes with different anode materials.

Instrumental parameters

Currently available maximum X-ray tube voltage 40-50 kV is quite adequate for the analysis of photographs but it is a good idea to test the stability of the X-ray tube when working at its maximum voltage. Photographs usually contain very small concentrations of analyzed material, so XRF spectrometers that provide higher maximum and very stable emission current are better suited for the analysis of photographs.

Vacuum / Helium flush

Several available XRF instruments provide arrangements for the analysis of lighter elements (sodium to potassium) using various configurations of vacuum or helium flush options. Some of these options are a little more difficult to use so additional testing of these options is highly recommended.

Filters

Several available instruments are equipped with replaceable filters for the primary X-ray beam or filter wheels. The use of filters of primary X-ray radiation is very useful when developing any screening or specific analytical procedure. The ability to remove filters is crucial for the use of a vacuum or helium flush when analyzing lighter elements.

Internal Standards

An internal standard that does not interfere with the analysis of photographs (we are using yttrium) allows for the normalization of several sets of measurements. This is very important in the case of the quantitative analysis where there is a need to eliminate, as much as possible, the effect of emission current variation of the X-ray tube on the measured result.

Live view and XRF software

A number of the XRF spectrometers work with PDA displays, but these are just a poor substitute for a live viewing of the ongoing analysis on a computer screen. The quality, flexibility and friendliness of the software provided with XRF spectrometers is also a very important issue. We have tested, over the years, a number of the handheld XRF spectrometers available on the market but we were not able to find one existing software package that met all our needs. Instrumental companies often promise future improvements of available software packages but that does not happen very often. Any good software should allow for good automation of analysis but also for a complete manual override of all instrumental parameters and a clear visibility and control of all spectral processing steps (spectra smoothing, background elimination, artifact removal, de-convolution, etc.) when processing recorded spectra.

Thin film software package

The quantitative XRF analysis of photographs should be done using a thin film analysis method. Several, but not all, XRF manufacturing companies provide a thin film method software package with their instrument or as an option.

Qualitative analysis of photographs

The XRF spectrum is composed of a number of both sharp and broader peaks superimposed on a spectrum background. Besides the characteristic X-ray peaks of the element, sometimes overlapping one another, the spectrum also can contain some other spectral features such as escape peaks, Compton scatter peaks, sum peaks (double photon counting events, see Chapter 1 for the definition of this term), etc., which are products of X-ray/matter interactions or of a particular instrumental configuration (X-ray source, detector, environment, etc.). In the special case of photograph analysis the problem of sum peaks is not very important due to the often low concentration of inorganic material in an analyzed photograph. The low concentration of analyzed elements makes both qualitative and quantitative

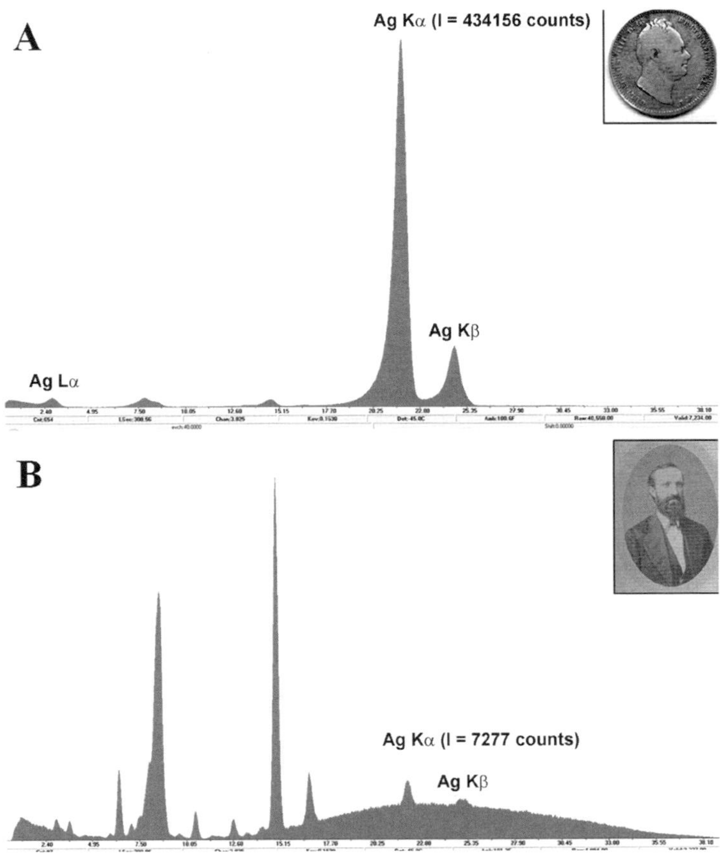

Figure 3.3:
A- XRF spectrum of an 1836 William (IV) shil ing silver coin; B- XRF spectrum of a 19th century albumen photograph (obtained using a TRACeR III portable XRF with a Re tube, and an Al/Cu filter operated at 40kV and 12μA).

analysis of photographs much more difficult and often very challenging. Figure 3.3 illustrates an issue of the intensity of analytical signal by the comparing the XRF spectrum of an 1830's silver coin and the XRF spectrum of a 19th century albumen silver photograph measured using identical instrumental parameters, sample-instrument geometry and time of measurement.

The qualitative interpretation of the measured spectrum is based on the careful assignment of all spectral peaks above the spectral background to individual chemical elements and in doing so deducing which elements are present in the sample. A good understanding of the chemical composition of photographs and of the historical use of different chemicals in creating, processing and post-processing treatments of photographs is very important for a proper interpretation, without over-interpretation, of the XRF spectra of photographs.

Detection limits in XRF analysis of photographs

The qualitative XRF analysis of photographs (and any other material) does not shows the presence of all chemical elements in an analyzed material but rather the presence of all elements that can be detected and are present in concentrations above the detection limits of the instrument and instrumental setting. Detection limits for different elements are functions of many physical factors (excitation and absorption processes) and of many instrumental factors (the anode material, X-ray tube operational parameters, instrumental geometry, spectral sensitivity of a detector, etc.).

Knowing and understanding detection limits of different elements is a great help to any analyst who has to deal with the analysis of minor and trace elements in bulk material and with the analysis of material containing small amounts of analyzed elements. The analysis of photographic materials is an important example of an analysis that deals with both low concentration of analytes and with the low sensitivity of XRF measurements for some key elements present in analyzed materials.

Table 3.3 shows detection limits for an "elemental survey analysis" using our TRACeR III Re tube XRF instrument operated at 40kV, 12μA, without vacuum, with an Al/Cu filter (70μm thick Al / 70μm thick Cu) for 300 dead time corrected seconds. The detection limits presented in Table 4 were obtained experimentally using a series of thin film standards containing single elements in different concentrations. Detection limits were determined by analyzing a series of quantitative ashless filter papers soaked in distilled water solutions of metal nitrates of different dilutions. The concentration of metal elements in the filter papers were determined using inductively coupled plasma – mass spectrometry (ICP-MS) calibrated with a series of standard solutions of known concentration. Dried filter papers were analyzed using XRF and the concentration of metal in the paper, for which the analytical signal was lower than 3x the standard deviation of the background was identified as the detection limit for that element. We have assumed that, in the first approximation, the analytical signal of the measured element was not diminished by the presence of thin cellulose, ~200 μm, of the filter paper. Following the methodologies applied for thin film XRF analysis, the concentrations of the analyzed elements were expressed in μg/cm^2.

Element	Primary Line	Line Energy (kV)	Detection Limit (μg/cm^2)
Sulfur	Kα	2,308	31.0 / 1.0*
Chlorine	Kα	2,622	24.0 / 1.0*
Barium	Lα	4,467	5,5
Chromium	Kα	5,414	1,5
Manganese	Kα	5,898	1,5
Cobalt	Kα	6,93	1,5
Gold	Lα	9,711	1,5
Uranium	Lα	13,613	5
Palladium	Kα	21,175	4,5
Cadmium	Kα	23,172	5,5
Iodine	Kα	28,61	45
Barium	Kα	32,191	1720

*Measured under vacuum

Table 3.3:
XRF detection limits of selected elements (to the nearest 0.5μg/cm^2).

Actual detection limits will differ for different XRF instruments and different experimental conditions but data in Table 3.3 may aid any analyst in assessing problems of detection limits and interpreting the presence or absence of different elements in analyzed photographs.

Knowing if some elements are present or absent in an analyzed photograph is only a partial answer to understanding the material makeup of a photograph. Every element present in the photograph may have its "photographic function" or it can be incorporated into a photograph during the manufacturing of photographic materials (paper, chemicals) or during darkroom processing or post-processing treatment. In permanently mounted photographs the analytical signal from certain elements may not come from the photograph itself but from the mounting material.

Our XRF analysis of a large series of Edward Weston's photographs from the JPGM collection showed that almost all palladium prints contained a small concentration of mercury (Fig. 3.4).

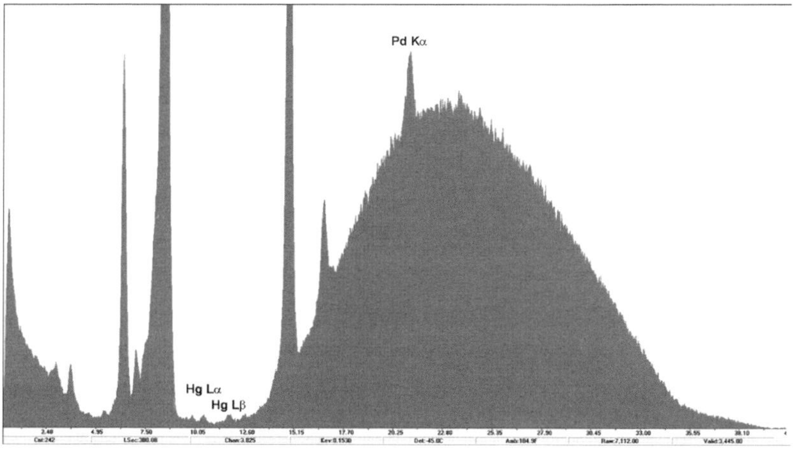

Figure 3.4:
XRF spectrum of an Edward Weston palladium print showing the presence of mercury (Hg) and palladium (Pd) (obtained using a TRACeR III portable XRF with a Re tube and an Al/Cu filter operated at 40kV and 12μA).

Interpreting these finding is not so easy and straightforward. It is well known that many brown toned platinotype photographs were toned using mercury. Toning of palladium photographs does not make too much sense because palladium prints have a brown tonality on their own. At the same time the amount of mercury detected in these photographs (approximately 5μg/cm^2) is too low to have any perceivable visual effect on the tonality of the final photograph. Our worldwide search for an unopened package or can of palladium photographic paper, which could have been used by Weston when making his palladiotype prints, has been unsuccessful so far and we were not able to identify and fully explain the discovery of mercury in his prints.

Spectral interferences in the analysis of photographs

In comparison with many other instrumental methods of elemental analysis (emission spectrometry, ICP-MS etc.), XRF exhibits a much smaller number of potential spectral interferences. A number of elemental spectral interferences and instrumental spectral interferences (interferences due to the material of a spectrometer) are still important to consider when interpreting the XRF spectra of photographs.

Numerous, already published tables of X-ray spectral lines (Thompson *et al.* 2009) show the entire selection of theoretical spectral lines and/or spectral lines detected using a very sensitive, high resolution X-ray spectrometer. Many of these spectral lines cannot be detected when analyzing the low concentration of elements that are present in photographs. To assess this analytical situation we have recorded XRF spectra of thin film, pure metal standards (50 $\mu g/cm^2$) and prepared a table of all recorded spectral peaks and some potential interferences (Table 3.4).

Element (Lines)	Possible Interferences
Al (K)	Br (L)
Cl (K)	Pd (L)
K (K)	Ar (K)
Ca (K)	Cd (L); I (L)
Ti (K)	Ba (L); I (L)
Cr (K)	Mn (K); Ba (L); Fe (Ka)
Mn (K)	Ba (L); Fe (K)
Fe (K)	Mn (K); Co (K)
Co (K)	Fe (K); Ni (K)
Ni (K)	Co (K); Cu (K), Pt (L); Au (L)
Cu (K)	Ni (K); Zn (K); Ir (L)
Zn (K)	Cu (K); Pt (L); Au (L); Hg (L)
Se (K)	Ir (L); Pt (L); Au (L); Hg (L)

Element (Lines)	Possible Interferences
Br (K)	Pt (L), Ir (L), Hg (L); Pt (L)
Sr (K)	U (L)
Pd (K)	Rh (compton scatter)
Ir (L)	Pt (L); Pb (L)
Pd (L)	Cl (K)
Cd (L)	Ca (K)
Ba (L)	Ti (K); Cr (K); Mn (K)
Pt (L)	Ir (L); Au (L); Pb (L); Ni (K)
Au (L)	Pt (L); Hg (L); Ni (K)
Hg (L)	Br (K); Zn (K); Au (L)
Pb (L)	Ir (L); Pt (L); As (K); Br (K)
U (L)	Sr (K)

Table 3.4:
Partial list of potential spectral interferences.

Table 3.4 shows a list of possible spectral interferences, but the occurrence of these interferences often depends on the concentration ratio between elements and on the intensity of interfering spectral lines. The most important spectral interferences when analyzing photographs are spectral interferences between the Kβ lines and Kα spectral lines of manganese, iron, cobalt, copper, nickel and zinc. The XRF spectra of pure manganese and iron and a combined spectrum of both manganese and iron together (Fig. 3.5) shows that the spectral peak around 6.40 keV represents a spectral superposition of the Kβ peak of manganese and Kα peak of iron. A spectral deconvolution or spectral subtraction is often needed when analyzing material containing both elements.

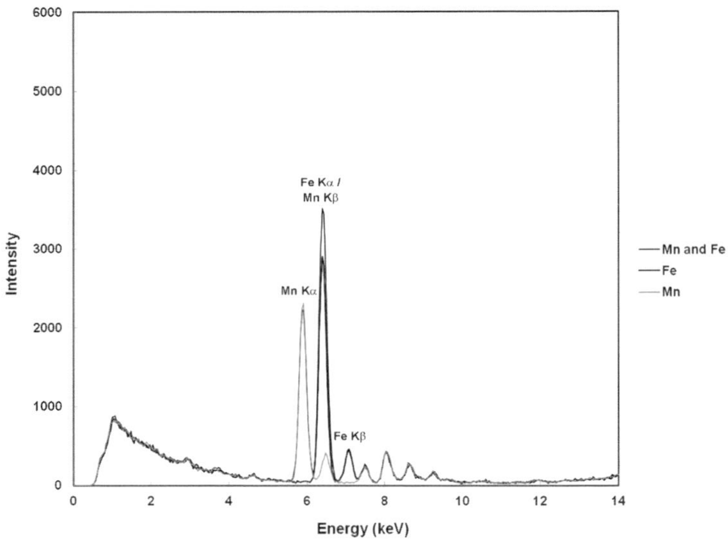

Figure 3.5:
A- XRF spectrum of pure manganese; B- XRF spectrum of pure iron; C- XRF spectrum of manganese and iron together. Spectra also show two yttrium peaks (14.96keV / 16.74keV) used as an internal standard in our instrument (obtained using a TRACeR III portable XRF with a Re tube and an Al/Cu filter operated at 40kV and 12μA).

Many 20th century photographs were printed on baryta coated, fiber based photographic paper. During the second part of the 20th century a new kind of photographic paper was introduced under the general name of RC paper. In the internal structure of RC paper (Fig. 3.1) the paper substrate is sealed between two thin layers of polyethylene. The PE layer facing the emulsion layer was usually made using a titanium dioxide filled PE. Both barium and titanium exhibit a very strong spectral interference and it is sometimes rather difficult to identify which element is present from the region of the spectra where the barium L peak and titanium K peaks overlap. Barium has some higher energy, low intensity L peaks that together with the presence of the Kα peak of barium at 32.191keV indicate the presence of barium (Fig. 3.6). To date we have not experienced a situation where an analyzed photograph contained both barium and titanium. Dealing with such a situation would require the use of spectral subtraction procedures based on the peak ratio values for the individual elements. Instances have occurred where a photograph containing barium is mounted to a board containing titanium. This situation can often be resolved by measuring the board alone followed by analysis of the mounted photograph and then performing a mathematical subtraction.

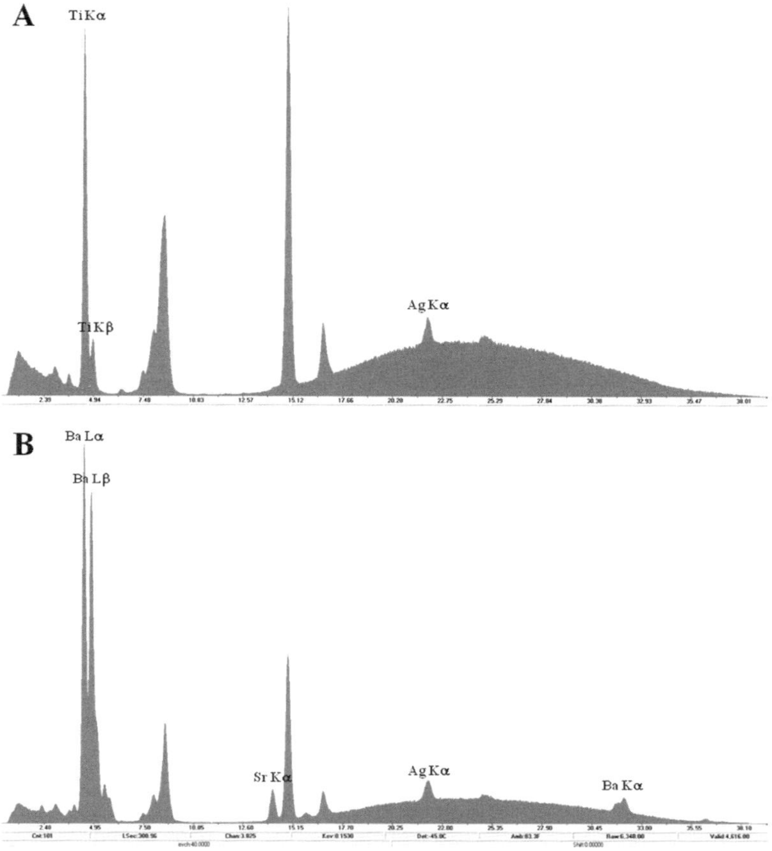

Figure 3.6:
A- XRF spectrum of a gelatin silver print on RC photographic paper; B- XRF spectrum of a gelatin silver print on baryta coated, fiber based photographic paper (obtained using a TRACeR III portable XRF with a Re tube and an Al/Cu filter operated at 40kV and 12μA).

In general both K and L spectral peaks are key spectral features used in the qualitative analysis of photographs. The M spectral peaks may be used to confirm the presence of certain elements but the spectral region of most M spectral lines have too many potential interferences to allow for the reliable identification of an element based on M lines alone.

Another important spectral interference is between lead and arsenic. This type of interference is much more important in paintings then in the case of photographs. Many photographs, papers and matting materials contain lead and lead compounds were also often used during the development or toning of photographs but the combined occurrence of both lead and arsenic in photographs is very rare. Higher L peaks of lead (Lβ-12.611keV and Lγ-14.762keV) have to be used to differentiate between both elements. The spectral interference of the Pb Mα_1 line (2.35keV) with the S Kα_1 line (2.31keV) makes it difficult to confirm the presence of low concentrations of sulfur when lead impurities are present in a photograph. Only when the concentration of sulfur is much higher, spectral subtraction can be used to confirm the presence of sulfur.

When dealing with low amounts or concentrations of analyzed elements even the XRF instrument itself can be a source of so called instrumental interferences.

Two types of instrumental interferences are of importance when analyzing photographs:

a - Interferences due to the anode material. Even when selecting an anode material very carefully there may still be a potential spectral interference which is important to account for and understand when interpreting the spectra of photographs. Figure 3.7 shows the XRF spectrum of zinc recorded using an XRF spectrometer equipped with a Re anode.

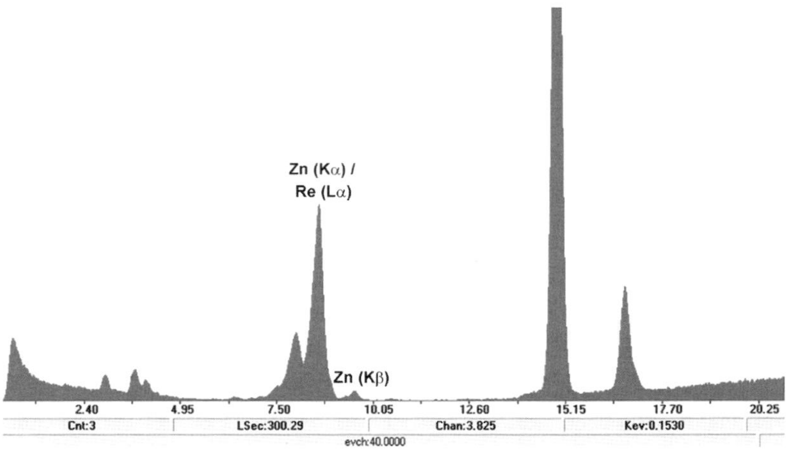

Figure 3.7:
A detail of the XRF spectrum of a zinc compound exhibiting an interference between the Zn Kα and Re Lα spectral peaks (obtained using a TRACeR III portable XRF with a Re tube and an Al/Cu filter operated at 40kV and 12µA).

Zinc does not occur in photography as an imaging metal or toner but it does occur as a white colorant in the manufacturing of paper materials or as a white pigment (ZnO) in some photograph tinting materials or retouching colors.

b - Interferences due to the material of the XRF spectrometer. The X-ray beam generated in the X-ray tube of an XRF spectrometer may interact with and be scattered within the head of the spectrometer creating a low intensity signal that can be detected by the detector. Some XRF manufacturing companies selected some elements as "sacrificial elements" and built X-ray source and detector compartments using these materials. (The Niton radioactive source based handheld XRF spectrometer uses nickel as a "sacrificial element" and informed customers about potential problems when analyzing nickel based material).

Figure 3.8 shows the XRF spectrum of a thick sample of pure cellulose (several layers of ashless quantitative filter paper) that does not contain any inorganic material.

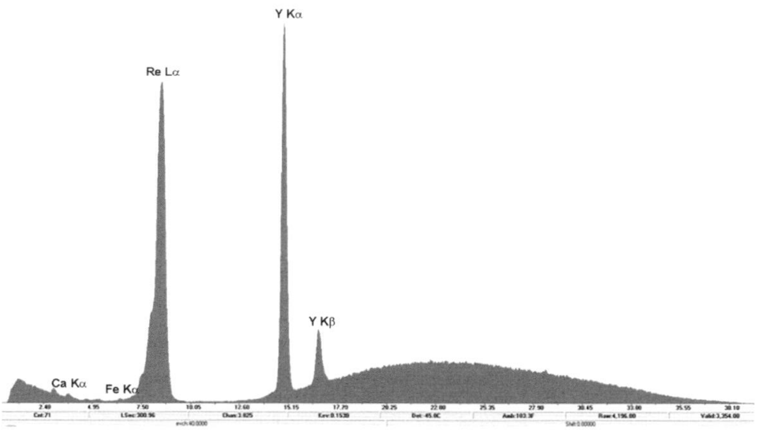

Figure 3.8:
Spectral artifacts due to instrumental interferences (obtained using a TRACeR III portable XRF with a Re tube and an Al/Cu filter operated at 40kV and 12μA).

A small signal of iron, copper and nickel recorded in the spectrum is due to instrumental interferences from the material used in the manufacturing of the XRF spectrometer. These spectral artifacts cannot be completely eliminated and it is important to consider them when interpreting the spectra of low or trace concentrations of elements in question. The analysis of low concentrations of iron in platinotype photographs is a very illustrative example of such an analysis.

Qualitative analysis: case studies

The main application of qualitative XRF analysis in the study and investigation of photographs is to identify all major and minor inorganic elements in an analyzed photograph. The process of the identification of a photograph usually starts with a detailed visual and microscopic observation of the visible photograph structure (one, two, three or multiple layer structure, microscopic structure of imaging material and of the photograph surface) and if such an investigation does not provide conclusive clues towards the identification of a photograph the investigation usually continues with the identification of both elemental (usually XRF analysis) and organic components (analysis of the organic substrate, binder and surface coatings using FTIR and/or Raman spectrometry). The combination of the visual and microscopic investigation with the results of the elemental and organic molecular analysis allows for a highly accurate and reliable identification of a photographic process, in most cases. The qualitative analysis of photographs can also lead to new information related to a photographer's darkroom techniques as well as provide new information related to the technical art history of photography.

The Durieu Album: the discovery of an early use of platinum and combined platinum and gold toning of albumen photographs

The qualitative XRF analysis of 115 photographs in the personal album of French photographer Eugene Durieu from the Cromer collection of the George Eastman House in Rochester showed (Stulik *et al.* 2002: 658-663) that 70% of photographs in the album were toned using platinum (Fig. 3.9).

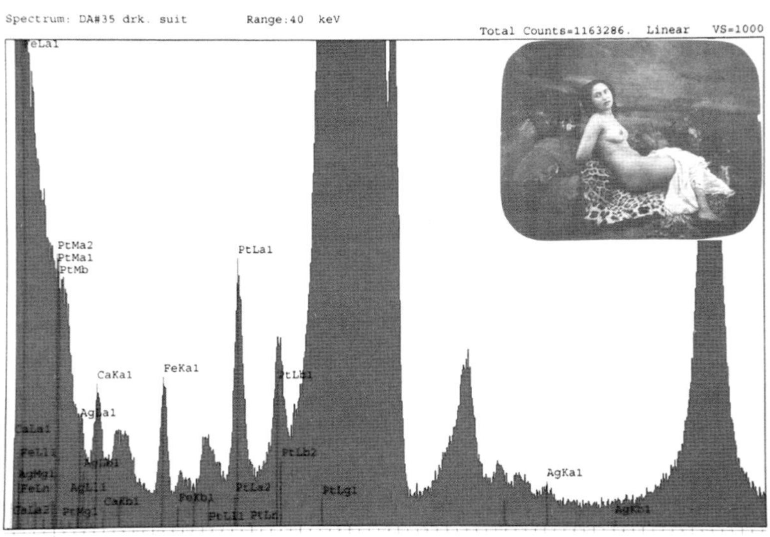

Figure 3.9:
XRF spectrum of a platinum toned image by Eugene Durieu (obtained using a Kevex 0750A XRF with a molybdenum tube and Ba/Sr secondary target operated at 55kV and 3.24μA).

A number of these photographs were made in 1854 when Durieu worked together with the famous French romantic painter Eugene Delacroix, who used Durieu's photographs of female nudes as models for several of his paintings. These analytical results were quite unexpected because until that time the earliest known use of platinum toning was described in 1856 (de Caranza 1856: 82-84). Several photographs in the album were also combination toned using both platinum and gold which was not known to be used during this time at all.

The analytical investigation of photographs by William Willis from the collection of the National Media Museum (NMeM) in Bradford, UK

Our project partners questioned an architectural plan that had been kept at the NMeM collection together with a number of other important photographs and photographic experiments of William Willis, the inventor of platinotype printing. The architectural print in question has on its back a pencil written note stating: "The first platinotype print by William Willis."

The XRF analysis of the architectural plan (Fig. 3.10) did not show the presence of platinum, silver or any other imaging material used around the 1870s when Willis researched his platinotype process.

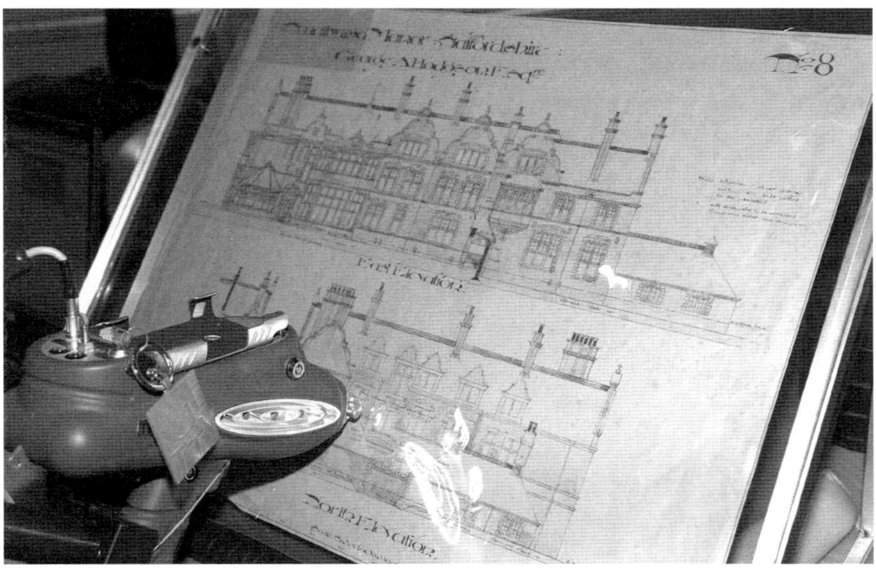

Figure 3.10:
XRF analysis of an architectural plan by William Willis from the NMeM collection (the analysis was performed using a mylar foil with a 2 inch diameter opening in order to keep the architectural copy flat for the analysis).

The only inorganic element detectable in the dark purple lines of the analyzed plan was chromium (Fig.3.11).

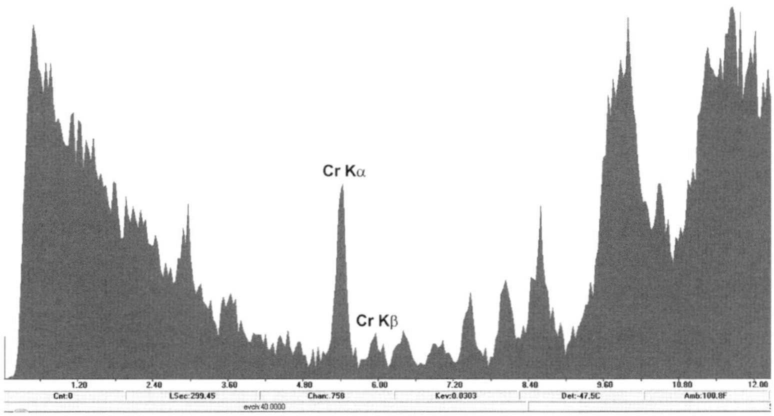

Figure 3.11:
XRF spectrum of Willis' Smallwood Manor showing the presence of chromium and no platinum (obtained using a TRACeR III portable XRF with a Re tube and an Al/Ti filter operated at 40kV and 12μA).

The identification of chromium in combination with our visual observations of the dark purple lines of the architectural plan calls for a quite different interpretation of the analytical findings. William Willis, the inventor of the platinotype process in 1873 (Nadeau 1994: 19-21) was the son of William Willis Sr., who also experimented with photographic processes. In 1868 he invented and patented his aniline reprographic process for making copies of architectural plans. His aniline process (Nadeau 1989: 32) was based on the photochemical sensitivity of chromium salts and on the formation of a purple dye material due to the reaction of aniline vapors with chromium (VI+).

At the conclusion of our scientific investigation we stated that the architectural plan in question was not an example of an early platinotype process but rather was produced using a reprographic process invented by William Willis Sr. in 1868.

Identification of rubidium in cyanotype photographs toned by a tea extract

Our collaborative project with members of the Alternative Process Photography community provides us with an opportunity to work with recreated, modified and newly developed photographic processes that, after the almost complete transition to digital photography by both photographic professionals and amateurs, represent the last bastion of chemical photography. Our XRF analysis of a series of untoned and toned cyanotype processes detected the presence of rubidium in a cyanotype photograph that was toned light brown using a strong tea extract (Fig. 3.12).

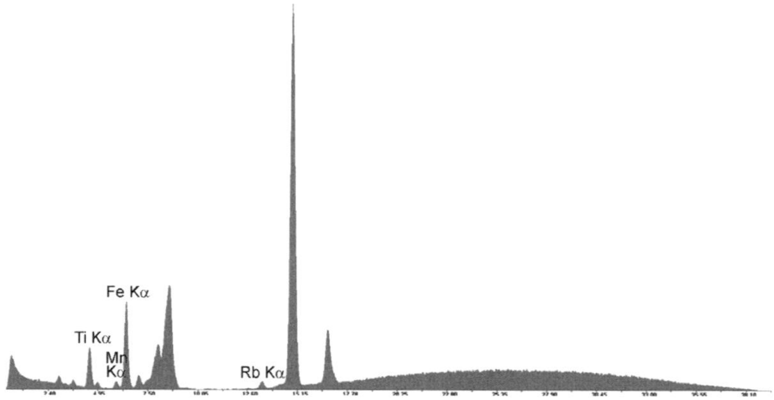

Figure 3.12:
XRF spectrum of a tea extract toned cyanotype showing the presence of rubidium (obtained using a TRACeR III portable XRF with a Re tube and an Al/Cu filter operated at 40kV and 12μA).

We have detected some small amounts of rubidium in some photographs before but were not sure, not ever seeing the presence of rubidium in photographs before. The photographer who created the toned cyanotype did not keep any sample of the tea used for toning and does not remember the make and source of the tea she was using. Our own toning experiments also showed the presence of rubidium in our tea extract toned photographs but the amount of rubidium detected was much lower. We are still working on a full explanation of the analytical findings.

Quantitative analysis of photographs

Quantitative XRF analysis has found a great many applications in art conservation research and archaeometry (Potts and West 2008). Most of these applications deal with the quantitative analysis of three dimensional objects such as metals, glass and ceramics. The primary X-ray beam of the XRF spectrometer penetrates deep into the analyzed material, generates the secondary X-ray fluorescence of elements present in the sample and the X-ray radiation of the sample atoms leaves the sample and is measured using an X-ray detector (Fig. 3.13).

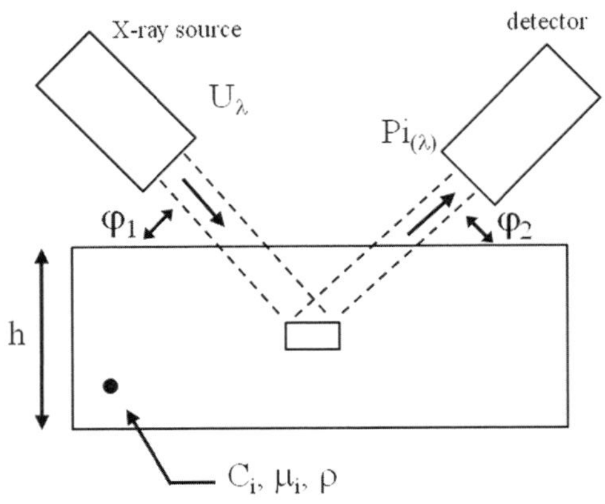

Figure 3.13:
XRF analysis of bulk samples (Ci – concentration of element i, U_λ - intensity of incident X-ray radiation for λ portion of the spectrum, φ_1 - angle of primary X-ray beam. φ_2 – angle of detector towards sample, h – sample thickness, $Pi_{(\lambda)}$ – detection efficiency for λ portion of the X-ray spectrum of element i, ρ - sample density, μi – mass absorption coefficient) .

To provide a quantitative interpretation of the measured X-ray spectrum it is important to first remove the spectral background, the secondary spectral features and by deconvoluting spectral overlaps to obtain "net intensities" of the spectral peaks of all chemical elements identified in the sample or elements that need to be quantified. In the second step of the quantitative interpretation of the measured X-ray spectrum it is necessary to find a relationship between net intensities of spectral peaks and concentration of chemical elements within the analyzed sample. The reasons why there is not a direct proportionality between the number of X-rays detected by the XRF instrument and the concentration of chemical elements in the analyzed sample are due to both instrumental and matrix effects.

The most important instrumental effects are geometric factors, effects of spectral resolution (spectral overlap) and the detector dead time. Some of the instrumental effects can be minimized by a careful selection of instrumental conditions, using mathematical corrections, or by using calibration standards.

Matrix effects are result of sample - X-ray interaction. Both primary and secondary X-ray radiation is partially absorbed in the sample matrix. The different chemical elements in the sample also have different X-ray absorption properties, and the peak intensity for a given chemical element will be affected by all of the chemical elements present in the sample. This situation can be further complicated by so-called multiple fluorescence when the fluorescent X-ray from one atom in the sample has enough energy to excite X-ray radiation from another element also present in the sample. In such a case the intensity of the peak of the element causing secondary fluorescence would be diminished and the peak intensity of the other element would be enhanced. Matrix effects have to be dealt with by using sophisticated mathematical models or by using calibration standards with a chemical composition and physical state very similar to the chemical composition and the physical state of the analyzed sample (Thomsen 2007).

Photographs as thin films

One of the experimental situations where matrix effects do not play such an important role is in the analysis of thin films or materials which resemble thin films. The vast majority of photographs have a very pronounced layered structure (Fig. 3.1) with imaging elements or components contained in a well defined layer that closely resembles a thin film. In such a case absorption of both primary and fluorescent X-rays beams in the analyzed material is negligible and we can assume a direct relationship between the magnitude of the analytical X-ray signal and concentration of the element within the analyzed thin film (Fig. 3.14).

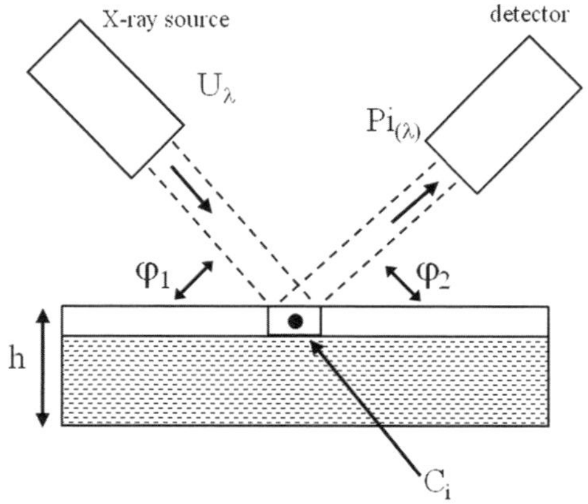

Figure 3.14:
XRF analysis of thin film samples (Ci – concentration of element i, U_λ - intensity of incident X-ray radiation for λ portion of the spectrum, φ_1 - angle of primary X-ray beam. φ_2 – angle of detector towards sample, h – sample thickness, $Pi_{(\lambda)}$ – detection efficiency for λ portion of the X-ray spectrum of element i).

The analogy between a photograph and a thin film makes the quantitative analysis of photographs much easier and straightforward. As so often is the case in the physical sciences the concept of the photograph as a thin film does not hold for all analytical situations. To understand when such a concept should and should not be used requires an introduction to so called "thin film criterion".

The expression for the intensity of fluorescent radiation from a flat homogeneous sample of thickness h, containing the concentration of chemical element C_i under the general experimental geometry depicted in Figure 16 was derived by Muller (Muller 1972) and later adjusted by Tertian and Claisse (Tertian and Claisse 1982). Based on this concept the fluorescent intensity, I_h, of a sample of finite thickness, h, can be expressed as:

(1) $$I_h = \frac{q}{\sin\phi_1} C_i \left\{ 1 - \exp\left[-\varrho h \left(\frac{\mu_{s,\lambda}}{\sin\phi_1} + \frac{\mu_{s,\lambda i}}{\sin\phi_2} \right) \right] \right\} \frac{E_i \mu_{i,\lambda} U_\lambda}{\frac{\mu_{s,\lambda}}{\sin\phi_1} + \frac{\mu_{s,\lambda i}}{\sin\phi_2}}$$

where:

(2) $$q = \frac{\sin\phi_1}{\sin\phi_2} \cdot \frac{d\Omega}{4\pi}$$

C_i	= concentration of element i
ϱ	= sample density
h	= sample thickness
$\mu_{s,\lambda}$	= mass absorption coefficient of sample s at wavelength λ
$\mu_{s,\lambda i}$	= mass absorption coefficient of sample s at wavelength λi
E_i	= excitation factor of element i
$\mu_{i,\lambda i}$	= mass absorption coefficient of element I at wavelength λi
U_λ	= intensity of the incident X-ray radiation

writing

(3) $$\bar{\mu}^* = \frac{\mu_{s,\lambda}}{\sin\phi_1} + \frac{\mu_{s,\lambda i}}{\sin\phi_2}$$

for the mean absorption coefficient
and

(4) $$k = \varrho h \bar{\mu}^*$$

then

(5) $$I_h = \frac{q}{\sin\phi_1} C_i \, 1 - e^{-k} \frac{E_i \mu_{i,\lambda} U_\lambda}{\bar{\mu}^*}$$

Similarly the fluorescent intensity from an infinitely thick sample is:

(6)
$$I_{\infty}=\frac{q}{\sin\phi_1}C_i\frac{E_i\mu_{i,\lambda}U_{\lambda}}{\bar{\mu}^*}$$

and the intensity ratio is:

(7)
$$\frac{I_h}{I_{\infty}}=1-e^{-k}$$

A thick sample can be defined as one in which:

(8)
$$\frac{I_h}{I_{\infty}}\geq 0.999$$

Thus, a value of $k = 6.91$ is the minimal condition under which the analytical signal emitted from the analyzed sample is of about the *same magnitude as the analytical signal from the infinitely thick sample*.

For small values of k, the approximation can be made:

(9)
$$1-e^{-k}\simeq k$$

which allows us to write:

(10)
$$\frac{I_h}{I_{\infty}}=k$$

The calculation of the relative error based on the magnitude of the factor k shows that the relative error of the measured X-ray intensity will be 0.5% when $k = 0.01$, 1% when $k = 0.002$ and 5% when $k = 0.1$. Accepting the 5% error of XRF measurement the thin film sample can be defined as one with $k < 0.1$.

From a practical experimental point of view this means that the analyzed sample fulfills the "thin film criterion" when the analyzed sample of the pure metal thin film emits characteristic X-ray radiation for a given element which is less than 10% of the characteristic radiation emitted under the same experimental conditions from the infinitely thick sample of the same pure metal.

Table 3.5 shows results of our XRF measurements using both bulk single element standards and thin film metal standards (50 µg/cm^2) prepared by vacuum evaporation of pure metal on a 2.5 micron thick Mylar film.

Element	Analytical line energy (keV)	Analytical intensity [counts] for infinitely thick sample	Analytical intensity [counts] for (~50 mg/cm^2) tfs	Thin film criteria maximum [counts]
Ti	TiKα_1- 4.500	15.900	n/a*	<1500
Cr	CrKα_1- 5.380	26.700	2.160	<2670
Fe	FeKα_1- 6.380	16.140	2.780	<1615
Co	CoKα_1-6.900	24.600	n/a*	<2460
Ni	NiKα_1- 7.460	30.100	n/a*	<3010
Cu	CuKα_1-8.020	45.700	4.680	<4570
Zn	ZnKα_1-8.620	38.100	n/a*	<3810
Pd	PdKα_1-21.14	4.200	210	<420
Ag	AgKα_1-22.14	9.200	285	<920
Cd	CdKα_1-23.14	12.200	n/a*	<1200
Sn	SnKα_1-25.26	5.000	n/a*	<500
Pt	PtLα_1-9.420	9.200	2.380	<920
Au	AuLα_1-9.700	17.000	2.820	<1700

*n/a: standard was not available

Table 3.5:
XRF analysis of bulk and thin film standards (tfs).

Our qualitative and quantitative XRF analysis of thousands of both 19th and 20th century photographs has shown that the concentration of most chemical elements detected in photographs falls well within the thin film criteria. The only important exception is the quantitative analysis of the baryta layer of gelatin silver POP and DOP photographs and the analysis of the thinner baryta layer of many 19th and 20th century glossy and matte collodion photographs.

Thin film calibration standards for quantitative analysis of photographs

Knowing that most of the image layers of analyzed photographs fulfill the thin film criteria allows for the use of thin film analysis software provided by several XRF manufacturing companies or of the use of quantitative XRF procedures based on the use of calibration curves obtained by the XRF measurements of thin film standards.

Our NITON XL700 handheld XRF spectrometer, with ^{55}Fe, ^{109}Cd and ^{241}Am radioisotope sources, allows the use of both bulk and thin film procedures and it was custom calibrated by the company for the quantitative analysis of 25 chemical elements important for research in photographs.

Most of currently used handheld XRF spectrometers use a miniature X-ray tube as the source of the primary X-ray radiation and many instruments on the market can also be supplied with quantitative thin film analysis software. The instruments can be both factory or in-house calibrated. We are using our extensive collection of thin film calibration standards (Fig. 3.15) to calibrate our KeyMaster TRACeR III, Re tube instrument for all chemical elements we were able to detect in analyzed photographs.

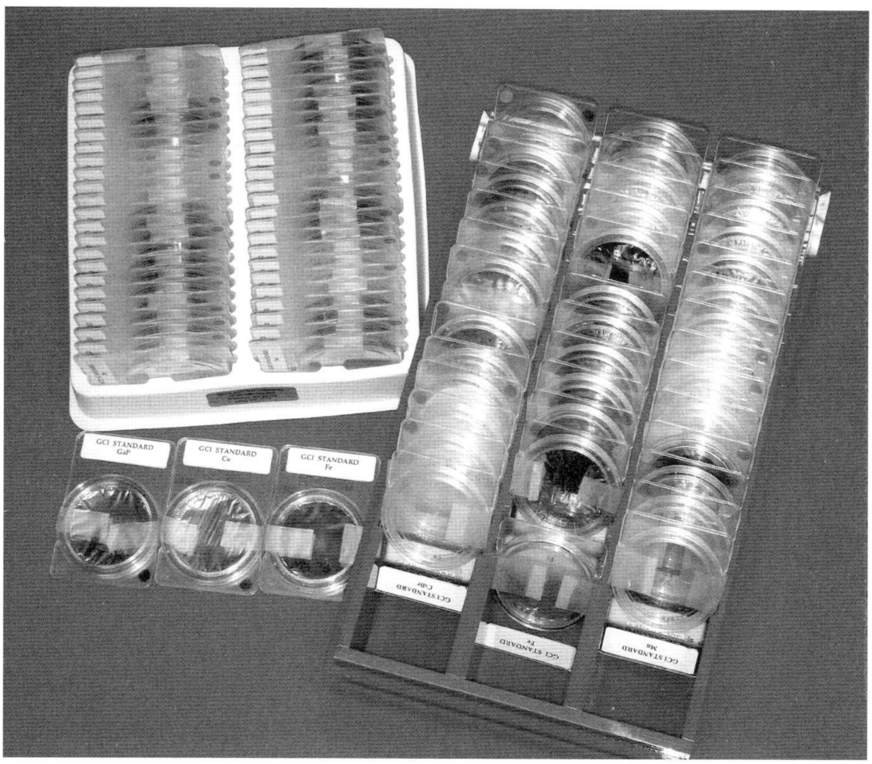

Figure 3.15:
GCI collection of thin film calibration standards.

These standards were prepared by vacuum deposition of pure metals, binary metal alloys or metal salts (usually fluorides). Such thin film standards are available in different sizes, deposited on Mylar foils of different thicknesses and in different amounts (2-5μg/cm^2, 15μg/cm^2, 50μg/cm^{2}and 150μg/cm^2, also we have added

300μg/cm² for selected elements like Ag and Au). The amount of deposited material was measured using a precise electronic quartz microbalance and is guaranteed with a precision of +/- 5%. Our set of standards ranges in concentration from 15 micrograms/square centimeter to 150 micrograms/square centimeter. For several important elements that usually constitute an imaging material of photographs (silver, platinum, iron) we also acquired thin film standards with concentrations of 300 micrograms/square centimeter.

Figure 3.16 shows calibration curves measured, using our set of thin film standards (Fig. 3.15), for several chemical elements often quantified by XRF analysis when analyzing photographs. The measurements were performed using a Keymaster TRACeR III, Re tube instrument operated with and Al/Cu (70μm thick Al/ 70μm thick Cu) primary beam filter at 40kV and 12μA in air and operated without a filter at 15kV and 12μA under vacuum.

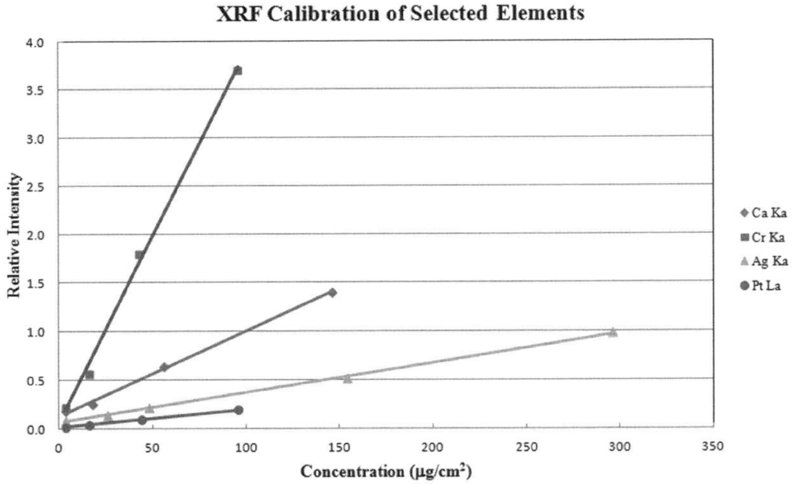

Figure 3.16:
XRF calibration curves for calcium, chromium, silver and platinum.

Calibration curves can be used to directly calculate the concentrations of analyzed chemical elements or they can be entered into the software of the XRF spectrometer for the automated quantitative interpretation of XRF measurements. Most of our measurements deal with material that corresponds well to thin film parameters. We also assume that the particle size of elements and compounds contained in analyzed photographs does not change the intensity of the analytical signal.

Application of quantitative XRF analysis in research in photographs

The intensities of spectral peaks in XRF spectra are directly related to the concentrations of individual elements that are present in the internal structure of photographs. In most cases, elements present in photographic images behave as thin films and the quantitative analysis of photographs does not require the use of complex quantitative models that include calculation of absorption of primary and fluorescent X-ray beams in the analyzed material. Most of the starting photographic material is also industrially produced and all layers of unexposed photographic material (both negatives and photographic papers) exhibit very uniform and homogeneous lateral distribution of all elements and chemical compounds in unexposed and unprocessed material. Light exposure in a camera, copy frame or in an enlarger followed by development, fixing and other processing and post processing steps result in a final image that exhibits a great variety in the concentrations of imaging elements (such as silver in silver based photographic processes, platinum in platinotypes etc.) or imaging compounds (various dye clusters in color photographs). Most of the imaging elements can be easily detected using XRF analysis. Important exceptions are photographic images formed using organic pigments, dyes and carbon based materials (pigment particles and inks).

The concentration of metal based (higher than sodium in the periodic table of chemical elements) imaging elements in photographs varies greatly from D-max (the darkest area of the final photograph containing the highest concentration of the imaging elements) to D-min (the lightest area of the final photograph containing the lowest concentration of the imaging elements) areas of the photograph. Many existing photographs contain a number of larger D-max and/or D-min areas that can be used when there is a need for quantitative analysis of the imaging material. The concentration of imaging elements, compounds or toning elements is not characteristic for any given photographic process but knowing concentrations of these elements can provide some important information on the unexposed material used or on some specific processing or post-processing treatments of photographs. The quantitative analysis of some chemical elements that are part of the internal structure of the photographic image and that do not change during processing or post-processing treatment of photographs (the presence of insoluble barium sulfate in the baryta layer of silver gelatin photographs or titanium dioxide permanently embedded in the white, reflecting, polymer layer of so-called RC (resin coated) modern photographic papers allows the use of the measured concentrations of these components for provenancing and authentication studies of photographs.

Single component analysis for provenancing of photographic material

Platinum photographic paper was invented and developed by the already mentioned William Willis, who later in 1879 established the Platinotype Company which produced several types of platinum photographic papers that were available commercially. Several other manufacturing companies also introduced platinum based photographic materials (Eastman, Ilford etc.). Each of these manufacturing companies used slightly different but chemically consistent photographic material that differed from one another by the amounts of platinum deposited on the paper substrate during the manufacturing process (Fig. 3.17).

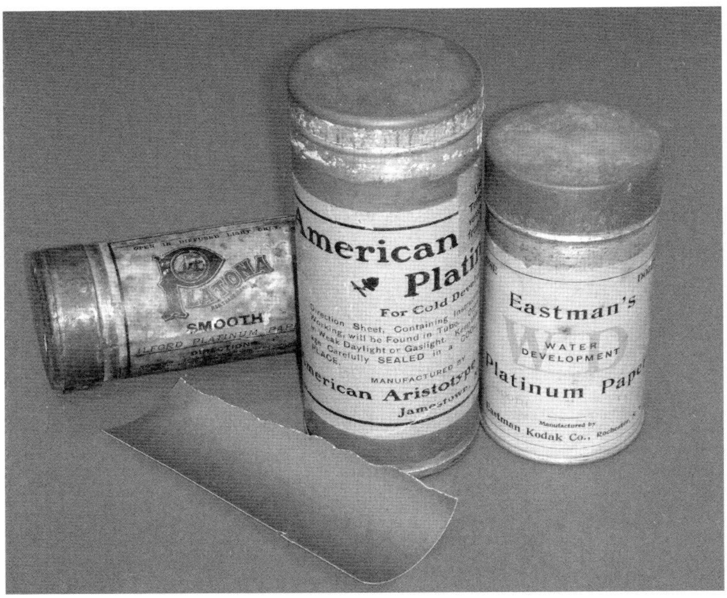

Figure 3.17:
Several examples of industrially produced platinum photographic paper that were available commercially before WWI.

The quantitative XRF analysis of unexposed and unprocessed platinotype paper from the American Aristotype Company from 1903 was analyzed using our handheld XRF spectrometer. The quantitative results of the analysis showed that the platinotype paper contained 66 $\mu g/cm^2$ of platinum.

Conservation scientists might be asked by curators or photograph dealers to answer some important questions related to provenancing or authentication of final photographs or questions related to the type of photographic material used by a photographer. The knowledge of the concentration of platinum in unexposed and unprocessed platinotype photographic paper is not very helpful because darkroom processing alters the amount of platinum and the concentration of platinum is usually lower than the concentration of platinum in the manufactured platinotype paper. The 66 $\mu g/cm^2$ of platinum in the photographic paper represents a theoretical upper limit for the concentration of platinum in any processed platinotype paper of this kind. If the XRF analysis of a series of platinotype photographs shows several prints with a concentration of platinum higher than 66 $\mu g/cm^2$ it is possible to conclude that these photographs could not be made using the 1903 American Aristotype Co. platinum paper[2].

Kinetics studies of toning processes

Art photographers of the second half of the 20th century started to be more conscious about the long term preservation of their photographs and developed the concept of archival processing and treatment of their exhibition photographs. Archival processing targeted both the elimination of fixer from photographic prints by a very thorough washing process as well as the use of the fixer eliminators and the improvement of the stability of silver particles by the application of selenium toning. Selenium toning was introduced during the first half of the 20th century producing a red-brown toning of gelatin silver photographs. The selenium toning of photographs during archival processing on the other hand should be done to protect silver particles against environmental pollutants but without any noticeable shift in the tonality of the photograph.

Our analytical investigation of selenium toned photographs and of the kinetics of selenium toning showed that archival selenium toning that cannot be detected by the visual inspection of photographs can be detected analytically using XRF spectrometry (Fig. 3.18).

The quantitative XRF of selenium toned photographs provides a more detailed insight into the kinetics of the toning process (Fig. 3.19) and allows the modification of the toning procedures for maximum protecting effects but without noticeable shifts in print tonality.

[2] Some earlier platinotype procedures used an addition of platinum in the developing solution. This procedure was not commonly used around 1903 but such a theoretical possibility has to be considered to avoid a possible over-interpretation of the results of the scientific investigation.

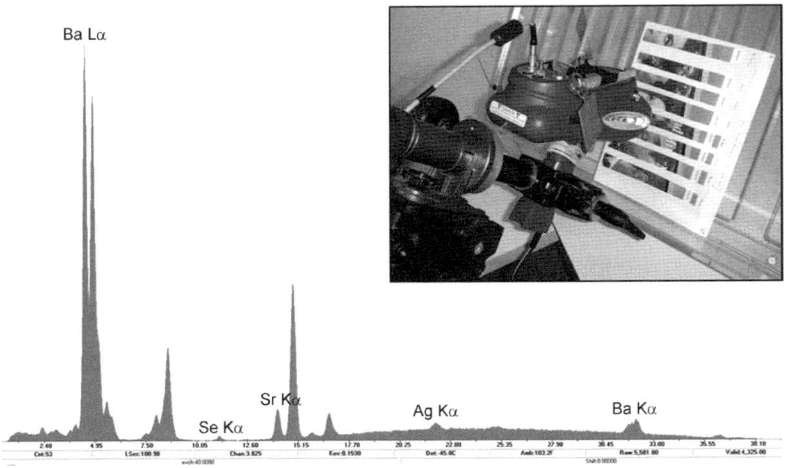

Figure 3.18:
XRF analysis of an archival processed, selenium toned photograph (obtained using a TRACeR III portable XRF with a Re tube and an Al/Cu filter operated at 40kV and 12μA).

The quantitative XRF analysis of a series of selenium toned photographs also provided data related to the kinetics of the selenium toning (Fig. 3.19).

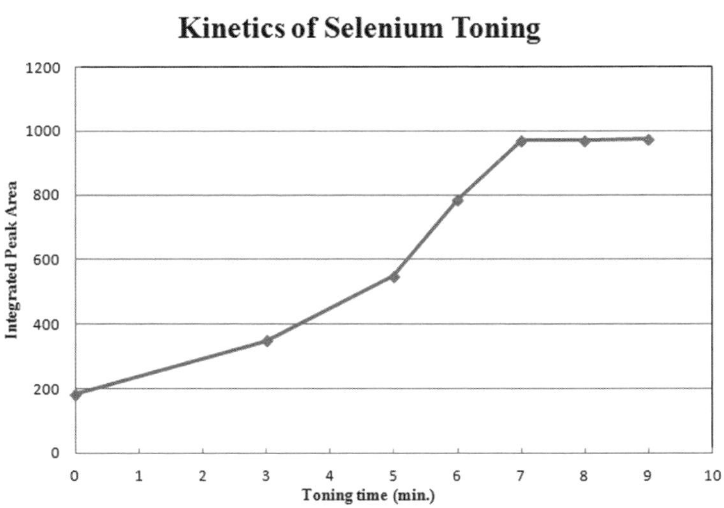

Figure 3.19:
Quantitative XRF study of the kinetics of selenium toning.

Provenancing and authentication of photographs based on quantitative XRF analysis of barium and strontium in photographs

Our quantitative XRF analysis of thousands of gelatin silver, fiber based photographs and photographic papers containing a baryta layer has shown that the concentrations of barium (from barium sulfate) and strontium (from strontium sulfate, which is a common impurity of barium containing material) can be used for the development of a new scientific methodology for the authentication and provenancing of photographs.

Provenancing and authentication of photographs is not an easy task. Photography is a medium of multiple originals. The photographer selects and records a photographed scene on a piece of film (in "classical" or "chemical" photography). From this moment the creation of a developed negative and the creation of a final photograph(s) can be performed by the photographer themselves, their assistants or printers, or by commercial photographic laboratories. The time period during which a photograph was printed is also important for art historical research.

"Vintage" prints are usually considered photographic prints made at the same time as the negative or within the next several years. So called "printed later" or "modern prints" were printed later by the photographer, his assistants, or printers under the direct supervision of the photographer. So called "posthumous" prints were made by a technician who never worked with the photographer, but they are still printed from the original negatives (Cartier-Bresson 2009: 138-157). At the same time all of the above mentioned types of photographs could be printed in different formats using different types of photographic papers.

Some curators, art dealers, and art collectors have developed a high level of visual connoisseurship that helps them deal with some more obvious authentication and forgery issues but several recently published cases of photographic forgeries (Robinson 2008, Woodward 2003: 67-72, 74-76) have shown that connoisseurship alone cannot solve all forgery, authentication and provenancing problems. Many problems related to forgeries of paintings, sculpture, drawings and other art and archaeological objects have already been solved using the methods of technical art history, conservation science, and forensic science (Ainsworth 2005).

Our research in photographs, conducted during the last several years, focused on the development of advanced and more objective methodology for the identification of photographic processes and process variants using both nondestructive and non-contact methods of chemical analysis and on the application of quantitative methods of XRF analysis in the development of new scientific methodologies for the provenancing and authentication of photographs.

Analyzing a number of 20th century, B&W, gelatin silver, baryta coated, fiber based photographs and a number of unexposed and unprocessed photographic papers used by important photographers of the "Chemical Photography Era" showed that even when the qualitative analysis of both photographs and photographic papers is very similar (Fig. 3.20), the quantitative analysis of each of the major inorganic components is often very different for material produced by different manufacturers, paper types and dates of the production (Stulik 2006, Kaplan 2006, Messier 2006, Miller 2006).

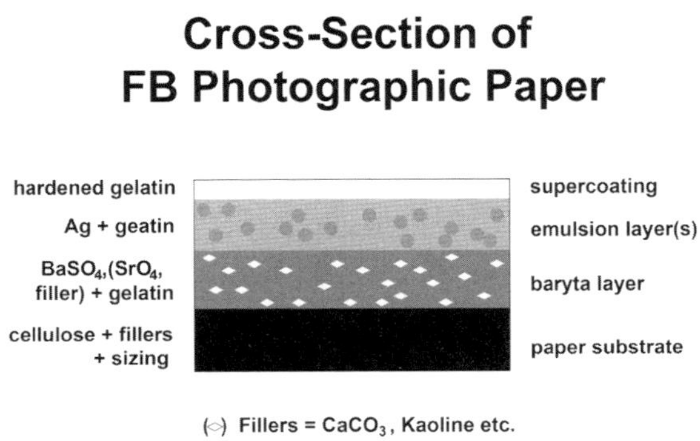

Figure 3.20:
Internal structure and chemical makeup of typical silver gelatin, fiber based, photographic papers.

The amount of elemental silver in analyzed photographs differs greatly but that information is not of primary importance when analyzing photographs because the final concentration of silver in a photograph depends heavily on darkroom processing procedures. Using XRF analysis we were able to determine that the concentration of barium and strontium in the baryta layer of gelatin silver photographs is almost identical (within fluctuations inherent to XRF measurements) when comparing photographs printed using the same photographic paper (the same box of paper) but varies substantially when the same photograph was printed using photographic paper produced by different photographic manufacturing companies. The baryta layer of photographic papers is made using barium sulfate (which often contains certain amounts of very stable strontium sulfate). Sandwiched between the paper base of the photograph and the emulsion layer of the photographic paper the concentration of both barium and strontium is invariant to any standard darkroom photograph processing or post-processing treatment. Based on this phenomenon, variation of the quantitative concentrations of barium and strontium in analyzed photographs became the basis for our quantitative XRF based methodology for provenancing and authentication of photographs.

Preliminary tests

A series of preliminary XRF measurements were used to prove that the concentration of both barium and strontium in machine-produced photographic paper is chemically and physically uniform for each emulsion run of the photographic paper. These test were conducted by quantitative XRF analyses of a number of analytical spots across a single photographic paper and by comparing multiple analyses conducted on the first and last photographic paper of a large volume packaging of the photographic papers (250 sheets of 8x10" photographic papers) and on several packages of photographic papers of the same and different emulsion runs (different emulsion numbers) (Fig. 3.21). The second battery of analyses tested the accuracy and reproducibility of XRF measurements. A series of consecutive analyses were conducted on a single sheet of photographic paper analyzed under several different operating conditions (accelerating voltage and emission current of the X-ray tube) without changing the geometry of the experiments. The identical analyses were repeated several times across several days after complete shut down and re-start of the XRF spectrometer. To conduct each of these measurements, the X-ray tube was warmed up and equilibrated for more than twenty minutes.

We have also studied the effect of the sample-XRF spectrometer geometry. These experiments have shown that even a small (1mm) increase of distance between the analyzed sample and the "nose" of the XRF spectrometer decreases the intensity of measured X-rays by about 15% (Barium Lα peak) and that a well defined and highly reproducible experiment geometry is critical for the success of actual measurements on real photographs. Our preliminary test led us to modify the XRF instrument and to develop a standard operating protocol for the quantitative analysis of the baryta layer of photographic papers.

Shaded areas across the analyzed sheet of the photographic paper (Fig. 3.21) represent areas where three consecutive analyses were conducted. Numerical values are presented together with calculated error values from three consecutive measurements that included removing the analyzer and repositioning the instrument for a new analysis. Measured values represent small local inhomogeneities in the concentrations of both Ba and Sr across the photographic paper. The bright grey rectangle area is a place where seven consecutive analyses were conducted without removing the instrument between individual measurement runs. These measurements represent X-ray instrument and counting errors that are inherent to all XRF measurements. Our experiments showed that the baryta layer is rather uniform and homogeneous and that quantitative data can be used when developing a working methodology for provenancing and authentication of photographs.

Sample # 1474 [ug/cm²]

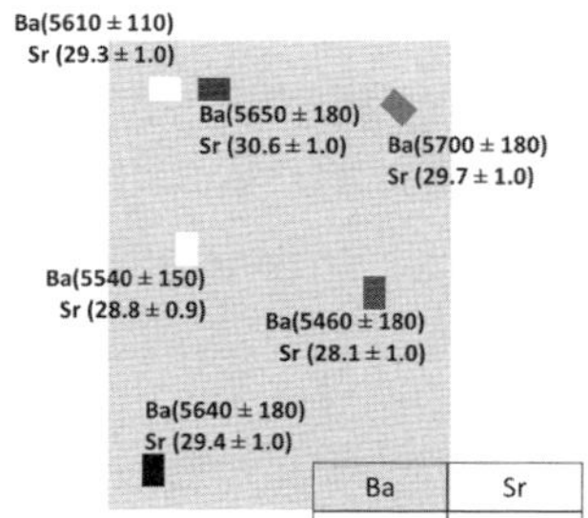

Ba	Sr
5630	30.2
5550	29.8
5580	29.8
5520	29.5
5570	29.4
5460	30.1
5600	28.9

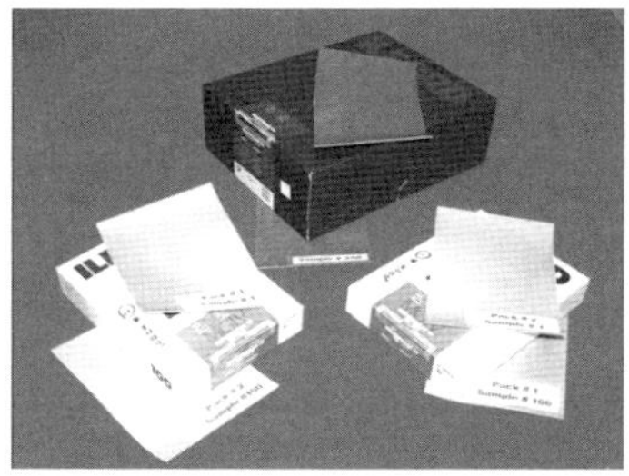

Figure 3.21:
Several experiments were conducted to determine the uniformity of mechanically coated photographic papers.

SOP (Standard Operating Protocol) for quantitative XRF measurements

Our laboratory has, at its disposal, several laboratory grade and handheld XRF spectrometers that are used for different aspects of GCI conservation science research. In our research in photographs our main instrument for the analysis of inorganic materials is the Keymaster TRACeR III XRF spectrometer equipped with a rhenium anode X-ray tube and an aluminum/copper (60μm/60μm thick) sandwich primary beam X-ray filter. We have modified the instrument for the quantitative analysis of photographs by attaching a 4x4" square 1mm thick geometry restrictor plate made of Lucite with a 1.5 cm diameter opening for primary and fluorescent X-ray beams. The geometry restrictor plate, when slightly touching an analyzed photograph keeps constant distance between a photograph and the XRF instrument. The geometry restrictor plate also allows one to keep the spectrometer perfectly perpendicular to the analyzed photograph as needed for quantitative analysis (Fig. 3.22). The X-ray tube is operated with an Al/Cu X-ray filter at 40kV and 12μA emission current in the air and at 15kV and 12μA with no filter under vacuum. Due to the low concentration of key elements in photographs we collect spectra for 300 seconds. The longer analysis time also provides for good statistics when interpreting the quantitative results of the analysis.

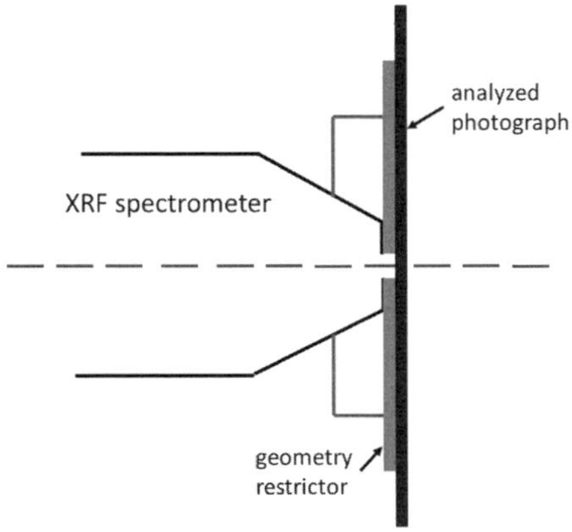

Figure 3.22:
Simple modification of the XRF spectrometer allows for well defined and reproducible analysis of photographs.

Preparation of XRF calibration standards

Knowing both current and future needs for a new scientific methodology for the provenancing and authentication of photographs we have designed and developed our working methodology in such a way that other conservation scientists and photograph experts would be able to fully adopt our working methodology in their work and research.

To achieve this goal it was important to prepare a set of high quality quantitative calibration standards that would potentially allow for the identical calibration of multiple XRF instruments.

We have prepared and analyzed a set of previously analyzed photographic papers from both the GCI's and Paul Messier's collections of photographic papers (Stulik 2006, Messier 2006). Photographic paper samples contained barium sulfate and strontium sulfate in varying ratios to one another as well as wide ranges in concentrations of both elements. Each photographic paper was thoroughly fixed to remove silver halide components of the photographic emulsion, leaving behind a paper substrate covered with a baryta layer and a thin layer of gelatin. Out of each fixed photographic paper we cut a circular (1-5/8" diameter) target for the XRF using a 1-5/8" metal punch. Five smaller, 5/16" diameter, samples were also removed from the area surrounding the larger target for neutron activation analysis (NAA). The NAA analysis was selected because of its high sensitivity, high precision and low level of spectral interferences. The fact that samples of photographic papers could be analyzed without any chemical treatments and preparation was also an important factor for the decision to use NAA analysis. Samples were sealed into special ultrapure polyethylene vials designated for NAA analysis and were loaded into the rotary rack of the nuclear reactor together with a number of NIST trace element standards and blanks. Samples were irradiated in a Triga nuclear reactor, at the Chemistry Department at the University of California at Irvine, at 250kW of power which generated a thermal neutron flux of 10^{12} neutrons/cm^2sec. Samples were irradiated for 30 minutes and allowed to decay for 30 min. prior to counting. The samples were counted for 4 minutes using a Canberra gamma ray spectrometer. Highly accurate analytical NAA data for both barium and strontium concentrations in analyzed photographic papers allowed us to use these samples as analytical standards for the calibration of our XRF spectrometer (Fig. 3.23). These XRF standards can also be used to calibrate any current or future XRF spectrometers that are or will be used in provenancing and authentication studies of baryta layer based photographs.

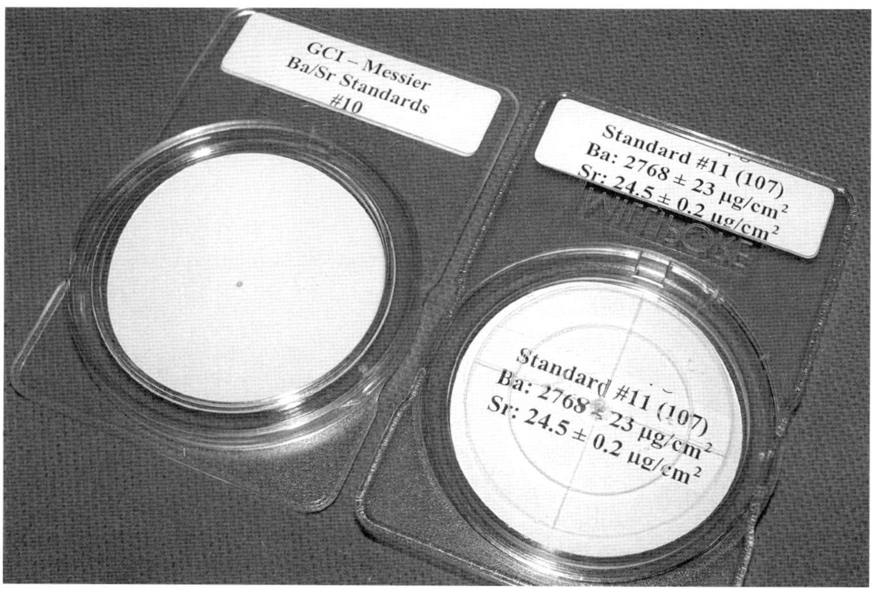

Figure 3.23:
Barium and strontium standards for the quantitative XRF analysis of baryta coated photographic papers.

Laboratory test of working methodology

To test our working methodology on real photographs we printed nine identical 8x10" photographs on several well known and widely used photographic papers that were commercially available in the year 2000. A randomly selected print from this test set was printed twice and it was used as an "unknown" for our laboratory test. Even when we tried to print all of the photographs to achieve identical contrast and tonality, there were some slight contrast and tonality differences between several prints in the sample set. To prevent any contrast or tonality clues when comparing the "unknown photograph" with a set of different photographs, the "unknown" print was toned dark brown using a selenium toner.

A group of leading photography conservators, museum curators, dealers of photographs, auction house experts and photography appraisers attending the 2006 Baryta Layer symposium held at the Getty Conservation Institute (Stulik 2006, Kaplan 2006, Miller 2006, Messier 2006, Duverne 2006) were asked to identify the "unknown" visually but none of the participants were able to correctly identify the paper the "unknown" photograph was printed on. All of the photographs in the test set were previously analyzed using XRF spectrometry following the above described standard operating protocol. Six consecutive analyses were performed on selected D-min areas of all photographs in a test set and the spectral intensities for both barium and strontium in each photographic paper were plotted on an X-Y graph together with corresponding standard deviation fields. The quantitative XRF analysis for barium and strontium in the "unknown" photograph when compared with analytical data of the test set of photographs allowed for the correct identification of the "unknown" photograph as being made on Forte Elegance Glossy FB Grade 2 B&W photographic paper (Fig. 3.24 a and b).

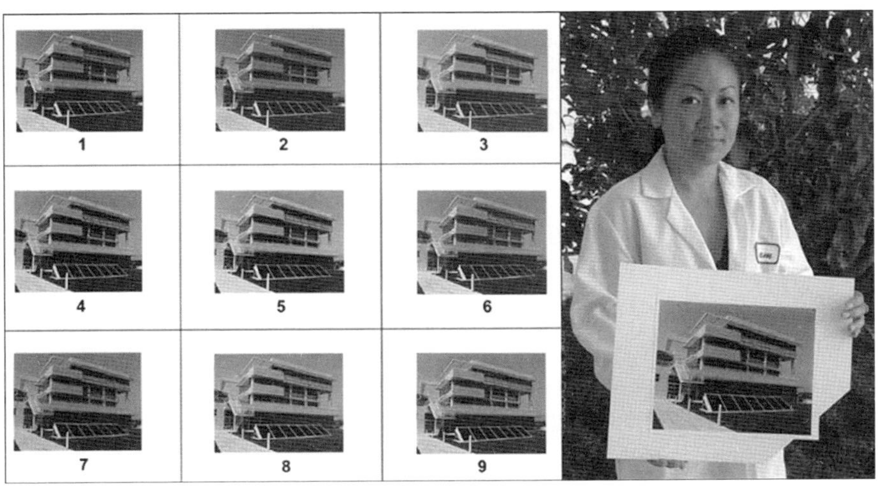

Figure 3.24:
A- Nine photographs were printed on different photographic papers;

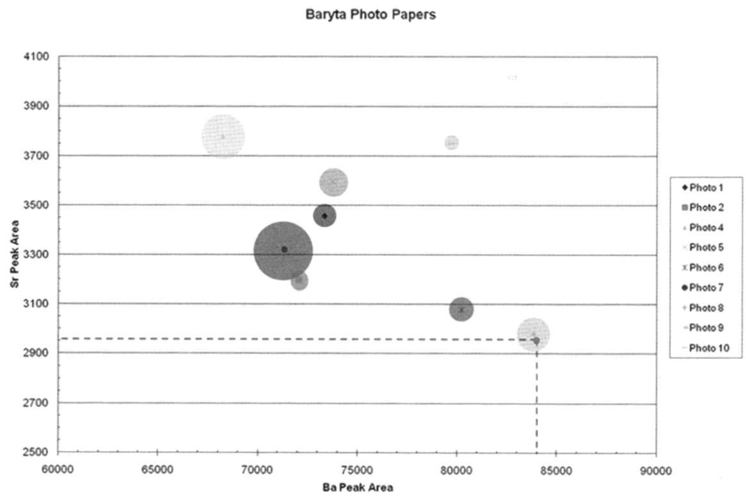

Figure 3.24:
B- XRF analysis correctly identified the "unknown" photograph (photo #8) based on the quantitative analysis of Ba and Sr in all photographs.

Reference material database

Working with samples of both historical and modern photographic papers we were able to plot the concentrations of barium vs. strontium in analyzed photographic papers into an X-Y graph (Fig. 3.25). The graph illustrates the range of concentrations of both elements in frequently used photographic papers. The same type of photographic paper produced in different years may have different chemical compositions or its composition may change due to changes in technology or due to changes in the sources of raw materials.

The graph in figure 3.25 is based only on the analytical data for concentrations of barium and strontium in the photographic papers. The inclusion of other chemical and physical parameters allows for more detailed separation of different paper types. For example, inclusion of paper thicknesses (a physical parameter measured using a micrometer) allows a separation of single and double weight photographic papers. So far we have identified eleven chemical and physical parameters that can be used, along with barium and strontium concentrations, in provenancing and authentication of photographs. This work is still in progress and will be published sometime in the future.

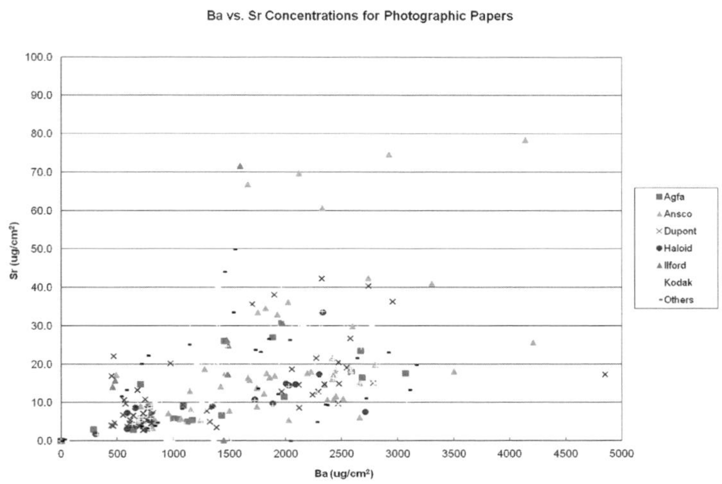

Figure 3.25:
Results of the XRF analysis of more than 500 different 20th century baryta based photographic papers.

To use our methodology for the identification of photographs without limitations (not counting concentration overlaps due to similar composition for both elements) would require access to the ultimate database of all photographic papers ever produced by all manufacturers during their manufacturing periods. A general strategy that can be applied in order to match analytical data from the analysis of an unknown photograph with analytical data from a database of "all" photographic papers is schematically depicted in figure 3.26. Figure 3.26 depicts a general concept of data matching between data obtained from the scientific examination of a photograph with the same data from a hypothetical database of photographic papers. Any chemical (ie. concentrations of barium and strontium) and physical (ie. paper thickness) parameters can be used for matching as long as the values of the parameters do not change during the processing or aging of the photographs. This is the reason figure 3.26 shows pictographic symbols (●, ■, ▲) rather than specific parameters. Our current research targets an inclusion of other experimental parameters (ie. trace elements, baryta layer thickness, etc.) and the use of chemometric software to aid in data matching.

A chance to build up a very comprehensive database of analytical data on photographic paper is very slim now and it will be even more difficult in the future. To provide an alternative solution to the problem of a comprehensive database of analytical data on photographic papers we developed the concept of Individual Database Segments (IDS).

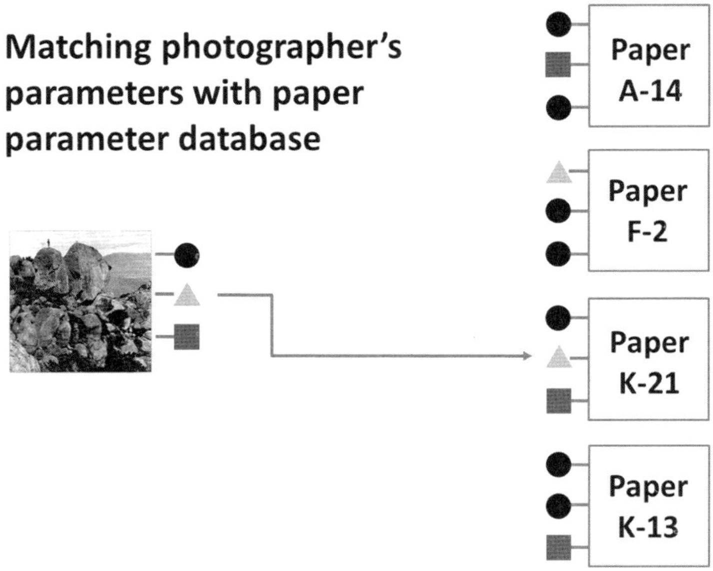

Figure 3.26:
General strategy for matching analytical parameters of unknown photographs with analytical parameters of the photographic paper database.

Working with "Individual Database Segments' (IDS)

Many photographers and printers liked to experiment with different photographic papers but many of them also developed a close attachment to certain kinds and types of photographic papers. Most art photographers opted for working with a small number of photographic papers and changed their selection only with changes in their "print aesthetics" or when their favorite papers were not available or did not provide good and consistent results. This observation, supported by our own photographic experience, has led us towards the working concept of the so called "Individual Database Segments" (IDS). Instead of having available a database of all photographic papers ever produced, to solve targeted provenancing and authentication problems dealing with the body of work of an individual photographer, the researcher can work only with a segment of a comprehensive paper data collection that relates to an individual photographer (Fig. 3.27). For example if there is a questionable Henri Cartier-Bresson (HCB) photograph on the market or in a museum collection, a researcher dealing with the authentication task would need to conduct XRF analysis of the photograph using a properly calibrated XRF instrument and to compare analytical results with data from HCB's IDS. Henri Cartier-Bresson's IDS would be created by analyzing many well authenticated and provenanced HCB photographs created throughout his photographic career. Some important information about the use of photographic paper by an individual artist could be also obtained in collaboration with art historians or historians when consulting photographers' records, business documents, photographer's writings or his/her darkroom notes.

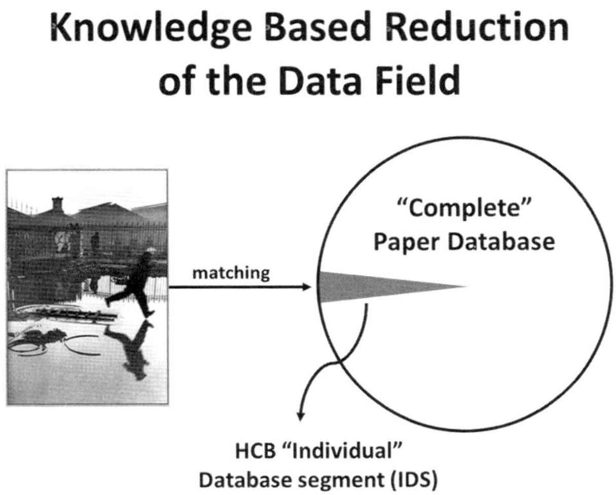

Figure 3.27:
Authentication of photographs using Individual Database Segments (IDS).

We have tested the IDS methodology for the body of work of the famous French photographer Henri Cartier-Bresson. Working very closely with the Atelier for the Restoration and Conservation of Photographs of the City of Paris (ARCP), Henri Cartier-Bresson Foundation (HCBF) and many museums in Paris we have selected and analyzed a number of photographs by Henri Cartier-Bresson printed both by himself and by a number of his printers who printed his B&W photographs after 1947. Analyzing many selected photographs from different periods of Cartier-Bresson's life we have built a starting (beta) version of the Individual Database Segment of analytical data for provenancing and authentication of Henri Cartier-Bresson's photographs.

Even when the work towards building HCB's IDS is not complete the information and data already available in our working HCB IDS can be used in supporting art historical research or in authentication and provenancing studies of Henri Cartier-Bresson's photographs (Stulik and Kaplan 2009: 158-173, Cartier-Bresson 2009: 138-157).

The data depicted graphically in figure 3.28, for example, show that the photographs from the Henri Cartier-Bresson "Master Collection" have very different chemical and physical signatures than so called "triage de presse" photographs. The XRF analysis of several visually identical photographs and the differences in their quantitative signatures showed that these photographs described in registrar information as prints from "1970-1980 decade" may have been made during at least three different printings during that time period.

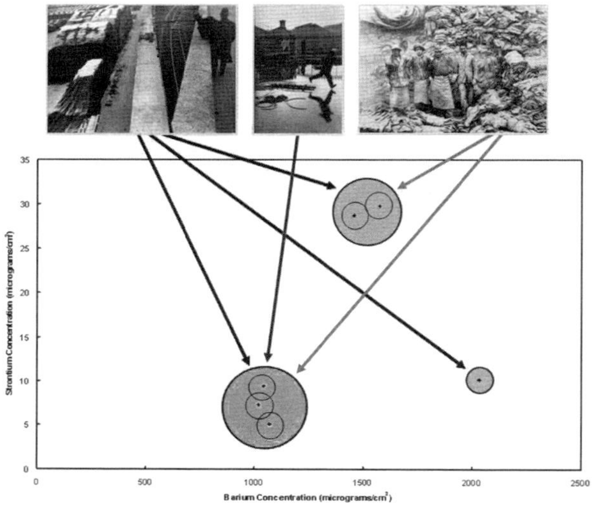

Figure 3.28:
Grouping of photographs according to their chemical signatures facilitates provenancing of photographs.

Our analysis of Henri Cartier-Bresson's photographs showed that all three photographs in figure 3.28 were printed when creating his "Master Collection". All three prints from the "Master Collection" group together nicely as depicted in the grey circle in the lower left part of the Ba-Sr 2D graph in figure 3.29. Two of the "triage de presse" photographs group together, in the top circle, and appear to have been printed on the same paper, possibly during the same printing session. The outlier is an additional "triage de presse" print listed as being from the same decade but based on the analysis appears to be printed on different paper during a different printing session. The information obtained from Henri Cartier-Bresson's printers on individual printing sessions and on the paper ordering practices of the printers was crucial in aiding the interpretation of the analytical data.

Matching of the chemical and physical signatures of photographs can also aid provenancing studies when dating photographs. Figure 3.29 shows a situation where the precise date of the making of a photograph is not known. The registrar information related to the photograph "*Quai St. Bernard, Paris*" provided only a rather loose time period of its printing (1940-1950). The analysis of the photograph and the data matching with other photographs in the current version of the IDS of Henri Cartier-Bresson's photographs showed a good match with a group of 1945 photographs from Dessau in Germany. Based on our findings we can suggest that this printed version of the "*Quai St. Bernard, Paris*" photograph was printed after 1945 when the Dessau prints had been produced.

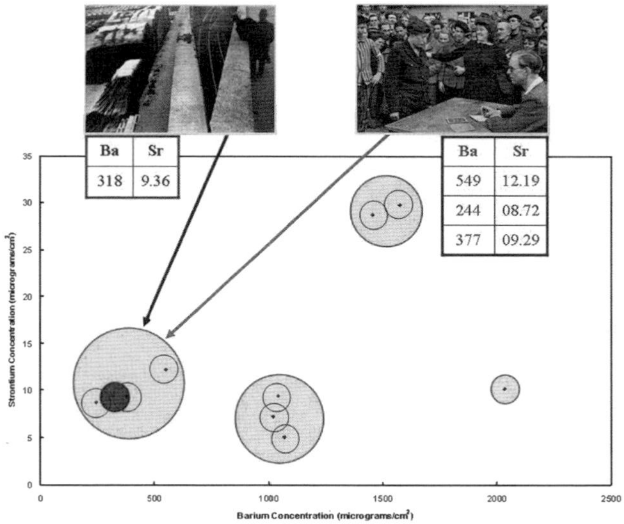

Figure 3.29:
Application of the scientific methodology in correcting a date assigned to a photograph.

The original form of the IDS of Henri Cartier Bresson's photographs was constructed using the analytical data obtained from the quantitative XRF analysis of 55 well provenanced and authenticated Henri Cartier Bresson photographs. The database later grew to 57 photographs. The results generated from the Henri Cartier-Bresson project showed that our quantitative XRF methodology for provenancing and authentication of photographs is applicable even when working with still incomplete IDSs and it is clear that the range of possible provenancing tasks and the quality of data matches will grow with expansion of the IDS. We can also see the clear value of working with collections of photographic papers, namely when these collections contain extensive holdings of photographic papers used by photographers for whom the IDS will be built.

XRF Investigation of the multilayer structure of photographs

It is a great advantage when a photograph can be separated from any frame, enclosure, matting assembly or album page for analysis. In many instances photographs were mounted permanently by photographers or photographic studios on mounting boards, into presentation folders or into albums.

When performing XRF analysis the primary X-ray beam penetrates deeply into the analyzed material and the resulting XRF spectrum of a mounted photograph usually represents a spectral superposition of the XRF spectrum of a photograph and all of the components of the mounting assembly.

To decipher the spectral contribution of different components it is advisable, if possible, to analyze both D-max and D-min areas of a photographic image, an area, if possible, of the matting or mounting assembly outside of the photographic image and sometimes even the matting assembly from behind (Fig. 3.30).

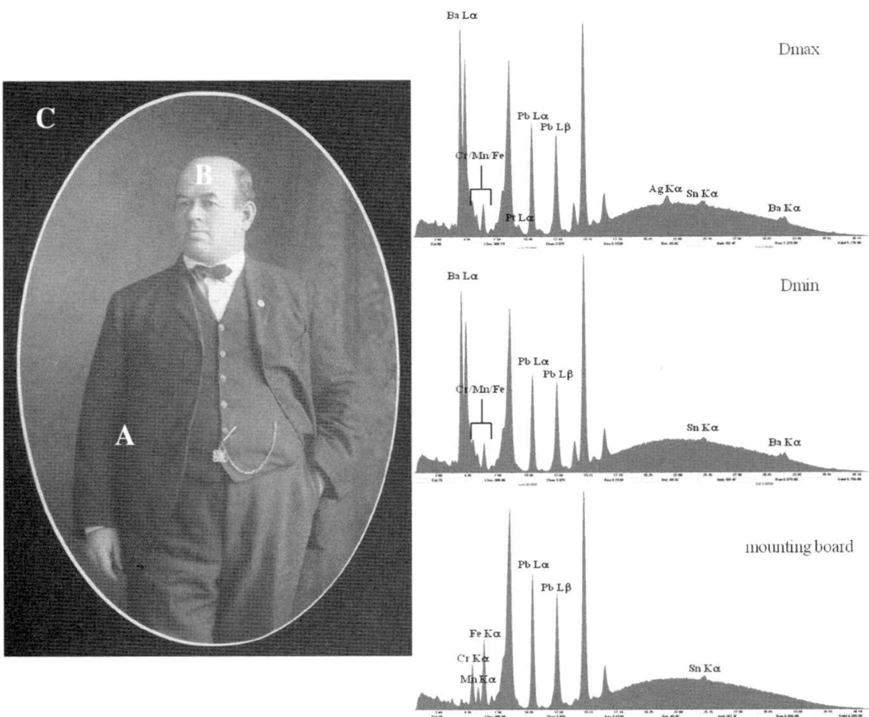

Figure 3.30:
XRF spectra of various spots of a platinum toned matte collodion photograph permanently mounted on a multilayer mounting board (letters A, B and C denote strategically selected analytical spots for the analysis). Obtained using a TRACeR III portable XRF with a Re tube and an Al/Cu filter operated at 40kV and 12μA).

Even relatively new archival and museum matting boards contain some inorganic materials that can complicate the XRF analysis. Figure 3.30 shows the XRF spectrum of a mounting board containing a relatively high concentration of titanium dioxide that was added to the board material to increase the whiteness of the matting material. Because titanium interferes strongly with barium the analysis of the baryta coated photograph mounted on such a support would be more complicated. The use of XRF quantitative methodology for provenancing and authentication of photographs described above may be very difficult or impossible.

The XRF analysis of photographs permanently glued to pages of photographic albums might also be more complicated (Fig. 3.31).

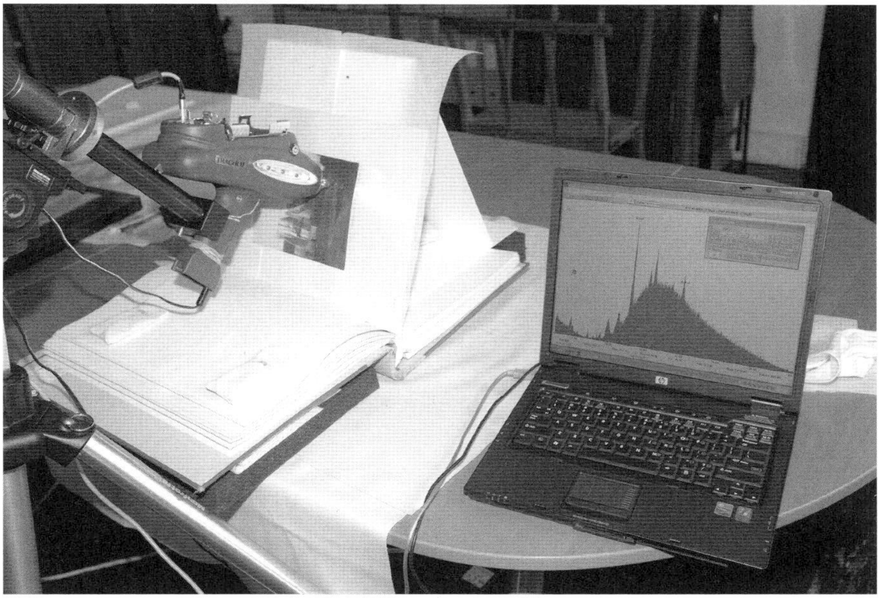

Figure 3.31:
XRF analysis of a 19th century photographic album by Brebisson.

To facilitate the analysis we have mounted our XRF spectrometer on a very stable tripod arm and we support a single page of an album using our custom made XRF easel that provides support for the album page as well as the ability to perform the analysis of a number of locations on a photograph where there are no other materials than the album page and page mounted photograph. In most cases we also analyze an album page alone so a spectral subtraction procedure can be conducted if needed.

Portable laboratory for research in photograph

The XRF as a primary component of the Portable Laboratory for Research of Photographs

The Getty Conservation Institute's project in research and conservation of photographs requires easy access to a variety of photographs and photographic processes, which is well beyond what can be found in any single photographic collection or museum. To facilitate access to a variety of photographs and photographic processes and at the same time to comply with current art conservation practices, which call for the restriction on movement and unnecessary transport of art materials, we have built a portable analytical laboratory for research in photographs (Stulik 2005: 556-564). Building the portable laboratory was made much easier by recent trends in modern analytical instrumentation. Miniaturization and computerization has made many instruments much smaller and often portable.

Using the portable laboratory allows us to modify the standard practice of art conservation research whereby the art was brought for analysis to the scientific or conservation laboratory: instead the laboratory goes to the art. The portable laboratory can travel worldwide, not only for the benefit of our research project but also for the benefit of many collaborating museums, collections, and institutions that have important questions about their photographic collections but which lack access to a conservation science laboratory or have limited financial resources to obtain analytical answers using commercial services. Table 3.6 shows the primary and secondary analytical and scientific instruments and tools included in the Getty Conservation Institute (GCI) Portable Laboratory.

Our portable laboratory needs two operators to work efficiently and can be transported in two standard Pelican cases. It can be checked-in when traveling by air or sent ahead by delivery services. The lab requires only a standard source of electricity (110/220V) and a desk area of approximately 3x6 feet. It is our experience that it takes about 40 minutes to go from the locked transportation cases to a fully functioning laboratory. As part of our continuous research we are actively looking and testing newly developed and commercially introduced portable instruments that might expand the working capabilities of the portable laboratory when analyzing and studying photographs and photographic materials.

Instrument	Manufacturer	Accessories	Task
S6D stereo-microscope	Leica	Fiber optic light, Pax-it image software, 2x close-up lens, Dell Latitude E6410 computer	Microscopic examination and photomicrography
DS-Fi1	Nikon		Image capture
DG-3 digital microscope	Scalar		Microscopic examination and photomicrography
TRACeR III	Keymaster	Dell Latitude D610 computer	XRF elemental analysis
TravelIR ATR-FTIR	Smiths Detection	Omnic FT-IR software, GCI Library of photographic processes, Dell Latitude D610 computer	FTIR organic analysis
TruDefender FT	Ahura Scientific		FTIR organic analysis
Digimatic Caliper	Mitutoyo		Physical measurements
Series 293 micrometer	Mitutoyo		Thickness measurements
Versalume UV lamp	Raytech		Detection of optical brighteners
Spectrocam	Avantes		Spectrophotometry, densitometry and colorimetry measurements

Table 3.6:
Components of the GCI portable laboratory.

References

Ainsworth, Maryan W. 2005. " From Connoisseurship to Technical History: The Evolution of the Interdisciplinary Study of Art." *Getty Conservation Institute Newsletter* Vol. 20 (Spring 1).

Bicchieri, M., S. Ronconi, F. P. Romano, L. Pappalardo, M. Corsi, G. Cristoforetti, S. Legnaioli, V. Palleschi, A. Salvetti, and E. Tognoni. 2002. "Study of foxing stains on paper by chemical methods, infrared spectroscopy, micro x-ray fluorescence spectrometry and laser induced breakdown spectroscopy." *Spectrochimica Acta Part B, Atomic spectroscopy* Vol. 57 (7):1235-1249.

Brown, B. 2002. "A brief interim report on the GCI/HRC collaborative project on the first photograph." *Western Association for Art Conservation Newsletter* Vol. 24 (3):13.

Brown, B., D. Stulik, and H. Khanjian. 2005. "Four metallic photographic prints from the Harry Ransom Center collection." *Topics in Photographic Preservation* Vol. 11:57-72.

Cartier-Bresson, A. 2008. *Le vocabulaire technique de la photographie*. Paris: Marval.

———. 2009. "Provenance et authentification des épreuves photographiques d'HCB." In *Revoir Henri Cartier-Bresson*, edited by Anne Cartier-Bresson and Jean-Pierre Montier, 138-157. Paris: Textuel.

Coe, Brian, and Mark Haworth-Booth. 1983. *A guide to early photographic processes*. London: Victoria & Albert Museum.

Daguerre, Louis Jacques Mandé, F. Arago, Joseph Louis Gay-Lussac, and Nicéphore Niépce. 1839. *Historique et description des procédés du daguerréotype et du diorama*. Nouv. éd., corr., et augm. du portrait de l'auteur. ed. Paris,: A. Giroux et cie.

de Caranza, E. 1856. "Procede de tirage des positives por le chlorure de platine." *Bulletin de la Société Française de photographie* Vol. April 15:82-84.

Duverne, R. . 2006. Structural Measurements of DOP Photographic Paper and Particle Size Analysis of Baryta Coating. Paper presented at the Understanding 20th Century Photographs: Baryta Layer Research Symposium, January 24, at Los Angeles: The Getty Conservation Institute.

Eder, Josef Maria. 1978. *History of photography*. New York: Dover Publications.

Enyeart, J. L., A. B. Anderson, S.J. Perron, D.K. Rollins, and Q. Fernando. 1983. "Non-destructive elemental analysis of photographic paper and emulsions by x-ray fluorescence spectrometry." *History of Photography* Vol. 7 (2):99-113.

Eremin, K., J. Tate, and J. Berry. 2003. "Non-destructive investigation of 19th century Scottish photographs." In *Conservation Science 2002: Papers from the Conference Held in Edinburgh, Scotland 22-24 May 2002*, edited by Joyce Townsend, Katherine Eremin and Annemie Adriaens, 201-207. London: Archetype Publications Ltd.

Glinsman, Lisha Deming. 2005. "The practical application of air-path x-ray fluorescence spectrometry in the analysis of museum objects." *Reviews in Conservation* Vol. 6:3-17.

Grieten, E., and F. Casadio. 2010. "X-ray fluorescence portable systems for the rapid assessment of photographic techniques in notable art collections: the Alfred Stieglitz Collection." *X-Ray Spectrometry* Vol. 39 (3):221-229.

Hey, M., G. Pasquariello, F. Gallo, G. Guidi, and F. Pierdominici. 1988. "Paper analysis in relation to foxing." In *2nd international conference on non-destructive testing, microanalytical methods and environment evaluation for study and conservation of works of art, Perugia, 17-20 April 1988*. Istituto Centrale Per Il Restauro - Associazione Italiana Prove Non Distruttive, 9.1-9.10. .

Jones, Bernard Edward. 1911. *Cassell's cyclopedia of photography*. London: Cassell and Company, ltd.

Kaplan, A. 2006. Photographic Paper Musical Chairs: Where Does Each Element Sit? Paper presented at the Understanding 20th Century Photographs: Baryta Layer Research Symposium, January 24, at Los Angeles: The Getty Conservation Institute.

Kaplan, A., and D. Stulik. 2008. "Getty Conservation Institute portable lab for photograph conservation: the first three years." In *ICOM-CC Preprints of the 15th Triennial Meeting, New Delhi, 22 - 26 September 2008*, edited by Janet Bridgland. London: James and James.

Knodt, R., and K. Pollmeier. 1999. *Verfahren der Fotografie: Bilder der Fotografischen Sammlung im Museum Folkwang Essen*. Essen: Museum Folkwang.

Lavédrine, Bertrand. 2009. *Photographs of the past: process and preservation*. Los Angeles: Getty Conservation Institute.

Mantler, M., and M. Schreiner M. 2000. "X-ray fluorescence spectrometry in art and archaeology." *X-Ray Spectrometry* Vol. 29 (1):3-17.

Marignier, Jean-Louis. 1999. *Nicéphore Niépce 1765-1833: l'invention de la photographie*. Paris: Belin.

McCabe, C., and L. Glinsman. 1996. "Understanding Alfred Stieglitz' platinum and palladium prints: examination by x-ray fluorescence spectrometry." In *Proceedings of the conference held at the National Museum of Denmark, Copenhagen, 14-19 May 1995*, edited by Mogens S. Koch, Tim Padfield, Jesper Stub Johnsen and Ulla Bøgvad Kejser. Copenhagen: The Royal Danish Academy of Fine Arts.

Messier, P. 2005. "Notes on dating photographic paper." *Topics in Photographic Preservation* Vol. 11:123-130.

———. 2006. Photographic Paper Exposed: New Meaning and Possibilities for the Understanding of 20th Century Photographs. Paper presented at the Understanding 20th Century Photographs: Baryta Layer Research Symposium, January 24, at Los Angeles: The Getty Conservation Institute.

Miller, D. 2006. Minor and Trace Elements in Photographic Material: Analysis and Meaning. Paper presented at the Understanding 20th Century Photographs: Baryta Layer Research Symposium, January 24, at Los Angeles: The Getty Conservation Institute.

Moore, C. 2007. "An analytical study of Eugéne Atget's photographs at the Museum of Modern Art." *Topics in Photographic Preservation* Vol. 12:194-210.

Muller, R. O. 1972. *Spetrochemical Analysis by X-ray Fluorescence*. London: Adam Hilger.

Nadeau, Luis. 1989. *Encyclopedia of printing, photographic, and photomechanical processes: a comprehensive reference to reproduction technologies containing invaluable information on over 1500 processes*. 2 vols. Fredericton, N.B.: Atelier Luis Nadeau.

———. 1994. *History and Practice of Platinum Printing*. 3rd rev. ed. Fredericton, N.B.: Atelier L. Nadeau.

Pailet, Joshua M. 2010. *Buy, Sell, Trade: Collecting Photographs* 1999 [cited October 25 2010]. Available from http://www.agallery.com/Pages/departments/buy_sell_trade.html.

Potts, P. J., and Margaret West. 2008. *Portable x-ray fluorescence spectrometry: capabilities for in situ analysis*. Cambridge, UK: Royal Society of Chemistry.

Reilly, J.M. 1986. *Care and Identification of 19th Century Photographic Prints*. Rochester N.Y.: Eastman Kodak Co.

Robinson, Walter. 2010. *Man Ray Forgeries Exposed* 2008 [cited October 25 2010]. Available from http://www.artnet.com/Magazine/news/robinson/robinson12-2-97.asp.

Rogers, David. 2007. *The Chemistry of Photography: from classical to digital technologies*. Cambridge, UK: Royal Society of Chemistry.

Stulik, D. 2005. "Getty Conservation Institute portable lab for photograph conservation: the first three years." In *ICOM-CC Preprints of the 14th Triennial Meeting, The Hague, 12 - 16 September 2005*, edited by Janet Bridgland, 556-564. London: James and James.

———. 2006. Research in Identification and Provenancing of 20th Century Photographs. Paper presented at the Understanding 20th Century Photographs: Baryta Layer Research Symposium, January 24, at Los Angeles: The Getty Conservation Institute.

———. 2008. "Une nouvelle methodologie pour l'identification des photographies." In *Le vocabulaire technique de la photographie*, edited by Anne Cartier-Bresson, 436-451. Paris: Marval.

Stulik, D. C., and A. Kaplan. 2008. "A New Scientific Methodology for Provenancing and Authentication of 20th Century Photographs: Nondestructive Approach." In *9th International Conference on NDT of Art, Jerusalem, Israel, 25-30 May 2008.*

Stulik, D., and A. Kaplan. 2009. "Nouvelle méthodologie scientifique d'analyse des photographies: son application aux tirages d'Henri Cartier-Bresson." In *Revoir Henri Cartier-Bresson*, edited by Anne Cartier-Bresson and Jean-Pierre Montier, 158-173. Paris: Textuel.

Stulik, D., H. Khanjian, A. de Tagle, and A. Botelho. 2002. "Investigation of Jean-Louis-Marie-Eugene Durieu's toning and varnishing experiments: a non-destructive approach." In *ICOM-CC Preprints of the 13rd Triennial Meeting, Rio de Janeiro, 22 - 27 September 2002*, edited by Roy Vontobel, 658-663. London: James and James.

Tertian, R., and F. Claisse. 1982. *Principles of Quantitative X-ray Fluorescence Analysis*. London: Heyden.

Thomsen, Volker. 2007. "Basic Fundamental Parameters in X-Ray Fluorescence." *Spectroscopy* Vol. 22 (5):46-50.

Woodbury, Walter E. 1898. *The encyclopædic dictionary of photography*. The S. & A. photographic series, no. 45. New York: The Scovill & Adams company.

Woodward, Richard B. 2003. "Too Much of a Good Thing." *The Atlantic* Vol. 291 (5):67-72, 74-76.

Wright, P., and D. Stulik. 2007. Challenging the Museum's Collections: The research project between the Getty Conservation Institute and the National Media Museum. *Archive (Magazine of the National Media Museum)*, 28-32.

Chapter 4

Handheld XRF for the examination of paintings: proper use and limitations

Chris McGlinchey

Introduction

When investigating paintings, examination of pigments often commences with X-ray fluorescence (XRF) even though this method only infers pigments via the elemental substituents identified. It is often amongst the first tools used because sometimes elemental data is sufficient for pigment characterization and it can be obtained without sampling or contact. However, the interpretation of data from layered structures typical for paintings is more complex than other applications such as metal alloy or soil analysis, where portable XRF is also used and where it is assumed elements are uniformly distributed and not stratified. Artists are of course free to mix pigments as they desire in whatever concentration they deem appropriate and blend them in a paint's organic binder which only contributes to the X-ray scatter observed from the low Z containing materials in the painting. Consequently, the assumptions that are reasonable for alloys or soils are not applicable to paintings.

A painting's structure, support, ground preparation, paint thickness, and pigment heterogeneity will influence how much useful information XRF can provide. Since the method only identifies elements, pigments with similar elemental compositions or different crystal structures cannot be differentiated. The two different crystal structures of titanium dioxide, anatase and rutile; the various lead or iron oxides; and copper carbonates or acetates that may be difficult to assess visually, are some examples where it is often important to pursue additional analytical methods. Furthermore, one element or a combination of elements may conceal the presence of additional elements due to peak overlaps, while others, notably those below aluminum, will go undetected because of their absorption and scattering during air-path analysis leaving these lower atomic number elements (notably those found in modern organic pigments and traditional natural dyes) only to contribute to the resulting spectrum as broadly scattered radiation. Due to the penetrating nature of X-rays (but depending ultimately on the materials used to construct the painting in addition to the experimental conditions that influence the excitation spectra) data will potentially be collected from all paint layers simultaneously. For all these reasons, the effort to report findings in paintings quantitatively should never be attempted. A telltale sign of poorly understood data is when the painting's identified elements are presented with concentrations that add up to 100%; giving a false sense of completeness and demonstrating a misunderstanding of what X-ray analysis is capable of when examining complex layered structures of potentially organic rich paint.

There are numerous questions XRF can help address regarding artists' pigments. They range from the straight-forward, identifying elements which can lead to pigment identification : is the white pigment titanium dioxide (TiO_2), zinc white (zinc oxide, ZnO), or white lead (lead carbonate hydroxide, $2PbCO_3 \cdot Pb(OH)_2$); is the isolated red pigment iron earth (iron oxide, Fe_2O_3), cadmium red (cadmium sulpho-selenide, Cd(S,Se)), vermilion (mercuric sulphide,HgS), or red lead Pb_3O_4; to the more complex: can the pigment creating the faint blue tint in the sky be identified; does the barium identified pigment also contain titanium; is there a little orpiment (arsenic sulphide, As_2S_3) pigment in a lead white and vermilion sample? The more straightforward type of inquiries may only require examination of a single representative site for a brief period of time. The latter type of investigations, which deal with pigments that might be in low concentration, fluoresce weakly, or have overlapping lines, will require collection of data from multiple sites, careful instrumental setup, standardized collection conditions, and will involve post-acquisition processing and statistical treatment of data.

XRF is frequently a good screening tool for the examination of paintings and may be well suited for the identification of elements characteristic of pigments from certain periods. For example, the transition of lead-tin yellow to lead antimonate that occurred in the 18th century; the increased use of zinc oxide, which established prominence in the latter half of the 19th century; the emergence of pure titanium dioxide or selenium containing cadmium pigments, which became common in the early 20th century; to more modern inorganic pigments like molybdate red, which became available in the 1930s. Properly gathered data may help link a painting undergoing technical examination to one that has been previously examined. If an artist is known to have worked with different ground preparations or was active during a time when certain pigments were introduced the method can be exceptionally useful at demarcating which working period the painting belongs to (if it belongs to any at all). The method can also be useful for identification of restoration areas and, since the method does not require sampling or contact, it is a useful way to examine possible acquisitions where sampling restrictions are often imposed.

Importance of raw data plots

Some paintings can be as X-ray transparent as works on paper while others nearly as radio-opaque as some bronze sculptures. In fact, there is often a broad range of X-ray density within a single painting. That property produces informative X-radiographs with regions of contrast originating from areas of high and low X-ray density (Figure 4.1) and it also influences X-ray spectroscopy data. With the spectroscopic examination of a painting, an instrument capable of raw data output in the form of energy vs. X-ray detector counts is the most useful because the intensity of background scatter is revealed. Instruments that only output elemental data as peaks 'auto-identified' make certain assumptions about the background scatter and can yield false positives with meaningless quantified values. Figure 4.2 shows why looking at the raw data is important: only one of the three spectra illustrated contains both molybdenum and cadmium but an additional one (the green paint) could be falsely reported if its background intensity was assumed to be the same as the background in the spectrum for the ground preparation.

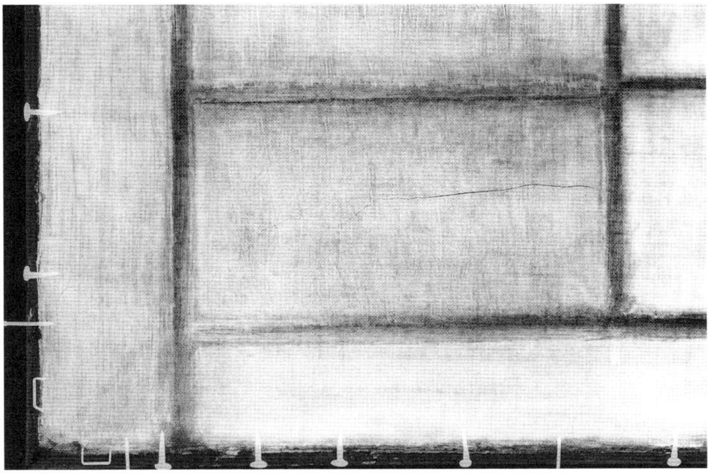

Figure 4.1:
X-radiograph detail of Mondrian's *Composition C* (1920), 257.1948 The Museum of Modern Art. The strong contrast shows there is a significant variation in radio-opacity. Light areas are radio-opaque and black lines, rich in low Z carbon containing black pigment, are relatively transparent. Conditions: 35 kV 45 μA, 80 second exposure.

One method that can be used to resolve this issue is to calculate the net area under the peak for each element of interest via the proper region of interest (ROI) along the x-axis and locally correcting the baseline. By doing this the misidentification of peaks can be avoided. To ensure that the spectra were accurately identified, all of the data presented in this chapter was collected in a raw x-y format and net area counts for elements' characteristic emission lines that did not overlap with other elements were used. Identification of elements difficult to resolve due to their overlapping spectroscopic lines and the role instrument selection and experimental conditions have on the signal response for individual elements is illustrated throughout the chapter. The three different instruments used in this study were manufactured by Bruker: two were Tracer III-V instruments, one with a rhenium (Re) target and the other with a Rhodium (Rh) target, both having Si PIN diode detectors with a resolution of 205 eV (FWHM at 6.4 keV); and a Tracer III SD with a rhodium (Rh) target and silicon drift detector with a resolution of 152 eV (FWHM at 6.4 keV). Data was analyzed using Bruker AXS PXRF and ARTAX software and files were exported in comma separated variable (CSV) format and graphed using Microsoft Excel.

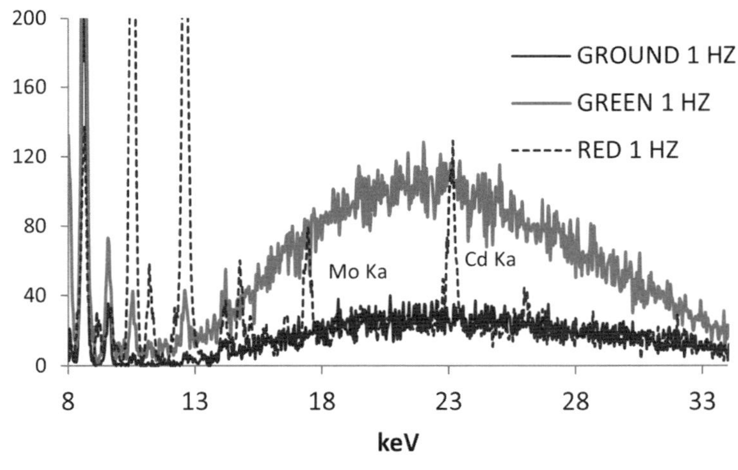

Figure 4.2:
This XRF data collected from de Kooning's painting *Woman II* (1952), 478.1953 The Museum of Modern Art, illustrates the importance of looking at raw data. The three spectra were taken from the ground and two different painted areas: one green and the other red. While the red contains both molybdenum and cadmium, the intense scatter in the green spectrum has about the same intensity in the channels associated with Mo (Kα=17.5 keV) and Cd (Kα=23.2 keV). In looking at the background scatter of the green sample it is obvious this equivalence is coincidental. However, for programs that do not look at local baseline effects but rather assume a certain background, the output could indicate the green contains Mo and Cd when in fact it does not. Instrumental conditions: rhenium (Re) anode, Si PIN diode with CuTiAl (25, 25, 250 μm thickness) filter. Collection conditions were 40 kV and 1.3 μA for 60 seconds. For further information about these conditions see the final section of this chapter regarding operating conditions.

Benefits of multiple XRF protocols

Prior to an analysis, one should ideally have a sense of what pigments to expect, choose the most appropriate XRF conditions for examination, and be aware of those that are undetectable by XRF. When XRF is suitable, multiple protocols optimized for specific regions of the XRF emission spectrum can be very helpful; one protocol may be acceptable for a broad range of elements but the gain in sensitivity for some will be at the expense of others.

Generally, it is useful to start with a voltage between 30 and 40 keV and current below 5 microamps with filtering chosen to minimize or eliminate the lines from the X-ray tube, such as the conditions described in Figure 4.2. While these conditions are a good place to start, the sensitivity of low Z elements is not optimized. Low Z elements are best excited by activation conditions set at no more than 15 keV and about 15 microamps. Whether or not a filter is used will depend upon the elements of interest and the type of X-ray tube used in the XRF instrument. The benefit of collecting conditions with and without filtration is illustrated in Figure 4.3. One can observe the increased signal response for low Z elements when the exciting voltage is set lower and the filters are removed. Since this example shows data from an instrument with a rhenium (Re) X-ray tube, the inelastic scatter of the X-ray target does not overlap with low Z elements; the downside is that the lines do not play a significant role in exciting the K lines of low Z elements the way a molybdenum (Mo), rhodium (Rh), or silver (Ag) tube would. However, with Mo L lines at 2.29 keV and 2.40 keV and Rh L lines at 2.7 keV and 2.83 keV, Ag L lines at 2.98 keV and 3.15 keV, while excellent at exciting aluminum (Al: Kα = 1.49 keV, Kβ = 1.56 keV) and silicon (Si: Kα = 1.74 keV, Kβ = 1.83 keV), these specific tube anodes begin to obscure or weakly excite elements above phosphorous (P: Kα = 2.02 keV, Kβ = 2.14 keV). Another way to improve sensitivity of low Z elements is to consider the instrument's detector used: for further discussion on this topic see text related to Figure 4.14 in this chapter.

In Figure 4.3, conditions optimized for high-Z detect the lead and some calcium but when low Z conditions are used, Ca and P are more clearly identified. The calcium and phosphorous come from a black area and confirm that bone black was used. One would be far less certain about the use of bone black if only one general purpose XRF protocol was used.

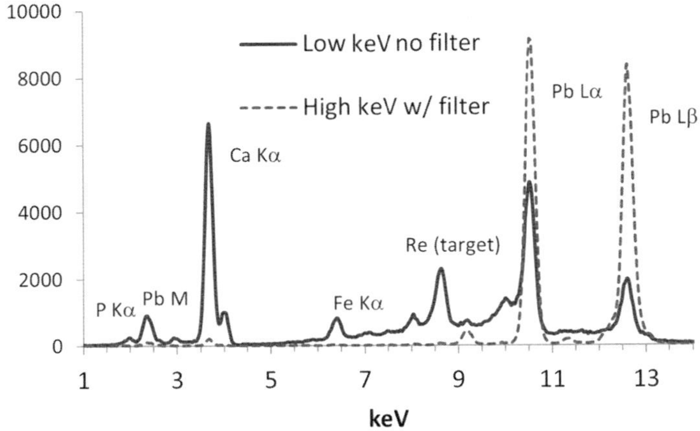

Figure 4.3:
XRF spectra collected under two different conditions from the same spot in a black passage in Matisse's *Piano Lesson* (1912), 125.1946 The Museum of Modern Art. One spectrum is optimized for high Z element identification (dotted line) and one optimized for low Z element identification (solid line). Low Z conditions show both Ca Kα (3.69 keV) and Kβ (4.04 keV) and P Kα,β (2.04, 2.14 keV). Instrument conditions: rhenium (Re) target with Si-PIN diode. Collection conditions: 60 second acquisition time for both methods, high Z: 40 kV and 1.3 μA with CuTiAl filter (25, 25, 250 μm thickness); low Z: 15kV and 15 μA with vacuum (between 3 and 10 mm Hg), no filter. For further information about conditions see the final section of this chapter regarding operating conditions. Note: potential interference from Re M lines (1.83 keV) with P Kα (2.04 keV), being separated by 210 eV, was ruled out by examining low z settings reflected from both a sheet of acrylic plexiglass and lead foil separately; Re M lines were minor only when reflected from lead foil and were resolvable from P Kα.

The effects of complex stratigraphy

The stratigraphic nature of paintings is one of the major reasons why data from XRF measurements will prohibit even semi-quantitative analysis. Both the incoming exciting radiation from the X-ray source and the emitted radiation from the fluorescing elements are attenuated so signals from lower layers can be reduced or undetected. They are not however reduced uniformly: lines at lower energies are attenuated more, and the greater the separation is between the lines of a particular shell for a given element the easier it is to see the effect. An example of significant attenuation from paint layer stratigraphy is shown in Figure 4.4. The top paint layer, purported to be aluminum radiator paint in the literature, was found to contain aluminum by XRF. The lead ground is attenuated significantly, as Pb L_α is shown to be much lower than the first Pb Lβ and only slightly more intense than Pb Lβ_2. When lead white is at the surface (as it is with Figure 4.6) the XRF data shows the Lβ_2 peak is typically 1/10 the height of Lα.

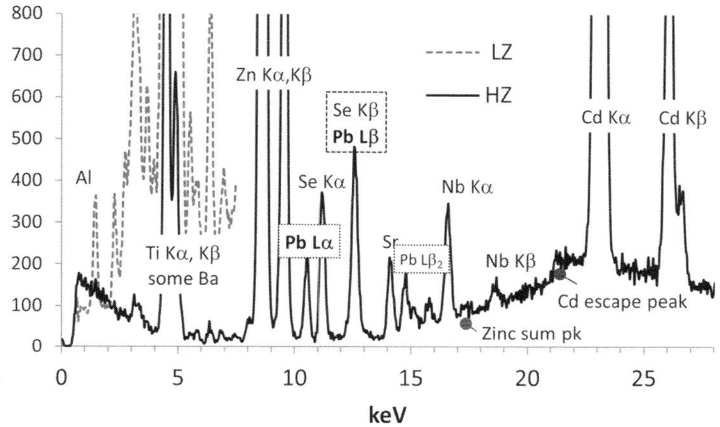

Figure 4.4:
XRF spectra from an area that appears silver in de Kooning's *Woman I* (1950-52), 332.1955 The Museum of Modern Art. The spectrum optimized for low Z (LZ) analysis (dotted line) confirms the top paint layer contains aluminum. For the high Z (HZ) analysis (solid line), the lead in the ground preparation is attenuated by paint layers and the attenuation of the relatively weaker Pb Lα compared to Pb Lβ and Pb Lβ2 confirm the lead is coming from a lower layer. The presence of selenium complicates this observation because the Se Kβ peak overlaps with Pb Lβ but since Pb Lα is normally much greater than Pb Lβ2 (which does not overlap with elements identified in this painting) we can assume the contribution of Se Kβ to Pb Lβ is minor. The weak zinc sum peak at 17.2 keV appears to the right of Nb Kα and the weak cadmium escape peak is noted be ow Cd Kα. (Instrument conditions: see figure 4.3).

The lead ground identification in Figure 4.4 was supported by examining an unpainted passage where the ground was exposed. For paintings that do not have exposed ground and are accessible from their verso, confirmation of the lowest layers may be more suitable from the back. However, complementary investigative tools, outlined further on in this chapter, will always be required to confirm what the XRF data suggests when complex mixtures are being examined.

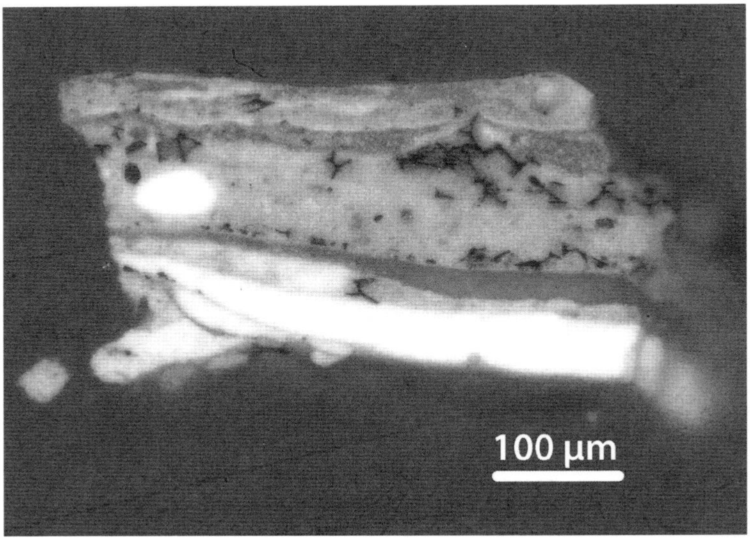

Figure 4.5:
Cross-section from de Kooning's *Woman I* (1950-52), 332.1955 The Museum of Modern Art. Interpretation of upper layers will likely be complicated by pigments in lower layers and emission from elements in lower layers will be attenuated.

In addition to the multi-layered stratigraphies of paintings, their intralayer heterogeneity and variable paint-layer thicknesses are also contributing factors that can make XRF analysis difficult. One only needs to be reminded how complex a painting's stratigraphy can be by looking at a painting's cross-section (see Figure 4.5) to recognize how intricate these structures are. With variable densities, thicknesses, compositions, and organic binders, it becomes clear that the analyst is restricted to confirming elements that are present and possibly their relative abundance and maybe even their relative position in the stratigraphy, but determining their absolute concentration in a layer is, in practical terms, impossible. The best approach in confirming elements related to certain pigments is to collect data from multiple locations with similar appearances of paint treatment and look for the strongest signal for a particular element that matches the expected pigment. One's expectations of plausible pigments are based on the date of the painting, the artist's preferential use of pigments previously documented, and the visual assessment by the curator or conservator. This combination of evidence and background information can strengthen a hypothesis of a particular pigment being present.

Some areas where information about the matrix stratigraphy can be determined are when one of the layers is a homogeneous element or alloy. For example, Cesareo (2004) found it was possible to measure the thickness of gold leaf in a wall painting by measuring its attenuation effect on lower layers of lead and Ida (2005), using handheld XRF, found that paints containing iron and titanium pigments up to about 0.5 mm thick allowed elements from the steel substrate to be detected for the purpose of alloy sorting. Measuring the attenuation of radiation from an underlying layer is one way to measure the coating's film thickness (Carpenter 2000, Piorek 2008). In all of these instances however, the success of this application relies on the layers in question not sharing any of the same elements in common. Given the nature of painting these applications might be used to determine the thickness of metal-leaf layer but they are rarely applicable to complex layers of paint. Furthermore, if the painting is on a metal support containing iron, copper, or tin, in many instances the ability to associate those elements with pigments in the paint layer will be compromised. However, using both a highly collimated synchrotron X-ray source and detector the noninvasive elemental depth profiling to make virtual paint cross-sections of elemental data via confocal XRF of a 17th c. wood panel painting was demonstrated (Woll 2008).

Missed elements and instrument artifacts

Interpretation of elements is not always a straightforward process: one must be cautious to avoid generating findings that will produce false negative or false positive results. With the former, certain elements or combinations of elements may be overlooked if they have overlapping emission lines with another element making the element in lesser concentration or weaker emission efficiency difficult or impossible to confirm by XRF alone. Some elemental combinations that are challenging to identify in paintings are mixtures of barium with titanium (discussed shortly) and mixtures of lead white, vermilion and orpiment, where the Pb Lα, Hg Lα, and As Kα lines overlap as illustrated in the next example (see Shugar and Sirois chapter 9 of this volume for an extended discussion about these elements). It is useful to work with mockup samples to determine the limitations of the instrumentation being used. One such mockup, used for the data shown in Figure 4.6, consists of a blend of the three pigments to achieve a convincing flesh-tone tinted paint.

The primary peaks for identifying barium and titanium also overlap closely (Figure 4.7). With a difference of only 40 eV, no energy dispersive detector currently available possesses a resolution good enough to resolve the Ti Kα (4.51

keV) from the Ba Lα line (4.47 keV) although the Ba Lβ (4.83 keV) and Ti Kβ (4.93 keV) are slightly easier to see because they differ by 100 eV. For the latter, there is peak broadening that is easier to observe when the two pigments are both well represented. In comparing a series of mixtures of the two from a 10:1, 1:1, and 1:10 it is clear that the peaks shift with these compositional differences but the lesser element can be easily overlooked. This is particularly true for small amounts of titanium because it doesn't have other emission lines that can confirm the way barium can be using BaKα (32.19 keV) and BaKβ (36.38 keV) lines.

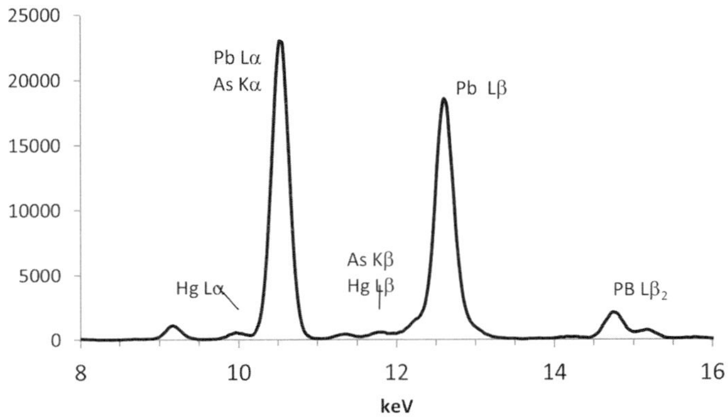

Figure 4.6:
XRF spectrum of a laboratory prepared flesh tone paint consisting of lead white, vermilion, and orpiment. At first glance, Hg from vermilion and As from orpiment could be misidentified either as minor Pb Lα and Lβ lines or as escape peaks. To complicate matters, both lines of arsenic are obscured when Pb and Hg are present: As Kα (10.54 keV) overlaps Pb Lα (10.56 keV) and As Kβ (11.73 keV) is close to Hg Lβ (11.82 keV). Instrument conditions: rhenium (Re) anode Si PIN diode detector. Collection conditions: 40 kV, 1.3 μA with CuTiAl (25, 25, 250 μm thickness) filter and 120 seconds collection time.

Szokefalvi-Nagy (2004) has used a curve-fitting program known as AXIL to suggest low quantities of titanium in barium based on the band broadening in the 4.5 keV range. Similar results can be obtained with the Bruker AXS ARTAX program to arrive at a value of corrected net counts for both elements. In cases where the Ti concentration is too low and identification is near impossible by handheld XRF, one must turn to investigating the crystal structure of the oxide using XRD or Raman spectroscopy. In matters of attribution for early 20th C. paintings, methods that identify the crystal structure of TiO_2 are particularly important to use when the element is detected: the anatase form was first readily available after World War I (1914-1918) while the white commercial form of rutile first appeared in 1941 (Patton 1973).

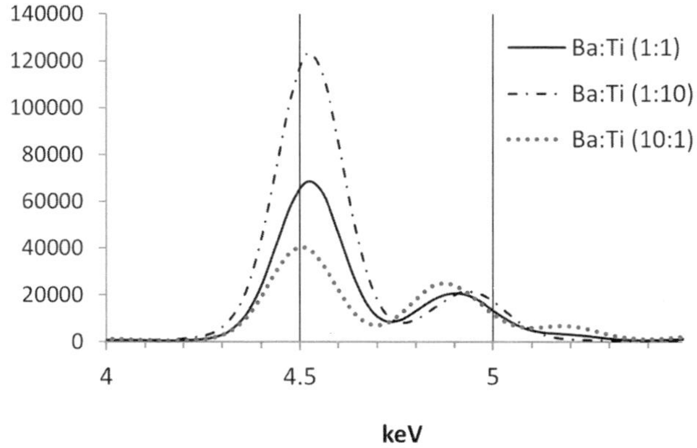

Figure 4.7:
Laboratory prepared mixtures of baryta and titanium dioxide. For Ba:Ti mixtures in 1:10 and 10:1 the spectrum approximates the element in high concentration for the range indicated. From left to right, Ba Lβ (4.47 keV); Ti Kα (4.51 keV); Ba Lβ (4.83 keV); Ti Kβ (4.93 keV). Instrument conditions: rhenium (Re) anode Si PIN diode detector. Collection conditions: 15kV and 15 μA with no filtration for 120 seconds.

For elements bromine (Z = 35) and below, K-lines become the only viable means to characterize the peaks by XRF. Without the aid of supplemental lines, certain combinations of elements in this range may be particularly difficult to identify. For example, for Z = 19 to 29 (K to Cu), low amounts of element Z are difficult to detect in the presence of large amounts of element Z-1 combined with element Z+1; that is, within this region $K\alpha_Z$ is close to $K\beta_{(Z-1)}$ and $K\beta_Z$ is close to $K\alpha_{(Z+1)}$. (This is easy to surmise by looking at a periodic chart of X-ray energies and comparing the Kβ and Kα energies of adjacent elements.) Theoretically, combinations are possible where the middle element would be impossible to detect. For example, it is not implausible that small amounts of Mn would be difficult to see when both Cr and Fe are confirmed. Realistically speaking, it would also be difficult to detect small amounts of cobalt blue in an earth pigment since the Co Kβ would be the only peak visually evident. With cobalt-containing pigments, since some are amorphous glasses, crystallographic methods may not always confirm the structure but a method like SEM-EDS or polarized light microscopy (PLM) might be a suitable approach.

With respect to false positives, confusing sum peaks for elements may be problematic for certain pigment combinations found in paintings. Sum peaks are typically easy to spot because they appear in clusters but, for example, the three different sum peaks from the lead L lines (α+α: 21.1 keV, α+β: 23.6 keV, β+β: 25.22 keV) can be confused for the Kα lines of palladium (Pd): 21.18 keV, cadmium (Cd): 23.17 keV, and tin (Sn): 25.27 keV, respectively. It would be very rare to find these three elements in combination because palladium red (ammonia-perchloride of palladium) and white tin (IV) oxide, yellow tin (IV) sulfide and green tin-copper oxide are all obscure pigments but the fact they do exist means their combination, as unlikely as it is, may need to be addressed.

Need for complementary methods

As previously mentioned, one must recognize that XRF alone will only provide information that is limited by its fundamental physics and instrument design. Portable XRF instruments examine sites on the order of 15 to 20 mm^2 which may be too large for analysis of small details. When additional methods are sought they may be different spectroscopic tools perhaps to identify a crystal or molecular structure, an elemental method capable of providing the same data but on a finer scale, or it may be an imaging method to examine the entire painting and possibly use it to identify other areas of the painting to investigate. Overall, complementary methods should form an integrated approach that is logical and considers any limitations caused by sampling restrictions, the work's condition, or intrinsic properties such as heavy impasto or mixed-media that may restrict access to some areas of interest.

One a finer scale, some complementary methods to consider include optical microscopy (both transmitted polarized light for particle analysis and reflected light for cross-section examination), X-ray diffraction (XRD), Raman spectroscopy, scanning electron microscopy with energy dispersive spectroscopy, separation sciences, and mass spectrometry methods. These methods examine much smaller areas than handheld XRF instruments are capable of identifying but some of those described below require sampling. Recently, Gatto-Rotondo (2010) has demonstrated a portable XRD system for the characterization of crystalline pigments and Mendoza-Cuevas (2011) describes a multi-functional XRF-XRD-radiography device for elemental, molecular analysis, and X-ray imaging from one device (!). Sawczak (2009) and Rosi (2009) use a statistical approach involving XRF with Raman and mid and near-FTIR, respectively, to examine artists' pigments. Burgio (2001) demonstrates laser induced breakdown spectroscopy (LIBS) and Raman spectroscopy for the identification of pigments in paintings, and Leona (2009) demonstrates *in situ* microanalysis of organic pigments using Raman. Lomax (2007) investigated modern organic pigments using FTIR and direct temperature-resolved mass spectrometry (DTMS). Fiber-optic reflectance spectroscopy (FORS) of regions slightly smaller than those examined by XRF was used by Picollo (2007) to identify modern white pigments in paintings.

Imaging methods such as UV examination, X-radiography, or multispectral imaging can help locate unique areas to examine with XRF or help with the interpretation of XRF data. Multispectral methods may be appropriate to identify certain pigments. For example, Berns (2001) demonstrates multi-spectral imaging for the examination of pigments in paintings and Delaney (2005) showed how multispectral imaging of paintings allows for the mapping of blue pigments characterized by XRF. These methods are particularly useful because they place emphasis on issues related to condition and artist's technique.

This is not an exhaustive review of pigment identification methods: it should give the reader the appreciation for the depth of literature and serve as a reminder that no single tool will answer every question. One source to consider for information on complementary techniques on specific pigments is the multi-volume set *Artists' Pigments* (Feller 1986; Roy 1993; FitzHugh 1997; Berrie 2007). *The Pigment Compendium* (Eastaugh *et al.* 2004) provides an encyclopedic list of pigments from common to obscure that may also be helpful. The reader is strongly advised to survey the relevant conservation and scientific literature related to previous studies of relevant paintings to determine how to integrate XRF with other currently available investigative tools.

Optimizing Instrumental Conditions

Setup

While the XRF is a handheld instrument, the most common and practical setup is to mount the instrument on a heavy duty tripod. Actual handheld operation might be suitable for cursory examination but may be risky for works on canvas although somewhat safer for paintings on a wood or metal support where gentle contact is acceptable. (Applying pressure against a canvas painting may cause the stretcher bar to press against and damage the paint.) For tripod operation, the instrument must be well secured and the tripod base weighted for stability as shown in Figure 4.8. Before beginning any work or placing equipment near the painting it is important to become completely familiar with the tripod, armature, and clamping mechanism and make sure that they are both stable and secure. For large paintings with a smooth flat appearance, pointing the instrument forward is the most common approach. For paintings with heavy impasto, it may be advantageous to use a 3-axis geared head to permit the instrument to be adjusted to variable angles. A geared head also allows for small paintings to be examined flat.

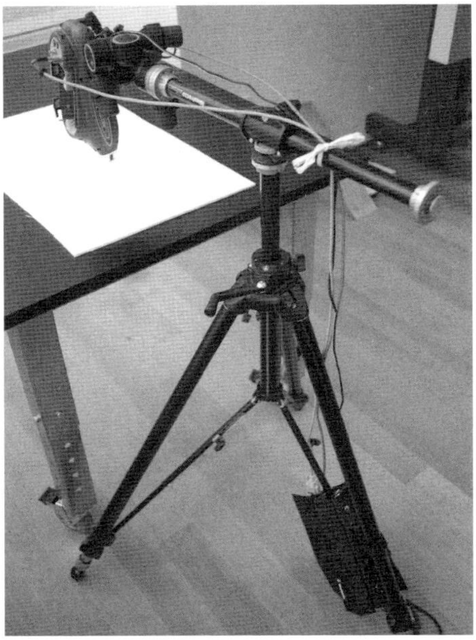

Figure 4.8:
XRF setup showing downward oriented operation on heavy-duty tripod with geared boom. The XRF can be safely backed away from the surface when it is necessary to change filters. Note weighted sandbag secured to leg opposite XRF. Cables should be secured away from examination area. Manfrotto equipment: tripod (058B), lateral side-arm (131D), geared head (405), and clamp (035RL).

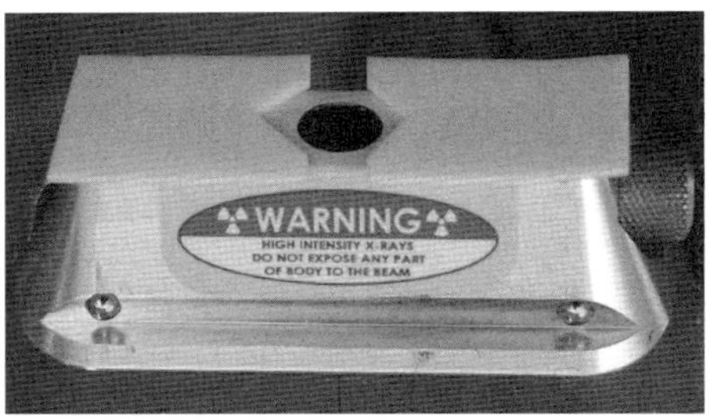

Figure 4.9:
2mm thick Ethafoam™ spacers affixed with transparent double-sided tape can be used to set a fixed gap if gentle contact with the surface is acceptable.

Ideally data should be collected under two conditions (e.g. optimized for high and low Z). This is best done by keeping the instrument at a specific location until both runs have been collected. It will take a little extra time for the instrument to stabilize between conditions and to change the filter but there won't be any ambiguity about the region analyzed.

For surfaces that can tolerate light contact the use of 2mm thick Ethafoam™ blocks on the left and right side of the instrument's nose (Figure 4.9) will both help establish a fixed gap between the object and the instrument, and also further reduce the risk of hard edges from accidentally denting the painted surface. One should not contact a painted surface that is fragile or easily burnished with foam or any other material; instead the gap should be set visually. Those responsible for the care and preservation of a specific painting should be consulted and any necessary permission obtained prior to proceeding with any technical examination.

Anodes

Miniature X-ray tubes for the handheld instruments used in cultural heritage science typically have either rhodium (Rh) or rhenium (Re) targets. Silver (Ag) anode targets are more common to areas of research other than art conservation since Ag targets are not ideal for the examination of materials that contain low quantities of silver such as gelatin silver print photographs and Renaissance bronzes.

Within the instrument housing, the X-ray tube is fixed so that when the instrument nose is parallel to the object the tube is oriented at an oblique angle (~62° for the instrument used here), allowing the detector to collect the emitted fluorescence oriented in the specular direction; traditional secondary target geometry found in earlier and larger instruments is not used in handheld XRF because the tube for the handheld instruments is not as powerful. In general, with proper settings, all anodes are good for the identification of the major elements found in pigments, but the Rh tube (Figure 4.10) is better suited for detecting small quantities of pigments containing elements between Z=13 (Al) to Z=16 (S) via excitation with the Rh L lines starting at 2.7 kV. The Re tube (Figure 4.11) is efficient at fluorescing elements between Z=17 (Cl) and Z=28 (Ni), partly because the baseline is smooth in this region (compared to Rh) and also from the Re L lines at 8.65 and above.

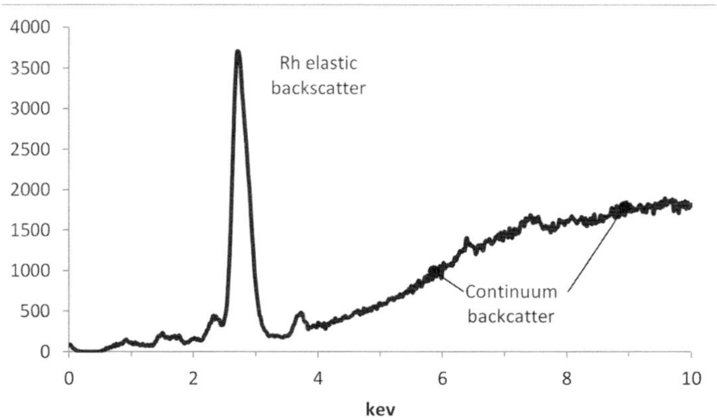

Figure 4.10:
Rhenium (Re) tube X-ray lines reflected off of acrylic plexiglass indicating elastic and continuum backscatter detected by a Si PIN diode detector. These conditions are suited for examination of a broad range of low Z elements. Collection conditions: 15 kV and 17 μA for 60 second.

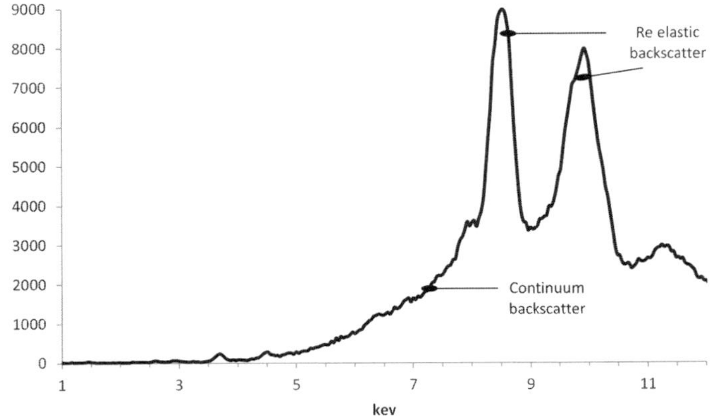

Figure 4.11:
Rhodium (Rh) tube X-ray lines reflected off acrylic plexiglass indicating primarily elastic and continuum backscatter detected by a Si PIN diode detector. These conditions are excellent for excitation of Al (Kα,β = 1.49, 1.56 keV) and Si (Kα,β = 1.74, 1.83 kV) but not for low z elements above P (Kα,β = 2.02, 2.14 keV). Collection conditions: 15 kV and 13 μA for 60 seconds.

Filters

Filtering out the L lines of Rh or Re will allow for the sample fluorescence to be largely driven by Bremsstrahlung scatter at higher keV settings. This accomplishes the excitation of elements that fluoresce above the L lines of either Rh or Re and makes elements with emission lines in this region easier to identify. A properly selected filter arrangement also mitigates the anode emission lines allowing for a smoother baseline in this region.

Detectors

The two most commonly used detectors in handheld XRF spectrometers are Si-PIN diode and Si drift detector (SDD). The first thermoelectrically cooled miniature detectors were developed in 1991 (Pantazis 2009) using Si-PIN diode detectors and by 1996 a Si-PIN diode with resolution of 186 eV FWHM at 5.9 keV was available. Leutenegger (2000) was first to apply a thermoelectrically cooled silicon drift detector (SDD) based XRF to the examination of art and cites a resolution approximately equivalent to a cryogenically cooled SDD detector [140 eV (cryo) vs. 150 eV (Peltier) FWHM at 6 keV]. In addition to the higher resolution ideal for identification of low-Z elements, SDD detectors permit lower dead time and allow for higher count rates so data can be collected more quickly with fewer spectral artifacts. Faster collection is achieved with more powerful energy settings so precautions must be taken to avoid the increased associated potential risks to human health.

The difference in detector sensitivity can be illustrated by examining an applicable combination of materials plausible for Old Master paintings: the use of ground glass mixed into paint medium. It will depend upon the amount of ground glass present, but the SDD detector gives is a significant increase in sensitivity for low Z elements (Figure 4.12).

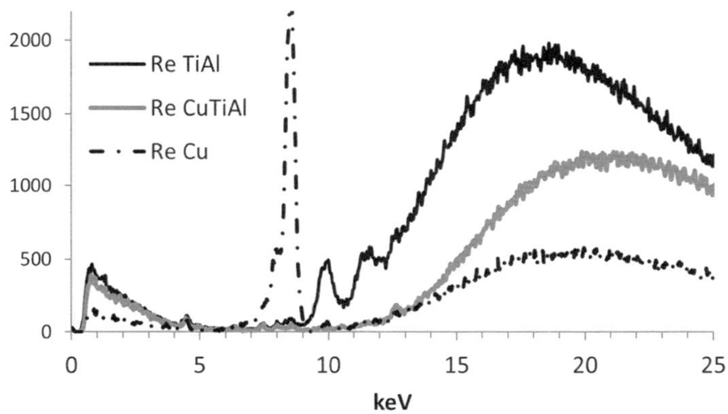

Figure 4.12:
Filters installed in a rhenium (Re) anode instrument. Cu filtered conditions being the most monochromatic may result in spurious diffraction peaks when crystalline structures are analyzed. Collection conditions: 40 kV and 1.3 μA of current, thicknesses of foil filters: Cu 25 μm, Ti 25 μm, and Al 250 μm, 60 seconds collection time.

Operating conditions

A) Knowing your instrument

It is good practice to know the range of excitation conditions an instrument is capable of achieving. This will confirm a stable baseline and appropriate excitation for specific elements of interest. If one archives the data, it can also be consulted if doubts arise about signal contamination or equipment malfunction. The quickest way to collect these reference spectra is to follow the manufacturer's recommendations for voltage, current, and filter settings and use them to collect data from a sheet of clean acrylic plexiglas (as illustrated in Figures 4.10, 4.11, 4.13 and 4.14) or distilled deionized water in an X-ray cell cup fitted with a thin polymer film of Ultralene® (Spex). The resulting spectra will either show a broad polychromatic curve of scattered radiation or one superimposed with sharp peaks related inelastic and elastic scatter from the anode target. The data will also reveal any weaker peaks that originate from the instrument that may overlap with trace elements in the sample. Such instrumental peaks could be related to collimator impurities, solder, or trace elements some manufacturers use to normalize data.

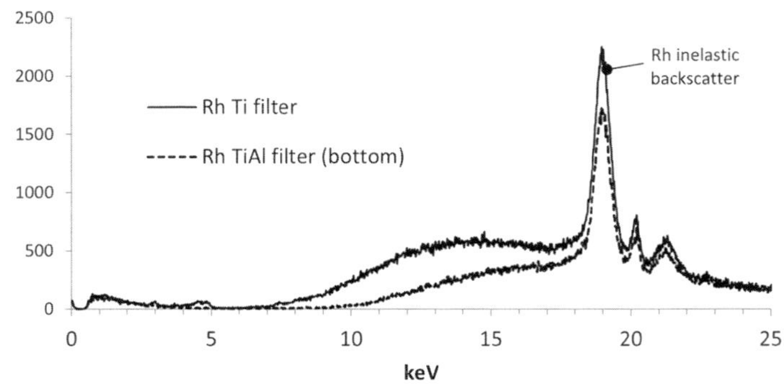

Figure 4.13:
Impact of Ti and TiAl filters on rhodium (Rh) anode detected with Si PIN diode detector. Collection conditions: 40 kV and 1.9 μA for 60 second. See figure 4.13 for filter information.

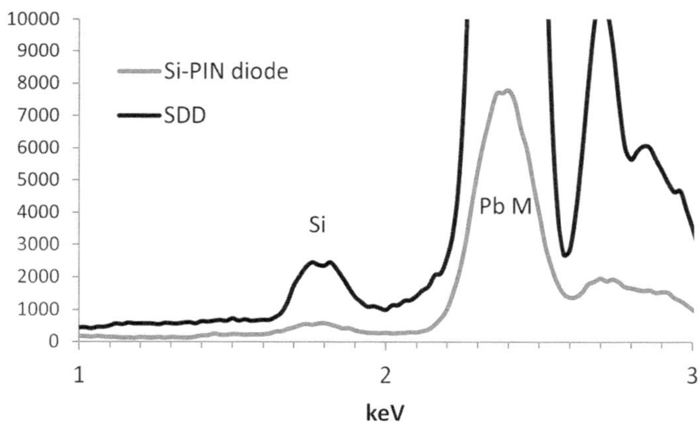

Figure 4.14:
Comparison of rhodium (Rh) Si-PIN diode to SDD detector for examining a laboratory prepared paint sample of lead white mixed with a low concentration of ground glass. The higher count rate and superior resolution makes the SDD detector ideal for low Z elements. Collection conditions: both instruments, no filtration for 120 seconds. Si PIN diode instrument: 15 kV and 13 μA; SDD instrument: 15 kV and 18 μA.

In addition to this basic data about an instrument, it is also worthwhile to examine some fabricated test panels with specific materials of interest at effective concentrations to determine which instrumental method most efficiently excites a particular element and if the method is sensitive enough to the concentration range that might be used by the artist. It also helps establish conditions that strike a compromise for general excitation conditions. Some examples to consider are tints of pigments; particularly for pigments that have strong tint-strength where low concentrations of pigment can have a significant impact on color such as Prussian blue and chromium-based greens. Another test panel might involve small spots of paint over a field of different color. Handheld XRF instruments examine areas around 3x5 mm but are capable of identifying elements in smaller passages. It would be useful to know how small a spot an instrument is capable of identifying for a specific pigment.

B) Maintaining consistency

The variables that will influence the intensity and signal to noise ratio of an emission spectrum are: the gap spacing between the instrument and the object, the voltage and current settings used, the attenuation from any filter inserted, and the length of time data is collected. For bulk assessment the latter factor is less critical for collection times beyond 60 seconds. An additional parameter critical for the examination of low-Z elements is whether or not a vacuum is drawn or a helium purge is applied; this will help increase the probability of characteristic but weak X-rays from low-Z elements reaching the detector and not being scattered or absorbed by air.

These above factors (with the exception of length of data collection time) will influence the intensity of emission spectra reaching the detector at a given moment. For detectors to function correctly they must rapidly switch between time collecting data to time converting it to an electronic signal. (Detectors cannot do both processes simultaneously.) The amount of information received between intervals will depend upon the type of detector: typically, a SDD instrument can optimally process data up to 50,000 events/second while a Si-PIN diode instrument should not exceed 5,000 events/second. When these values are exceeded the detector is more likely to produce anomalous signals such as escape peaks and sum peaks. It is particularly important to reduce this risk when elements in low or trace levels are of interest since they may be difficult to differentiate from anomalous peaks. In these circumstances, an improved approach would employ conditions that detect a less powerful response and employ longer collection times. It is possible that future generations of detectors will be able to exceed these conditions so it is prudent to consider the manufacturer's current recommendations on the signal strength optimal for a specific detector.

When conducting a comparative investigation of multiple paintings it is imperative to maintain the same conditions for all paintings studied, particularly the instrument settings and, when low Z elements are important, gap and purge conditions. For optimized low Z conditions, one should keep an eye on the argon (Ar) K lines to make sure they remain constant: they can be an indicator of some parameter having been modified. If a vacuum is used for low Z conditions only the instrument is evacuated and the vacuum typically drops to below 50 mm of Hg. Over time the polymer window protecting the instrument cavity from contaminants will stretch and ultimately cause the air gap between the object and the window to increase over time, resulting in a slight increase in Ar signal.

In comparing data from similar paintings but collected under different conditions, one may be at risk of false conclusions about the materials used and any possible relationship between the two paintings. This would be especially true for low concentration elements, but not necessarily true for bulk elements.

c1) Settings for general purpose, high Z, screening

As stated previously, general purpose and high Z screening methods typically employ a kV setting around 35 to 40 kV and 1 to 4 microamps of current for either a rhodium or rhenium anode. For a rhenium anode instrument CuTiAl or TiAl foil filter (see Figure 4.12) are effective at removing Re L lines. (The foil thicknesses used in these studies are Cu: 25 μm; Ti: 25 μm; and Al: 250 μm) The benefit of a CuTiAl filter with a Re target is the inelastic and elastic scatter from the tube is effectively filtered leaving the excitation to be driven by the polychromatic Bremstrahlung.

c2) Settings for low Z investigations

If low-Z elements (calcium or below) are important it is essential to collect data under conditions optimized for their detection. Typically, low-Z conditions will be collected with all filters removed from the X-ray path and a low voltage (e.g. 15 kV), high current selected (Figure 4.15). In addition, an SDD detector is significantly more sensitive for low Z elements and should be considered for low Z critical work.

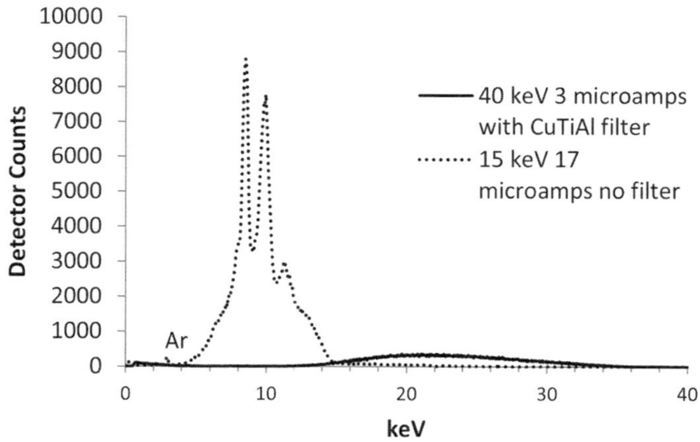

Figure 4.15:
Emission spectra from a rhenium (Re) target instrument with Si PIN detector. The 40 kV conditions are good for exciting elemental lines below 20 keV but diminish where the 15 kV conditions improve, typically 5 keV and below. Other targets and filter conditions will produce different conditions that may be more suitable for other queries. See figure 4.3 for further information on collection conditions. Note that the weak argon peak in the 15 kV data is evidence that an air gap between the instrument and acrylic plexiglass exists and/or the instrument cavity was not under vacuum.

Collecting data with the instrument under vacuum or flushing the beam path with helium will displace the argon (Ar) in air. (Ar Kα = 2.96 keV, Kβ = 3.2 keV) While an air path might not have an impact on excitation via the Rh Lα line, it will attenuate the emission from the sample site to the detector.

Interpreting and reporting XRF results

Easily detected elements

Identification of most bulk elements is easily achieved by using the instrument manufacturer's peak identification tools to identify the elements via K and or L lines present from the raw data. For clearly defined peaks, confirmation of Kα, β and or Lα, β lines are sufficient if the peaks representing the different elements are isolated.

For paintings, some elements even when in high concentration will have Kα:Kβ ratios that do not generally fit the 5:1 ratio encountered with alloys. Matrix attenuation from oxides, sulfates, and carbonates in addition to the paint binder will cause some pigments to have and α:β ratio closer to 1:1 even when the pigment is in the top layer. The L lines of lead from lead white or mercury from vermilion are good examples to illustrate this. With this in mind, peak ratios that do not follow this reduced ratio may be indicative of overlapping elements. If data was collected from multiple passages that look similar it may be possible to observe small fluctuations of characteristic intensities that could allow one to determine if the anomaly is due to a layering effect or a weakly detected element. These ambiguities should be pointed out when reporting findings and, if time permits, complementary analytical methods used to reach more conclusive findings.

Weakly detected elements

Weakly detected elements are either in trace level concentrations or only weakly detected because of the experimental conditions used or the limitations of XRF. Prior to assessing whether or not a peak is present one needs to examine background noise under conditions that approximate the sample being examined. It is possible that background data can be extracted from a painting's data but only if one is confident there are no additional trace elements that happen to occur in the area being identified as background. Using identical experimental conditions, the background selected should appear to have the same signal to noise as the region adjacent to the baseline around the element of interest.

Most programs will automatically determine the average area-counts in defined regions by selecting either emission regions characteristic of a specific element or background noise. This can also be accomplished manually by outputting data to a spreadsheet. By collecting multiple spectra from the same sample and location one can measure the standard deviation of the background. Typically, if the net area counts in the region associated with an element are three times the background's standard deviation then the peak is confirmed.

How the baseline gets treated will vary and depend upon the slope of the baseline or lines nearby associated with other elements within the set of paintings being examined. Jenkins (1999, p163) reviews the treatment of background to obtain net counts manually. Once proper baseline areas have been established on either side of the peak, these counts can be averaged together and subtracted from the gross channel counts for the element. Confirming an element is accomplished by carrying this out for the two main peaks. When this cannot be accomplished one can only suggest the element is present and recommend further analysis by other means or state the reasoning why one peak is sufficient.

Case Study: Pigment Analysis of Palette used by Chaïm Soutine (1893-1943)

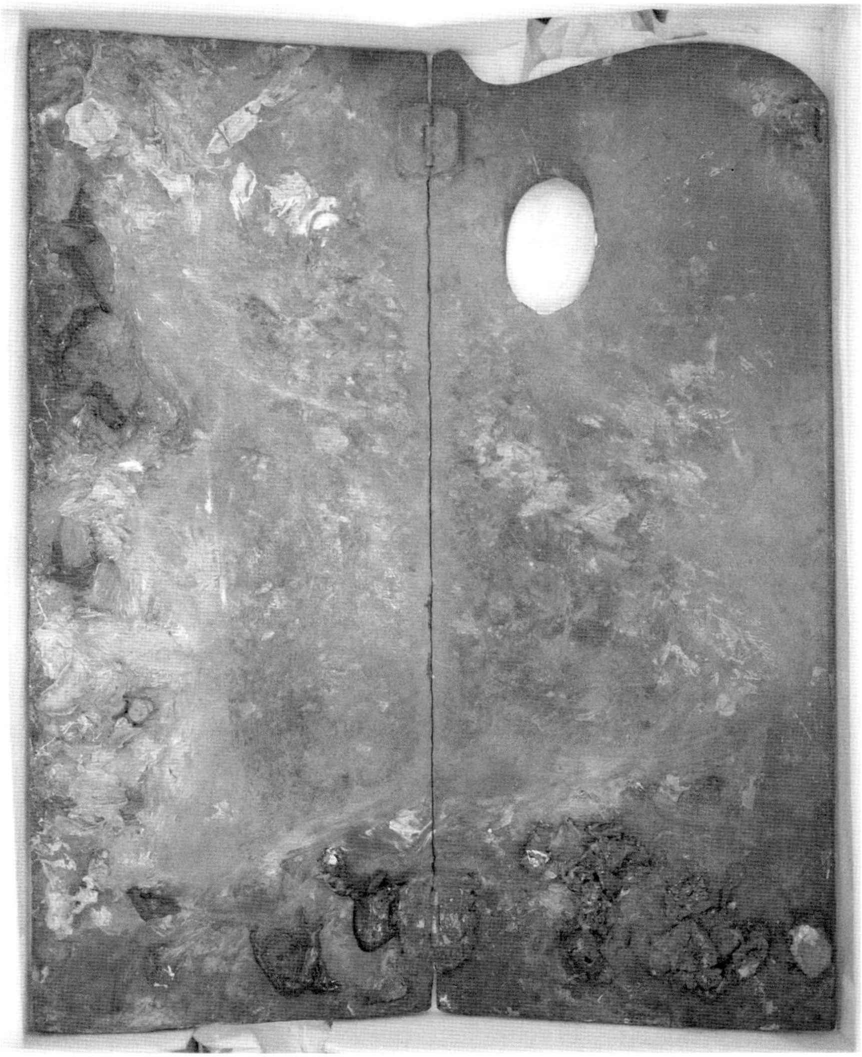

Figure 4.16:
Palette used by Chaïm Soutine (1893-1943), considered the artist's last palette. (Accession number: SC1243.1964) The Museum of Modern Art.

XRF analysis of Soutine's palette provides some examples of the method's strengths and weaknesses when the method is applied to the identification of pigments in paintings (Figure 4.16). The palette, a gift to the Museum of Modern

Art by Pierre Loeb in 1964 is considered to be his last palette used and is part of larger study- a collaborative investigation initiated by the Barnes Foundation being carried out by Barbara Buckley, Lauren Bradley, and Jennifer Mass. Soutine was an early twentieth century painter who developed a highly individual style of painting typically using a heavily loaded brush. From an XRF perspective, the palette can be viewed as a painting on a rigid support; this particular palette could describe a painting containing varied impasto but in terms of the traditional ground preparation that quality is absent here. From an art historical perspective the palette is considered an important artifact that can help provide information about the artist's materials, technique, and approach to painting. Analysis of the palette can help support connections to paintings that would otherwise be difficult to prove.

Nineteen different locations were analyzed by XRF with each site analyzed twice: once under conditions optimal for high Z analysis and another for conditions more suitable for low Z element identification. The instrument used has a rhenium (Re) anode excitation source and Si PIN diode detector with a resolution of 205 eV (FWHM at 6.4 keV). One location on the palette (a dark blue) was examined using an instrument with a rhodium (Rh) anode excitation source and silicon drift detector with a resolution of 152 ev (FWHM at 6.4 keV). Data was collected for 60 seconds according to conditions described in figure 4.3 and 4.14 of this chapter.

Lead was identified in the area of white paint by its characteristic L lines ($L\alpha = 10.56$ keV, $L\beta = 12.61$ keV) and its M line (Pb M = 2.39 keV) was noted. This suggests the use of lead white. Zinc [Zn ($K\alpha = 8.64$ keV, $K\beta = 9.57$ keV)] was detected in the lighter shades of some colors suggests zinc white was used to create some lighter shades of paint. An intense signal for mercury ($L\alpha = 9.99$ keV and $L\beta$ 11.82) was detected in the red paint to suggest vermilion while the yellow paint contained cadmium ($K\alpha = 23.17$ keV, $K\beta = 26.09$ keV) and is indicative of cadmium yellow. X-ray diffraction (XRD) would be an ideal method to use to declare with certainty that the pigments suggested by XRF are indeed those that were used. Since the palette measures 36.5 cm x 29.4 cm open, about the size of a small panel panting, it could fit into most open architecture XRD instruments and thus be analyzed without sampling.

One dark blue paint contained cobalt [Co ($K\alpha = 6.93$ keV, $K\beta = 7.65$ keV)] indicative of cobalt blue, but a different dark blue paint did not contain any elements that could serve as a pigment marker. This paint was reexamined with a silicon drift detector (SDD) and a rhodium (Rh) anode source utilizing low Z conditions. With these conditions and equipment, it was possible to detect aluminum [Al ($K\alpha,\beta =$ 1.49 keV, 1.56 keV)], silicon [Si ($K\alpha,\beta = 1.74$ keV, 1.83 keV)] and possibly sulfur S [($K\alpha,\beta = 2.31$ keV, 2.47 keV)] although this last peak overlaps with the M line of lead which was also detected in the sample. It could not be confirmed but a possible weak peak at 1.0 keV was observed and may suggest sodium, Na (Kα = 1.04 keV). A helium purge of the instrument to object air-path was not used but could be considered in the future to see if the signal intensity increases to 3σ above the baseline to indicate Na with greater confidence. (For further information on this approach see Chapter 14 by Kaiser and Shugar on the examination of glass. Note: exposing a paint film to dry helium, as Kaiser and Shugar demonstrate for glass, may cause stresses that could lead to paint film cracking.) The intensity of the blue hue in conjunction with the confirmed identification of aluminum and silicon suggests the dark blue paint contains ultramarine. Raman spectroscopy could also be used to identify ultramarine and would have been necessary if it was only present on the palette as a tint.

Some green paints appear to be prepared from green pigments and others are optical green pigments- blue and yellow pigment mixtures. Copper [Cu (Kα = 8.05 keV, Kβ = 8.9 keV)] and arsenic [Ar (Kα = 10.54 keV, Kβ 11.73 keV)] were present in some of the green paint and suggest a copper arsenite based pigment such as emerald green [$Cu(C_2H_3O_2)_2{\cdot}Cu(AsO_2)_2$]. The darkest green paint only gave a very weak signal for copper and arsenic but a strong signal for barium and strontium; low z conditions indicate this sample contains chlorine [Cl (Kα,β = 2.62 keV, 2.82 keV)] and since no lead is present in this sample, sulfur [S (Kα,β = 2.31 keV, 2.47 keV)] could be confirmed. Barium and strontium and sulfur suggest barium sulfate. (For a discussion on the relationship between barium and strontium see Chapter 3 by Stulik and Kaplan on the examination of photographs.) Chlorine, being an auxochrome to some organic pigments suggests such an organic pigment may be present but is insufficient information to confirm. And since copper is present in many of the samples, phthalocyanine pigments which contain chlorine auxochromes on a porphryin structure that chelates copper, one can't indicate with any confidence what the source of the copper is in the XRF signal. Subsequent analysis of the dark green by Raman spectroscopy does confirm that the sample is an organometallic pigment, specifically phthalocyanine green (PG 7). The X-ray signatures suggestive of cobalt blue and cadmium yellow appear in some of the greens and one green sample contains cadmium with some iron [Fe (Kα = 6.40 keV and Kβ = 7.65 keV)]. The iron may indicate Prussian blue but complementary methods are essential to confirm the pigment's presence since iron is so ubiquitous.

These findings demonstrate the value of multiple XRF protocols and how they can guide what type of subsequent tests are needed for a particular paint being analyzed. Together, these tools when used wisely provide data that efficiently dovetails to form an accurate understanding of the materials present.

Final Remarks

There is no doubt XRF data can contribute to a deeper understanding of an artist's materials and technique: in many instances the method can provide useful elemental data to help confirm that a particular pigment, filler, or ground preparation was used. More carefully conducted studies could provide information to help understand how several paintings relate to each other. Like all analytical tools, each has their own limitations and a strong understanding of XRF's limitations applied to paintings is essential. Full grasp of the strengths and weaknesses of any method will increase the confidence one can invest in reporting data in addition to guiding the analyst to adopt an approach to technical investigations that relies on a rational use of complementary techniques and the establishment of productive collaborations.

Acknowledgements

The author would like to thank the editors for helpful comments to earlier drafts of the manuscript and to Karen Trentelman for valuable advice on the chapter's structure. Marco Leona kindly identified the phthalo green pigment from Soutine's palette using Raman spectrometry. Cindy Albertson provided the radiograph illustrated in Figure 4.1. Staff, fellows and interns in the Conservation Department at MoMA provided practical suggestions to potential applications.

Further reading

Buxbaum, Gunter. 1998. *Industrial inorganic pigments*. 2nd ed. New York: Wiley-VCH.

Hou, X., and B.T. Jones. 2000. "Field instrumentation in atomic spectrometry." *Microchemical Journal* Vol. 66:115-145.

Janssens, K., F. Adams, and A. Rindby. 2000. *Microscopic X-ray Fluorescence Analysis*. Chichester: John Wiley & Sons Ltd.

Namowicz, Carole, Karen Trentelman, and Christopher McGlinchey. 2009. "XRF of cultural heritage materials: round-robin IV - paint on canvas." *Powder Diffraction* Vol. 24 (2):124-129.

Patton, Temple C. 1973. *Pigment Handbook*. Vol. 1. New York: Wiley.

Potts, P.J., and P.C. Webb. 1992. "X-ray fluorescence spectrometry." *Journal of Geochemical Exploration* Vol. 44:251-261.

References

Berns, Roy S., and Francisco H. Imai. 2001. "Pigment identification of artist materials via multi-spectral imaging." In *Proceedings of IS&T and SID's 9th Color Imaging Conference: Color Science, Systems and Applications*, 85-90. Scottsdale, Arizona.

Berrie, Barbara. 2007. *Artists' Pigments: A handbook of their history and characteristics*. Vol. 4. Washington: National Gallery of Art.

Burgio, L., K. Melessanaki, M. Doulgeridis, R. J. H. Clark, and D. Anglos. 2001. "Pigment identification in paintings employing laser induced breakdown spectroscopy and Raman microscopy." *Spectrochimica Acta Part B: Atomic Spectroscopy* Vol. 56 (6):905-913.

Carpenter, Donald. 2000. "Industrial Applications of u-XRF." In *Microscopic X-ray Fluorescence Analysis*, edited by Koen H.A. Jansens, Freddy C.V. Adams and Anders Rindby, 348. New York: John Wiley.

Cesareo, R, A Castellano, G Buccolieri, S Quarta, M Marabelli, P Santopadre, M Leole, and A Brunetti. 2004. "Portable equipment for energy dispersive X-ray fluorescence analysis of Giotto's frescoes in the Chapel of the Scrovegni." *Nuclear Instruments and Methods in Physics Research Section B: Beam Interactions with Materials and Atoms* Vol. 213:703-706.

Chaplin, Tracey D., Robin J. H. Clark, and Marcos Martinón-Torres. 2010. "A combined Raman microscopy, XRF and SEM-EDX study of three valuable objects - A large painted leather screen and two illuminated title pages in 17th century books of ordinances of the Worshipful Company of Barbers, London." *Journal of Molecular Structure* Vol. 973 (1-3):350-359.

Delaney, John K., Elizabeth Walmsley, Barbara H. Berrie, and Colin F. Fletcher. 2005. "Multi-spectral Imaging in the Infrared to Detect and Map Blue Pigments." In *Scientific Examination of Art: Modern Techniques in Conservation and Analysis*, 120-136. Washington: National Academies Press.

Eastaugh, Nicholas, V. Walsh, T. Chaplin, and R. Siddal. 2004. *The Pigment Compendium: A dictionary of historical pigments*. Boston: Elsevier Butterworth-Heinemann.

Feller, Robert. 1986. Vol. 1, *Artists' Pigments: A handbook of their history and characteristics*. Washington: National Gallery of Art.

Ferretti, Marco. 2004. "Fluorescence from the collimator in Si-PIN and Si-drift detectors: problems and solutions for the XRF analysis of archaeological and historical materials." *Nuclear Instruments and Methods in Physics Research Section B: Beam Interactions with Materials and Atoms* Vol. 226 (3):453-460.

FitzHugh, Elizabeth West. 1997. *Artists' Pigments: A handbook of their history and characteristics*. Vol. 3. Washington: National Gallery of Art.

Gatto Rotondo, G., F. P. Romano, G. Pappalardo, L. Pappalardo, and F. Rizzo. 2010. "Non-destructive characterization of fifty various species of pigments of archaeological and artistic interest by using the portable X-ray diffraction system of the LANDIS laboratory of Catania (Italy)." *Microchemical Journal* Vol. 96 (2):252-258.

Glinsman, L. 2004. *The Application of X-ray Fluorescence Spectrometry to the Study of Museum Objects*. Ph.D., University of Amsterdam.

Ida, Hiroyuki, Takao Segawa, Shigeo Tohyama, and Jun Kawai. 2005. "Analysis of painted steel by a hand-held X-ray fluorescence spectrometer." *Spectrochimica Acta Part B: Atomic Spectroscopy* Vol. 60 (2):249-252.

Jenkins, Ron. 1999. *X-ray Fluorescence Spectrometry*. 2 ed, *Chemical analysis*. New York: John Wiley & Sons, Inc.

Leona, Marco. 2009. "Microanalysis of organic pigments and glazes in polychrome works of art by surface-enhanced resonance Raman scattering." *Proceedings of the National Academy of Sciences* Vol. 106 (35):14757-14762.

Lomax, Suzanne Quillen, Michael R. Schilling, and Tom Learner. 2007. "The Identification of Synthetic Organic Pigments by FTIR and DTMS." In *Modern Paints Uncovered, London, May 2006*, edited by Mark Greenberg, 105-117. Los Angeles: Getty Conservation Institute.

Mendoza-Cuevas, Ariadna, and Homero Perez Gravie. 2011. "Portable energy dispersive X-ray fluorescence and X-ray diffraction and radiography system for archaeometry." *Nuclear Instruments and Methods in Physics Research Section A: Accelerators, Spectrometers, Detectors and Associated Equipment* Vol. 633 (1):72-78.

Pantazis, T., J. Pantazis, A. Huber, and R. Redus. 2010. "The historical development of the thermoelectrically cooled X-ray detector and its impact on the portable and hand-held XRF industries " *X-Ray Spectrometry* Vol. 39 (February):90-97.

Picollo, M., M. Bacci, D. Magrini, B. Radicati, G. Trumpy, M. Tsukada, and D. Kunzelman. 2007. "Modern White Pigments: Their Identification by Means if Noninvasive Ultraviolet, Visible, and Infrared Fiber Optic Reflectance Spectroscopy." In *Modern Paints Uncovered, London, May 2006*, edited by Mark Greenberg, 118-128. Los Angeles: Getty Conservation Institute.

Piorek, S. 2008. "Coatings, Paint and Thin Film Deposits." In *Portable x-ray fluorescence spectrometry: capabilities for in situ analysis*, edited by P. J. Potts and Margaret West, 56-82. Cambridge, UK: Royal Society of Chemistry.

Rosi, Francesca, Aviva Burnstock, Klaas Jan Van den Berg, Costanza Miliani, Brunetto Giovanni Brunetti, and Antonio Sgamellotti. 2009. "A non-invasive XRF study supported by multivariate statistical analysis and reflectance FTIR to assess the composition of modern painting materials." *Spectrochimica Acta Part A: Molecular and Biomolecular Spectroscopy* Vol. 71 (5):1655-1662.

Roy, A. 1993. *Artists' Pigments: A handbook of their history and characteristics*. Vol. 2. Washington: National Gallery of Art.

Sawczak, M., A. Kamińska, G. Rabczuk, M. Ferretti, R. Jendrzejewski, and G. Śliwiński. 2009. "Complementary use of the Raman and XRF techniques for non-destructive analysis of historical paint layers." *Applied Surface Science* Vol. 255 (10):5542-5545.

Szökefalvi-Nagy, Z., I. Demeter, A. Kocsonya, and I. Kovács. 2009. "Non-destructive XRF analysis of paintings." *Nuclear Instruments and Methods in Physics Research Section B: Beam Interactions with Materials and Atoms* Vol. 226 (1-2):53-59.

Woll, A.R., J. L. Mass, C. Bisulca, M. Cushman, C. Griggs, T. Wazny, and N. Ocon. 2008. "Appropriation or Collaboration? The Unique Construction of The Armorer's Shop." *Studies in Conservation* Vol. 53 (2):93-109.

Chapter 5

XRF analysis of manuscript illuminations[1]

K. Trentelman, C. Schmidt Patterson and N. Turner

1. Introduction

Illuminated manuscripts, bound books consisting of pages (folios) of either parchment or paper containing written text, painted illustrations (often decorated with gold, from which the term "illumination" is derived), decorative initial letters and borders or margin designs, are among the most prevalent object types to survive into the modern era (Brown 1994; de Hamel 2001; Clemens and Graham 2007). Illuminated manuscripts were produced by nearly every culture, and date from antiquity through the present day. The examples discussed in this chapter draw primarily from medieval manuscripts, but the techniques described for their study using handheld XRF would be equally applicable to any manuscript.

Although largely understudied, manuscripts and manuscript illuminations offer tremendous opportunities for expanding scholarly knowledge and understanding of painting materials, artists' working methods, and art technology. Many illuminated manuscripts remain intact, but over the years it has unfortunately been common practice to cut out important illuminations or other decorative elements. Consequently, many manuscript collections may also include detached folios or excised images, which, having been removed from their original surroundings, may particularly benefit from study in order to better contextualize them. Frequently, images were excised for the purposes of display, and as such may have been subjected to the deteriorating effects of light or other environmental agents. By contrast, illuminations that have been protected within the pages of a closed manuscript may offer the rare opportunity to study pigments in a state close to their original condition. Therefore, the study of illuminated manuscripts can provide important insight into the interpretation and conservation of related works, such as large-scale paintings on canvas or panel.

1.1 Role of XRF in the study of manuscript illuminations

Removed samples from a painted object can reveal the stratigraphy of the paint layers present (if prepared as a cross-section) and often this is also the only way to determine the composition of binders or other organic materials present (Pinna, Galeotti, and Mazzeo 2009). While samples can be removed from large-scale paintings or polychrome sculpture with relatively little impact on the integrity of the image, because of the small scale of most manuscript illuminations, the removal of even small samples would be visually conspicuous and difficult to conceal. Therefore, the removal of samples from manuscript illuminations for analysis is generally severely limited, or more typically, completely prohibited. Occasionally, "volunteer" samples can be obtained from areas of pigment offset (onto a facing page) or from areas of pre-existing flake loss, but the location of such samples cannot be controlled and therefore may or may not be from an area of interest. Consequently, the study of manuscript illuminations is largely limited to those analytical techniques that are non-invasive and can be applied *in situ* – such as X-ray fluorescence (XRF) spectroscopy.

Illuminated manuscripts may be held in the collections of a variety of institutions in addition to fine art museums – most notably libraries, archives and private collections. Therefore, the development of handheld XRF spectrometers (Moioli and Seccaroni 2000; Cesareo *et al.* 2008; Pessanha, Guilherme, and Carvalho 2009; Bonizzoni *et al.* 2010; Pantazis *et al.* 2010) was an important factor in opening up illuminated manuscripts in particular, but also works of art in general, to wider study (Mantler and Schreiner 2000; Michael Mantler and Schreiner 2000; Mantler and Schreiner 2001; Mantler and Klikovits 2004). By allowing manuscripts to be studied in their current environments, the need to subject rare and valuable objects to the risks associated with transporting them to an off-site location for analysis is eliminated. Furthermore, studies need not be limited to works in a single collection: equipped with a handheld XRF spectrometer a single researcher may conduct analyses on works in many different collections/locations.

XRF spectroscopy provides a means of obtaining, relative quickly and non-invasively, the elemental composition of an area under investigation, from which the pigments may be inferred. The number of pigments traditionally used in the painting of manuscript illuminations is relatively limited, and many of the inorganic pigments may be inferred from their XRF elemental signature. However, in many cases the results can be ambiguous, either because the elements detected are not unique to a single pigment, because multiple pigments are present in complex mixtures or layers, or because materials comprised of low-Z elements (such as organic dyes) are present that cannot be inferred from the XRF data alone. In these cases, the elemental data obtained using XRF can help guide the selection of appropriate complementary analytical techniques.

Providing compound specific identification, Raman spectroscopy is a particularly good complement to XRF, and also being non-invasive, is well-suited to the study of manuscript illuminations (Clark 1995b, 1995a; Clarke 2001a). Raman spectroscopy can definitively identify many pigments (including some organic pigments) (Burgio and Clark 2001), and, when coupled to a microscope, can examine single pigment particles, enabling the identification of the individual components comprising mixtures. Elemental data provided by XRF is helpful in the interpretation of Raman spectra since it effectively narrows the number of possible compounds. Similarly, recently developed spectrometers (Duran *et al.* 2009; Duran A 2009; Sarrazin, Chairi, and Gailhanou 2009) simultaneously collect XRF and X-ray diffraction (XRD) data at each examined location; the elemental data can again be used to refine the interpretation of the diffraction pattern. Because XRF, Raman and XRD are all primarily spot analysis techniques, techniques that provide a more global assessment of the object, such as multispectral imaging, ultraviolet illumination, infrared reflectography and X-radiography, are useful in helping determine the overall distribution of pigments or other materials within an illumination (Bruni *et al.* 2008; Ricciardi *et al.* 2009). XRF, by providing a completely non-invasive means for rapidly obtaining the elemental composition of a specific area, is a good first choice in the study of pigments used in manuscript illuminations. However, by employing multiple analytical techniques (along with close visual examination) when possible, a more complete understanding of the materials and techniques used to create manuscript illuminations may be obtained.

1.2 Topics covered in this chapter

This chapter will discuss the use of handheld XRF spectroscopy as applied to the study of manuscripts, and specifically, manuscript illuminations and the pigments that comprise them. Section 2 discusses the elemental signatures of typical pigments found in manuscript illuminations, as well as the paper or parchment support and inks.

Manuscript illuminations may be found in their original context within a bound manuscript, or as individual folios that have been cut out and mounted for display. In developing an analytical approach for the XRF analysis of manuscript illuminations, one therefore first must consider the physical form of the object to be studied. If it is a bound book, what are the limitations with respect to opening the book to the desired pages (i.e. how widely can the manuscript binding be opened safely; can it be stood vertically or must it lay flat or at a particular angle if it is on display?). If the investigation is of a detached folio, is it loose or adhered/attached to a secondary support? Section 3 describes strategies for examining bound manuscripts versus individual folios.

Illuminated manuscripts typically have text and imagery on both sides of the folios. Because of the penetrating nature of X-rays, the researcher needs to consider what may be contributing to the total XRF signal, from all sources. Section 4 discusses the interpretation of data collected from folios with text/illustrations on both sides, and how to minimize contributions from underlying pages.

Lastly, another distinguishing characteristic of manuscript illuminations is that they often contain finely detailed and elaborate features on the scale of a millimeter or less. The recent development of capillary optics for XRF spectrometers has made it possible to analyze these fine features, and furthermore, when combined with motorized stage control, generate detailed maps of the distribution of elements over a region. Section 5 discusses the benefits of spot, line and area mapping for the study of manuscript illuminations.

2. Illumination materials: pigments on paper/parchment

2.1 Elements in paper or parchment support

Most medieval illuminated manuscripts – such as prayer books, lectern missals or choir books - were written and painted on parchment (prepared sheets of goat/sheep skin). Paper was also used, and with the rise of printing in 16th century, parchment was largely supplanted by paper (Brown 1994). Whether the manuscript under study is written on paper or parchment, an XRF spectrum from an undecorated area of the underlying support material always should be collected, if possible, to provide a background reference. Since these substrates are organic, and thus will not be significant contributors to the overall XRF spectrum, elements detectable by XRF in the paper/parchment supports most likely will be indicative of materials used in their production. Elements detected in paper may include Ca (and associated Sr) from chalk fillers, and Fe (or sometimes other metals) imparted into the paper from screens or other metal-containing devices used in the manufacturing process or other elements introduced as impurities in the water (Hunter 1947; Manso, Costa, and Carvalho 2007; Baker 2010). In parchment, elements commonly found include residual Ca from the lime baths used to prepare the hides, chalk used to whiten the surface, and Fe (or other metals) from the tools used to scrape and prepare the parchment, and impurities in the water used to soak the hides (Reed 1972; Haberditzl *et al.* 1984; Reed 1991; Larsen 2002). In addition, Ca and Fe also are found in many pigments (i.e. iron earth pigments), colorants (i.e. chalk lake substrates) and inks (i.e. iron gall), or even surface dirt contamination, and thus a spectrum of the bare support is necessary to determine whether the origin of these elements is in the painted illumination or from the underlying support. Along with Ca and Fe, other trace elements may also be detected, and may be significant (Dietz *et al.* 2011). Studies conducted by Tim Barrett *et al.*, for example, demonstrated that trace element signatures in European handmade papers samples dating from 1300 to 1800 measured by XRF could be correlated to changes in production technology (Barrett 1989; Barrett *et al.*, Chapter 6 this volume).

When collecting spectra from undecorated, or bare, areas of support, care should be taken to make sure that the area is truly bare, and that there are no decorations on the reverse or subsequent pages. In some cases, particularly for illuminations that have been cut out of manuscripts, there may not be any undecorated support available for analysis. If this is the case, the interpretation of the results must consider that elements that appear consistently across the illumination may be due to the underlying support.

2.2 Elements in inks

Inks are ubiquitous in manuscripts. A number of analytical studies have been published on inks, with regards to their composition, production, deterioration and conservation (Kolar 2003; Hahn *et al.* 2004; Kanngießer *et al.* 2004; Budnar 2006; Kolar 2006; Hahn 2010). In illuminations, inks may be encountered as underdrawing materials, in nearby written text, or as part of the illumination itself. Illuminations frequently have lines of text (or musical notation) on the reverse side of the folio, and therefore it is important to be able to recognize the elemental signatures of inks. In medieval manuscripts, the most common type of ink encountered is iron gall. Iron gall ink is readily identified by the detection of elevated levels of iron, sometimes accompanied by zinc and copper (Malzer, Hahn, and Kanngiesser 2004; Cechák *et al.* 2010) (incorporated into the ink from brass vessels used in the preparation process), or other trace elements from the metal ores. Carbon-based inks are also frequently encountered, but since carbon cannot be detected via XRF, their presence may be inferred from the lack of heavier elements, and confirmed using another analytical technique, such as Raman spectroscopy. Similarly, brown or yellow inks made from organic colorants are not detectable via XRF, although, if they are prepared as a lake, elements associated with laking agents (such as Ca) may be. Metallic inks - gold, copper or brass - are readily detected by XRF because of their high-Z elemental content. Although not strictly inks, musical stave lines painted in vermilion (characterized by mercury) or other heavy metal-containing pigments are similarly easily detected through their elemental signature.

Applied with a pen or other fine drawing tool, inks generally are localized in discrete areas. For very fine lines, depending on the aperture of the XRF spectrometer, the ink may only represent a small proportion of the signal collected, and thus the composition of the ink may be masked by a large background component from the underlying support. Similarly, if the ink has been applied over a large area as a thin wash, the signal from the ink may be quite weak in comparison to the background. In both cases, subtraction of a background spectrum collected from a nearby, undecorated, area may help to determine the composition of the ink.

2.3 Elements in pigments and colorants

Painting treatises and manuals, such as Cennini's famous *Il libro dell'arte* (The Craftsman's Handbook) (Cennini 1954), and the anonymous *De Arte Illuminandi* (Thompson 1933) as well as other alchemical, medicinal, and pharmacological treatises, books of secrets and pigment recipe collections dating from the period during which the object under study was painted (Merrifield 1967) provide a useful starting point for identifying the pigments one might expect to find in manuscript illuminations (Clarke 2001b). Most illumination pigments are also common to the painting palette of monumental painters and were in use throughout the middle ages and Renaissance (Thompson 1956; Gettens and Stout 1966; Feller 1986; Roy 1993; FitzHugh 1997; Berrie 2007). Some pigments, however, have special importance in the study of manuscript illuminations. In particular, the pigment mosaic gold (SnS_2), is frequently found in medieval and early Renaissance manuscripts (Turner 1998). Cennini specifically identified mosaic gold with the work of manuscript illuminators, and it has only rarely been found in monumental paintings (Smith, Reeve, and Roy 1981; Wallert 1996; Speelers 1999). By contrast, identified occurrences of the pigment lead tin yellow in monumental painting and sculpture has its use confined from c. 1300 – c. 1750 (Kuhn 1993); yet recent identifications of this pigment by XRF in Byzantine manuscripts dating to as early as the 12th century may challenge this understanding of its use (Oltrögge 2011). Similarly, bismuth black was recently discovered in the work of a French illuminator, Jean Bourdichon, in illuminations dating to as early as 1485 (Burgio *et al.* 2009; Trentelman and Turner 2009); prior to this discovery bismuth black had only been identified in monumental paintings by early 16th century Italian artists such as Raphael and Fra Bartolomeo (Spring, Grout, and White 2003). Therefore, the most frequently identified or typical pigments should be used as a

Pigment/ material	Chemical formula	Elements (*lines*) detectable by portable XRF	Potential interferences/ identification notes
Lead white	$2PbCO_3 \cdot Pb(OH)_2$	Pb (*L,M*)	Ubiquitous, often admixed with other pigments.
Red lead	Pb_3O_4	Pb (*L,M*)	Difficult to detect in presence of lead white; confirmation using complementary analysis recommended.
Vermilion	HgS	Hg (*L, M*) S (*K'*)	Unique element: presence of vermilion may be assumed based on XRF analysis alone.[2]
Iron earths (red and yellow)	Fe_2O_3 / FeOOH in clay/silicate matrix	Fe (*K*) Mn (*K*)	Fe (K) may be confused with Cu escape peak if Cu is present. Presence of Mn (K) indicative of umber; exact form of Fe-based colorant (hematite vs goethite) requires analysis via molecularly specific technique.
Orpiment	As_2S_3	As (*K*) S (*K'*)	As Kα may be difficult to detect in presence of lead white: As Kα overlapped by Pb Lα. As Kβ (close to Hg Lβ) necessary for positive identification of As. Exceedingly difficult to detect if both Pb and Hg are present.
Realgar/ Pararealgar	As_4S_4	As (*K*) S (*K'*)	As Kα may be difficult to detect in presence of lead white: As Kα overlapped by Pb Lα. As Kβ (close to Hg Lβ) necessary for positive identification of As. Exceedingly difficult to detect if both Pb and Hg are present. Raman particularly useful to identify specific form.
Azurite	$2CuCO_3 \cdot Cu(OH)_2$	Cu (*K*)	Molecularly specific technique generally required for complete identification.
Malachite	$CuCO_3 \cdot Cu(OH)_2$	Cu (*K*)	Molecularly specific technique generally required for complete identification.
Lead tin yellow	Pb_2SnO_4 (type I) $PbSn_{1-x}Si_xO_3$ (type II)	Pb (*L, M*) Sn (*K,L*)	Coincident occurrence of Pb and Sn necessary, although additional presence of lead white may change observed ratios. Sn Kα overlapped by Pb Lβ sum peak; Sn Kβ necessary for positive identification. Raman spectroscopy can distinguish type I from type II.
Emerald green	$Cu(C_2H_3O_2)_2 \cdot 3Cu(AsO_2)_2$	Cu (*K*) As (*K*)	Coincident occurrence of Cu and As suggestive of presence; confirmation by complementary technique recommended.
Ultra-marine	$3Na_2O \cdot 3Al_2O_3 \cdot 6SiO_2 \cdot 2Na_2S$	S (*K'*) Si (*K*) Al (*K*)	Si and Al detectable by XRF if instrument equipped with vacuum/purge capability.
Terra verte	Celadonite/glauconite Fe, Mg, Al, K hydrosilicate	Fe (*K*) Si (*K*) Al (*K*)	Level of Fe typically very low, may be difficult to distinguish from background. Si and Al detectable by XRF if instrument equipped with vacuum/purge capability.
Prussian blue	$Fe_4[Fe(CN)_6]_3 \cdot xH_2O$	Fe (*K*)	Due to high tinting strength, often present in low quantity and Fe signals may be weak; complementary technique necessary for positive identification.
Mosaic gold	SnS_2	Sn (*K, L*) S (*K'*)	Sn Kα overlapped by Pb Lβ sum peak; Sn Kβ necessary for positive identification if Pb is also present.

Gypsum	$CaSO_4•2H_2O$	Ca (*K*) S (*K*[1])	S K lines are overlapped by Pb M; complementary technique necessary for positive identification when lead is present to distinguish from chalk.
Calcite/chalk	$CaCO_3$	Ca (*K*)	May be present as component of pigment or substrate. Small amounts in pigment can be difficult to discern from background.
Gold	Au	Au (*L, M*)	Unique element. Small amounts difficult to identify if in combination with Pb, Hg, As.
Silver	Ag	Ag (*K*)	For Rh anode instruments, overlapped by elastic scatter from the Rh Kα line. Must be confirmed by the presence of the Ag Kβ line.

[1] S K line may be detectable in instruments equipped with capability to purge/evacuate area between sample and detector, although if Pb is present, the Pb M lines overlap S K.

[2] Hg may also be detected in items that have been amalgam gilt (typically only metal surfaces), and thus in manuscript illuminations the presence of Hg can confidently be attributed to the presence of HgS.

Table 5.1:
Elements detectable by handheld XRF for pigments most commonly encountered in manuscript illuminations.

starting point for the interpretation of XRF spectra, but one should never preclude the expectation of finding unusual pigments or pigments not typically associated with manuscript illuminations.

Before beginning XRF analysis, close visual examination and photo-documentation of the painted surface under binocular magnification (employing various lighting conditions, such as raking and transmitted light, ultraviolet illumination and infrared reflectography) can help identify areas for study by revealing the presence of mixtures, multiple layers or previous restoration campaigns. Pigments cannot be positively identified using visual inspection alone, but, with experience, one can use visual observations to form a hypothesis that can help guide in the collection of XRF data (Porter 1995). At the very least, close visual inspection can help establish the minimum number of materials that might be present in an area – e.g. are the green areas composed of a mixture of blue and yellow pigments, or of a green pigment only?

Pigments most commonly encountered in medieval and Renaissance manuscript illuminations, along with their elemental signatures (and possible interferences), are listed in Tables 5.1, while Table 5.2 list the possible sources of detected elements. The pigment vermilion (HgS, mineral equivalent cinnabar) is ubiquitous in manuscript illumination, and is easily inferred from the detection of a single element, Hg. Many other commonly encountered pigments, even though they do not have unique elemental signatures, when coupled with visual analysis often also may be inferred from the detection of a single element: lead white (Pb), red lead (Pb), iron earths (Fe), orpiment (As), azurite (Cu) and mosaic gold (Sn).

For some pigments, the simultaneous detection of two elements is characteristic - notably lead tin yellow (Pb and Sn) or emerald green (Cu and As). However, since each partner of these pairs of elements also occurs alone in other pigments (e.g. lead white and mosaic gold, or malachite and orpiment) the possibility of mixtures must always be considered. Many pigments share common elemental signatures, and as such, are difficult to identify based on XRF data alone, as highlighted in Table 5.2. The detection of Cu in particular often is difficult to interpret. For example, the presence of Cu in a green colored area could indicate a number of materials including the green pigments malachite or verdigris, a glaze of copper resinate, or the blue pigment azurite mixed with a yellow pigment. Again, visual analysis often can aid the interpretation, but additional analysis using complementary, molecularly specific, techniques may also be necessary.

Elements detected by portable XRF	Possible source(s)
Al	ultramarine, terra verte, organic lake pigment with alum substrate
Si	ultramarine, iron earths (red, yellow), terra verte, lead tin yellow (type II)
S	vermilion, orpiment, realgar/pararealgar, mosaic gold, gypsum, misinterpreted Pb M lines
K	terra verte
Ca	gypsum, calcite/chalk, bone white, organic lake pigment with calcite substrate, component of parchment/paper substrate, constituent in gold ground prepatory layer
Ti	titanium white, incidental component of iron-containing pigment
Mn	umber, incidental component of iron-containing pigment
Fe	iron earths (red, yellow), terra verte, Prussian blue, component of parchment/paper substrate, iron gall ink, constituent in gold ground prepatory layer, misinterpretated Cu escape peak
Cu	azurite, malachite, emerald green, verdigris, copper resinate, incidental component of iron gall ink, copper or brass "ink," constituent in gold ground prepatory layer
Zn	zinc white, incidental component of copper-containing pigment, incidental component of iron gall ink
As	orpiment, realgar/pararealgar, emerald green
Sr	component of parchment/paper substrate, incidental to presence of Ca
Ag	silver leaf/mordant gilding
Sn	lead tin yellow (type I or type II), mosaic gold, brass "ink", misinterpreted Pb sum peak
Au	gold leaf/mordant gilding, gold "ink"
Hg	vermilion, constituent in gold ground prepatory layer
Pb	lead white, red lead, lead tin yellow (type I or type II), litharge, massicot, constituent in gold ground prepatory layer

Table 5.2:
Possible source(s) of elements commonly detected in manuscript illuminations.

Other common pigments, such as ultramarine, indigo, and organic reds (such as madder and cochineal), consist of low-Z elements that are generally not readily detectable with a handheld XRF spectrometer. Likewise, some pigments, such as terra verte and Prussian blue may contain such small quantities of the elements detectable by XRF (Fe/Si and Fe, respectively) such that their fluorescence signal is difficult to distinguish from the background. The presence of these pigments might be suspected based on the lack of higher-Z elements that would be expected if other likely (i.e. similarly colored) pigments were present. For example, if no Cu (indicating azurite) is detected in a bright blue area, one might reasonably assume ultramarine is present. However, the converse is not true: if Cu is detected in a bright blue area, azurite might reasonably be assumed to be present, but the additional presence of ultramarine cannot be excluded. Identification should never be made on the lack of evidence. Instead, as will be shown in the example discussed in section 4.2, such data should be used to inform subsequent analyses, such as optimizing XRF conditions to enhance the detection of low-Z elements or using a complementary technique, such as Raman microscopy.

2.4 Gold and other decorative metals

Illuminations take their name from the use of decorative metals – typically gold and silver – to embellish the illustrations to provide a luminous quality. The identification of the type of metal used is easily accomplished using XRF, and the detection of impurity elements, such as silver and copper within gold, may be useful in characterizing the materials used by a particular artist or workshop. The composition of the adhesive ground under metal leaf may also be characteristic of particular artists, workshops, or the location in which it was made. In contrast to the typical iron-rich clay bole used in the gilding of panel paintings, a wider range of elements (including copper, calcium, mercury, lead and iron) have all been found as major constituents in gold grounds in manuscript illuminations (Turner 2009). However, these are intrinsically layered structures, which, as will be discussed in section 4, necessitate a consideration of absorption of fluorescence emission from different strata in the interpretation of the spectra.

3. Experimental setup

There is no single ideal configuration for the XRF examination of manuscripts. Each setup must be tailored to each individual project, depending on the particular characteristics of the manuscript or individual folio to be studied (i.e. bound vs.

unbound, matted vs. unmatted), and should be designed in consultation with a conservation professional. In general, the XRF spectrometer should be positioned as close as possible (ideally no more than 1-2 mm) from the surface, but should never come into direct contact with the illumination. The use of protective sheets (preferably of glassine to avoid the possibility of detaching loose pigment via static attraction) with windows over the area to be analyzed may be used at the discretion of the conservator; if there is any concern about the stability of the pigment, it is best to avoid any contact with the surface of the illumination whatsoever. Regardless of the mounting scheme utilized, another important consideration is the possibility that the parchment/paper will flex with minor air currents and/or fluctuations in temperature or humidity during the course of the analysis. Whether bound in the manuscript or detached, folios may be strapped or weighted, at the discretion of the conservator, to limit such unintended motion, but the experimental setup should be able to accommodate and adjust for such motion if it occurs. In all cases, unnecessary handling of the manuscript should be avoided, and the folios and binding should never be put under strain.

Detached folios that are pasted down or otherwise securely affixed to mat board or other supports can be examined using a similar configuration to that commonly used for the examination of paintings (Figure 5.1a): the portable instrument mounted on a tripod with an adjustable arm to adjust the distance of the nose of the instrument from the illumination, which will be vertically positioned on an easel or gallery wall. Unmounted folios generally cannot be safely positioned vertically without a special mounting system, and instead are best examined laid flat on a horizontal surface (Figure 5.1b). This necessitates the instrument being positioned over the manuscript, which may require different mounting hardware to securely couple the spectrometer to the tripod and maintain control over the distance to the illumination. Whether vertical or horizontal, particular care and attention must be given to securing the instrument, such as making sure all fittings are tight and applying appropriate counter-weights to prevent tipping, to insure there is no chance the instrument will accidentally come into contact with the work of art.

Because X-rays will penetrate through most manuscript folios and generate fluorescence from elements in underlying materials, such as a mat board or other mounting materials, care must be taken to reduce potentially interfering signals from these materials. Ideally, only air should be behind the area under examination. In practice, however, because the folio under investigation must be appropriately physically supported, this is not always possible. For an unmounted detached folio, an elevated platform with a small window over which the area of interest is placed (easily constructed out of Plexiglas) can provide adequate support while allowing the area of analysis to be free of underlying materials. However, if such a support is not available, the folio may be placed directly on a thick sheet of Plexiglas, Teflon or other low-Z element-containing material. As will be discussed below, while this will increase the background from scattered X-rays, it will largely prevent unwanted fluorescence signals from the table or other support. A reference spectrum of the low-Z material itself should be collected. Similarly, for folios that are pasted down, or otherwise cannot be removed from their mount, a spectrum of any underlying mounting material(s) should be collected, so that any contributions to subsequently acquired spectra may be evaluated. For folios that are hinge-mounted, a thin sheet of a low-Z material may be inserted between the folio and the mat board to filter out unwanted fluorescence signals.

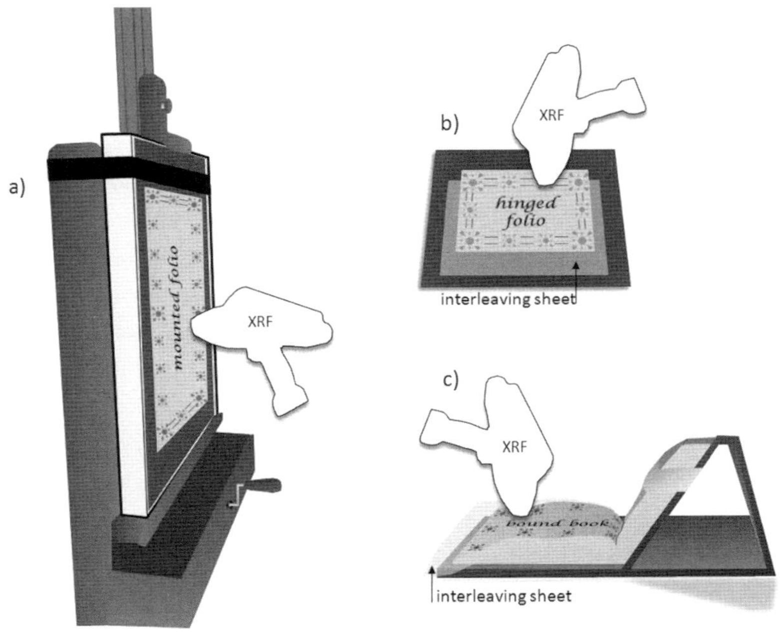

Figure 5.1:
Examples of configurations for performing XRF analysis on manuscript illuminations: a) detached folio mounted on mat board and positioned vertically and strapped to a painting easel for XRF examination; b) XRF spectrometer in position over detached hinge-mounted folio; note interleaving sheet positioned under folio to be studied; and c) bound manuscript securely attached to angled support such that folio to be examined is close to horizontal (an interleaving sheet has been positioned under the folio to be studied).

For folios bound in manuscripts, the manuscript should be positioned either on a horizontal surface or in an angled book cradle (Figure 5.1c), whichever is safest for the manuscript, such that the folio to be examined is as close to horizontal as possible. The manuscript should be secured to the cradle and properly supported to insure it will not shift position during the analysis. Adequate reduction of interference from elements present in underlying pages may be achieved by inserting a sheet of low-Z material under the folio of interest, in the same manner as described above for hinge-mounted folios. The selection of material and thickness of these filter sheets will be described in detail in section 4.3.

The selection of instrument parameters – anode, voltage, current, filtering and acquisition time – should be made according to the specific instrument manufacturer's suggestions to optimize performance for the detection of the desired elements. In general, at least for the initial study of pigments in a manuscript illumination, parameters enabling screening of the widest range of elements should be employed. Subsequent analyses may be made using parameters selected to enhance a particular region of the spectrum, for example, employing a lower excitation voltage and applying a purge or vacuum to enhance the detection of low-Z elements. In all cases, it is advisable to understand the X-ray output of the instrument in order to make optimized choices regarding instrument parameters. A measure of the X-ray output can be obtained by recording a spectrum of a low-Z material such as Plexiglas or Teflon; the resulting spectrum will inform the user of possible interferences from the anode (Rayleigh and Compton scatter), elements in the instrument housing (typically Fe, Ni, Cu and Zn) or impurities in filters (if used).

4. Interpreting the results

Some images in manuscripts contain large passages comprised of a single pigment. If the image is also present on only one side, and the folio can be isolated from all underlying pages or other supports, interpretation of the XRF spectra acquired can be a relatively straightforward matter of inferring the pigment(s) present based the elements detected (see Section 2.3 Elements in pigments and colorants). However, most illuminated manuscripts do not present such an ideal situation for study by XRF. More typically, the illuminations and illustrations consist of complex, finely detailed images created using multiple pigments and colorants, either applied as mixtures or as layers of overlapping strokes to create subtle shading and tonal effects. In addition, the penetrating nature of X-rays will induce fluorescence not only from elements present in the topmost pigment layer, but also from inks or pigments on the reverse (section 4.1), from elements in underlying pigment layers (section 4.2), and on underlying pages as well (section 4.3).

4.1 The recto-verso problem: XRF emission from both sides of a folio

Illuminated manuscript folios typically contain text and imagery on both sides, so the interpretation of XRF spectra acquired from one side must always consider what is on the other side. If possible, a visual examination can help determine locations for analysis that are free from possible interferences. However, in cases where only one side of a folio is accessible (e.g. a detached folio pasted down to a secondary support), care must be taken in the interpretation of XRF spectra: the detection of elements inconsistent with the pigments visible on the surface may be an indicator of a hidden layer.

For example, as shown in Figure 2, in the study of *The Temperate and the Intemperate* (Master of the Dresden Prayerbook, Flemish, 1475-1480, J. Paul Getty Museum, 91.MS.81), the detection of a large amount of Hg in an area of black background (Figure 5.2a) could not be rationalized with the appearance of the surface. The collection of a second spectrum in another area within the black background (Figure 5.2b) did not contain Hg, suggesting the Hg-containing pigment was most likely present on the reverse side. Evaluating the spatial distribution of anomalous elements – either through multiple spot analyses as was done in this instance, or through mapping (see section 5) – is perhaps the easiest means of determining whether or not the element is present due to a feature painted on the other side. However, there are also clues within the XRF spectrum as to the relative location of the element within a layered structure.

In layered materials, X-ray fluorescence originating from lower layers can be absorbed by elements in upper layers. For an X-ray fluorescence photon emitted by one element to be absorbed by a second, different, element, the fluorescence photon must have sufficient energy to eject an electron from the absorbing atom (photoelectric effect). A detailed discussion of absorption effects is beyond the scope of this chapter, and the reader is referred to the many available reference books on XRF spectroscopy for more detail (Jenkins 1974; Tertian and Claisse 1982; Van Grieken and Markowicz 1993).

For the purposes of illustrating some of the challenges in interpreting XRF spectra from folios with pigments on both sides, a simplified model as a three-layer system consisting of a single sheet of parchment with a layer of the copper-based pigment azurite on one side and a layer of the lead-based pigment red lead on the other will be considered. Such as system may be encountered in an illuminated manuscript, if, for example, the first letter in a passage of the written text executed in red lead is on the reverse of an illumination containing a blue field painted in azurite.

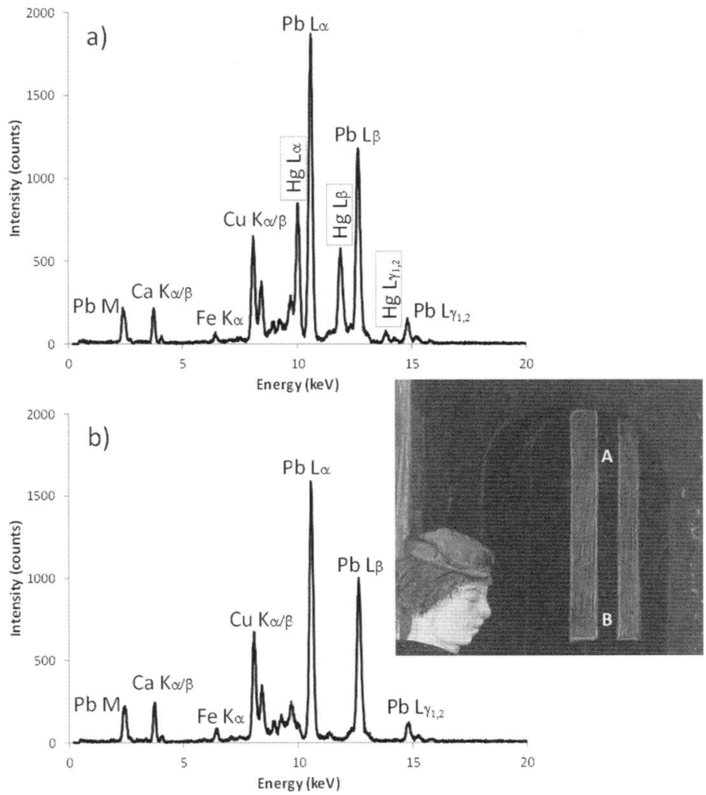

Figure 5.2:
XRF spectra acquired from a) spot A and b) spot B within grey-black area in illumination (see inset). Inset: detail from *The Temperate and the Intemperate* (Master of the Dresden Prayerbook, 1475-1480, J. Paul Getty Museum, 91.MS.81). Detection of Hg in spot A could not be rationalized with appearance of surface of illumination. XRF spectrum acquired from spot B is identical to that acquired from spot A except for the absence of Hg, suggesting Hg is likely present on the other side of the sheet. Spectral acquisition conditions: Bruker Artax, W anode, 40 kV, 600 μA, no filters, in air, 30 s live time.

As shown in Figure 5.3, an XRF spectrum acquired from the illumination (Cu-pigment azurite, black trace) side contains strong Cu K emission lines, along with Pb L lines. Close inspection of this spectrum can help determine whether the Pb is present in the upper surface (perhaps as lead white mixed with azurite) or, as is the case in this example, as red lead present on the other side. The first clue that Pb is not present on the surface is that the low energy Pb M lines (centered at ~2.4 keV) are absent, having been subjected to absorption by the parchment/Cu-pigment layers. The Pb L lines are also attenuated, but to different degrees: the lower energy Lα line is more strongly attenuated than the Lβ line, leading to the two lines being similar in intensity. A spectrum acquired from the side of the model sheet painted with red lead, showing the expected Lα/Lβ ratio for unattenuated Pb L lines as well as relatively strong Pb M lines, is shown for comparison (grey trace). In the spectrum acquired from the Cu-pigment side, the reduction of the Lα/Lβ ratio, coupled with the weak (or missing) M lines, indicates that the Pb fluorescence emission is being absorbed, suggesting the material is likely present in a lower layer.

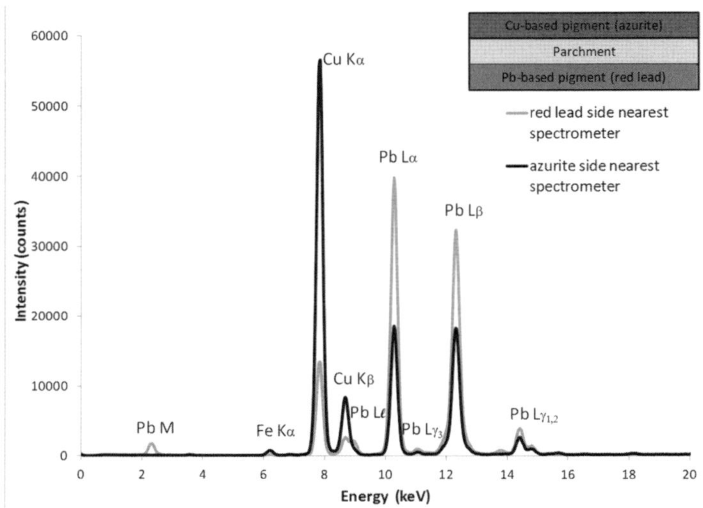

Figure 5.3:
XRF spectra from a sample of parchment with the Cu- based pigment (azurite, $2CuCO_3{\cdot}Cu(OH)_2$) on one side and the Pb-based pigment (red lead, Pb_3O_4) on the other. Black trace is the spectrum acquired from the azurite side; grey trace is the spectrum taken from the red lead side. Note the lack of Pb M lines and reduction in Pb Lα/Lβ ratio in the spectrum acquired from the azurite side in comparison to the spectrum acquired from the red lead side. Spectral acquisition conditions: Bruker Tracer III-V, Re anode, 40 kV, 4.5 µA, Al/Ti filter, in air, 120 s live time.

This reduction of the Lα/Lβ ratio is often, and most easily, observed for Au, Pb, and Hg, elements commonly found in illuminated manuscripts in the form of gold leaf/paint; lead white, lead tin yellow or red lead; and vermilion respectively. It should be noted that although the observation of a reduction of the Lα/Lβ ratio in the L series can indicate that the particular element may be present in a lower layer a similar effect can also be produced within a single layer if the element is present as part of a mixture (Namowicz, Trentelman, and McGlinchey 2009; Bonizzoni *et al.* 2011). De Viguerie *et al.* (De Viguerie, Sole, and Walter 2009; De Viguerie *et al.* 2010) have shown the possibility of exploiting the differential

X-ray attenuation, in particular the greater attenuation of lower energy Pb M lines relative to the higher energy Pb L lines, as a means to calculate the thickness of overlying varnish and glaze layers.

In illuminated manuscripts it is common to find lines of text, typically written either in iron gall or a carbon-based ink, on the backs of folios containing illuminations. While carbon-based inks would not produce interfering X-ray fluorescence, Fe K emission lines from iron gall ink might be misinterpreted as iron earth pigments. Therefore, the interpretation of Fe should always be considered in light of the possibility of the presence of iron gall ink (either under the pigment layer or on the other side of the folio). In Figure 5.4, XRF spectra acquired from both sides of a sheet of parchment painted with lead tin yellow on one side and a letter written in iron gall ink on the other are shown. As shown in the inset (in which the two spectra have been normalized to the Fe Kβ line for clarity), a slight attenuation of the Fe Kα/Kβ ratio can be seen in the spectrum acquired from the side painted with lead tin yellow, due to the differential attenuation of the Fe XRF by the parchment and paint layers.

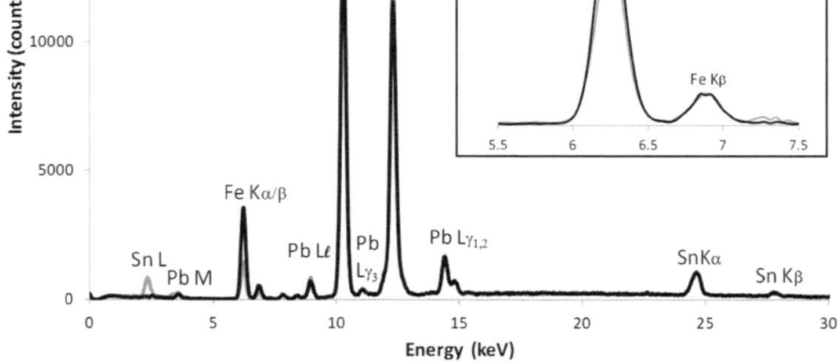

Figure 5.4:
XRF spectra acquired from a sample of parchment with the pigment lead tin yellow (Pb_2SnO_4) on one side and a letter written in iron gall ink on the other. In the spectrum acquired from the lead tin yellow painted side (grey trace), the Fe Kα/Kβ ratio is attenuated (see inset, spectra normalized to Fe Kβ line for clarity) with respect to that observed from the iron gall ink side, where Fe is on the surface. Spectral acquisition conditions: Bruker Tracer III-V, Re anode, 40 kV, 4.5 μA, Al/Ti filter, in air, 120 s live time.

Like the L series emission lines discussed above, differential attenuation of K emission lines also may be used to infer the presence of an element in a lower layer. In general, attenuation across a fluorescence emission series decreases with increasing energy (Jenkins 1974) (to be absorbed, a fluorescence photon must have sufficient energy to eject an electron from the absorbing atom, and the efficiency with which it can do this decreases the greater the difference between its energy and the absorption edge of the absorbing material). Therefore, because the Kα and Kβ lines of Fe (and other elements frequently encountered in illuminated manuscripts, such as Ca and Cu), are more closely spaced in energy than the Lα and Lβ lines of heavier elements such as Au, Hg and Pb, the reduction in the K α/Kβ ratio is smaller, and may not produce a noticeable change in the appearance of the XRF spectrum unless a spectrum of the pure material is available for comparison. Trojek *et al.* (Trojek, Čechák, and Musílek 2010) successfully employed calibration curves of Kα/Kβ and Lα/Lβ ratios of single-element pigments as a function of thickness to predict whether or not an element is on the surface, and the technique may find use in particular applications/analyses.

4.2 The layered paint problem: XRF from multiple overlapping layers of paint

In large scale paintings on canvas or panel, the application of multiple layers of paint to achieve desired effects is common, and, as discussed above, clues to the relative position of various elements within the paint stratigraphy often can be found in the XRF spectrum (Namowicz, Trentelman, and McGlinchey 2009). Some manuscript illuminations also contain passages in which two or more layers of paint have been applied.

One such example was encountered in a study of illuminated manuscripts by the 14th century Florentine artist Pacino di Bonaguida. In Pacino's illumination from the *Laudario of Sant'Agnese* depicting the Ascension (Italian, c.1340, J. Paul Getty Museum, 2005.26), the blue of the Virgin Mary's robe at first presented what appeared to be contradictory results: XRF analysis indicated the strong presence of Cu, but Raman analysis identified ultramarine – a sodium sulfur aluminosilicate – as the only blue pigment present. Since no copper-containing pigment was present on the back of the folio, the best explanation of the source of the Cu X-ray emission was a copper-containing pigment *underneath* the layer of ultramarine. Indeed, a small crack was discovered that revealed a layered structure, and Raman analysis confirmed the presence of azurite underneath the ultramarine layer.

As shown in Figure 5.5, the XRF spectrum of such a layered structure is dominated by the Cu Kα emission line; a small Fe Kα line is also present, which may be due to iron compounds sometimes found in azurite (Burgio, Clark, and Hark 2010). Evidence that a Cu-containing layer of azurite is concealed under a layer of ultramarine is reflected in the XRF spectrum, although the effect is subtle. The Cu K lines, upon passing through the low-Z ultramarine layer, are attenuated, with the Kα line attenuated slightly more than the Kβ line. As shown in inset (a) in Figure 5.5, comparison of the K lines in a layered structure to those of pure azurite (normalized to the Kβ line for clarity), reveals this slight reduction of the Cu Kα/Kβ ratio. However, because these ratios may also be affected by the material being present in mixtures, the presence of a layered structure should always be confirmed directly through visual analysis – either through a fortuitous crack as described above, or by the removal of a sample for preparation as a cross section.

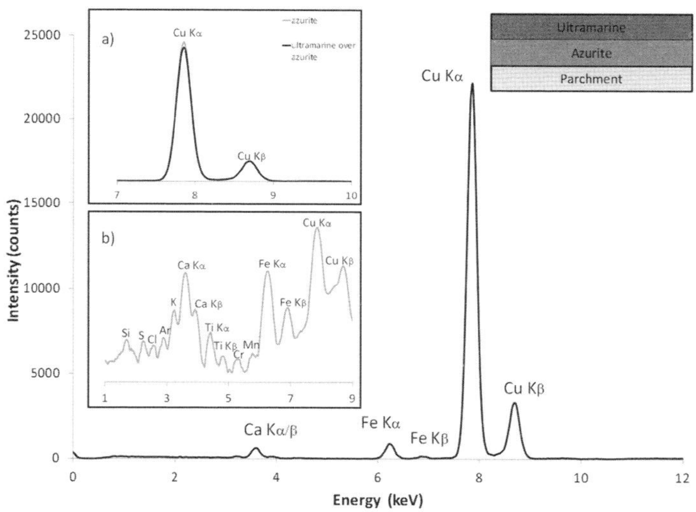

Figure 5.5:
XRF spectrum of area with a layer of ultramarine painted over a layer of azurite (Bruker Tracer III- V, Re anode, 40 kV, 4.5 μA, Al/Ti filter, in air, 120 s live time). Inset (a) shows comparison to spectrum of pure azurite (grey trace), normalized to the Cu Kβ line for clarity. Inset (b) shows a spectrum (log scale intensity) taken under conditions to optimize the detection of low Z elements (Re anode, 15 kV, 4.5 μA, no filters, vacuum between window and detector, 240 s live time). Indicated element lines are all Kα, unless otherwise noted. The detection of Cl, Ti, Cr and Mn are attributed to impurities in the detector window; the detection of Ar is due to the small air gap between the detector window and the illuminated manuscript page.

Evidence that ultramarine is present may also be obtained by collecting an XRF spectrum using instrument settings that enhance the detection of the low-Z elements present in ultramarine (Na, Al, Si and S). Evacuating or purging the region between the area to the studied and the detector will further enhance the detection of low energy emission lines. As shown in inset (b) in Figure 5.5, Si and S can be detected in a spectrum thus obtained (some instruments may also be able to detect Al), suggesting the presence of ultramarine. However, it must be noted that this spectrum was obtained on a mockup prepared using commercial pigments. In manuscript illuminations, Pb is often also present, and the Pb M lines would obscure the S K lines. Similarly, the presence of earth pigments might also contribute to the detection of Si. Therefore, although XRF may provide an indication that ultramarine is present, its identification should always be confirmed using another technique.

4.3 The bound manuscript problem: XRF emission from underlying pages

In bound manuscripts, XRF emission from elements present in text or painted decorations on underlying pages may also be detected. Depending on the energy of the emission lines and composition of intermediate folios, the interfering fluorescence emission may originate from pigments present multiple pages below the folio being examined. However, the depth from which interfering fluorescence may be detected will vary, depending on the thickness and composition of the parchment and decorations, and the characteristics of the individual XRF spectrometer. To avoid detecting fluorescence from underlying pages, ideally, each folio should be completely isolated for examination by suspending it such that there is nothing behind the area being excited. For bound manuscripts, this may not be feasible without subjecting the manuscript to unacceptable risk of damage. Fortunately, however, in most cases it is safe to slip a thin sheet of an absorbing material under the folio to be examined. Low-Z materials such as Plexiglas or Teflon are the easiest and most readily available materials to isolate the folio of interest, although the shielding is not uniform across the entire emission spectrum. Lower energy X-ray emission (<10 keV), including the K emission lines from elements such as Ca, Fe, and Cu are shielded quite strongly. Moderately energetic X-ray emission (10-15 keV), such as the L emission lines from Pb, Hg and Au, will not be attenuated as completely as element lines with lower emission energies, but should be noticeably reduced. Finally, highly energetic X-ray emission (>15 keV), such as the K emission lines of Ag and Sn, are not effectively attenuated by these filters.

To illustrate this differential absorption of fluorescence emission, the spectra shown in Figure 5.6 were acquired with equal thicknesses (3 mm) of air, Plexiglas and Teflon between the spectrometer and a Cu-Pb-Sn bronze reference standard. A bronze was employed as a proxy material representing elements with X-ray fluorescence emission over a wide range of energies (Cu Kα at 8.047 keV, Pb Lβ at 12.620 keV, and Sn Kα at 25.270 keV) that might be present in pigments used in illuminated manuscripts. The spectrum acquired with a 3 mm air gap, a representative distance at which XRF spectra on manuscript illuminations should be acquired, is shown as the baseline reference. The top set of spectra in Figure 5.6 shows the spectra of the bronze acquired through air and through 3 mm sheets of Plexiglas and Teflon for the low energy (<10 keV) region: the Cu K lines are considerably attenuated (~90%) by the Plexiglas, and virtually eliminated by the Teflon. The middle set of spectra examines the region between 10-15 keV: here again, the Plexiglas produces noticeable attenuation of the Pb L lines (for Pb Lβ the attenuation is ~30% in total counts, ~50% with background subtraction) while the same thickness of Teflon attenuates them to an even greater extent (80-90%). The bottom set of spectra show the behavior above 15 keV: in total counts, neither Plexiglas nor Teflon attenuates the Sn K lines (and in fact in the Plexiglas spectrum an increase in total counts is seen due to the increase in background). With background subtraction, the Plexiglas produces only a slight attenuation, while the Teflon attenuates the Sn Kα line on the order of 50%. Aluminum can also be used to filter out unwanted fluorescence from underlying materials, providing shielding efficiencies similar to that of Teflon, but, unless very pure aluminum is used the presence of elements commonly found in aluminum alloys - Fe, Cu, Mn, Mg or Zn - may contribute fluorescence signals, thereby complicating the analysis rather than simplifying it.

The significantly increased background, clearly apparent in the mid- and high-energy regimes of the spectra in Figure 5.6, is largely due to X-ray scatter from the underlying Plexiglas/Teflon (which, for low-Z materials is dominated by incoherent, or Compton, scattering) (Jenkins 1974; Van Grieken and Markowicz 1993). In illuminated manuscripts this scatter can also be produced from the low-Z parchment/paper substrates, but for example, as can be seen in Figure 5.4, usually is negligible with respect to the emission produced by the high-Z elements of interest. When selecting a material to be placed between the folio of interest and any underlying pages (or other material), one must keep in mind not only the desired filtering effect, but also the possibility of producing unwanted scatter that may interfere with the detection of elements present in lower concentrations.

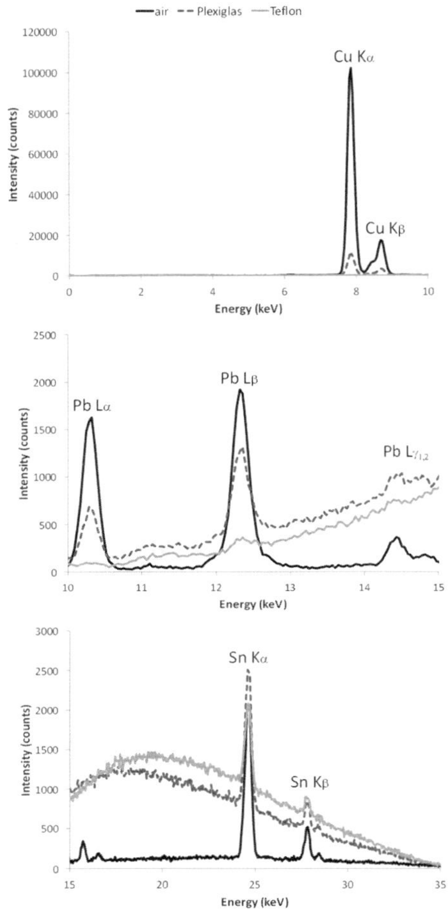

Figure 5.6:
XRF spectra of a Cu-Pb-Sn containing a bronze standard (BS544A) acquired through 3mm of air (black traces), Plexiglas (dashed traces) and Teflon (grey traces). Spectral acquisition conditions: Bruker Tracer III-V, Re anode, 40 kV, 1.5 μA, Ti/Al filter, in air, 240 s live time.

Absorption is a function of thickness ($I = I_0 e^{-\mu\rho x}$, where μ is the mass absorption coefficient for material of density ρ and thickness x), (Jenkins 1974) so the amount of attenuation can be increased by increasing the thickness of the interleaving layer. However, the contribution from scatter will increase as well. Therefore, the choice of interleaving material thickness should take into consideration the benefits of increased absorption against the impact of an increased background. In addition, when dealing with bound manuscripts, the effectiveness of shielding always must be balanced against the safety of the manuscript. In particular, any physical constraints, such as the angle to which the manuscript may be safely opened, of inserting a stiff sheet of material between manuscript pages must be carefully considered. In practice, sheets of Plexiglas or Teflon (wrapped in Mylar or glassine to protect the manuscript pages) 2-3 mm thick have been found to provide a good balance between shielding, scatter, and safety of the manuscript. In some cases, it may be preferable to place only a small piece of material directly under the area to be studied, although this has the disadvantage of presenting additional risk to the manuscript if it has to be repositioned multiple times during the analysis. Similarly, sheets of interleaving material may also be slipped under hinge-mounted illuminations to reduce detection of elements in the mat or other mounting material without having to remove the hinge. As always, the selection and placement of interleaving materials should be done in consultation with a conservator.

5. Spot versus line/area mapping

The small scale of features found on most manuscripts, many on the order of a few millimeters or less, cannot be adequately probed using the 3-6 mm beam diameter typical of most handheld XRF spectrometers. However, the recent incorporation of polycapillary optics into XRF spectrometers, coupled with improved hardware and software capabilities for creating element maps, (Janssens, Adams, and Rindby 2000; Janssens *et al.*. 2000; Bronk *et al.* 2001; Vittiglio *et al.* 2004; Longoni *et al.* 2005; Uhlir *et al.* 2008; Van Der Snickt *et al.* 2008) has proven to be an ideal match for the examination of manuscripts. While these mapping micro-XRF spectrometers cannot be classified as handheld, some are portable in the sense that they are modular (i.e. they can be disassembled and transported to a different location), open architecture, and mounted on wheeled tripod stands that can be moved from one location to another or within a gallery or laboratory space. As such, they offer many of the same benefits of handheld instruments (Trentelman *et al.* 2010).

5.1 Considerations in setting up line/area scans

The logistics – tube selection, page orientation, cradle setup, manuscript positioning, and selection of filters to shield unwanted fluorescence – of using a mapping micro-XRF spectrometer are the same as those described above for the use of a handheld instrument. One additional factor to take into consideration when generating XRF maps is that, since manuscript pages generally are not perfectly flat, care must be taken to insure that there is no chance of the instrument coming into contact with the manuscript over the course of acquiring the data. A good practice is to move the instrument to the extremes of the planned scan, and to various locations within the area to be mapped, watching for changes in manuscript-instrument distance to make sure there is no chance of accidental contact. Depending on the manuscript, it may be possible to apply weights to the edges of a folio to hold it more securely or create a mask that can be placed over the page, leaving the area to be mapped exposed. However, the placing of weights or any other material directly on a manuscript illumination should only be done with the guidance and approval of a manuscripts conservator.

Another consideration in the acquisition of line or area element maps is the total time involved. Depending on the desired resolution, the aperture size, and the live time for each individual spectrum, it may take several hours to collect a map over an area as small as a few millimeters on a side. Fortunately, because of the very small features on most manuscripts, in practice, it is usually possible to select a combination of spot size, step size, and live time that allows the feature of interest to be mapped with sufficient resolution in a reasonable length of time, typically over the course of a business day.

5.2 Area mapping to study fine details

One of the most dramatic examples of using area scans to examine fine details in a manuscript illumination is the examination of human faces. Faces were usually painted with the greatest care and attention, and many of these miniature paintings have extraordinary complex constructions, consisting of multiple thin layers of incredibly fine, overlapping strokes. One such example is a detached folio from the Turin-Milan Hours (Flemish, 1440-1445, J. Paul Getty Museum, 2000.33), thought to have been painted by a follower of Jan van Eyck. The face of Christ, shown in Figure 5.7a, measures a mere 15 mm from chin to forehead and has a depth of complexity that can best be appreciated under magnification. Element distribution maps of the face generated by micro-XRF mapping, besides being striking visual indications of the distribution of pigments, contain important information about the techniques used to paint this illumination. However, as with spectra collected individually, care must be taken when interpreting these maps. For instance, the Pb L and Hg L maps (Figures 5.7b and c, respectively), when plotted over the full intensity scale, would seem to suggest that the artist utilized two different red pigments: vermilion (HgS) to paint the lips, and red lead (Pb_3O_4) to give the rosy blush to the cheeks and forehead. Closer examination of the lower intensities of the Hg L map (plotted on a log scale, Figure 5.7d) reveals that mercury is also present in the cheeks and forehead, but at much lower concentrations than in the lips. Raman spectroscopy confirmed that vermilion is present in the rosy color in the cheeks and forehead. It is mixed with lead white ($2PbOH_2 \bullet Pb(CO)_3$), thus accounting for the increased level of Pb in these areas that was originally misinterpreted as indicating the presence of red lead. This example serves to underscore the fact that XRF analyses can provide a suggestion of the pigments present, but care should be taken to consider that pigments present in lower concentrations may be overlooked, and that complementary analytical techniques should be used to confirm the identity of pigments that do not have unique elemental signatures (such as lead white versus red lead).

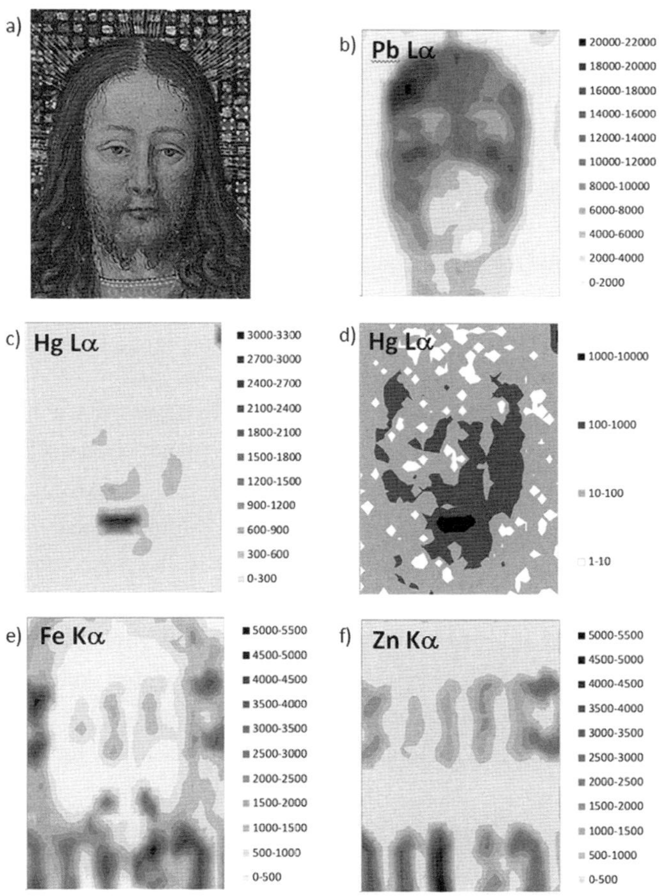

Figure 5.7:
a) detail from detached folio from Turin-Milan Hours (Flemish, 1440-1445, J. Paul Getty Museum, 2000.33); element distribution maps of b) Pb Lα, c) Hg Lα (linear scale), d) Hg Lα (log scale), e) Fe Kα and f) Zn Kα. Spectral acquisition conditions: Bruker Artax micro-XRF spectrometer, W anode, 40 kV, 600 μA, 0.65 mm collimator, no filter, in air, 0.45 mm step size, 10 s live time/spot.

Another benefit of XRF mapping is its ability to detect hidden features (Bergmann 2007; Dik *et al.* 2008). The Fe K map (Figure 5.7e) appears to contain features in addition to the hair and beard of the figure: the penetrating X-rays are also detecting the text on the reverse of the page written in iron gall ink. By mapping an element that is present only in the ink – in this case Zn – a clear map of the text emerges (Figure 5.7f). Although in this particular example the back of the page is accessible, and the text can be easily read simply by turning the folio over, this example serves to demonstrate the possibility of using XRF mapping to reveal features hidden on the reverse side of folios in a way that would not be possible with spot analyses alone. This ability may be particularly important when only one side of a folio is accessible, and there is scholarly interest in the materials – and their context – present on the hidden side.

6. Summary/conclusions

This chapter described the use of handheld and portable XRF spectroscopy for the examination of illuminated manuscripts. The same principles described here for the study of pigments can be applied to any of the components of a manuscript (i.e. text, support, illuminations), or almost any other painted work. As such, XRF analysis of manuscript illuminations is best used to assess the major elements present in different colored areas, from which the pigments used may be inferred. Since in most cases the elements detected are not unique to a particular pigment, definitive identification should be verified using a complementary, molecularly specific technique whenever possible. Because of its ability to perform analysis *in situ*, and with single particle resolution, Raman microspectroscopy provides a good complement to XRF spectroscopy. Recently developed XRF/XRD instruments similarly may be able to provide specific identification of crystalline materials. Additionally, imaging techniques such as infrared reflectography, X-radiography or multispectral imaging can be helpful in determining the distribution of materials across an illumination.

Despite its name, a handheld XRF spectrometer should never actually be handheld in the study of illuminated manuscripts; using a tripod or other secure mount, the spectrometer should positioned close to, but not touching, the surface of the manuscript in order to avoid any potential damage to the painted surface. Because manuscript illuminations may be found in a variety of physical situations, the XRF setup must be individually configured for each illumination, with the safety of the manuscript/illumination always given primary consideration. The use of shielding materials to reduce unwanted fluorescence from underlying pages may be used, but only if they can be placed without risk to the manuscript.

The results of any XRF study should always be interpreted in conjunction with a close visual analysis of the surface. If the elements detected cannot be rationalized with the surface appearance of the illumination, the XRF spectra should be analyzed carefully, looking for differential attenuation of the lower energy lines in an emission series that may be an indicator that the element is present in a lower layer. If a layered structure is suspected, additional analysis should be conducted to ascertain whether the elements might be present in underlying layers, on the reverse side the folio, or on underlying pages. Along with visual or microscopic inspection, this additional analysis might take the form of collecting spectra from several spots to ascertain the distribution of the anomalous element across the illumination. Finally, the acquisition of line or area scans using a mapping micro-XRF spectrometer, if available, may be helpful in probing the finely painted details found in many manuscript illuminations, or reveal the form of hidden features.

Handheld XRF spectrometers, by providing a relatively inexpensive means to obtain elemental analysis on works of art, have found great utility in the study of cultural heritage materials. As discussed in this chapter, they are particularly useful in the study of illuminated manuscripts, which generally only can be studied using non-invasive techniques such as XRF. However, perhaps the greatest advantage of using a handheld XRF spectrometer is the ease with which it may be transported, providing researchers with the ability to conduct studies on manuscripts regardless of where they are housed, which has greatly expanded the study of illuminated manuscripts and the materials that comprise them.

References

Baker, Cathleen A. 2010. *From the hand to the machine: Nineteenth-Century American Paper and Mediums: Technologies, Materials, and Conservation*. Ann Arbor, Mich.: Legacy Press.

Barrett, Timothy. 1989. "Early European Papers/Contemporary Conservation Papers: A Report on Research undertaken from Fall 1984 until Fall 1987." *The Paper Conservator* Vol. 13.

Bergmann, Uwe. 2007. "Archimedes brought to light." *Physics World* Vol. 20 (11):39-42.

Berrie, Barbara. 2007. *Artists' Pigments: A handbook of their history and characteristics*. Vol. 4. Washington: National Gallery of Art.

Bonizzoni, L., C. Colombo, S. Ferrati, M. Gargano, M. Greco, N. Ludwig, and M. Realini. 2011. "A critical analysis of the application of EDXRF spectrometry on complex stratigraphies." *X-Ray Spectrometry* Vol. 40 (4):247-253.

Bonizzoni, Letizia, Simone Caglio, Anna Galli, and Gianluca Poldi. 2010. "Comparison of three portable EDXRF spectrometers for pigment characterization." *X-Ray Spectrometry* Vol. 39 (3):233-242.

Bronk, H., S. Röhrs, A. Bjeoymikhov, N. Langhoff, J. Schmalz, R. Wedell, H.-E. Gorny, A. Herold, and U. Waldschläger. 2001. "ArtTAX - a new mobile spectrometer for energy-dispersive micro X-ray fluorescence spectrometry on art and archaeological artifacts." *Fresenius Journal of Analytical Chemistry* Vol. 371:307-316.

Brown, Michelle P. 1994. *Understanding Illuminated Manuscripts: A Guide to Technical Terms*. Los Angeles, CA: J. Paul Getty Museum.

Bruni, S., S. Caglio, V. Guglielmi, and G. Poldi. 2008. "The joined use of n.i. spectroscopic analyses - FTIR, Raman, visible reflectance spectrometry and EDXRF - To study drawings and illuminated manuscripts " *Applied Physics A: Materials Science and Processing* Vol. 92 (1):103-108.

Budnar, Milos; Ursic, Mitja; Simcic, Jure; Pelicon, Primoz; Kolar, Jana; Elih, Vid Simon; and Strlic, Matija. 2006. "Analysis of iron gall inks by PIXE." *Nuclear Instruments & Methods in Physics Research. Section B, Beam Interactions With Materials and Atoms* Vol. 243:407-416.

Burgio, L., R. J. H. Clark, and Richard Hark. 2010. "Raman microscopy and x-ray fluorescence analysis of pigments on medieval and Renaissance Italian manuscript cuttings." *PNAS* Vol. 107 (13):5726-5731.

Burgio, Lucia, Robin J. H. Clark, Richard R. Hark, Mike S. Rumsey, and Carlotta Zannini. 2009. "Spectroscopic Investigations of Bourdichon Miniatures: Masterpieces of Light and Color." *Applied Spectroscopy* Vol. 63 (6):611-620.

Cechák, Tomáa, Tomáa Trojek, Ladislav Musílek, and Hana Paulusová. 2010. "Application of X-ray fluorescence in investigations of Bohemian historical manuscripts." *Applied Radiation and Isotopes* Vol. 68 (4-5):875-878.

Cennini, Cennino d'Andrea. 1954. *The Craftsman's Handbook (Il Libro dell'Arte)*. Translated by D.V. Thompson. New York: Dover Publications.

Cesareo, R., S. Ridolfi, M. Marabelli, A. Castellano, G. Buccolieri, M. Donativi, G. E. Gigante, A. Brunetti, and M.A. Rosales Medina. 2008. "Portable Systems for Energy-Dispersive X-ray Fluorescence Analysis of Works of Art." In *Portable X-ray Fluorescence Spectrometry: Capabilities for In Situ Analysis*, edited by Philip J. Potts and Margaret West, 206-246. Cambridge: Royal Society of Chemistry.

Clark, R.J.H. 1995a. "Pigment identification on Medieval manuscripts by Raman microscopy." *Journal of Molecular Structure* Vol. 347:417-427.

———. 1995b. "Raman microscopy: Application to the identification of pigments on Medieval manuscripts." *Chem. Soc. Rev.* Vol. 24:187-196.

Clarke, Mark. 2001a. "The analysis of medieval European manuscripts." *Reviews in Conservation* Vol. 2:3-17.

———. 2001b. *The Art of All Colours: Mediaeval Recipe Books for Painters and Illuminators*. London: Archetype Publications.

Clemens, Raymond, and Timothy Graham. 2007. *Introduction to Manuscript Studies*. Ithaca: Cornell University Press.

de Hamel, Christopher. 2001. *The British Library Guide to Manuscript Illumination: History and Techniques*. Toronto: University of Toronto Press.

De Viguerie, Laurence, V. Armando Sole, and Ph. Walter. 2009. "Multilayers quantitative X-ray fluorescence analysis applied to easel paintings." *Analytical Bioanalytical Chemistry* Vol. 395:2015-2020.

De Viguerie, Laurence, Ph. Walter, Eric Laval, Bruno Mottin, and V. Armando Sole. 2010. "Revealing the sfumato technique of Leonardo da Vinci by X-ray fluorescence spectroscopy." *Angewandte Chemie* Vol. 49:6125-6128.

Dietz, Georg, Thomas Ketelsen, Manfred Hoss, Olaf Simon, Carsten Wintermann, Timo Wolff, Ira Rabin, and Oliver Hahn. 2011. "The Egmont Master phenomenon: X-ray fluorescence spectrometric and paper studies for art history research." *Analytical and Bioanalytical Chemistry*:1-11.

Dik, Joris, Koen Janssens, Geert Van Der Snickt, Luuk van der Loeff, Karen Rickers, and Marine Cotte. 2008. "Visualization of a Lost Painting by Vincent van Gogh Using Synchrotron Radiation Based X-ray Fluorescence Elemental Mapping." *Analytical Chemistry* Vol. 80 (16):6436.

Duran, A., J. Perez-Rodriguez, T. Espejo, M. Franquelo, J. Castaing, and P. Walter. 2009. "Characterization of illuminated manuscripts by laboratory-made portable XRD and micro-XRD systems." *Analytical and Bioanalytical Chemistry* Vol. 395 (7):1997-2004.

Feller, Robert. 1986. *Artists' Pigments: A handbook of their history and characteristics*. Vol. 1. Washington: National Gallery of Art.

FitzHugh, Elizabeth West. 1997. *Artists' Pigments: A handbook of their history and characteristics*. Vol. 3. Washington: National Gallery of Art.

Gettens, Rutherford J., and George L. Stout. 1966. *Painting Materials: A Short Encyclopedia*. New York: Dover.

Haberditzl, A., H. Stachelberger, G. Banik, F. Bauer, and F. Mairinger. 1984. Chemical characterization of parchment. Paper presented at the ICOM Committee for Conservation 7th Triennial Meeting, at Copenhagen.

Hahn, O. 2010. "Analyses of iron gall and carbon inks by means of x-ray fluorescence analysis: a non-destructive approach in the field of archaeometry and conservation science." *Restaurator: International Journal for the Preservation of Library and Archival Material* Vol. 31 (1):41-64.

Hahn, O., W. Malzer, B. Kanngiesser, and B. Beckhoff. 2004. "Characterization of iron-gall inks in historical manuscripts and music compositions using x-ray fluorescence spectrometry." *X-Ray Spectrometry* Vol. 33 (4):234-239.

Hunter, Dard. 1947. *Papermaking: The History and Technique of and Ancient Craft*. New York: Dover Publications, Inc.

Janssens, K., F. Adams, and A. Rindby. 2000. *Microscopic X-ray Fluorescence Analysis*. Chichester: John Wiley & Sons Ltd.

Janssens, K., G. Vittiglio, I. Deraedt, A. Aerts, B. Vekemans, L. Vincze, F. Wei, I. Deryck, O. Schalm, F. Adams, A. Rindby, A. Knöchel, A. Simionovici, and A. Snigirev. 2000. "Use of microscopic XRF for non-destructive analysis in art and archaeometry." *X-Ray Spectrometry* Vol. 29:73-91.

Jenkins, Ron. 1974. *An Introduction to X-ray Spectrometry*. London: Heyden & Son, Ltd.

Kanngießer, Birgit, Oliver Hahn, Max Wilke, Bettina Nekat, Wolfgang Malzer, and Alexei Erko. 2004. "Investigation of oxidation and migration processes of inorganic compounds in ink-corroded manuscripts." *Spectrochimica Acta Part B: Atomic Spectroscopy* Vol. 59 (10-11):1511-1516.

Kolar, Jana; Štolfa, Andrej; Strlič, Matija; Pompe, Matevž; Pihlar, Boris; Budnar, Miloš; Simčič, Jure; and Reissland, Birgit. 2006. "Historical iron gall ink containing documents: properties affecting their condition." *Analytica Chimica Acta* Vol. 555:167-174.

Kolar, Jana; Strlic, Matija; Budnar, Milos; Malesic, Jasna; Selih, Vid Simon; and Simcic, Jure. 2003. "Stabilisation of corrosive iron gall inks." *Acta chimica Slovenica* Vol. 50 (4):763-770.

Kuhn, H. 1993. "Lead-tin yellow " In *Artist's Pigments: A Handbook of their History and Characteristics*, edited by A. Roy. Washington, D.C.: National Gallery of Art.

Larsen, René. 2002. *Microanalysis of parchment*. London Archetype Press.

Longoni, A., C. Fiorini, C. Guazzoni, S. Buzzetti, M. Bellini, L. Strüder, P. Lechner, A. Bjeoymikhov, and J. Kemmer. 2005. "A novel high-resolution XRF spectrometer for elemental mapping based on a monolithic array of silicon drift detectors and on a polycapillary x-ray lens." *X-Ray Spectrometry* Vol. 34:439-445.

Malzer, W., O. Hahn, and Birgit Kanngiesser. 2004. "A fingerprint model for inhomogeneous ink-paper layer systems measured with micro-x-ray fluorescence analysis." *X-Ray Spectrometry* Vol. 33:229-233.

Manso, M., M. Costa, and M.L. Carvalho. 2007. "From papyrus to paper: Elemental characterization by X-ray fluorescence spectrometer." *Nuclear instruments & Methods in Physics Research. Section A* Vol. 580 (1):732-734.

Mantler, M., and M. Schreiner. 2000. "X-ray fluorescence spectrometry in art and archaeology." *X-Ray Spectrometry* Vol. 29 (1):3-17.

———. 2001. "X-ray analysis of objects of art and archaeology." *Journal of Radioanalytical and Nuclear Chemistry* Vol. 247 (3):635-644.

Mantler, Michael, and J. Klikovits. 2004. "Analysis of art objects and other delicate samples: Is XRF really nondestructive?" *Powder Diffraction* Vol. 19 (1):16-19.

Merrifield, Mary. 1967. Medieval and Renaissance Treatises on the Arts of Painting: Original texts and English translations. New York: Dover Publications, Inc.

Moioli, P., and C. Seccaroni. 2000. "Analysis of art objects using a portable x-ray fluorescence spectrometer." *X-ray spectrometry* Vol. 29 (1):48-52.

Namowicz, Carole, Karen Trentelman, and Christopher McGlinchey. 2009. "XRF of cultural heritage materials: round-robin IV - paint on canvas." *Powder Diffraction* Vol. 24 (2):124-129.

Oltrögge, Doris. 2011. "Byzantine recipes and book illumination." *Revista de História da Arte special issue: Medieval Colours. Between Beauty and Meaning* Vol. Série W (1):68.

Pantazis, T., J. Pantazis, A. Huber, and R. Redus. 2010. "The historical development of the thermoelectrically cooled X-ray detector and its impact on the portable and hand-held XRF industries " *X-Ray Spectrometry* Vol. 39 (February):90-97.

Porter, Cheryl. 1995. "You can't tell a pigment by its colour." In *Making the Medieval Book. Techniques of Production*, edited by Linda L. Brownrigg, 111-116. Los Altos: Anderson-Lovelace.

Reed, Ronald. 1972. *Ancient Skins, Parchments, and Leathers* London and New York: Seminar Press.

———. 1991. "Parchment -- Old and New." In *Pergament : Geschichte, Struktur, Restaurierung, Herstellung*, edited by Peter Rück. Sigmaringen J. Thorbecke.

Ricciardi, Paola, John K. Delaney, Lisha Glinsman, Mathieu Thoury, Michelle Facini, and E. Rene de la Rie. 2009. Use of visible and infrared reflectance and luminescence imaging spectroscopy to study illuminated manuscripts: pigment identification and visualization of underdrawings. Paper presented at the.

Roy, A. 1993. *Artists' Pigments: A handbook of their history and characteristics*. Vol. 2. Washington: National Gallery of Art.

Sarrazin, P., G. Chairi, and M. Gailhanou. 2009. "A portable non-invasive XRD-XRF instrument for the study of art objects." *Advances in X-ray Analysis* Vol. 52:1757-186.

Smith, A., A. Reeve, and A. Roy. 1981. "Francesco del Cossa's "Saint Vincent Ferrer." *National Gallery Technical Bulletin* Vol. 5:44.

Speelers, L. 1999. "An Early Example of the use of Mosaic Gold." *Zeitschrift für Kunsttechnologie und Konservierung* Vol. 13 (1):50-54.

Spring, Marika, Rachel Grout, and Raymond White. 2003. ""Black Earths": A study of unusual black and dark grey pigments used by artists in the sixteenth century." *National Gallery Technical Bulletin* Vol. 24:96-113.

Tertian, R., and F. Claisse. 1982. *Principles of Quantitative X-ray Fluorescence Analysis*. London: Heyden.

Thompson, Daniel V. 1956. *The Materials and Techniques of Medieval Painting*. New York: Dover.

Trentelman, Karen, Michel Bouchard, Monica Ganio, Carole Namowicz, Catherine Schmidt Patterson, and Marc Walton. 2010. "The examination of works of art using in situ XRF line and area scans." *X-Ray Spectrometry* Vol. 39 (3):159-166.

Trentelman, Karen, and Nancy Turner. 2009. "Investigation of the painting materials and techniques of the late-15th century manuscript illuminator Jean Bourdichon." *Journal of Raman Spectroscopy* Vol. 40:577-584.

Trojek, Tomáš, Tomáš Čechák, and Ladislav Musílek. 2010. "Recognition of pigment layers in illuminated manuscripts by means of $K\alpha/K\beta$ and $L\alpha/L\beta$ ratios of characteristic X-rays." *Applied Radiation and Isotopes* Vol. 68 (4-5):871-874.

Turner, N. 1998. The Recipe Collection of Johannes Acheriius and the Painting Materials use in Manuscript Illumination in France and Northern Italy, c. 1380-1420. Paper presented at the Painting Techniques: History, Materials and Studio Practice, 1998, at Dublin.

———. 2009. Illuminare: The uses and embellishment of gold and other metallic leaf and inks in medieval and Renaissance manuscript painting. In *Medieval Colours: between beauty and meaning: An interdisciplinary conference on the study of colour in medieval manuscripts*. Lisbon, Portugal.

Uhlir, K., M. Griesser, G. Buzanich, P. Wobrauschek, C. Streli, D. Wegrzynek, A. Markowicz, and E. Chinea-Cano. 2008. "Applications of a new portable (micro) XRF instrument having low-Z elements determination capability in the field of works of art." *X-Ray Spectrometry* Vol. 37:450-457.

Chapter 6

XRF analysis of historical paper in open books

Tim Barrett, Robert Shannon, Jennifer Wade and Joseph Lang

Introduction

In 2007 the Institute for Museum and Library Services funded research at the University of Iowa Center for the Book designed to gather data on previously unstudied components in 14th through 19th century western papers. The research employed non-destructive analytical techniques fairly new to the paper conservation field. Its purpose was to provide conservators and other preservation specialists with new information about paper stability, manufacture, and compositional changes during aging and conservation treatment.

During our main period of interest, between 1300 and 1800, all paper in Europe was made by hand. The raw material for writing and printing paper was old linen and hempen rags. The rags were fermented to leave them more receptive to beating with large wooden trip hammers or 'stampers.' The hammer heads were fitted with nail heads of various configurations which worked against a metal beadplate to cut the rags, and then shorten and fibrillate the fiber. Lime was sometimes used in the stamper pits, to help swell the cellulose fiber and ground chalk (calcium carbonate) may have been added to help lighten the color of the finished paper. (Drabowski and Simmons, 2003) The 'hollander' beater was developed in the low countries around 1650 to replace the stampers. Rags and water were added to an oblong tub fitted in the center with a 'midfeather' to promote circulation of the mixture. On one side of the midfeather iron bars on a spinning roll -- constructed like the paddle wheel on a steamboat-- pulled the rags beneath the roll where they were sheared against another set of bars mounted permanently in the bottom of the machine. The desired beating action was similar to that produced by the stampers: breaking the rags into threads and fibers followed by a degree of individual fiber shortening and fibrillation.

The prepared pulp was diluted in a large vat where the 'vatman' formed sheets one at a time using a rectangular, flat, sieve-like tool called a mould. The 'deckle,' an open wooden frame, was fitted to the top of the mould and served to contain a layer of the pulp as the vatman dipped the assembly into the vat and brought it up and out again in a horizontal position. When the excess water had drained through the porous surface of the mould, the vatman removed the deckle and passed the mould to the coucher, who inverted the mould and pressed it against a damp woolen blanket or felt, transferring the new sheet of paper to the felt. Another felt followed, then another sheet of paper and so on until a repeated sandwich or 'post' of 100 or more sheets had been built up. After squeezing out all possible excess water in a large screw press, the damp sheets could be separated from the felts and set aside in a pile by a third man, the layer. This 3 person team could make 1500-2000 or more sheets in a day. The paper was dried by hanging the still moist sheets over ropes stretched across upper floor lofts in the mill. By opening and closing specially designed shutters, the workers could dry paper in a wide range of weather conditions.

Later the sheets were taken down from the loft for gelatin sizing. The gelatin was prepared by heating 'various bits and pieces of 4 footed animals' in water (LaLande, 1761). The hot water dissolved the collagen, or white connective tissue in the bones, hooves, ears, and so on, and yielded gelatin in solution. This is effectively the same raw material used in making hide glue, or *Jello*™ dessert (without the flavoring and sugar). Alum was added to the gelatin to help harden it, and perhaps too, to help preserve the prepared solution. The size was applied, in theory and in practice, to leave the paper resistant to the bleeding of ink. But the gelatin sized sheet was also significantly tougher and stronger.

Additional details of 14th – 18th century papermaking are offered by Barrett (1989). For the present research, it is important to point out that some of the ingredients introduced by papermakers, such as calcium based carbonates added to the rags originally as lime or as a ground chalk white pigment, very likely had a positive impact on the long term stability of paper. Alum, on the other hand, is acidic and can have a detrimental effect on paper permanence.

Our key elements of interest were Al, K, and S for estimation of alum concentration, and Ca, Mg, Cu and Fe for data on related compound concentrations. In this project we considered the alum to be potash alum, $AlK(SO_4)_2$ prior to 1800 and aluminum sulfate, $Al_2(SO_4)_3$, sometimes called 'papermaker's alum,' after that date. Ca and Mg we associated with alkaline carbonates such as $CaCO_3$, $MgCO_3$ or dolomite $MgCa(CO_3)_2$ which are generally considered beneficial to paper stability. Cu and Fe have been associated with foxing and/or discoloration in historical papers (Tang 1978). Sources of the elements of interest could be rags, processing water, papermaking additives, or bits of metal from papermaking equipment. Our period of interest trails off at the beginning of the 19th century when the paper machine, chlorine bleach, and rosin and alum internal sizing were all introduced (Baker 2010).

The role of these and other metals or compounds in paper stability is generating increasing interest. Historically, metals in paper have been either ignored, or studied destructively using chromatography. More recently, Brunelle *et al.* (1971) used neutron activation analysis to characterize modern machine made papers by their elemental content. Tang's work on foxing employed flameless atomic absorption (1978). Shahani and Hengemihle (1986) also used atomic absorption methods to look at bleached kraft paper wash-treated with various Cu- and Fe-rich sulfate and other solutions. The papers were artificially aged and then tested for pH, fold endurance, and elemental composition. Calvini and Goriassini (2002) employed FTIR to consider the negative effects of iron and copper in paper. McCready (1999) gives an overview of conservation and paper industry literature on the role of metals in paper stability.

Quantitative data from paper artifacts [collected, for example in parts per million (ppm, mg/kg)] can still only be determined reliably by using destructive tests such as ICP-MS and -OES. In some instances, very small samples (e.g. 100mg, as used in this study, or even less) are adequate to run an analysis, but this requires some special sample-introduction equipment, and very sensitive instrumentation. In addition, small samples such as this reduce the likelihood that the sample represents the larger whole. For this reason, and for obvious preservation reasons, non-destructive analyses and tests are preferred in the conservation field. Strlič *et al.* (2006) give an excellent overview of the full array of analytical techniques available for micro-sample and non-destructive analysis of paper and ink.

Our research plan included analysis work at multiple institutions. Because of the availability of handheld XRF instrumentation and the long history of XRF use in metals analysis, we focused our efforts on XRF methodology. Analytical techniques can be considered in terms of the type of data they provide: qualitative or quantitative. Qualitative data from paper artifacts, such as XRF peak height or intensity, can be compared and used to identify artist's media, or differences in relative concentrations of artist's media between artifacts or in areas of the same artifact, and that information, in turn, can play a useful role in establishing provenance. (Noda 2002; ,Brostoff, Wade, and Klinger 2009; Brostoff, Maynor, and Speakman 2009; Sanchez and Valentinuzzi 2006).

XRF is especially effective where the density of elements of interest is high, as in the analysis of easel and mural paintings, mosaics, glasses, ceramics and bronzes. (Moioli and Seccaroni 2000; Dussubieux *et al.* 2005). Mantler and Schreiner (2001), and Janssens (2005) provide excellent overviews of the principals of XRF qualitative and quantitative analysis of multi layered materials with case studies describing analyses of paint layers, weathered medieval glass, corroded medieval coins and other artifacts.

The development and utilization of XRF quantification techniques for paper analysis is less common. Hanson (1981) surveyed elemental content in a range of historical papers using EDXRF and Manso *et al*. (2008) used the same method to investigate elemental content in modern and historical papers. XRF for analysis of elemental content in paper and inks is still being explored (Hahn *et al*. 2004, Hahn *et al*. 2005).

For our purposes, quantification (or estimation of ppm elemental content) was the only way to effectively compare over 1500 historical paper specimens we planned to analyze during the research. We decided a handheld XRF instrument and its spectrometric technique for the non-destructive analysis of paper in open books was our best approach.

XRF analysis of historical paper is made difficult by its minimal thickness, limited density, and the usually low concentrations of our elements of interest. Quantitative analysis is further complicated by the wide variation in thickness and density of papers made in Europe during our research period of interest: 1300-1800.

The transmission of X-rays is generally thought to be very good, however for lower energy X-rays the transmission is more like UV rays and can be stopped even by lightweight materials like paper. For example, based on our preliminary tests, potassium at 3.3KeV loses 72% of its signal going through a 20 lb (75 gsm) piece of photocopy paper and 90% going through two thickness of the same paper (thickness approximately 0.097 mm per sheet). So for elements like potassium, XRF reads more at the surface of the specimen, while heavier elements such as Fe are detected much more evenly throughout the thickness of the specimen. For lighter elements like Al and K, paper can get thick enough to be considered infinitely thick. In such cases as the paper gets thicker the signal does not increase. Therefore the resulting prediction in our desired ppm units would be inaccurate if the test (unknown) specimen is thicker than the thickness of the calibration specimens.

Because the calibration was based on a destructive weight percent basis assay of standards (ICP-OES ppm) and we planned to report the results in ppm, we knew the thickness and density of the specimen tested could be a factor in the calibration. A paper can be thick but not dense (blotting paper), thin but quite dense (tracing paper, glassine), thick and dense, thin but not dense, etc. In the transmission of X-rays, thickness and density together impact the results. We therefore wanted a means to ensure the weight percent results would be valid over a wide range of paper thickness/density.

We settled on a plan to create a calibration that would correct for thickness/density by using the CrBr background signal described below. In the end, the calibration standards did not vary much in thickness/density. (As a point of reference grammage in the standards averaged 81 gsm with a low of 59 and a high of 109 gsm). In the final calibration and analysis of unknowns the thickness/density data is used as a check to compensate for any specimen tested that is not in the same range as the thickness/density range in the calibration set.

Based on our experience, the use of handheld XRF instrumentation to estimate ppm elemental content in paper in open books requires the following 8 steps. This chapter proceeds by describing how we addressed each of these steps and expands on our 2009 AIC annual meeting presentation on the same subject (Barrett *et al.* 2009).

1. Select instrument and tube type.
2. Design instrument positioning device to facilitate analysis.
3. Normalize spectra.
4. Establish method to correct for the variation in paper thickness/density.
5. Select paper specimens for destructive testing and calibration, or create equivalent known samples to create secondary standards for quantitative analysis.
6. Create the calibration.
7. Estimate the overall predictive ability of the calibration by testing the calibration on known samples.
8. Acquire data; output results.

Procedures

1. Select instrument and tube type

2. Design instrument positioning device for analysis

We selected the Bruker TRACeR III-V instrument because it is used by many conservation laboratories in the US and because of its vacuum head, which minimizes the amount of air the X-rays must pass through and therefore increases its sensitivity to the lighter elements. The instrument was originally fitted with a rhodium tube to generate the excitation X-rays but we later switched to a silver tube manufactured by Moxtek because the original tube developed stability problems and failed calibration checks. The new tube was more consistent and stable at our chosen 15 kV and 25 microamp power settings while still retaining adequate sensitivity to our elements of interest. A 1 mil (25um) Ti filter was used in all subsequent work described here. The Ti filter removes low energy X-rays coming from the tube reducing background levels and hence improving the detection limits. In addition, the filter removed the L lines from both silver and rhodium tubes and at 15 kV the K lines are not fluoresced in the tube, so the target material was not critical for this calibration.

At the University of Iowa Center for the Book the TRACeR III-V was mounted upside down, in order to better position the instrument for accessory attachment and paper specimen testing (see Figures 6.1 and 6.2). A vertical slide allows graduated adjustments to accommodate artifacts of varying size. (Further plans of the positioning device and manufacturing details are provided at http://paper.lib.uiowa.edu/xrf.php.)

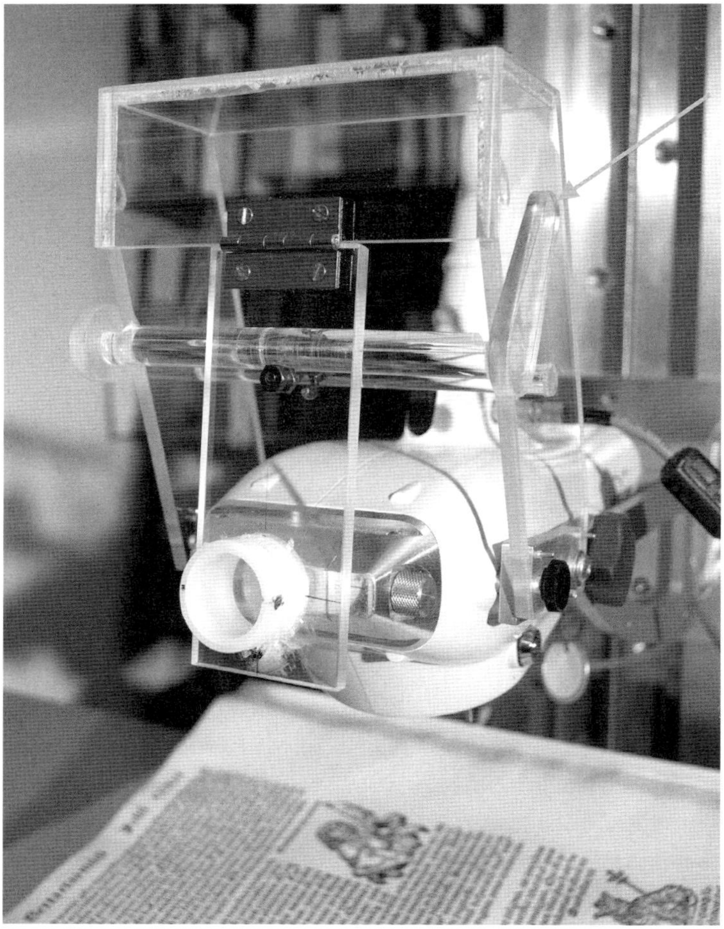

Figure 6.1:
Instrument mounted on positioner with calibration #2 resin Cr Br thin film seen within the cup as the cup is held in position on the nose of the instrument. Arrow indicates pivot point leaver.

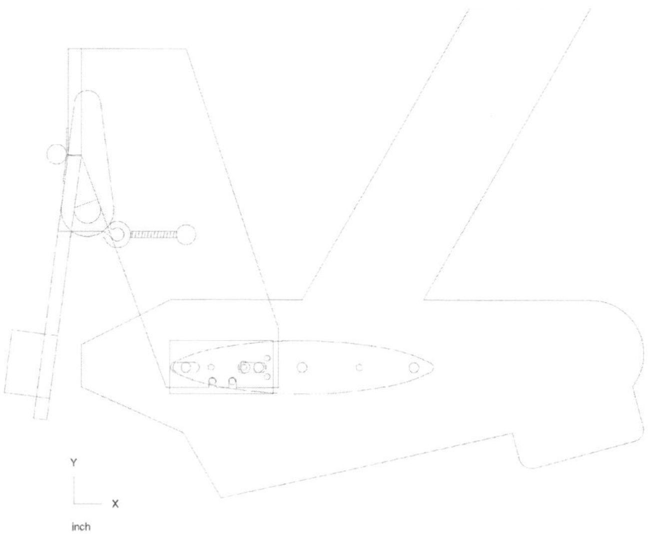

Figure 6.2:
Schematic drawing of the XRF with the acrylic accessory open.

3. Normalize spectra

Normalization is a way to compensate for sources of variation. There are a number of different sources of variation, and numerous approaches to normalization.

The simplest example is normalization by time. Time normalization allows results to be calculated from spectra collected for different scan times, and it can be accomplished simply by dividing by a time. In our calibration the scan time for the artifact signal was the same as the scan time for all other signals. We therefore normalized for scan time variations but we also did so for tube/excitation source variations. Many "normalization" methods try to correct for variations in the excitation source. The bromine artifact signal used in our normalization is a measure of the instrument's X-ray tube flux or output. The counts per second for a given XRF element line is a function of the elements concentration and matrix effects, but it is also a function of the excitation source. To control for long term variations in the tube output it is useful to normalize the spectra for the paper specimens tested relative to a known reference. We noted that the transmission of the bromine 12 KeV XRF line through paper is very high, making it a good artifact for normalizing the spectrum. (Emission line for Br Kα_1 is at 11.9242 and Kα_2 at 11.8776). We used the bromine artifact described below to remove variations due to changes in the excitation source.

4. Establish method to correct for the variation in paper thickness/density

To generate data to correct for the variation in paper thickness/density we likewise noted that the chromium 5.4KeV emission line (5.4147 for $K\alpha_1$ and 5.4055 for $K\alpha_2$) is significantly attenuated by paper, and we therefore considered it useful for generating a relative thickness/density number. In our work, we anticipated no Cr or Br compounds of interest in the historical paper specimens to be tested.

There were two calibrations used in our work and both required a Cr Br impregnated thin film artifact mounted behind the paper being tested. The first calibration was executed using the original rhodium tube. The first iteration of the thin film (film #1) employed a tissue paper impregnated with Cr and Br compounds. This tissue did not contain elements the instrument could detect prior to doping and weighed approximately 17 grams/square meter. (The tissue is the material supplied as a separator between the Proleene® XRF films described below). To create the doped tissue a solution of 5.0 ml isopropyl alcohol plus 0.100 ml of 50.9% Chromium(III) 2-ethylhexanoate in 2-ethylhexanoic acid [$Cr(C_8H_{15}O_2)_3$] [5.5% Cr] [Alfa Aesar #89304, CAS #3444-17-5] plus 0.010g of 3,3',5,5'-Tetrabromobisphenol A [$C_{15}H_{12}Br_4O_2$] [Albemarle SAYTEX_CP-2000, CAS #79-94-7] was made. Then 200ml of the solution was put on a 64mm diameter tissue sheet, allowed to dry and then placed between two Proleene® XRF films from Chemplex® Industries before being mounted onto an XRF sample cup. The resulting distribution of Br and Cr was tested and found to be slightly uneven. Therefore the special accessory described below was mounted onto the instrument to hold the XRF cup in the same position. The repeatability of results from the Br-Cr doped tissue was then tested and found acceptable. However, because of ongoing concerns about heterogeneity in distribution of the Cr Br compounds in film #1, when we later installed the X-ray tube with a silver anode, we prepared a new Cr Br thin film by using the same compounds dispersed in Elmer's All Purpose fiberglass resin. This second generation thin film (#2) was found to be much more homogenous. It too was placed between two Proleene® XRF films and again mounted in a XRF sample cup (See Figure 6.1). We estimate the final thickness of the hardened and sanded resin film inside the white cup to be less than .025mm. Concentrations of the two elements in the resin before hardening were 260 ppm (mg/kg) for the Br and 55 ppm for the Cr.

At the University of Iowa, the Physics and Astronomy Machine shop built the acrylic CrBr cup mounting accessory following our design (See Figures 6.1 and 6.2).

The cup is fixed on a spring loaded hinged arm and the acrylic rod and lever indicated by the red arrows raises and lowers the arm via cams machined into the rod ends. When the arm is lowered, a consistent light pressure keeps the paper specimen in its final position against the nose of the instrument (Figure 6.3). At the same time, the CrBr cup is kept in close contact with the back side of the specimen and in the same position from analysis to analysis (Drawings of the acrylic accessory are available at http://paper.lib.uiowa.edu/xrf.php). Our assumption of better homogeneity in the #2 film was based on rotating the cup in the accessory holder and observing fairly consistent spectra. Spectra were more irregular with the #1 tissue film. Homogeneity of the CrBr film, and lab prepared paper standards for XRF work is an area that deserves additional research attention. Billiet and Dams (1981) describe a method of producing thin film standards for XRF work using water soluble polymers. In both the case of CrBr thin films and paper standards, ICP-MS or other suitable methods must be used to quantify elemental concentrations.

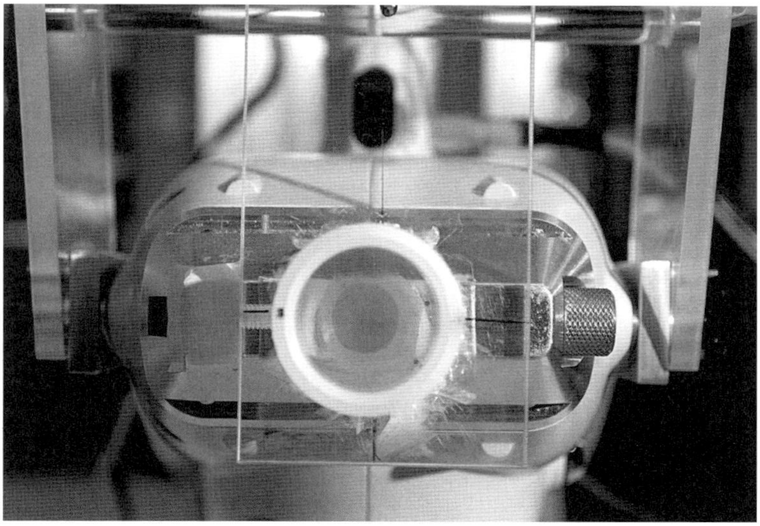

Figure 6.3:
Accessory with specimen in place and cup lowered.

Figure 6.4 shows the spectra from the Cr Br cup alone resting directly against the nose of the instrument. (Calibration #2: resin matrix for Cr Br thin film). The lines show the difference between the lowest and highest readings taken on different days after standard instrument warm ups, with multiple lifts and rests of the cup away from and back against the nose of the instrument. This constituted our 'Min Max notation.' Before any analysis of unknowns, the instrument is checked by taking a 'cup only' scan and comparing it to a library of earlier scans that include the ones shown in Figure 6.4.

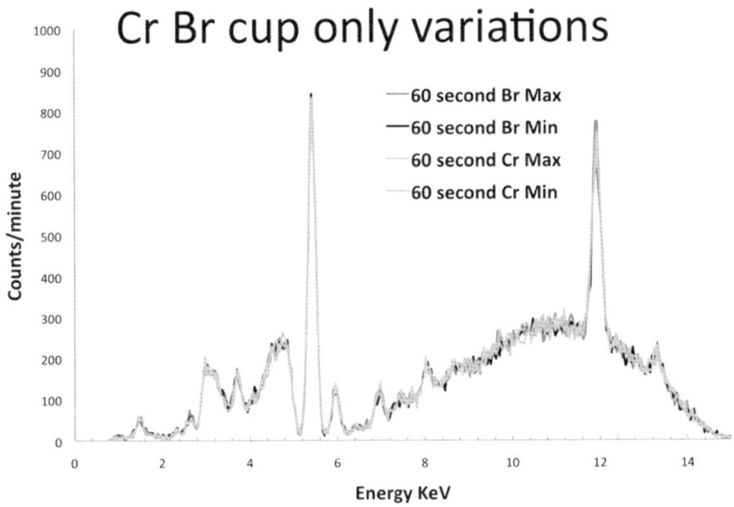

Figure 6.4:
Indication of instrument stability showing the slight variation in Br and Cr detection from the Cr Br cup upon various instrument startups and cup placements.

In summary, with regard to steps 3. and 4. above, variations in Cr and Br signal strength and backscatter in a selected region of the spectrum allowed us to normalize the spectra and generate a 'thickness/density' value that became part of the final calibration.

5. Select paper specimens for destructive testing and calibration

As part of an earlier project the 40-specimen set of historical papers shown in Figure 6.5 was available for study (Stevens *et al.*, 2008). These specimens, from books published between 1477 and 1793, served as our calibration set because they included the range of condition, age, and thickness that we considered representative of the unknown leaves the authors planned to analyze.

We also had on hand ppm data on all our elements of interest as a result of ICP-OES analyses executed at the Library of Congress Division of Preservation Research and Testing. From each specimen, samples were cut with a diamond blade from ten locations, and totaled ~ 0.1 g of sample for each specimen. The locations were adjacent to sampling sites where material was taken for prior determination of gelatin concentration, etc. Sample papers were prepared for analysis following 1:10 HF-HNO_3 digestion in Teflon screw-top vials on a constant-temperature hot plate, and resulting solutions were diluted with MilliQ H_2O to ~ 400 x the original paper weight. [Note: work with strong acids should be undertaken only by trained laboratory technicians]. Seven elements were measured in ppm in these solutions using the Thermo iCAP 6500 ICP-OES: calcium (Ca), magnesium (Mg), potassium (K), aluminum (Al), sulfur (S), copper (Cu), and iron (Fe). The wavelength used to measure the presence of each element was 370.6, 285.2, 769.8, 237.3, 180.7, 324.7, and 259.9 nm, respectively. The detector on the iCAP 6500 is a charge injection device (CID) camera detector, with continuous detection of wavelengths 166 to 847nm, and Fullframe capability. Raw counts were blank-subtracted, calibrated using synthetic solution standards, and corrected for the dilution volume. Each sample was run in triplicate, and reproducibility of replicate analyses is generally < 1% rsd.

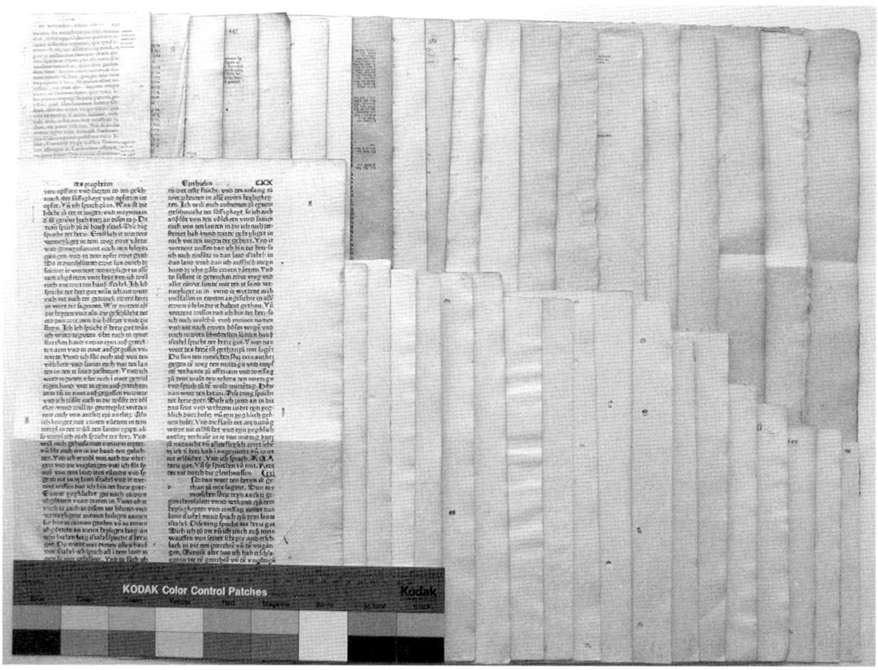

Figure 6.5:
Set of 40 historical paper specimens available for testing.

6. Create the calibration

To undertake the XRF calibration work, one 60 second XRF scan was gathered as close as possible to the each of the 10 ICP sampling locations in each specimen. In addition, 10 replicate 120 second scans at a single location on each specimen were collected to test for precision. From the full set of 40 historical specimens, four were selected to validate the accuracy of the calibration and were therefore not used in the calibration calculations. The 4 reserved specimens were selected because they represented the high and low values for most of the elements of interest.

We used an empirical calibration similar to the one proposed by Lucas-Tooth and Price (Jenkins *et al.* 1995). XRF scans near each of the 10 spots used for the ICP-OES testing were collected and then summed before being used in the calibration. This summed scan should correspond to the average result obtained through ICP-OES.

A normalization of the intensity in accordance with the Lucas-Tooth and Pyne model (Jenkins *et al.* 1995:366) was performed by dividing total raw intensities by a Br artifact intensity introduced into the spectrum as described above.

The standard Lucas-Tooth equation is;

(1) $$W_{nm} = I_{nm}(K_0 + \Sigma_x K_{nm} I_{xm}) + b_n$$

Where: W is the concentration or chemistry result, in our case the concentration in ppm, I is the intensity or normalized XRF signal, $W_{nm,}$ is the concentration and I_{nm} is the intensity or normalized XRF signal of the nth element in the mth sample, The K_{nm}'s, K_0, and b_n are the fit parameters. K_0 is the sensitivity to element n. K_{nm} are secondary fluoresces/absorption amplitudes and b_n is a background correction and Σ represents the sum of all 'x' elements.

For the case of no interferences all the K_{nm}'s are zero and the above equation reduces to a linear equation with K_0 as the slope and b_n as the y-intercept. The K_{nm}'s are sometimes referred to as slope corrections as they modify the slope of the uncorrected linear equation. In addition to the slope corrections we use constant corrections which modify b_n in the same way that the slope corrections modify K_0. Furthermore, we sometimes allow x to equal n in the above Standard Lucas-tooth equation, which makes the fit quadratic as opposed to linear. This possibility of a quadratic fit is accomplished by allowing a slope correction for the nth element in the calculation of the nth element's concentration.

The bromine artifact signal can be seen in Figure 6.4 above to be on top of the backscatter. A backscatter range of the same energy width used for Br but at an energy just below that of Br was defined. This backscatter intensity was subtracted from the signal measured in the BrKα_1 XRF peak before it was used to normalize the XRF line intensities in the Lucas-Tooth model. This backscatter range could also be used as a "correction" in the Lucas-Tooth type model. For XRF lines like FeKα_1 which are on top of the Bremsstrahlung backscatter this correction was useful.

As described above, a chromium artifact was introduced at the same time the Br artifact was introduced. CrK α_1 at 5.4 KeV is significantly attenuated by paper, and is used to estimate a relative thickness number. The XRF signal transmission through the paper is given by the Beer-Lambert law (Jenkins *et al*. 1995:15),

(2) $$I(E) = I_o e^{-\mu\varrho t}$$

Where: I= measured intensity, E= energy of the XRF line of interest. The intensity is a function of energy because μ is a function of energy, I_o= the intensity when thickness is zero or no paper was analyzed (cup only scan), *e* is the exponential function, *t* is the paper thickness, ϱ is the density, and μ is the mass attenuation coefficient.

From the Cr signal relative to (I_0), the Cr signal when only the doped artifact is present, we calculate $-\ln(I/I_0)$ where ln is the natural logarithm which is equal to ($\mu\varrho t$). Assuming the composition of the papers is similar so that mass attenuation coefficient is the same for the paper being tested then $-\ln(I/I_0)$ is proportional to a density-thickness number, and provides some information about relative thickness.

The corrections used for each of the fits are shown in Table 6.1 as well as the goodness of fit parameter, chi squared which is 1 for a perfect fit and 0.9 or more for a good fit.

	Corrections		
XRF Line	Slope	Constant	chi squared
Al Kα_1	S		0,5145582
S Kα_1			0,9272355
K Kα_1	Cr		0,9237389
Ca Kα_1			0,953417
Fe Kα_1	Fe	Ca, Cr, Bg	0,9129584

Table 6.1:
The elemental slope and constant corrections used for the X-ray lines of interest.

In general Slope Corrections are due to absorption or secondary fluorescence, while Constant Corrections are due to line overlap, where two lines that are close to each other directly contribute signal in each other's energy range. Higher chi squared values were found using additional corrections but it was determined that the results of some tests were becoming negative. These clearly low results were improved when some of the corrections originally used were taken out. Additionally we calculated a propagated variance, or estimated precision, associated with the measured peak intensity and this estimated precision was looked at as a function of the corrections used. Any corrections which significantly increased variance were not used.

Precision

The statistical probability associated with counting photons, as is done in XRF, has an intrinsic variation. This variation is governed by a Poissonian distribution (Jenkins *et al.* 1995:170) where the estimated standard deviation of counts n is $\sqrt{n}$. The propagation of this variation through equations involving Addition/ Subtraction and Multiplication/Division as in the Lucas-Tooth like model used here is well known. The propagation of this variation though equations like the Beer-Lambert law (equation 2) is more complicated. The relative thickness number as defined in the text is the inverse of the Beer-Lambert law, or $-\ln(I/I_0)$. The propagated error assuming I_0 is a constant is, $1/\sqrt{I}$, where a negative sign has been dropped. Substituting $I=I_0e^{-T}$ gives $1/\sqrt{I_o e^{-T}} = e^{\left(\frac{T}{2}\right)}/\sqrt{I_o}$, where I_0= (64.390 Counts/Second)*(the scan time used in seconds) and T is the relative thickness number. This equation or (Table 6.2) can be used to estimate one standard deviation for the relative thickness number we have calculated as a percentage of the number.

	Scan Time (sec)				
Relative thickness number	60	120	180	300	600
0,1	16,91%	11,96%	9,76%	7,56%	5,35%
0,2	8,89%	6,29%	5,13%	3,98%	2,81%
0,3	6,23%	4,41%	3,60%	2,79%	1,97%
0,4	4,91%	3,47%	2,84%	2,20%	1,55%
0,5	4,13%	2,92%	2,39%	1,85%	1,31%
0,6	3,62%	2,56%	2,09%	1,62%	1,14%
0,7	3,26%	2,31%	1,88%	1,46%	1,03%
0,8	3,00%	2,12%	1,73%	1,34%	0,95%
0,9	2,80%	1,98%	1,62%	1,25%	0,89%
1	2,65%	1,88%	1,53%	1,19%	0,84%

Table 6.2:
Indication of the estimated standard deviation based on scan time and the relative thickness number (based on the Beer-Lambert law).

Figure 6.6 shows results for iron. The dotted line fit through the diamond points represent the uncorrected fit, the solid line through the square points the corrected fit. Error bars indicate the error in XRF intensity.

The plot for Fe shows the improvement using 4 different corrections. It was not possible to sufficiently correct the calibration for Al so it was excluded from final calibration set. Mg was also too light to be included. Cu had ICP values only in the single or double digits ppm and was likewise excluded. Of the calibrations used, the Fe calibration shown above has the lowest R^2 at 0.888. The rest had R^2 values slightly over 0.92. See Figure 6.7 where the calibration for potassium had an R^2 of 0.88 before corrections and 0.924 with corrections. Ca and S did not require corrections and had R^2 values of 0.953 and 0.927 respectively.

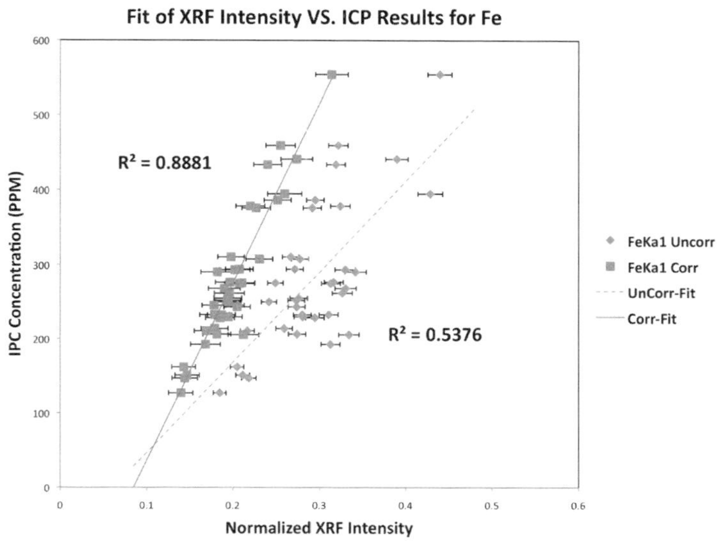

Figure 6.6:
Fit plot for iron comparing the XRF data versus ICP data.

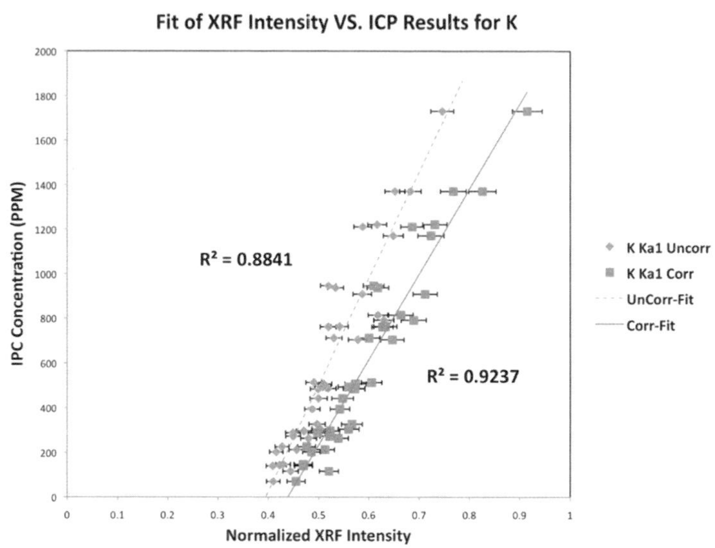

Figure 6.7:
Fit plot for potassium comparing the XRF data versus ICP data.

For the IMLS analysis of historical papers we were interested in estimating alum concentration. The loss of Al from the calibration will hopefully be offset by the stronger calibrations for K and S, the other two elements in alum during the period of interest. In the future we hope that another non-destructive procedure will be developed for estimation of Al at low concentrations in paper.

7. Estimate the overall predictive ability of the calibration

Sources of uncertainty impacting accuracy and precision in our technique include

1) Instrument error (which incorporates the precision and accuracy of both the XRF and ICP);
2) Specimen placement/geometry;
3) Specimen thickness and;
4) Calibration accuracy (from the 4 samples not used in the calibration).

The first of these uncertainties we consider to be addressed by our precision value (2 x standard deviation) while the second and third are included in the calibration.

		Sample Accuracy					**Sample Precision**			
	Sample Label	D16	D5	L19	L2	Sample Label	D16	D5	L19	L2
Al $K\alpha_1$	Conc.(ppm)	1300	1200	160	240	2STDEV	170	334	114	130
	Bias (ppm)	-258	142	100	350	Est. Prec.	147	191	75	115
S $K\alpha_1$	Conc.(ppm)	2620	3220	361	665	2STDEV	240	522	173	255
	Bias (ppm)	-120	283	-114	174	Est. Prec.	292	366	160	212
K $K\alpha_1$	Conc.(ppm)	1670	1260	282	468	2STDEV	163	261	211	180
	Bias (ppm)	-119	251	-96	48	Est. Prec.	250	234	154	182
Ca $K\alpha_1$	Conc.(ppm)	759	1140	5280	19900	2STDEV	119	109	514	796
	Bias (ppm)	-26	-169	326	-3982	Est. Prec.	134	149	411	1190
Fe $K\alpha_1$	Conc.(ppm)	188	106	291	302	2STDEV	36	42	37	26
	Bias (ppm)	19	-45	-35	15	Est. Prec.	76	76	84	100

Table 6.3:
Both the accuracy and precision of analysis are presented for Al, S, K, Ca, and Fe. The Conc. result is from ICP data and the Bias result is the difference between the XRF and ICP data. The Precision was determined by calculating 2*standard deviation from 10 measurements. The estimated precision is provided as well.

Table 6.3 documents our test of the technique using the 4 reserved paper specimens. The left part of the table shows Accuracy and the two numbers are the Conc. (ICP-OES result) and the Bias (XRF result minus the ICP-OES result). The right part of the table shows Precision data. The first number is 2 times the standard deviation as measured by 10 repeat measurements on the same spot. The second number is the estimated precision number we calculated for these 4 test samples, as we did for all results. It shows how good the assumption of the source of the precision is.

From our test using the 4 reserved specimens we feel our calibration allows us to estimate ppm concentrations of K, S, Ca, and Fe with a precision of ± 20% at a 90% confidence level when concentrations are close to 1000 ppm. Precision improves at higher concentrations, and is poorer at lower concentrations. Elemental variation must also be considered. For instance, Fe was present in the calibration specimens in a narrower range making the accuracy outside of this range poorer. The precision for S is not as good as for K due to instrument sensitivity.

To establish our limits of detection and limits of quantification from the calibration data we found the low and high concentrations for each element and then calculated a slope y-intercept and x-intercept (counts for blank). From this data we calculated a LOD using 3 sigma assuming the square-root of counts was the standard deviation. Results are shown in the table 6.4 below.

	Scan Time (seconds)				
LOD(ppm)	30	60	120	180	300
Al $K\alpha_1$	3065	2167	1533	1251	969
S $K\alpha_1$	459	324	229	187	145
K $K\alpha_1$	279	197	140	114	88
Ca $K\alpha_1$	243	172	122	99	77
Fe $K\alpha_1$	56	39	28	23	18

The LOQ is (10/3) times the LOD.
Note that our Al calibration samples had concentrations near our limit of detection.

Table 6.4:
Limit of detection of Al, S, K, Ca, and Fe based on scan time.

8. Acquire data; output results

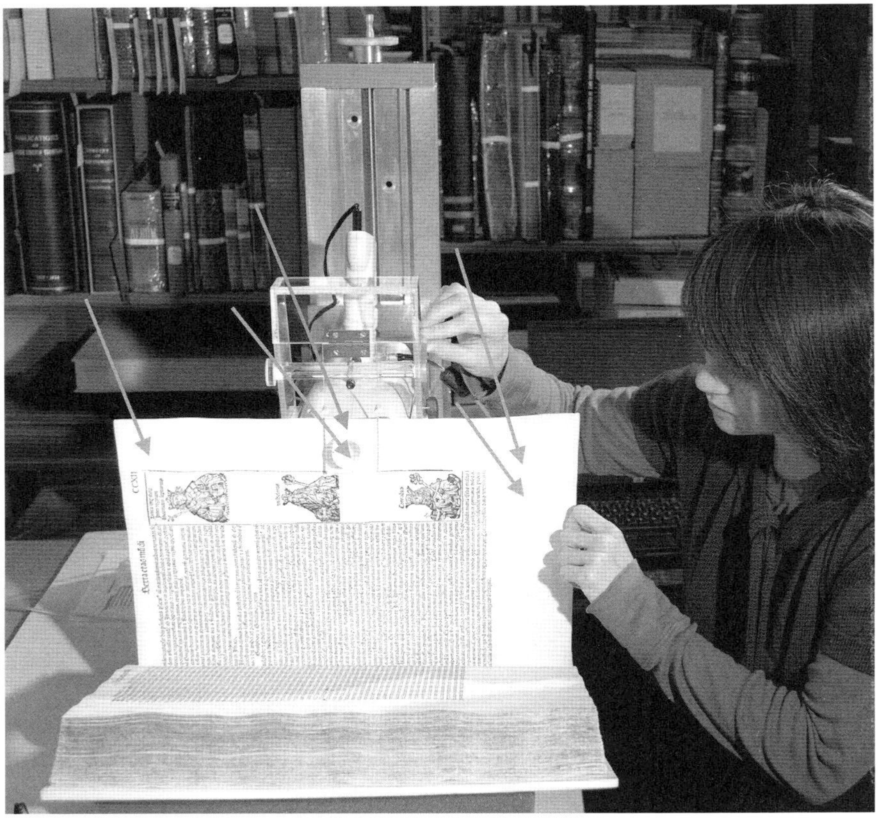

Figure 6.8:
Analysis locations on each sheet of paper.

Five locations on each specimen were analyzed (see Figure 6.8); one 120 second scan at the center of the fore edge of the leaf as close to the edge as possible (central upper arrow). That was followed by four 30 second scans in the margins: one midway between the fore edge center and the printed area, one midway between the upper right hand corner and the printed area, likewise at the lower right hand corner, and finally, one in the lower margin midway between edge and printed area as far from the lower right corner as the cup holder assembly permits– about 5 cm (remaining arrows). The 4 "interior" margin analyses were later averaged into a single "interior" value for a total of 120 seconds of analysis time which could then be compared to the single 120 second analysis at the very edge of the leaf. This is to investigate whether airborne components such as sulfur accumulate at the edge of the leaf relative to the interior of the sheet. The 4 interior measurements were done to minimize the possible lack of homogeneity in paper or the impact of random concentrations of compounds containing elements of interest such as small bits of iron from papermaking equipment. The single edge determination carried with it the chance of a skewed result due to such contamination. Averaging data from more analyses is generally better but requires more time. We decided on the 4 interior 1 edge analyses regime because of our plan to test hundreds of samples overall, and the time available to implement the project.

A version of the Bruker empirical method modified to handle the artifact normalization and thickness-density number was then used. The calibration file outputs data for all of our elements of interest in ppm values along with a precision value calculated by propagating the error in XRF intensities through the calibration. As discussed above, the final precision values are an estimate of two standard deviations.

Results

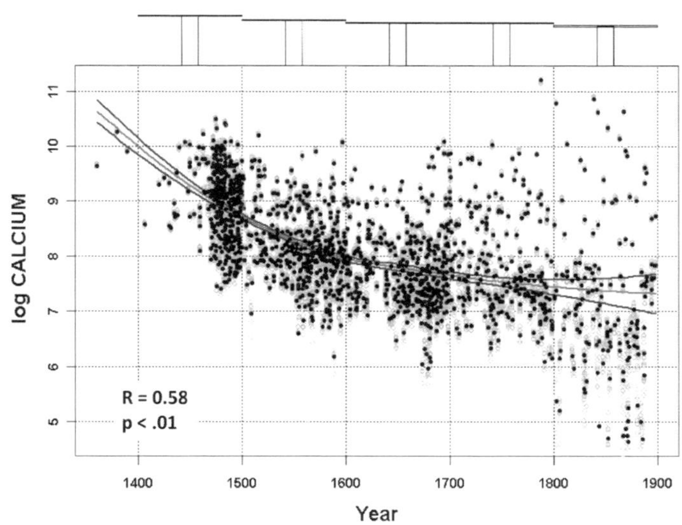

Figure 6.9:
Plot of log estimated ppm calcium versus year.

In Figure 6.9, we see the calcium data (solid black symbols) for 1578 specimens plotted over time. The vertical scale displays the natural log of calcium. The table-top display across the top of the graph depicts the mean levels across the centuries. The thickness of the table top gives the 95% confidence interval for the mean. Non-overlapping table tops imply that the difference between the means is statistically significant.

The center curve represents a locally-smoothed estimate of the mean, as a function of year. The outer curves give the point-wise 95% confidence intervals for the means over the years. We used loess (locally weighted scatterplot smoothing) in R to compute the smoothed estimate of the mean. The confidence bands were derived using Monte Carlo simulation. The confidence bands account for the measurement error as described in the next paragraph.

The black points in the scatter plot represent the observed values. These observed values were measured with some error, so if we had carried out the same measurement procedures on the same samples, we would have come up with slightly different values. As discussed earlier, precision in our work, calculated by propagating the variation in the measured intensities through the calibration equations, is equivalent to 2 standard deviations. Using these precision parameters, corresponding to each observed (solid black symbol), we generated 10 auxiliary data points (grey circles) that represent data we would expect to see if the same measurement process was repeated 10 times. The light gray points represent potential values under replicate measurements. The spread of these potential values reflects the lack of precision in the measurements. The more spread out these potential values are, the less precision.

The statistic R is a generalization of the Pearson correlation coefficient. Whereas the correlation measures the strength of the linear relationship between X and Y, the statistic R measures the strength of the functional (linear or non-linear) relationship between X and Y. If the observed (X, Y) values fall close to a smooth, non constant, function of X then R will take on a value close to 1. If X and Y are linearly related then R will be numerically identical to the absolute value of the correlation coefficient. Numerically, R is the correlation between the smoothed estimates of the Y means and the observed Y values. We used the function "loess" in R to compute the smoothed estimates of the Y means, and the p- value was computed using a non-parametric bootstrap approach. P-values less than .05 indicate the probability the results in the plot could occur by chance is very low. Thus the P-value of .01 for Figure 6.9 indicates the declining relationship between Log Ca and year is real.

Discussion

Regarding the change around 1500 in the plot for calcium in Figure 6.9, we found similar significant decreases around 1500 for gelatin concentration, paper thickness and paper color (lighter to darker; Barrett *et al.*, 2010). We believe one cause of the apparent change around 1500 in Ca concentration could be bias in the specimen sample pool. For instance, many books from the 15th century were available for testing from Italy and Germany, but very few were available from England. During later centuries we analyzed significantly more books from England and relatively fewer from Italy and Germany. Thus the trend seen in Figure 6.9 could be a comparison of papermaking in different countries as much or more than it is an indication of trends throughout Europe over the centuries. However, the other possible reason for the trend seen is that higher levels of Ca compounds were used in the process during a period when superior quality paper capable of competing with parchment was the goal. In this view, after 1500 the demand for books exploded and along with it a demand for cheaper paper which meant lower ingredient concentrations (Ca compounds and gelatin) and thinner paper overall. The details of the broad IMLS funded research project are available at http://paper.lib.uiowa.edu.

We believe the level of precision obtained using the methods we have described here is adequate for documenting trends in the concentrations of alum (potassium aluminum sulfate), and Ca and Fe compounds over the centuries when hundreds of specimens are tested. Likewise the technique may also be useful documenting the average concentrations in large collections. However, our degree of precision needs to be improved substantially before this method can be used to monitor elemental content in paper artifacts during conservation treatment. To that end, we are now analyzing data from an experiment documenting XRF and ICP-OES changes in single artifacts during aqueous treatment. We hope the precision will improve over what we obtained when analyzing widely differing unknowns because single artifacts will provide the same or a very similar paper thickness/density from analysis to analysis. Nine paper specimens (6 historical specimens and 3 controls) were subjected to seven treatments as follows:

1 Deionized water cold (tap, ca 15 C)
2 Deionized water warm 30 C
3 Calcium hydroxide, low concentration (pH 8)
4 Calcium hydroxide, low concentration (pH 8), 3 times with intermediate drying
5 Calcium hydroxide, higher concentration (pH 9.5)
6 Calcium hydrogen carbonate (also known as calcium bicarbonate)
7 Calcium phytate / calcium hydrogen carbonate

These results are available at http://paper.lib.uiowa.edu/experiments.php, Experiment A: XRF as a Technique for Monitoring Changes in the Elemental Content of Paper During Washing.

Conclusions

Handheld XRF instrumentation equipped with the proper accessories and software holds promise for the quantitative estimation of the concentration of elements of interest in historical paper specimens in open books or other works of art on paper. As with most studies, predictive power is enhanced as the number of unknowns analyzed increases. Thus surveys of large numbers of items in collections result in the most reliable data.

Despite these limitations, XRF surveys could help pinpoint why some books or drawings in a collection are browned and weak while others are not. If, for example, XRF analysis shows that the discolored papers are much higher in concentrations of elements associated with alum (K, Al, and S, or Al and S post 1800), or Fe, that information could help a conservation or collections care team decide whether or not aqueous neutralization and deposition of alkaline reserve or treatment with a chelating agent is warranted. Alternatively the same data might help the team recommend new alkaline buffered housing for the artifacts with problems, as a cheaper alternative.

More research is necessary before the method described here can be recommended for monitoring elemental changes during single item treatment. However, predictive capability may improve during treatment because single artifacts consist of essentially the same substrate (with similar thickness and density) during treatment steps. Again, this method, instrumentation, and all our results are best evaluated in the context of the broader IMLS funded research of which it was a part (Barrett *et al*. 2010).

Acknowledgements

We wish to thank the following for their support of this research: the Institute for Museum and Library Services, the University of Iowa Center for the Book, Bruker Elemental, Library of Congress, Preservation Research and Testing Division and The Kress Foundation.

References

Baker, Cathleen A. 2010. *From the hand to the machine: Nineteenth-Century American Paper and Mediums: Technologies, Materials, and Conservation*. Ann Arbor, Mich.: Legacy Press.

Barrett, T. 1989. "Early European papers, contemporary conservation papers: A report on research undertaken from fall 1984 through fall 1987." *The Paper Conservator* Vol. 13:1–108.

Barrett, T., M. Ormsby, I. Brückle, J. Lang, J. Mazurek, M. Schilling, J. Wade, and J. White. 2010. Non-destructive analysis of 14th- 19th century paper. Paper presented at the American Institute for Conservation Annual Meeting, Joint RATS and BPG session., at Milwaukee, WI.

Barrett, T., R. Shannon, J. Wade, and J. Lang. 2009. XRF analysis of paper in open books. Paper presented at the American Institute for Conservation Annual Meeting, RATS Session, at Los Angeles, CA.

Baty, John. 2011. Personal communication, 27 January.

Biliet, J., and R. Dams. 1981. "Multi-elemental thin film standards for XRF analysis." *Mikrochimica Acta* Vol. Wien (II):37-44.

Brostoff, L., J.A. Wade, and J. Klinger. 2009. Handheld XRF analysis of written material from the United States Holocaust Memorial Museum Archives. Paper presented at the American Institute for Conservation Annual Meeting, at Los Angeles, CA.

Brostoff, L.B., C.I. Maynor, and R.J. Speakman. 2009. "Preliminary Study of a Georgia O'Keeffe pastel drawing using XRF and μXRD." *Powder Diffraction* Vol. 24 (2):116–123.

Brunelle, R., W. Washington, and M. Pro. 1971. "Use of neutron activation analysis for the characterization of paper." *Journal of the Assoc. Off. Anal. Chem.* Vol. 54 (1):920-924.

Calvini, P., and A. Gorassini. 2002. "The degrading action of iron and copper on paper- a FTIR-deconvolution analysis." *Restaurator: international journal for the preservation of library and archival material* Vol. 23 (4):205-221.

Dussubieux, L., S. Eleni Pinchin, Jia-sun Tsang, and C. S. Tumosa. 2005. "Non-destructive elemental analysis: reliability of a portable x-ray fluorescence spectrometer for museum applications." In *14th triennial meeting, The Hague, 12-16 September 2005: preprints (ICOM Committee for Conservation)*, edited by Isabelle Verger, 766-773. London: James & James (Science Publishers) Ltd.

Hahn, O., B. Kanngießer, and W. Malzer. 2005. "X-ray fluorescence analysis of iron gall inks, pencils and coloured crayons." *Studies in Conservation* Vol. 50 (1):23-32.

Hahn, O., W. Malzer, B. Kanngiesser, and B. Beckhoff. 2004. "Characterization of iron-gall inks in historical manuscripts and music compositions using x-ray fluorescence spectrometry." *X-ray spectrometry* Vol. 33 (4):234-239.

Hanson, Victor F. 1981. "Determination of trace elements in paper by energy dispersive x-ray fluorescence." In *Preservation of Paper and Textiles of Historic and Artistic Value II*, edited by J. Williams, 143-168. Washington, DC: American Chemical Society.

Janssens, K. A. 2005. "Survey of the recent use of x-ray beam methods for non-destructive investigations in the cultural heritage sector." In *Cultural heritage conservation and environmental impact assessment by non-destructive testing and micro-analysis*, edited by René Van Grieken and Koen Janssens, 265-308. New York: A.A.Balkema.

Jenkins, Ron, Dale Gedcke, and Robert William Gould. 1995. *Quantitative X-ray Spectrometry*. 2nd ed, *Practical spectroscopy*. New York: M. Dekker.

Lalande, Joseph Jérôme Le Français de. 1761. *Art de faire le papier*. Paris: Académie royale des sciences, (Description des arts et métiers, vol 4). Translated by Richard MacIntyre Atkinson as *The Art of Papermaking*, Kilmurry, Sixmilebridge: The Ashling Press, 1976.

Manso, M., M. Costa, and M. Carvalho. 2008. "Comparison of elemental content on modern and ancient papers by EDXRF." *Applied Physics A Materials Science & Processing* Vol. 90 (1):43-48.

Mantler, M., and M. Schreiner. 2001. "X-ray analysis of objects of art and archeology." *Journal of Radioanalytical and Nuclear Chemistry* Vol. 247 (3):635-644.

McCrady, E. 1999. "Effect of metals on paper: A literature review." *Alkaline Paper Advocate* Vol. May 9 (1).

Moioli, P., and C. Seccaroni. 2000. "Analysis of art objects using a portable x-ray fluorescence spectrometer." *X-ray spectrometry* Vol. 29 (1):48-52.

Noda, Y., and S. Shimoyama. 2002. "Non-destructive analysis of ukiyo-e, traditional Japanese woodblock prints, using a portable x-ray fluorescence spectrometer." In *Dyes in History and Archaeology: Vol. 18: Papers Presented at the 18th Meeting, Brussels, 1999*, 73-86. London: Archetype.

Sánchez, H. J., and M. C. Valentinuzzi. 2006. "Authentication of postal pieces by spatially resolved x-ray fluorescence analysis." *X-ray spectrometry* Vol. 35 (6):379-382.

Shahani, C., and F. Hengemihle. 1986. "The influence of copper and iron on the permanence of paper." In *Historic textile and paper materials: conservation and characterization*, edited by Howard L. Needles and S. Haig Zeronian, 387-410. American Chemical Society.

Stevens, Catherine H., T. Barrett, P. Whitmore, J. Wade, J. Mazurek, and M. Schilling. 2008. "Composition and condition of naturally aged papers." *Journal of the American Institute of Conservation* Vol. 47:201-215.

Strlič, Matija., J. Kolar, V. S. Šelih, M. Budnar, J. Simčič, P. Kump, M. Nečemer, M. Marinšek, and B. Pihlar. 2006. "Chemical analysis of metals in paper and ink, in iron gall inks." In *Iron Gall Inks: On Manufacture, Characterisation, Degradation and Stabilisation*, edited by J. Kolar and M. Strlič. Ljubljana, Slovenia: National and University Library.

Tang, L. 1978. "Determination of iron and copper in 18th and 19th century books by flameless atomic absorption spectroscopy." *Journal of the American Institute for Conservation* Vol. 7 (2):19-32.

Chapter 7

Quantitative non-destructive analysis of historic silver alloys: X-ray fluorescence approaches and challenges

Jennifer Mass and Catherine Matsen

Introduction: the quantification of silver alloys at winterthur - methodologies and rationales

Winterthur Museum's Scientific Research and Analysis Laboratory has used open architecture energy-dispersive X-ray fluorescence (XRF) to study historic silver alloys since 1970. These alloy analyses are carried out to answer questions about an object's provenance, authenticity, technology of manufacture, and state of preservation. The origins of the metal and the identification of later modifications and pastiches are of particular importance for American silver alloys. The advent of handheld XRF has allowed even greater access to historic silver objects from the museum's collections and also from related collections. However, the quantitative and non-destructive analysis of historic silver alloys using XRF is challenging due to a number of factors: element segregation in copper-silver alloys; irregular surface structures with small radii of curvature (absence of ideal and reproducible geometries); limited availability of alloy standards for instrument calibration; post-manufacturing acid treatments that dissolve surface copper from the alloy (pickling); and the presence of corrosion/tarnish layers. This chapter represents an attempt to quantify these sources of error, and, where possible, to provide analytical protocols that can minimize or even eliminate them. It is hoped that this chapter can serve as a reference for obtaining meaningful quantitative data from historic silver alloys using handheld XRF. To place the analytical data presented here in context, the origins of American silver alloys are discussed, along with their relationship to contemporary European silver alloys. The microstructures of American silver alloys are also discussed, in conjunction with the effects of these microstructures on the quantitative data that can be obtained from them. Examples of objects from the New York Historical Society are used to illustrate the quantification issues and phenomena addressed.

Winterthur Museum's first open architecture XRF for silver analysis was a home-built radioisotope-based system built by Victor Hanson. This was followed by a Kevex spectrometer equipped with secondary targets, a rhodium X-ray tube, and fundamental parameters quantification software; next a laboratory-based Bruker Artax spectrometer outfitted with a molybdenum X-ray tube with a 50 μm nickel filter, empirical quantification software, and a polycapillary focusing optic was employed; then a Bruker Tracer III-V handheld system with a rhenium X-ray tube, 304.8 μm aluminum and 25.4 μm titanium filters, and empirical quantification software; and finally a Bruker Tracer III-SD rhodium tube handheld system with empirical quantification software and a 304.8 μm aluminum/25.4 μm titanium filter. Advantages and limitations have been encountered with each system, including the recently acquired handheld spectrometers. The advantages and limitations of the laboratory's current lab-based and handheld spectrometers will be discussed in particular, as they are most representative of the XRF spectrometers employed in cultural heritage research today. The unique difficulties of conducting nondestructive quantification of historic silver alloys will also be reviewed. Some of the problems encountered for generating high quality, reproducible quantitative data include: the presence of variable thickness copper-depleted/silver-enriched surface layers due to cycles of pickling (acid dipping) and polishing during and after manufacture, the presence of (often undocumented) tarnish prevention coatings such as cellulose nitrates, residues of surface gilding, microstructural heterogeneity (a particular problem when using a polycapillary focusing optic), surface or buried firescale (also known as firestain, a black copper oxide layer formed during manufacture), and layered microstructures resulting from various plating technologies (Tate 1986). The most important consideration is an inherent limitation for the nondestructive alloy analysis of silver objects - *there can be no guarantee that the surface of the object does not have a layered, silver-enriched surface structure that is not representative of the bulk alloy*. This limitation does not prevent XRF from being a useful tool for the study of historic silver, but without proper surface preparation it does limit the accuracy of the results that can be obtained. As a result, the analyst must be aware of both the problem and how it can be minimized or eliminated. A full discussion of the approaches and challenges for the quantitative XRF analysis of historic silver alloys will be followed by a brief historical overview of New World silver. This metal, so important to early American public and private life, will be examined from a number of perspectives. These include: the ore sources that supplied New World silver; the processes for winning silver from these ores (and how they affect impurity levels found in silver alloys that can be used for dating); the laws governing the degrees of fineness in Old and New World alloys, and how these laws are used to identify countries of origin, repairs, repurposed wares, pastiches, and period forms prepared from modern alloys. Specific examples of these issues, highlighted in wares from the New York Historical Society, will be discussed.

While the primary concerns of museum curators and conservators are typically the provenance, state of preservation, and technology of manufacture of collection objects, invariably during the analysis of colonial-era American silvercan, the question arises of whether or not the piece is what it purports to be (especially when analyses are conducted for potential acquisitions). The identification of pastiches and later modifications is of particular importance, as silver objects are frequently reworked and repaired over the course of their lifetimes. Common examples of this are canns being transformed into tankards by the addition of lids, tankard barrels or bases being replaced and re-marked, and tankards being given pouring spouts and then these spouts later removed to make presentation pieces.

Elemental analysis cannot be used to precisely date a silver object, but can instead be used to determine whether the alloy dates to pre- or post-electrolytic refining production (see below for a full discussion on assigning a date to this nineteenth-century process for purifying silver). Victor Hanson's XRF studies of early American silver at Winterthur in 1970 found that pieces made before electrolytic refining contain trace amounts of gold and lead (up to 1.2 wt. % lead and up to 0.1 wt. % gold (Hanson 1970:27). One can often observe the presence of these elements with qualitative analysis but, in practice, quantitative analysis is required so that a definitive statement can be made about lead and gold concentrations, and so that questions of provenance can be addressed simultaneously with questions of dating. One may gain insight into the country of origin of silver wares through quantitative XRF since many countries had specific silver fineness standards in place at different times. For the Winterthur collection the determination of imported British wares versus colonial American silver is particularly important, especially given the rarity of colonial American silver and that British wares were sometimes given American marks after import (Barquist 2001:42). Colonial silversmiths such as Myer Myers (New York, 1723-95) and Joseph Richardson, Sr. (Philadelphia, 1711-84) were both known to be regular importers and retailers of English wares (Barquist 2001). Quantitative XRF analysis of historic silver is also used to identify changes made to wares such as the insertion of false marks and acid dipping or plating to disguise repairs. Silver solders can be dated using XRF by identifying the presence or absence of cadmium, which was first introduced into these solders c. 1920. All research into alloy compositions and subsequent interpretation is done in conjunction with a conservation assessment of the piece's construction, as well as a connoisseurship assessment of the piece's salient stylistic attributes. Since old metal and/or imported metal can be (and was) melted down and reformed into new objects, it is critical that this holistic evaluation occur along with alloy analysis for a full understanding of the piece.

Part I. Quantitative handheld XRF of historic silver alloys

Pickling, Polishing, and Corrosion: Layered Surface Structures in Historic Silver Objects and their Impact on Quantitative XRF

Pickling

When working or raising a silver object by hammering, twisting, rolling and drawing, the alloy work hardens, becoming brittle as stresses are created in the grain structure due to the deformation of the crystal lattice. The alloy's crystal lattice must be relaxed to avoid stress-corrosion cracking and allow for continued working. It is therefore annealed after working, heating it to an approximate temperature of 1200°F (650°C). When heated during the annealing process, the copper present in the silver alloy oxidizes to form a black oxide scale called firescale or firestain (see Figure 7.1 Step 1).

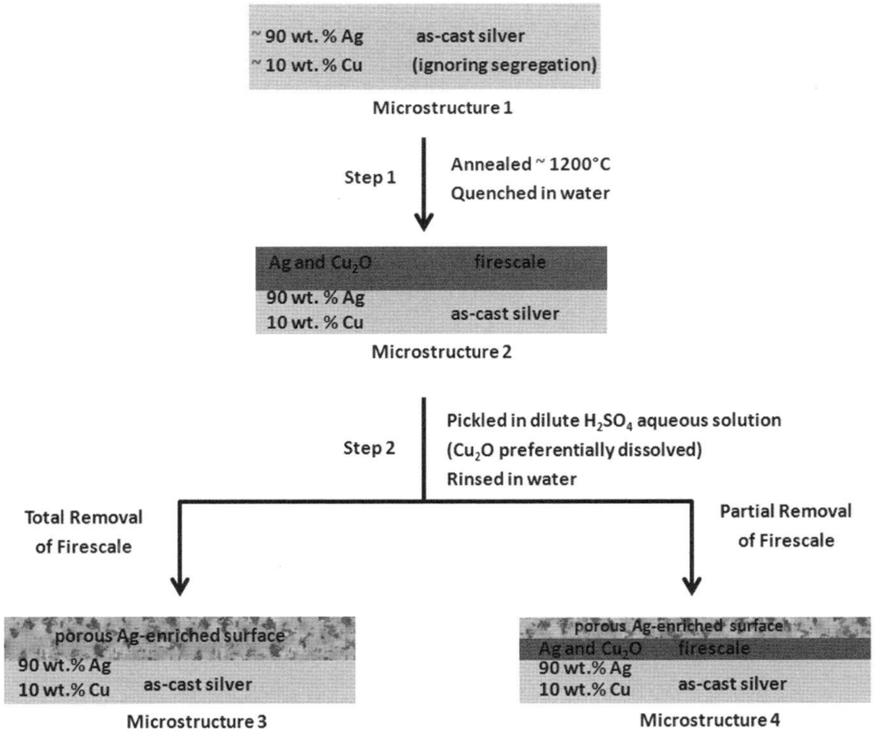

Figure 7.1:
Microstructures resulting from the casting and pickling of a 90/10 silver/copper alloy (90 wt. % silver and 10 wt. % copper bulk alloy composition).

This black copper oxide layer is removed from the surface by "pickling" the object in a dilute sulfuric acid (H_2SO_4) bath (approximately one part acid to between ten and twenty parts water) (Holden 1954:14; Bovin 1963:27). The copper oxide is solubilized and either completely or partially removed by the sulfuric acid (see Figure 7.1 Step 2), the dissolved copper ions turning the pickling bath blue. Pickling is a generic name given to the process of removing the oxide layer by means of acid treatment, and historically many different types of acids have been used for this process. Removal of the copper oxide scale produces a silver-rich surface of nearly pure silver (this surface is porous and has a whitish haze). Pickling is repeated throughout the process of raising a silver piece, and it is also performed following completion of manufacture and after any repairs (Tate 1986). After pickling the metal object is rinsed with water to remove all traces of acid, and burnished to consolidate the porous surface and to restore a bright, reflective surface. XRF analysis of the object at this point will provide the composition of the silver-enriched surface only, or, depending upon the thickness of this surface layer, a mixture of the compositions of the silver-enriched surface and the bulk.

The exact thickness of the silver-enriched surface on a given object will depend on: the initial composition of the alloy, how long the metal was heated during annealing and the annealing temperature (these will control the thickness of the firescale layer), how long the piece was dipped in acid and the composition of the acid, and any subsequent polishing that takes place. However, a layer thickness of 14 μm, observed for a c. 1771-1791 Philadelphia silver tankard, can be used as a starting point for the calculations below, and for the interpretation of the data collected on historic silver objects at Winterthur (Hanson 1970:24-27). XRF is a surface analysis technique, and the silver Lα X-ray line (2.984 keV) commonly used for analysis and quantification can escape from only the top 8.5 μm of a pure silver object (see Equation 1 below). As a result, it may provide data either from the silver-enriched surface only or from a combination of the silver-enriched surface and the bulk alloy [this emission depth is for the emission of silver Lα X-rays from pure silver, and is calculated using the following equation defining infinite thickness of an alloy (thickness at which 99% of the fluorescent X-rays can no longer escape from the surface):

(1) $$x_{(\text{emission depth})} = -\ln(I/I_o)/[(\mu\rho)]$$

where ρ is the density of silver, 10.5 g/cm^3, and μ is the X-ray mass attenuation coefficient (see Table 7.1). I_o is equal to the initial intensity of the emitted X-rays and I is the final intensity of the X-rays (intensity at the detector, in this instance I_o would equal 100 and I would equal 1, leading to a $-\ln(I/I_o)$ of 4.605). Note that for emission depth calculations the density of the alloy is approximated as the density of pure silver, whereas the actual compositions of historic American alloys will vary between 86 wt. % silver and 97 wt. % silver. Escape depths that include both the bulk alloy and the silver-enriched surface layer are even more likely for the copper, lead, and gold X-ray lines of interest (the Kα line for copper and the Lα lines for lead and gold), where we can calculate an emission depth of 20.3 μm for copper and 36.7 μm for lead and gold (see Table 7.1).

X-ray line	Energy (keV)	Mass attenuation coefficient (from NIST)* μ (cm2/g)	(density × X-ray mass attenuation coefficient) (μρ)	Calculated emission depth (μm)
Cu Kα	8,047	216,4	2272,2	20,3
Pb Lα	10,551	119,3	1252,7	36,7
Au Lα	9,988	119,3	1252,7	36,7
Ag Lα	2,984	513,6	5392,8	8,5
Ag Kα	22,163	12,6	193,2	348

* From (Hubbell and Seltzer, 2011)

Table 7.1:
X-ray line energies and approximate emission depths from a pure silver matrix relevant to silver alloy quantification.

Laboratory pickling experiments in addition to Hanson's (1970) measurements have revealed a wide range of silver-enriched surface thicknesses that depend upon the exact pickling procedures used. One study measured enriched surface thicknesses of only 2-3.5 μm for coins cleaned using commercial cleaners containing hydrochloric acid and sulphamic acid for 48 hours [to simulate incompletely cleaned surfaces produced by dipping, (Beck *et al.* 2008)]. A stable composition was reached on an 80/20 silver/copper alloy's surface after two hours of this treatment, that of almost pure silver. PIXE analysis of these silver-enriched coins initially revealed a composition between that of the bulk and the enriched surface, and only after removal of the enriched surface by mechanical abrasion did PIXE produce surface analysis data that matched the bulk. Corrosion layers also play a critical role in the depth of silver-enriched surfaces (copper can be leached from the surface during corrosion), and for archaeological copper-silver alloys studied by depth-profiling methods this silver-enriched layer can be as large as 200 μm (Tate 1986; Beck *et al.* 2004:157).

Polishing

Polishing can be defined as the action by which metal is removed from an object by an abrasive, and it is typically done to remove corrosion layers from a metal object's surface. The effect of polishing a historic silver object on its composition as measured by XRF depends on the presence of any buried firescale in its surface structure. It is important to note that often only the surface copper oxide scale is removed by pickling, and that a layer of firescale will be present beneath the silver-enriched surface. Intensive polishing of a silver object after pickling and burnishing will bring layers of firescale previously buried beneath the pickled surface closer to the surface or reveal them at the surface. If no buried firescale is present in the object, the effects of polishing are more straightforward. Polishing may result in the total removal of the silver-enriched layer to reveal the bulk alloy (see Figure 7.2 microstructure 7). Alternatively, polishing may simply reduce the thickness of the silver-enriched layer (see Figure 7.2 microstructure 8). If buried firescale is present, the polishing may expose the buried firescale (see Figure 7.2 microstructure 9 and Figure 7.3). Alternately, the reduction of the silver-enriched surface layer by polishing may leave any buried firescale intact (Figure 7.2 microstructure 10). The extent of polishing will thus determine whether, in the case of buried firescale, the resulting silver surface has a two- or three-layered microstructure in which the varying silver and copper contents present will interact with the X-ray beam during analysis (Figure 7.2 microstructures 9 and 10). If the firescale is completely removed by pickling, either a double layered microstructure with a silver-enriched surface will be revealed from polishing (Figure 7.2 microstructure 8), or the bulk alloy will be revealed (Figure 7.2 microstructure 7). Which of these four possible outcomes is achieved for a polished silver object's surface depends on the extent of polishing. As a result, the object's surface microstructure, its surface appearance, and its apparent composition as measured by XRF will change over its lifetime. Any firescale brought to the surface of an object will appear as an irregular dull grayish color (see Figure 7.3).

These multilayered microstructures make the simplified approach to determining the X-ray emission depths described above less accurate, and also increase the likelihood that matrix effects will complicate both the quantification of individual objects, and the comparison of quantitative data between objects.

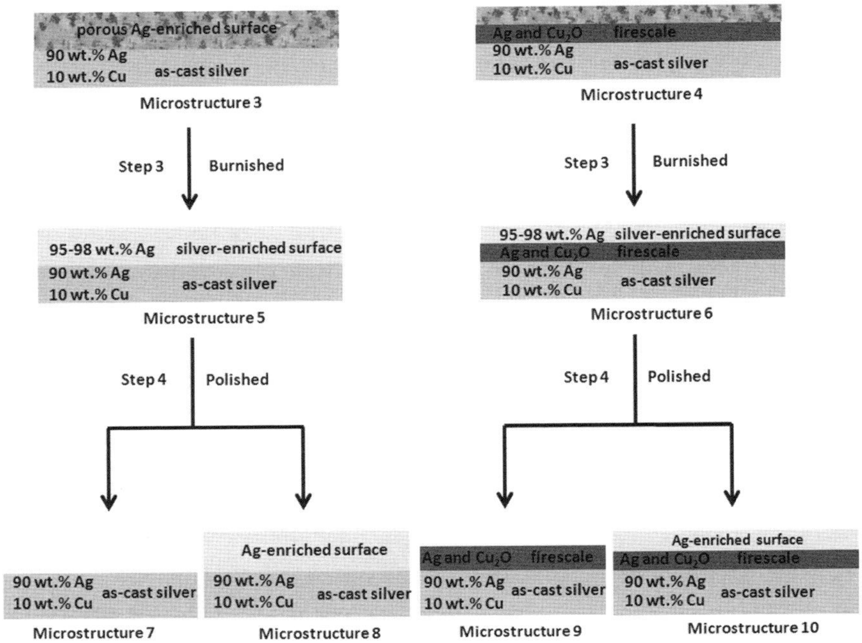

Figure 7.2:
Microstructures produced for a 90/10 silver/copper alloy following the pickling process, both after burnishing (microstructures 5 and 6) and after polishing (microstructures 7-10).

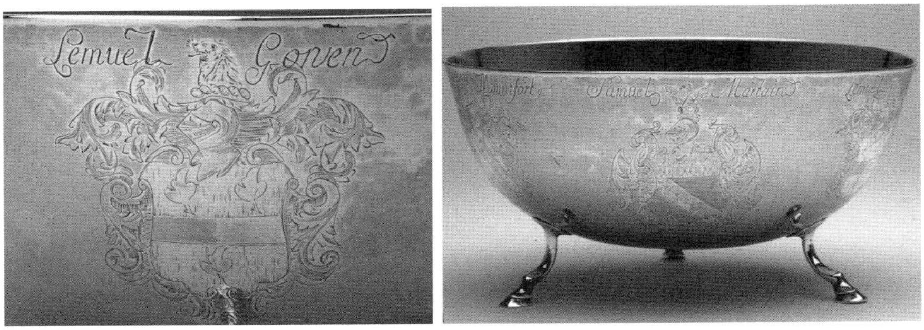

Figure 7.3:
Left: detail of firescale; Right: Firescale on the surface of a Boston footed bowl, ca. 1740-1745, John Burt. Winterthur 2004.0052.

Compositions closest to the as-pickled compositions of hollow wares can often be measured on their undersides, where measured compositions are highly enriched in silver compared to the pieces' presentation surfaces, the undersides often showing between a 0.5 to 5 wt. % increase in silver concentrations (see Table 7.2). This suggests that hollow wares are seldom polished on their undersides, which would be consistent with the tarnish patterns observed on these objects. Table 7.2 documents this trend for Colonial-era American tankards from New York, Philadelphia, Boston, and Newport.

Object	Body (wt. %)	Underside (wt. %)	Handle (wt. %)	Lid (wt. %)	Quimby catalogue number
Elias Pelletreau tankard 1760-1790 New York	91,5	95,5	90,7	91,1	237
John Coney tankard c. 1690 Boston	91,3	95,1	92,7	89,1	37
Richard Conyers tankard 1696-1708 Boston	89,1	94,2	89,9	92,6	39
Jeremiah Drummer tankard c. 1676 Boston	92,4	94,8	93,1	93	44
David Jesse tankard c. 1698 Boston	95,2	96,7	91,3	94,1	95
Paul Revere Jr., 6 tankards 1772, Boston	88,7	93,1	94,4	93,8	111
Daniel Russell, Jr. tankard ca. 1735 Newport	90,7	92,5	90,1	91,1	118
Edward Winslow Boston tankard ca. 1730	90,1	95	91,3	-	141
Cornelius Kierstede New York tankard ca. 1720	90,3	92,9	91,6	90,2	221
John David Philadelphia tankard ca. 1765	90,6	93	90,4	92,8 (replacement)	336
Johannes Nys Philadelphia tankard ca. 1720	88	89,4	87,3	88,8	418
Philip Syng, Jr. Philadelphia tankard 1741-7	93,4	94,1	93,3	93,1	472
Philip Syng, Jr. Philadelphia tankard 1750-52	92,5	93	90,8	93,3	473
Philip Syng, Jr. Philadelphia tankard 1740-1772	90,8	92,1	87,3	94,3	474
Joseph Lownes Philadelphia tankard c. 1800	93,1	93,4	93,1	92,6	389

Table 7.2:
Compositions for the undersides of silver tankards from Winterthur's collection compared to their presentation surfaces. Analysis was conducted using a Kevex model #9020 XRF spectrometer with a silver secondary target at 40 kV and 1.7 μA. Data collected by Kate Duffy and Janice H. Carlson, and compiled from American Silver at Winterthur, Ian M.G. Quimby, University Press of Virginia, 1995. Errors based on limitations of secondary standards for empirical instrument calibration (rather than counting statistics) are on the order of ± 0.1 wt. %.

This same phenomenon has also been observed in a recent case study of hollow wares from the collection of the New York Historical Society. This case study will be examined in detail after the effects of silver alloy microstructures, element segregation, secondary standard empirical calibrations, corrosion, and non-ideal geometries are considered. American silver sources and ores, silver refining methods, and their impact on alloy compositions will also be discussed to place the New York Historical Society data in context.

Silver-Copper Alloy Microstructures and Phase Diagrams: The Problem of Element Segregation

To understand the microstructural properties of silver-copper alloys and how they impact alloy quantification by XRF it is necessary to consider the silver-copper phase diagram, keeping in mind that phase diagrams are representations of the alloy system under equilibrium (a thermodynamically controlled system where the alloy atoms have sufficient time to reach their equilibrium positions), and that ancient and historic alloys are typically cast from the melt in a manner that leads to rapid cooling and situations that are kinetically rather than thermodynamically controlled (exceptions to this are bivalve casting and sand casting, which can produce microstructures close to those predicted at equilibrium conditions) (Ottaway and Yang 2004). Regardless, the phase diagram remains a useful resource for understanding ancient and historic alloy microstructures even if the alloy compositions as a function of temperature may not precisely match those predicted at equilibrium. Silver and copper are completely miscible in the liquid state with pure silver melting at 961.93°C and pure copper melting at 1084.5°C. When the two metals are alloyed, the eutectic (minimum melting point) occurs at an alloy composed of 28.1 wt. % copper in silver. In the solid state silver and copper can exist as: the solid solution α (a solution of silver atoms in copper lattice), the solid solution β (a solution of copper atoms in the silver lattice), and the eutectic mixture of finely divided α and β grains. American silver alloys typically range in composition from 5 to 15 wt. % copper with coin silver (90 wt. % silver) being the most commonly observed alloy. When a 90 wt. % silver melt is cooled during the casting process, the alloy starts to crystallize at approximately 890°C, forming dendritic crystals of β that are enriched in the lower melting metal, nearly pure silver (97 wt. % Ag, and 3 wt. % Cu, see Figure 7.4).

As the temperature decreases further, solid solutions of copper in silver continue to solidify, increasing the dimensions of the original dendrites but with compositions progressively depleted in silver as dictated by the shape of the phase diagram solidus and liquidus lines. These dendritic crystals, with composition gradients from their centers to their exteriors, are referred to as cored dendrites and are commonly observed in cast silver objects. When the temperature drops to the eutectic temperature of 779 °C the solidification completes, with the cored dendrites being surrounded by the eutectic mixture, a finely divided mixture of α and β solid solutions with compositions found at each end of the horizontal component of the solidus (see Figure 7.4). As the alloy approaches room temperature the compositions of the α and β phases will be reflected in the shape of the solvus lines, leading to an α composition of less than 0.1 wt. % silver and a β composition of 0.5 wt. % copper. However, the exact curvature of the solvus lines is not completely understood for this alloy system, and solid state diffusion of atoms from the enriched silver dendrites into the eutectic is thought to occur. In non-equilibrium conditions copper atoms trapped in the dendrites can also be responsible for precipitation of nearly pure copper at the dendrite grain boundaries. If the object is sufficiently annealed, then the cored structure of the dendrites will be lost due to diffusion. The result will be equiaxed grains of the silver-copper alloy.

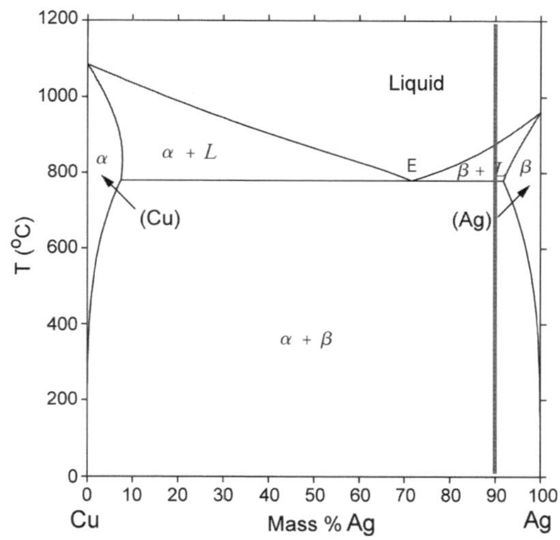

Figure 7.4:
Copper-Silver Phase Diagram. After F.H. Hayes, H.L. Lukas, G. Effenberg and G. Petzow, *Z. Metallkde.* 77 (1986) 749-754. Coin silver, the most commonly observed early American alloy, is composed of 90 wt. % silver and is noted with the vertical line.

Alternately, if the starting alloy is less than 8.8 wt. % copper, such as a sterling silver alloy with 7.5 wt. % copper, the phase diagram predicts that the alloy will be composed of predominately cored grains of β solid solution, and then as the alloy reaches room temperature small particles of α solid solution (nearly pure copper) will precipitate at the grain boundaries of the β grains. This has been observed in the metallographic analysis of a Genoese Renaissance silver basin from c. AD 1625 (Scott 1991). In the case of a copper-silver alloy such as coin silver (90 wt. % silver) being cast into a mold, one would expect the crystals to nucleate from the melt at the mold walls first, leading to segregation, an enrichment of silver, at the object's exterior. In other words, the highest temperature solid state material that can precipitate from the melt is pure silver, which should make up the majority of the chill crystals (Hanson 1970). However, silver-rich alloys of the copper-silver system, such as coin silver and sterling silver, can undergo reverse segregation, being enriched instead in copper at their exteriors (Lejcek, Kovac, and Vanickova 2010). This is particularly true in the presence of lead (or other low-melting constituents) (Watson 1932). Lead is present at between 0.1 and 1.2 wt. % in ancient and historic silver alloys (alloys prepared prior to electrolytic refining in the 19th century, see below). As a result, inverse segregation of copper-silver alloys has been reported in the metallurgical literature since the sixteenth century. An early 20th century study of this phenomenon revealed that over a 10 mm depth into a test ingot, the silver content can increase by 5 wt. % (Bauer and Deiss 1933). This phenomenon can be greatly reduced in a rapidly cooled alloy, however, and when silver alloys have greater than 72 wt. % silver normal segregation can also occur (Beck *et al.* 2004). In the use of a non-invasive surface technique such as XRF for quantitative analysis, segregation effects must be considered but are expected to be minor in comparison to surface treatments such as pickling, wear due to use, corrosion, burnishing, and polishing. *Segregation effects can predominate, however, in cast copper-silver alloy standards employed for empirical calibrations.* As a result, several mm of surface material should be removed from alloy standards by polishing prior to their use in a calibration. To determine if segregation is taking place in an alloy, quantification should be carried out using both the silver L and the silver K X-ray lines. The emission depth for the silver Kα line is much larger than the Lα line (238.4 µm versus 8.5 µm), and so if the alloy compositions obtained are different for the two quantifications, then segregation is occurring.

Raised or wrought objects formed from rolling, hammering, or forging can be expected to have a more uniform composition than those of cast objects. Objects that have undergone a substantial degree of deformation have a lower activation energy for recrystallization upon annealing, and repeated mechanical working followed by annealing will produce a finer and more uniform grain size and composition. As a result, nondestructive surface analysis methods like XRF should produce more reliable results for wrought than cast objects. This issue is particularly relevant for the laboratory-based XRF instruments with polycapillary focusing optics that are commonly used in art museums, where the beam size can be on the order of 70 µm. The grain sizes and shapes in ancient and historic silver/copper alloys will be highly variable depending upon: the alloy's solidification rate, any working and annealing, and in particular any recrystallization. Published microstructures of ancient and historic copper-silver alloys display grain sizes ranging from 250 µm to 25 µm (Scott 1991). As a result, multiple replicate measurements are required when doing quantitative analysis of alloys with microbeam instruments to account for the inherent inhomogeneity of the system being studied. The handheld spectrometers currently being used for cultural heritage applications have the advantage of having beam sizes of several millimeters, and therefore require fewer replicate measurements to obtain representative results.

Instrument calibration

The lack of primary calibration standards for copper-silver alloys has led researchers in art conservation to develop sets of secondary standards purchased from metallurgical corporations. The standards employed by Winterthur for a calibration suitable for American and European alloys of the 18th and 19th centuries are shown in Table 7.3

SILVER STANDARD	ELEMENT (wt.%)				
	Ag	Cu	Au	Pb	Zn
S-1W	92,5	7,5	—	—	—
S-2W	92,5	7,4	0,1	—	—
S-3W	92,5	7,3	0,2	—	—
S-4W	92,5	7	0,5	—	—
S-5W	90	10	—	—	—
S-6W	95	5	—	—	—
S-1H	99,06	0,49	0,5	—	—
S-2H	98,02	1,51	0,51	—	—
S-3H	94,94	2,04	0,97	0,96	1,05
S-4H	92,51	7,49	—	—	—
S-5H	92,54	7,42	0,10	—	—
S-6H	92,56	6,98	0,52	—	—
S-7H	90,03	4,01	1,98	1,86	2,16
0X	100	—	—	—	—
1X	92,5	7,5	—	—	—
2X	80,0	20,0	—	—	—
3X	99,4	—	0,2	0,3	0,1

Table 7.3:
Secondary calibration standards used to produce Winterthur's copper-silver alloy empirical XRF calibrations.

Any standards should be polished and analyzed in an iterative fashion until consistent compositions are obtained to avoid the element segregation problems discussed above. It is also important that the standard surface be flat, polished and free of any corrosion products to ensure a reproducible geometry. The standard should also be large enough to cover the spectrometer's whole incident beam, and thick enough to exceed the infinite thickness value for each element included in the calibration. Note that the silver, copper, lead, and gold composition ranges spanned by the calibration standards exceed the expected values of our collection (American silver alloy contents can range from 86.0 wt. % to 97.0 wt. %, copper contents range from 15.0 wt. % to 5.0 wt. %, lead contents range from 0.1 wt. % to 1.2 wt. %, and gold contents range from 0.0 wt. % to 1.0 wt. %). These standards provide silver concentrations within the range expected for eighteenth- and nineteenth-century silver, and also contain the minor amounts of gold and lead that would be present in pre-electrolytically refined alloys. Ideally a bulk elemental analysis method, such as ICP-MS, should also be carried out on one's secondary standards (after removal of any surface segregation layer) so that independent verification of the standards' stated compositions is available.

Table 7.3 shows the secondary silver standards used to develop our initial calibration curves and their concentrations for the Bruker Tracer III-V and the Bruker Artax systems. Based on the copper-silver phase diagram, ultimately it would be ideal to have two sets of calibrations, one with alloys containing less than 8.8 wt. % copper and one with alloys containing more than 8.8 wt. % copper. This will help to address issues related to microstructural differences in eutectic versus non-eutectic alloys. The standard concentrations are programmed into the software to the hundredth decimal place and the calculated results are also provided to the hundredth decimal place, but results are only reported to the tenth decimal place, because several of the secondary standards are known only to this lesser degree of accuracy. For empirical XRF analysis the spectrometer software plots each elemental fluorescence intensity in the standard against the corresponding composition and fits a regression to the known points. The fluorescence intensity of the unknown is then compared to the calibrated regression, and the composition is derived. Empirical models typically account for absorption, secondary fluorescence and other matrix effects using empirically derived correction coefficients based on regression analysis (Vrebos 2002).

Before analysis with either the handheld or laboratory-based instrument, the Winterthur silver standard S-2 with 92.5 wt.% silver, 7.4 wt.% copper and 0.1 wt.% gold is analyzed as a spot scan (for the handheld) or a line scan (sum over ten measurements) with the microfocus instrument. If necessary, the calibration files are then modified slightly to minimize the effects from instrumental drift and produce values closer to the reported standard values. Once acceptable values are obtained spot analyses are carried out on the object being studied. Instrumental drift has been more of a problem with the microfocus laboratory-based XRF than the handheld spectrometers, and so the calibration curve modifications are carried out on the handheld spectrometers only immediately following the completion of the initial calibration curve. The X-ray lines used for quantification are the copper K lines, the lead L lines, the gold L lines, and both the silver K and L lines. Using both the silver K and L lines in the calibration can alert one to any compositional heterogeneities in the calibration standards (such as element segregation) as described above. The reported data is from the silver L lines though, so that the emission depth values between the copper and silver lines are relatively similar.

Spectrometers and spectrometer settings

A Bruker Tracer III-V rhenium tube handheld spectrometer (with a SI-PIN diode detector) calibration was carried out using a Ti /Al 25.4 micrometer/ 304.8 micrometers filter, a 40 kV tube voltage, a 1.8 μA current, and 90 seconds collection times. Three analyses were carried out on different spots for each standard to provide a representative composition. A Bruker Tracer III-SD rhodium tube handheld spectrometer was calibrated using a Al/Ti filter (304.8 μm aluminum/25.4 μm titanium), a 40 kV tube voltage, a 9.6 μA current, and 90 secondscollection times. A Bruker Artax molybdenum microfocus laboratory-based XRF spectrometer was calibrated using 50 kV, 600 μA, 300 seconds data collection time, and a 50 micron nickel filter to reduce background and remove Bragg peaks from the spectra. Some instrument manufacturers provide a manufacturer-developed silver quantitative calibration file with their software, but the standards used are not in the composition ranges relevant to historic alloys. Therefore, in-house calibration files were created using the seventeen standards listed above.

Corrosion

It has long been stated that the presence of any corrosion layers will have a deleterious effect on attempts to perform quantitative analysis of alloy surfaces (Vittiglio *et al.* 1999; Canovaro *et al.* 2011). While silver collections in museums are sometimes lacquered to prevent tarnish or displayed in inert atmosphere cases, it has become increasingly common for museums to display objects with a thin film of incipient tarnish intact, typically resulting in interference colors ranging from golden yellow to a deep purple. To demonstrate the effect these incipient tarnish layers has on quantitative XRF analysis of these alloys, quantitative analysis was performed on a sterling silver coupon that had a silver sulfide film visible only as a golden tint to the alloy's surface (Table 7.4).

Unpolished sterling coupon			Polished sterling coupon		
Analysis Number	Wt. % Cu	Wt. % Ag	Analysis Number	Wt. % Cu	Wt. % Ag
1	7,3	92,4	1	7,3	92,4
2	7,4	92,3	2	7,2	92,6
3	7,4	92,3	3	7,3	92,4
4	7,4	92,4	4	7,2	92,5
5	7,4	92,4	5	7,3	92,5
Mean (1 SD)	7.4 (0.04)	92.4 (0.05)	Mean (1 SD)	7.2 (0.06)	92.5 (0.05)
Range	7.36-7.44	92.35-92.45	Range	7.14-7.26	92.45-92.55

*Note data is presented to a tenth of a percent to reflect the accuracy with which the calibration standards are known, but the hundredth of a percent data was used for the standard deviation calculations, which were calculated according to the following standard method:

$$\sigma=\sqrt{\frac{1}{N}\sum_{i=1}^{N}(x_i-\mu)^2}, \text{ where } \mu=\frac{1}{N}\sum_{i=1}^{N}x_i$$

Table 7.4:
Quantitative XRF results for lightly tarnished silver coupon (incipient tarnish) versus freshly polished silver coupon.* All data was collected using a Bruker TRACER III-V with a Re tube, a 304.8 μm/25.4 μm Al/Ti filter, 300 second collection times, 40 kV, and 1.2 μA. Quantification was carried out using the lead Lα line, the gold Lα line, the copper Kα line, and the silver Lα line.

In the case of this lightly tarnished coupon, the measured silver content was reproducibly 0.1 wt. % lower than that of a freshly polished silver coupon (see Table 7.4), revealing that even incipient tarnishing can affect quantitative XRF data. This data was then compared to heavily tarnished (matte brown/black) silver (tarnish caused by a lacuna in a cellulose nitrate lacquer) on the underside Winterthur collection object, a diminutive Elias Pelletreau tankard (c. 1760-90, Long Island, New York, height 11.7 cm, 1959.2514). In the case of the heavily tarnished area, there was a reproducible 0.5 wt. % decrease in the measured silver content versus an adjacent clean region. In the case of an N. and S. Richardson tankard lid (Winterthur Museum 1964.52), the decrease in the silver content of a black tarnished spot (again caused by a lacuna in a cellulose nitrate coating) after polishing to a bright surface was a reproducible 0.9 wt. %. The increased silver content observed upon polishing for the lightly tarnished silver coupon can be explained by the removal of a silver sulfide layer on the surface. However, the silver content of the heavily tarnished historic silver *decreased* reproducibly after polishing. This phenomenon can be explained by the polishing of the Pelletreau and Richardson surfaces removing part of the silver-enriched (pickled) layer, and thus the analysis reveals a composition closer to the bulk alloy. This result emphasizes the complex nature of historic silver surface structures, and reveals that further quantification experiments on heavily tarnished surfaces before and after polishing should also be carried out on modern alloy coupons.

In the case of both incipient tarnish and fully tarnished surfaces, it is evident that the measured silver content using XRF will not reflect the true composition of the bulk alloy. A 0.1 wt. % error in silver content may not be acceptable when performing analysis to determine the provenance of a piece of hollow ware, and a 0.5 – 0.9 wt. % error could easily prevent the researcher from distinguishing between an alloy that exceeds the sterling standard and one that does not. The situation with the lead and gold contents measured for tarnished versus untarnished silver is more complex. In the case of silver won before electrolytic refining, with minor amounts of gold and lead present, the measured lead and gold contents were almost totally unaffected by the presence of tarnish/polishing. For example, the Pelletreau tankard underside was found to have 0.2 wt. % Au and 0.3 wt. % lead both for the untarnished and the heavily tarnished surface. The lack of variation with surface corrosion is due to the relatively large emission depths of the Au Lα X-rays and the Pb Lα X-rays used to quantify these elements (36.7 μm emission depth from pure silver).

Reproducible geometry

Whether analyzing a tea caddy, the barrel of a tankard, a dram cup, or a chocolate pot, the majority of silver hollow wares have curved surfaces. In addition, because of the high cost of silver, silver wares tend to be smaller than their copper alloy counterparts, and as a result have surfaces defined by smaller radii of curvature. To determine the extent of the effect of surface geometry on quantitative alloy analysis using XRF, two sets of five measurements were performed on the side of the Pelletreau tankard (Winterthur Museum 1959.2514) in the same location, one in which the spectrometer head was parallel to the tankard side, and one that was misaligned by 2°. The mean silver content measured using the parallel measurement head was 91.9± 0.06 wt. %, and the mean silver content of the misaligned measurement head was 91.4 ± 0.09 wt. %. This demonstrates that 0.5 wt % error or more will be built into analyses performed on surfaces that are not flat enough to be parallel to the measurement head. Given the high degree of curvature of major components of historic silver objects such as cylindrical candlesticks, tankard handles, teapot handles, teapot spouts, and the bowls of spoons, it is important to select areas of analysis to minimize this source of error whenever possible. While it is impossible to completely avoid curved surfaces for many objects such as a tankard, for example, analyses conducted on the barrel should be more accurate than those conducted on the handle with its smaller radius of curvature. When trying to identity replacement parts on a complex object, it is particularly important to determine if the difference between, for example, a teapot wall composition and a teapot spout composition exceeds the 0.5 wt. % error that could be ascribed to geometry differences, and so multiple measurements on each component are crucial. The same principle holds true for spoons – the data collected from the handle of a spoon (typically flat) will be more accurate than the data collected from the bowl of the spoon, whether the convex or concave surface is measured. Ultimately, the loss in accuracy that comes from performing measurements on any surface other than a flat and polished surface is an unavoidable consequence of doing one's analysis nondestructively, and thus an intrinsic source of error. The best that can be done is to quantify the error using experiments such as the one described above. Replacement components on historic silver hollow wares are most easily identified by their lack of trace lead and gold contents (in the cases where the components have been fashioned from modern silver).

Part II. Historic silver sources, dating silver based on chemical composition, and case studies from the New York Historical Society

Origins of American silver and impact on alloy compositions

As critical as silver was to colonial American economic, domestic, and religious life, the raw materials of the colonial American silversmith were far from readily accessible (Ward and Ward 1979:xiii; Mulholland 198:xiii). The incredible mineral wealth of the Spanish colonies in Mexico and Bolivia made the search for similar precious metal ores an important driving factor in the establishment of the English New World colonies in Virginia and New England (Ward and Ward 1979:xiii; Mulholland 1981:3). Tales of finding rich silver and gold mines in the early settlements were common, and Gilbert's 1583 voyage to the New World optimistically included both a "Saxon assayer" and "mineral men and refiners" (Mulholland 1981:XI, 14). However, ore samples from the 1606 Jamestown settlement returned to the Old World for assaying revealed no silver and gold (ibid:20). Rumors of gold and silver deposits in the colonies continued throughout the seventeenth and eighteenth centuries, although none were revealed to be founded on truth (ibid:37). Throughout the eighteenth century the English colonists' metallurgical experience remained limited (with the exception of Cornish miners), and German miners and refiners were repeatedly brought over to identify ores, to build smelters, and participate in an ongoing transfer of knowledge to the colonists (Cronau 1916; Mulholland 1981:13; Herbert, Knapp, and Pigott 1998:174,106). While Europeans had mastered the mining, assaying, and refining of ores by the sixteenth century, with this knowledge being catalogued in such texts as Biringuccio's *Pirotechnia* and Agricola's *De Re Metallica* (c. 1556), these works were not translated into English until the nineteenth century and the early twentieth century respectively (Mulholland 1981:12).

In 1680 a lead mine was operated in Southhampton, Massachusetts with the intent of extracting silver from its ores, and in the early 1770s Alexander Henry began to study the north shore of Lake Superior to consider extracting silver from the copper ores there (ibid:53). These ventures failed to produce appreciable silver for use by the colonial American silversmith, and in fact it was not until the discovery of Nevada's great Comstock lode in Virginia City in 1859, followed by the massive native silver and silver ores at Cobalt, Ontario in the 1890s, that substantive sources of silver north of Mexico could be economically exploited. *As a result, even though silver plate and coin was critical to the economic lives of the colonists, all silver used by seventeenth- and eighteenth-century American silversmiths was imported.*

The lack of a local silver supply for the colonial American silversmith presented a substantial difficulty, but raw metal to be recycled by melting in crucibles was obtained in the form of plate brought over by the colonists from the Old World, which at times could be substantial (Falino and Ward 2008:3). The settlers who arrived in Massachusetts before 1640 were from wealthy families and arrived with their own silver. Silver also arrived in the form of gifts from England, or English purchases for either personal or ecclesiastical use (Mulholland 1981:87). Robert Sanderson began work as a silversmith in Boston in 1634, and the earliest wares from his shop likely were prepared from melted down English and Dutch wares (Mulholland 1981:77-78; Falino and Ward 2008). While this practice has been often considered the "ruthless destruction of early masterpieces", in fact the recycling of out of fashion or no longer functional metal wares has been common practice since the dawn of metallurgy (Tylecote 1976:101; Forbes 1999:15-16). Philadelphia silversmith Joseph Richardson Sr.'s day book (dating from 1744-48) documents this practice by noting when patrons would provide him with the silver necessary for the requested wares (Kauffman 1995:11). The practice is also suggested by the c. 1795 trade sign of the New York silversmith William V. Bueren, which notes the highest prices offered for "old gold and silver" (Fales 1973:228). One consequence of this practice is that a number of well documented and provenanced colonial American silver objects, bearing the appropriate American marks of colonial silversmiths, adhere to the English sterling standard of 92.5 wt. % silver

(Forbes 1999:16-18) (a standard in place since 1328, see Table 7.5) because they were prepared, often in their entirety, from imported English silver. The sterling standard remained in place in England (and for imported English wares) until 1697 when the Britannia act raised the fineness of England's plate to 95.8 wt. % silver (in an effort to prevent coins from being consigned to recycling for the production of plate) (Fales 1973:232; Forbes 1999:159-184). The Britannia act was repealed shortly thereafter, in 1719, restoring the sterling standard. However, the Britannia standard has not been completely abolished, and remains in limited use up to the present day. To correctly interpret the alloy composition of early American silver in light of the re-use of English and continental alloys, it is necessary to understand the evolution of silver alloy standards from several countries, in particular during the eighteenth and nineteenth centuries (see Table 7.5 below).

Alloy Standard Name	Location	Time of Use	Standard
Sterling	England	1328-1697 and 1719 to present	92.5 wt. % silver
Britannia	England	1697-1719 exclusively and up to the present day	95.8 wt. % silver
Coinage Act	U.S.A.	1792-1837	89.2 wt. % silver
Congressional Act	U.S.A.	1837	90 wt. %
Coin, coin silver or pure coin	U.S.A.	1825-1860s	90 wt. % silver
Sterling standard	U.S.A.	1850s onward	92.5 wt. % - no government oversight until 20th c.
First (Parisian) Standard	France	1838 onward	95 wt. % silver

Table 7.5:
Relevant silver alloy standards for eighteenth- and nineteenth-century silver made and used in America.

Other potentially relevant alloy standards in effect in the eighteenth and nineteenth centuries include the German standard – 81.25 wt.% silver (13 lötiges), the Austrian standard (Viennese) – 81.25 wt.% silver, and the Italian standard – 91.7 wt.% silver (in effect late eighteenth century) (Duffy and Carlson 2000:224-243).

A second major source of silver for colonial American silversmiths was silver coin. This source included English coins, coins from the prolific Mexican and Bolivian silver mines governed by Spain, and continental European coins (Mulholland 1981:87; Falino and Ward 2008:4). The enormous native silver and surface silver ore deposits in Mexico and Bolivia created incredible wealth for the Spanish Conquistadors, and the first Spanish-controlled silver mine was opened in Mexico just a year after Cortes's victory in 1522 (Mulholland 1981:5). After the establishment of mines in Zacatecas, Mexico and Potosí, Bolivia, shipments of silver would leave the New World for Spain annually (*ibid*). One major mechanism for this silver to reach the British colonies to the north was through privateering, and Spanish ships bearing New World silver were waylaid by English privateers such as Sir Francis Drake on numerous occasions, a practice continuing up until 1859 with the discovery of silver in Nevada (Fales 1973:229; Mulholland 1981:13).

Silver also arrived in the colonies as specie as a result of direct trade. Spanish dollars (Reales) arrived in the colonies as a result of trade with Southern Europe, and these coins were circulated widely throughout the western hemisphere (Ward and Ward 1979:57-58). Spanish dollars were in fact legal currency in the United States until 1857. The Ordinance of Medina del Campo (1497) set the fineness of the Reale at 93 wt. % silver, but subsequent regulations relaxed this standard and by the eighteenth century the fineness of the Reale had fallen below 90 wt. % (Sumner 1898; Desaulty *et al*. 2011). Spanish coins remained the dominant source of silver for American silversmiths into the late 1860s, and the 1844 Hartford Courant notes that "10,000 American or Mexican half-dollars are melted in Boston every week. New York and Philadelphia must use still more" (Venable 1994:20). The minted coins that could be found in the American colonies included coins from Mexico, Peru, Brazil, France, Holland, Spain, Portugal, and England (Fales 1973:230). These diverse coins arrived in the colonies as a result of complex and far-reaching trade routes (Fales 1973:231). That these coins were melted down to produce hollow ware is documented by George Washington's 1777 order of twelve camp cups from silversmith Edmund Milne, noting that Milne was being supplied with "16 silv.^{r} Doll.^{s}" (Fales 1973:231). Silver was also directly obtained from Mexico after the end of Spanish rule in 1821, and silversmith Sidney Gardiner is documented as having traveled to Mexico twice, once in 1823 and again in 1827, to trade small items such as sealing wax, quills, lamps, and topaz seals for silver (Fennimore and Wagner 2007:51).

To commission a silver object, a person in colonial America would save their coins and any scrap/outmoded plate (silver wares were referred to as plate in colonial English from the Spanish word for silver, *plata*) until they had sufficient weight for a piece of hollow ware, and then take these materials to the smith, who would weigh them and melt them in a small furnace (forge) (Falino and Ward 2008:4). The number of pieces and their design would be agreed upon, and the silver would then be cast into ingot or billet molds. Hollow wares would then be shaped/raised by repeated cycles of heating, hammering, and annealing to prevent work hardening. The majority of first generation colonial silversmiths apprenticed in England, Holland, or France, and so the earliest colonial wares were similar in design to common English or Dutch silver forms (Mulholland 1981:90; Falino and Ward 2008:4-5).

Silversmithing was carried out in New York prior to 1650 by Dutch smiths for the Dutch estates on the Hudson (Falino and Ward 2008:6) and by French Huguenot protestants in New York in the second half of the seventeenth century (for example, Bartholomew Le Roux who arrived in 1687)(Mulholland 1981:88). Silversmithing was carried out in Boston prior to 1652 (the date when Hull and Sanderson began to operate a Colonial mint, but as noted above Sanderson was working as a smith from the 1630s). It had begun in Maryland by 1668, in Virginia (in Norfolk, Williamsburg, Alexandria, and Fredericksburg) by the last decade of the seventeenth century and by 1707 in Charleston, South Carolina (Mulholland 1981:88). In the eighteenth century the rise of both population and affluence in the colonies meant that silver wares could now be produced in some quantity (Falino and Ward 2008:6-7). The wares produced included ecclesiastical wares, presentation pieces, and table wares such as flatwares and hollow wares. Specific items included porringers, dram cups, beakers, chalices, candlesticks, spoons, bowls, snuffers, salt cellars, and tankards. Not only were such wares a reflection of the owner's sophistication and affluence, but these wares, especially when marked with the maker's mark and the owner's initials or a cipher, were a safer method of storing household silver than as coin, which, if stolen, could not provide proof of ownership. Silversmiths also offered small items such as pins, brooches, mourning rings, chains, combs, buckles, and Indian trade goods (items for political exchange such as peace medals or in exchange for animal pelts), as well as stock directly imported from England. Ecclesiastical wares included flagons, chalices, cups, and patens (Kauffman 1995:17).

The early American silversmith's crucible, then, contained melted down and recycled English wares, Spanish coins, English coins, and continental coins, and could be expected to have an overall composition varying from the purity of coin silver (approximately 90 wt. % silver but occasionally as low as 85 wt. % silver), up to, depending on the time of production, the English Britannia standard (95.8 wt. % silver) or the French 1838 standard of 95 wt. % silver. That alloys below the sterling standard were in use and that the sterling standard was both acknowledged and desired is demonstrated by the following notice from the May 2nd 1743 South Carolina Gazette "...whereas 'tis complained that the Silver which is worked up here is not true Sterling Standard, this is to acquaint all Gentlemen and other, that the said Wright will warrant it to be true Sterling Standard, he will finish his work with the utmost Dispatch" (Kauffman 1995:30). The Americans did not adopt the sterling standard after the revolution, but rather in 1792 Congress passed the Coinage Act, which set a silver standard at 89.2 wt. % silver (see Table 7.5), the alloy being defined as 1485 parts pure silver to 179 parts copper (this standard lasted until 1837). There was no guild in the colonies or during the federal period (c. 1790-1828) to govern the fineness of silver employed by American silversmiths, although for a short time in Baltimore an assay office did exist (1814-1820) (Kauffman 1995:30), and after 1837 a higher purity standard of 90 wt. % silver was adopted (Waters 1977). Starting in the second quarter of the nineteenth century, silver objects marked "coin", "coin silver", or "pure coin", were typically 90 wt. % silver or less. Specie continued to be the dominant source of silver into the 1850s, at which point the largest makers of plate changed from "coin" silver to the sterling alloy (Tiffany and Co. in 1851 and Bailey and Co. of Philadelphia in 1855) to improve their ability to compete with wares imported from England.

New world silver sources, and how trace metals left from silver refining methods are used to date silver

Unlike the agentiferous galena silver ores that are so prevalent in southern Europe, the silver deposits of Bolivia, Mexico, and the American Southwest are predominately comprised of proustite (Ag_3AsS_3), a red silver ore containing approximately 65.4% silver; argentite (Ag_2S), a black silver sulfide ore containing approximately 87% silver, and horn silver (AgCl), a colorless silver chloride (Ward and Ward 1979:50). Polybasite, ($Ag_{16}Sb_2S_{11}$), containing approximately 75% Ag, was first recognized in Nevada in 1859 and was a common silver ore in this locale. Argentiferous lead carbonate ores were identified in 1874 in Leadville, Colorado (*ibid*:51), and they have become the dominant source of silver in the American west (*ibid*:52).

Silver has been separated from its ores since antiquity using an oxidation process known as cupellation. This method of refining argentiferous lead ores starts with the lead-silver alloy formed from the smelted ore (typically argentiferous galena, PbS). This alloy is placed on a porous hearth liner, or cupel, typically fashioned from powdered bone ash, and then heated under an oxidizing atmosphere to between 730 and 1100°C (using a forced air furnace)(Tylecote 1976:38,79-80). These conditions cause the lead to be oxidized to lead oxide, predominately litharge (PbO), a yellow compound that is largely absorbed by the cupel (as shown in Agricola's *De Re Metallica* and in Biringuccio's *Pirotechnia*). The silver and any gold present in the ore remain as a "button" on the surface of the cupel (Agricola 1556). This process is not 100% effective, leaving trace amounts of gold and lead in the silver (along with copper and bismuth) that can be detected by XRF (Hanson 1973; Tylecote 1976:123-124; Gale and Stos-Gale 1981; Rehren, Hauptmann, and Muhly 1998; Weber *et al.* 2000). This ancient technique was modified in the beginning of the eighteenth century (1705 at the London Lead Company) when a coal-fired reverberatory furnace replaced the wood fuel for cupellation, and after the 1760s when fern ash, pearl ash, and bone ash mixtures were used for the cupel. By the 1830s the bellows traditionally used had been replaced by blowing cylinders (Day and Tylecote 1991:119).

The Pattinson process of refining silver was developed in 1829 (*ibid*:120-1, 96) and introduced in 1833, after which cupellation continued to be used only for assaying and final, post-Pattinson refining. The Pattinson process is a crystallization process that works by exploiting the different melting temperatures of lead and silver (327.5°C and 961.78°C respectively). A central iron pot in a row of eight or nine pots heated by coal fires is charged with argentiferous lead alloy smelted from the ore to be refined (*ibid*:121). As the alloy cools a lead-enriched phase crystallizes first, and these crystals are skimmed off and moved to an adjacent pot. The remaining silver-enriched melt is moved in the opposite direction, and the process is repeated until the pots on either end contain (relatively) pure lead and silver. *Like cupellation, minor amounts of lead, typically between 0.1 and 1.2 wt. %, are retained in the silver and these are detectible by XRF.* By 1850-1852 the Pattinson process was replaced by the even more efficient and economical Parkes process (Tylecote 1976:121), which involved adding zinc metal to form a silver-rich alloy that floated to the surface of the melt and could then be removed by skimming. Silver could then be separated from the zinc by volatilization. *It is important to note that cupellation, the Pattinson process, and the Parkes process were all not able to fully separate gold from silver because of their close chemical properties, and so each of these refining techniques leaves a minor amount of gold in the silver alloy, typically between 0.01 and 0.1 wt %, which can, like the lead, be identified using XRF.*

The amalgamation process, commonly used to extract silver from its ores in Mexico from 1557 to the end of the nineteenth century, involved, like cupellation, relying on the nobility of silver to isolate it from more reactive metals (Ward and Ward 1979:4-5). This process involved crushing the ore and then grinding it with mercury, salt, and copper sulfate. The gold and silver is amalgamated with the mercury, and the amalgam collected into a canvas bag. The bag is then squeezed to remove excess mercury as liquid through the fabric, and a dark paste remains. This paste is then heated to liberate the mercury as a vapor, leaving any gold and silver behind (*ibid*:5). Like cupellation, the Parkes process, and the Pattinson process, trace amounts of gold and lead identifiable by XRF will be contained in silver alloys prepared from silver isolated by amalgamation.

Electrolytic refining of silver ores and dating silver alloys

The electrical industry of the mid-1800s drove the demand for higher purity metals – specifically undoped copper for electric wire – than could be obtained through the cupellation, Pattinson, or Parkes processes. The first commercial application of electrolysis to copper refining was granted in an 1865 patent (Gore 1887:1894). The invention of the dynamo, a powerful electrical generator, in 1869 made commercial electrolytic refining of metals possible on an even larger scale (Tylecote 1976:151).

The electrolytic refining of silver is a secondary process to the recovery of associated metals such as copper, lead, or zinc. Through the electrolytic refining of 97-99% "blister copper" produced by smelting, impurities such as silver, gold, and platinum accumulate at the bottom of the refining tank. These impurities are roasted to produce oxides of all metals except silver, gold and platinum. The silver, gold and platinum metals are then recovered and cast as an ingot to be used as the anode in the electrolytic refining of one of these particular metals.

The multiple advances and improvements in the electricity-producing machines that made large-scale metal refining possible in the mid-1800s mean that there is no single year to delineate between the pre- and post-electrolytic refinement of metals. Furthermore, patents were usually granted years after an invention was made and in use. The March 1903 publication of *Mines and Minerals* reports the first electrolytic silver refinery in the United States was erected in 1886 by the Pennsylvania Lead Company near Pittsburg, PA (Mines and Minerals, March 1903:353). However, the majority of electrolytic refineries in the United States were built in the early 1890s in the Perth Amboy, NJ and Long Island, NY area of the Greater New York harbor region. As previously noted , silver wares made before electrolytic refining reproducibly contain trace amounts of lead and gold, between 0.1 and 1.2 wt. % lead and 0.01 and 0.1 wt. % gold (Hanson 1970:27; Hanson 1973). Therefore, while it is commonly stated that "clean" silver alloys lacking these trace amounts were made after the mid-nineteenth century, the conservative conclusion is that American silver alloys free of lead and gold impurities date after 1890 (although it is certainly possible that such alloys could date to the second half of the nineteenth century). Any wares prepared from recycled silver that had been won prior to electrolytic refining would, of course, have the trace impurities expected for an earlier ware.

Eighteenth-century hollow wares alloy analysis case study: New York Historical Society silver

Table 7.6 illustrates the phenomena of trace amounts of gold and lead present in eighteenth-century hollow wares from the collection of the New York Historical Society: a wrought silver caster attributed to Abraham Poutreau, ca. 1730, New York; a silver porringer attributed to Peter Van Dyk, ca. 1705-1724, New York; a silver tankard attributed to Stephen Curtis, ca. 1730, Bristol, England ; and a silver teapot attributed to Van Voorhis and Schanck, 1791-1793, New York (See Figure 7.5).

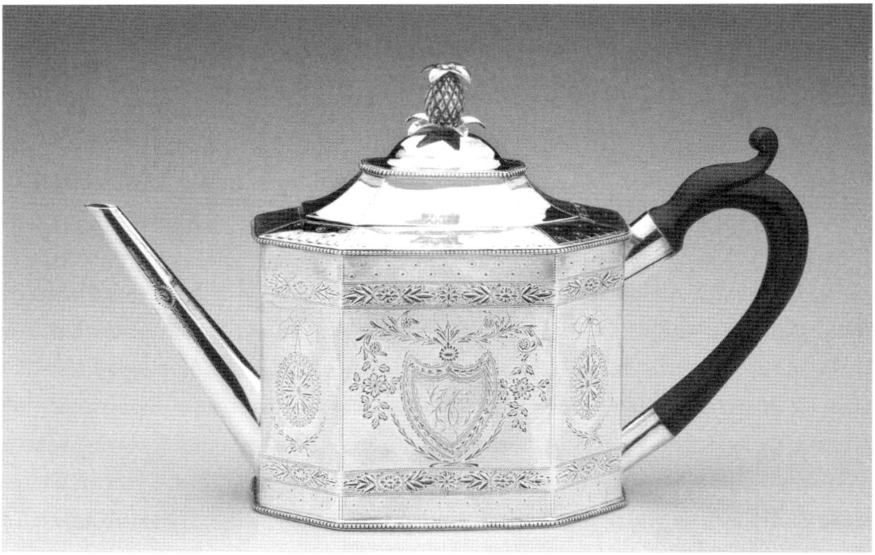

Figure 7.5:
New York Historical Society 1950.357, a silver teapot attributed to Van Voorhis and Schanck, 1791-1793, New York. Photo courtesy of the New York Historical Society.

The Poutreau caster was analyzed because the finial appeared uncharacteristically plain and the cap showed evidence of possible repair, suggesting that perhaps the entire cap was a replacement. The data in Table 7.6 however, suggest that the cap is not a later replacement, even a replacement prepared from period silver, based upon the similarity of the cap and body alloy compositions including trace components. The Van Dyk porringer was analyzed to determine whether its handle was a replacement, since several pierces were missing for this handle design, and the solder line adhering the handle was exaggerated and sloppy, suggested a possible repair. The high lead content of the handle suggests, but does not prove, that although the handle was prepared with a period alloy, it may not be contemporary with the construction of the bowl.

The Curtis tankard was analyzed to determine if the spout was a later addition, and also possibly the lid and handle, which are somewhat out of proportion to the rest of the piece. The tankard body conforms to the English sterling standard, with an alloy content of 92.9 wt. % silver. Likewise, the lid and the foot conform to the sterling standard, with alloy contents of 92.9 wt. % silver and 95.9 wt. % respectively. However, the spout and the handle both appear to be coin silver additions, with average silver compositions of 91.4 wt. % and 89.4 wt. % respectively. The dimensions of the lid with respect to the body suggest that it may not be original to the piece. This may be the case (the lid could be an English replacement) but the alloy analysis did not significantly distinguish between the two pieces. The anomalous zinc contents of the handle suggest that it may not contemporary with the rest of the piece, but all of the components predate electrolytic refining.

The Van Voorhis and Schanck teapot (see Figure 7.5) was analyzed to determine if it had been prepared from English or American silver, since it appeared as if the underside contained traces of English hallmarks that had been removed by abrasion. The sterling silver alloy composition of the teapot is consistent overall with an English origin for the piece, suggesting that Van Voorhis and Schanck did remove English marks and replace them with their own before resale. Colonial American silversmiths such as Myer Myers are known to have engaged in the removal of marks from imported English silver followed by the addition of their own marks prior to resale (Barquist 2001:42). While the 92.4 wt. % silver measured for the lid is 0.1 wt. % below the sterling standard, this anomaly is not sufficient to call into question an English origin overall, particularly considering that an ideal measurement geometry could not be achieved for this component. The underside of the teapot has a composition of 95.6 wt. %, which is indicative of a pickled surface rather than the bulk alloy composition. Copper-depleted pickled surface compositions are observed on the undersides of hollow wares only because polishing has removed (or reduced) this thin layer from the other surfaces. The alloy of the screw in the finial reveals a post-1850 manufacture, with only 0.1 wt. % lead and no gold. In addition, the alloy of the finial leaf does not meet the sterling standard, with only 91.0 wt % silver. This suggests that the finial is not original to the piece, and was perhaps added after the piece was imported to the colonies. The finial screw may represent a modern replacement or repair. Daniel Van Voorhis spoon alloys have been analyzed and found to have the more variable alloy compositions characteristic of early American silver, ranging from 88.2 to 92.5 wt. % silver (Quimby 1995:315).

New York Historical Society Accession Number / Object Description	Wt. % Silver	Wt. % Copper	Wt. % Gold	Wt. % Lead	Wt. % Zinc
1967.44ab Wrought silver caster attributed to Abraham Poutreau, ca. 1730, New York					
Tip of finial	91.2±1.3	8.1±0.2	0.2±0.03	0.4±0.04	0.3±0.04
Side of finial	90.7±1.2	8.5±0.2	0.1±0.02	0.4±0.04	0.3±0.04
Pierced cap	91.7±1.3	7.9±0.2	0.2±0.03	0.3±0.03	n.d.*
Body above contraflexure	90.0±1.2	9.5±0.2	0.2±0.03	0.3±0.03	n.d.
Body below contraflexure	92.1±1.2	7.4±0.2	0.2±0.03	0.3±0.03	n.d.
Underside of foot	94.5±1.5	4.8±0.2	0.2±0.03	0.4±0.04	0.2±0.04
1962.35 Silver porringer attributed to Peter Van Dyke, ca. 1705-1724, New York					
Underside of bowl	92.0±1.2	6.1±0.2	0.1±0.02	0.3±0.04	0.6±0.06
Handle	92.6±1.2	6.3±0.2	0.1±0.02	0.8±0.05	0.1±0.02
1922.100 Silver tankard attributed to Stephen Curtis, ca. 1730, Bristol, England					
Body	92.9±1.2	6.8±0.2	0.1±0.02	0.2±0.02	0.1±0.03
Underside of foot	95.9±1.3	6.4±0.2	0.1±0.02	0.1±0.02	n.d.
Lid	92.9±1.3	6.4±0.2	0.1±0.03	0.3±0.03	0.4±0.04
Spout	91.4±1.2	8.1±0.2	0.1±0.03	0.3±0.03	0.1±0.03
Handle	89.4±1.2	5.7±0.2	0.2±0.03	0.5±0.04	4.3±0.1
1950.357, a silver teapot attributed to Van Voorhis and Schanck, 1791-1793, New York					
Side	93.9±1.2	5.3±0.1	0.1±0.02	0.2±0.02	0.5±0.04
Spout	93.3±1.2	5.7±0.1	0.1±0.02	0.3±0.03	0.6±0.05
Top of finial	92.2±1.2	7.7±0.2	n.d.	0.1±0.02	n.d.
Finial leaf	91.0±1.2	7.7±0.2	0.2±0.03	0.3±0.03	0.8±0.05
Underside of bottom	95.6±1.3	3.6±0.1	0.1±0.02	0.3±0.03	0.4±0.04
Lid	92.4±1.2	6.3±0.1	0.1±0.02	0.4±0.04	0.8±0.05

*not detected

Table 7.6:
XRF Analysis of New York Historical Society hollow wares, Bruker Keymaster III-V (40 kV, 1.2 μA, 180 seconds collection time, calibration file Winterthur Ag Cal Rev 1, 304.8 μm/25.4 μm Al/Ti filter).

The compositions measured for the presentation surfaces of these wares reflect the four different microstructures outlined in Figure 7.2 - the bulk composition may be represented (if the entire pickled surface has been polished away) or a composition intermediate between the bulk and the pickled surface (if remnants of the silver-enriched pickled surface layer remain). The inclusion of buried or surface firescale in the interaction volume of the X-ray beam will further complicate the interpretation of the resulting data. *The presence of pickled surface layers and buried firescale reveal an inherent limitation of nondestructive alloy analysis for silver objects; there can be no guarantee that the surface of a historic silver object does not have a layered, silver-enriched surface structure*. There is also, because of uneven polishing, no guarantee that the thickness of any silver-enriched surface layer is uniform throughout the object. This problem could be addressed by performing destructive sampling allowing for bulk study using metallography and SEM-EDS, for example. However most museums, including Winterthur, have policies prohibiting destructive sampling of their silver collections. The current protocol to obtain the most representative and accurate data sets on hollow wares then, is to conduct multiple replicate analyses on each component of the ware not only to account for any inherent inhomogeneity of the alloy (due to element segregation), but also to identify any areas of the ware in which silver-enriched regions remain. Regions that tend to receive more extensive polishing due to handling and so have more homogeneous surfaces include the sides and lids of hollow wares. There are also some areas that get polished rarely if at all, including crevices or engraving. Such areas could be studied using microbeam XRF to obtain information about the alloy's as-pickled surface, but access to many of these areas is severely restricted by geometry, and the data that would be obtained from them is challenging to interpret also due to geometry considerations. A protocol for obtaining information about the bulk alloy would be to conduct analysis of an unobtrusive area, subject the area to brief polishing with a slurry of precipitated chalk in an ethanol/water mixture, and repeat the analysis. If the silver content decreases, then the silver-enriched layer was still present for the first analysis, and this process should be repeated until consistent results are obtained (Tate 1986). Once the measured concentrations are reproducible, the bulk alloy has been reached. In interpreting quantitative data from copper-silver alloys that were collected without following this protocol (the majority of the data available in the literature), identifying the presence of any silver-enriched layer on these objects is critical. The majority of hollow wares in Winterthur's and related collections retain consistent evidence of a pickled surface remaining on their undersides and variable evidence of this surface treatment elsewhere. The undersides commonly display silver contents 0.5 to 5 wt. % greater than the bulk compositions. The presentation surfaces of these wares more commonly reflect coin silver or sterling silver compositions, suggesting that the enriched pickled surface has been greatly reduced or even completely removed by polishing. There are, however, intermediate cases where presentation surfaces reveal silver compositions between 92.5 and 95 wt. %, and one must consider that these values reflect a mixture of the depleted surface and bulk compositions.

The presence of "islands" of residual silver-enriched surfaces is also suggested by the observation of silver contents that are highly variable within a given hollow ware component (such as a tankard barrel) and also above the sterling standard. While the penetration depth of the incident X-ray beam in an XRF experiment can be controlled by adjusting the angle of incidence to obtain depth-dependent data (Awane *et al.* 2009), this is not possible with the handheld spectrometers currently on the market.

Conclusions

Information about historic silver alloys is of value for answering questions about an objects' state of preservation, authenticity, technology of manufacture, and provenance. However, the surface compositions of historic silver objects differ substantially from their bulk compositions, and this is only one of the potential pitfalls encountered when attempting nondestructive XRF analysis of these wares. A more complete list of the problems encountered for generating high quality, reproducible quantitative data nondestructively on these wares includes: obtaining ideal and reproducible measurement geometries, the presence of variable thickness copper-depleted/silver-enriched surface layers due to cycles of pickling (acid dipping) and polishing, surface or buried firescale (copper oxide) layers, residues of surface gilding, microstructural heterogeneities (a particular problem when using a polycapillary focusing optic), layered microstructures resulting from various plating technologies, elemental segregation within the alloy, surface corrosion layers, and obtaining appropriate standards for empirical quantification methods. We have seen that problems of silver-enriched surfaces and firescale can only be addressed by polishing the region of analysis until reproducible results are obtained, and by using carefully prepared secondary standards to remove element segregation effects. Knowledge about the origins and compositions of colonial American silver alloys, as well as how American silver alloy compositions change over time, are also critical for interpretation of quantitative XRF data.

Acknowledgements

The authors would like to thank Winterthur Museum Metals Curator Ann Wagner and Winterthur Museum Metals Curator emeritus Don Fennimore for their many helpful discussions during the preparation of this chapter, in particular with their assistance in identifying numerous highly valuable references. Margi Hofer,

Curator of Decorative Arts of the New York Historical Society, is thanked not only for providing generous access to her collection, but also her valued insights on the interpretation of our findings. The authors are also grateful to Bruno Pouliot and Lauren Fair for their assistance in providing interesting objects/silver surfaces for research, and for their insights about silver conservation. Finally, the authors are grateful to Aaron Shugar for his many helpful metallurgical insights.

References

Agricola, Georgius. 1556. *De Re Metallica*. Translated by Herbert Clark Hoover and Lou Henry Hoover. New York: Dover.

Awane, T., S. Fukuoka, K. Nakamachi, and K. Tsuji. 2009. "Grazing Exit Micro X-ray Fluorescence Analysis of a Hazardous Metal Attached to a Plant Leaf Surface Using an X-ray Absorber Method." *Analytical Chemistry* no. 81 (9):3356-3364.

Barquist, David. 2001. *Myer Myers: Jewish Silversmith in Colonial New York*. New Haven: Yale University Press.

Bauer, O., and E. Deiss. 1933. "Seigerung in Einen Silberblock." *Z. Anal. Chem.* no. 92:161-164.

Beck, L., E. Alloin, C. Berthier, S. Réveillon, and V. Costa. 2008. "Silver surface enrichment controlled by simultaneous RBS for reliable PIXE analysis of ancient coins." *Nuclear Instruments and Methods in Physics Research Section B: Beam Interactions with Materials and Atoms* no. 266 (10):2320-2324. doi: 10.1016/j.nimb.2008.03.084.

Beck, L., S. Bosonnet, S. Réveillon, D. Eliot, and F. Pilon. 2004. "Silver surface enrichment of silver-copper alloys: a limitation for the analysis of ancient silver coins by surface techniques." *Nuclear Instruments and Methods in Physics Research Section B: Beam Interactions with Materials and Atoms* no. 226 (1-2):153-162. doi: 10.1016/j.nimb.2004.06.044.

Bovin, Murray. 1963. *Silversmithing and Art Metal for Schools, Tradesmen, and Craftsmen*. Forest Hills, New York: Bovin Publishing.

Canovaro, C. Canovaro C., I. Calliari, S. Gottardello, and M. Asolati. 2011. "Microstructural characterization of official and imitative nummi of Vth century AD." *Metallurgia Italiana* (2):21-25.

Cronau, Rudolph. 1916. *German Achievements in America*. New York

Day, Joan, and R. F. Tylecote. 1991. *Industrial Revolution in Metals*. London: Institute of Metals.

Desaulty, Anne-Marie, Philippe Telouk, Emmanuelle Albalat, and Francis Albarede. 2011. "Isotopic Ag-Cu-Pb record of silver circulation through 16th-18th century Spain." *Proceedings of the National Academy of Sciences* no. 108 (22):9002-9007.

Duffy, Kate, and Janice Carlson. 2000. "A Scientific Assessment of Silver: XRF Analysis." In *Campbell Collection of Soup Tureens*, edited by Donald L. Fennimore and Patricia A. Halfpenny. Lebanon, New Hampshire: University Press of New England.

Fales, Martha Gandy. 1973. *Early American Silver*. New York: Dutton.

Falino, Jeannine, and Gerald W.R. Ward. 2008. *Silver of the Americas, 1600-2000*. Boston, Massachusetts: MFA Publications.

Fennimore, Donald L., and Ann K. Wagner. 2007. *Silversmiths to the Nation: Thomas Fletcher and Sidney Gardiner 1808-1842*. New York, New York: Antique Collectors Club.

Forbes, John S. 1999. *Hallmark: A History of the London Assay Office*. London: Unicorn Press in Conjunction with the Goldsmiths Company.

Gale, N. H., and Z. A. Stos-Gale. 1981. "Cycladic Lead and Silver Metallurgy." *Annuals of the British School at Athens* no. 76:169-224.

Gore, George. 1887. *The Theory and Practice of Electro-Deposition*. London: Charles and Co.

Hanson, Victor F. 1970. The Curator's Dream Instrument. Paper read at Application of Science in Examination of Works of Art: Proceedings of the Seminar, 15-19 June 1970, at Museum of Fine Arts, Boston.

———. 1973. "Quantitative Elemental Analysis of Art Objects by Energy-Dispersive X-ray Fluorescence Spectroscopy." *Applied Spectroscopy* no. 27:309-34.

Herbert, Eugenia W., A. Bernard Knapp, and Vincent C. Pigott. 1998. *Social Approaches to an Industrial Past: the Archaeology and Anthropology of Mining*. London: Psychology Press.

Holden, Geoffrey. 1954. *The Craft of the Silversmith*. New York: Studio Publications.

Hubbell, J. H., and S. M. Seltzer. 2011. *Tables of X-ray Mass Attenuation Coefficients and Mass Energy-Absorption Coefficients from 1 keV to 20 MeV for Elements Z = 1 to 92 and 48 Additional Substances of Dosimetric Interest*. The National Institute of Standards and Technology 2011 [cited December 2011]. Available from http://www.nist.gov/pml/data/xraycoef/index.cfm.

Kauffman, Henry J. 1995. *The Colonial Silversmith: His Techniques and His Products*. Lakeville, Minnesota: Astragal Press.

Lejcek, Pavel, Janez Kovac, and Jaroslava Vanickova. 2010. "Copper Surface Enrichment of Ag-Cu Alloys." *Surface and Interface Analysis* no. 42 (6-7):662-665.

Mulholland, James A. 1981. *A History of Metals in Colonial America*. Tuscaloosa, Alabama: University of Alabama Press.

Ottaway, Barbara S., and Quanyu Yang. 2004. *Casting Experiments and Microstructure of Archaeologically Relevant Bronzes*. Vol. 1331, *BAR International Series* London: Archaeopress.

Quimby, Ian M. G. 1995. *American Silver at Winterthur*. Charlottesville: The University Press of Viriginia.

Rehren, Thilo, Andreas Hauptmann, and James D. Muhly. 1998. *Metallurgica Antiqua: In Honor of Hans-Gert Bachmann and Robert Maddin*. Bochum: Deutsche Bergbau Museum.

Scott, David A. 1991. *Metallography and microstructure of ancient and historic metals*. Los Angeles: Getty Conservation Institute.

Sumner, W. G. 1898. "The Spanish Dollar and the Colonial Shilling." *American Historical Review* no. 3:607-619.

Tate, James. 1986. "Some problems in analysing museum material by nondestructive surface sensitive techniques." *Nuclear Instruments and Methods in Physics Research Section B: Beam Interactions with Materials and Atoms* no. 14 (1):20-23. doi: 10.1016/0168-583x(86)90417-9.

Tylecote, R. F. 1976. *A History of Metallurgy*. London: The Metals Society.

Venable, Charles L. 1994. *Silver In America 1840-1940: A Century of Splendor*. New York: Harry N. Abrams, Inc.

Vittiglio, G., K. Janssens, B. Vekemans, F. Adams, and A. Oost. 1999. "A compact small-beam XRF instrument for in-situ analysis of objects of historical and/or artistic value." *Spectrochimica Acta Part B-Atomic Spectroscopy* no. 54 (12):1697-1710. doi: 10.1016/s0584-8547(99)00100-7.

Vrebos, DeVries and. 2002. "Quantification of Infinitely Thick Specimens." In *Handbook of X-ray Spectrometry*, edited by Rene Grieken and Andrzej Markowicz. New York: Marcel Dekker, Inc.

Ward, Barbara McLean, and Gerald W.R. Ward. 1979. *Silver in American Life: Selections from the Mabel Brady Garvan and Other Collections at Yale University*. Boston: David R. Godine.

Waters, Deborah Dependahl. 1977. "From Pure Coin: The Manufacture of American Silver Flatware 1800-1860." *Winterthur Portfolio* no. 12:19-33.

Watson, J. H. 1932. "Liquation or "inverse segregation" in the silver-copper alloys." *Journal of the Institute of Metals* no. 49:347-362.

Weber, G., J. Guillaume, D. Strivay, H. P. Garnir, A. Marchal, and L. Martinot. 2000. "Is the external beam PIXE method suitable for determining ancient silver artifact fineness." *Nuclear Instruments and Methods in Physics Research Section B: Beam Interactions with Materials and Atoms* no. 161-163:724-729.

Chapter 8

The analysis of porcelain using handheld and portable X-ray fluorescence spectrometers

Anikó Bezur and Francesca Casadio

Introduction

Porcelain objects are most often curated as part of decorative arts collections or those highlighting the arts of Asia. Unless pursued as part of archaeometric studies that involve object fragments or production-related debris, the compositional study of porcelain bodies presents a special challenge due to the need that sample removal be avoided. Energy-dispersive X-ray fluorescence spectrometers have the potential to meet this challenge due to their ability to be used in a non-destructive and non-invasive fashion. While scientists have explored the use of XRF for the analysis of ceramic remains, most recent studies have focused on relatively low-fired, iron-rich matrices rather than porcelain. Researchers have compared XRF with more traditional analytical methods for these types of ceramics, such as inductively coupled plasma optical emission spectroscopy (ICP-OES), inductively coupled plasma mass spectrometry (ICP-MS), and neutron activation analysis (NAA). This chapter reviews the extant literature on the analysis of ceramics using XRF, and it also surveys the use of Raman spectroscopy, which can yield complimentary, molecular information regarding the components of the ceramic matrix. Handheld XRF results and methodologies will be detailed and also compared to results and methodologies from laboratory-based XRF systems. The literature review portion of the chapter will emphasize the role of instrument configuration, operating parameters, and the use of standard reference materials in order to optimize the quantitative analysis of ceramic matrices. Practical issues relating to the analysis of porcelain bodies will be discussed in the context of recent work by the authors, which explored compositional similarities between the products of the first two true, hard-paste, porcelain factories in Europe: Meissen and Du Paquier. Topics of discussion will include the relationship between instrument aperture size and the area of unglazed porcelain bodies to be analyzed, depth resolution, and the availability of suitable reference materials.

Introduction to porcelain

Porcelain as a material may be described as a dense, translucent, vitreous ceramic, with a white or off-white color that is produced at high firing temperatures (around or above 1200-1300 °C). Fired porcelain pastes have very low porosity; Maggetti *et al.* (2009:20) cite that water adsorption of less than 2% distinguishes porcelains from white earthenware. Dinsdale (1986:45) discusses porcelain as a type of whiteware system, having a triaxial composition of clay, flux, and filler. When sufficiently thin, porcelain is translucent and vessels may be sonorous when struck. Curators and collectors frequently use these properties, especially translucency, which can be assessed by directing a bright light at the wall of a vessel, as criteria for the identification of porcelains (Kingery and Vandiver 1986b: 282). In the field of cultural heritage studies, the word "porcelain" typically refers to porcelain bodies that have a partial, most often clear, glaze coating which has partially diffused into the body during high-temperature firing. The term "biscuit" or "bisque" may refer to fired but unglazed objects, or to unglazed portions of pieces. Colored decoration may be applied to the unglazed body (e.g. enamel on biscuit), can be present at the body-glaze interface (underglaze enamels), or applied on top of the glaze layer (fired overglaze enamels or unfired, "cold colors" consisting of pigments in an oil medium). Gilding and silvering may also be present at the glaze-air interface.

History of Asian and European porcelain technology prior to the twentieth century

Porcelain technology developed in China during the Tang period, 618-906 (Vandiver and Kingery 1984; Rice 1987; Finlay 2010), though some argue that the first true porcelains were produced considerably earlier, during the Han period, 206 BCE-220 CE (Jiazhi 1984). The technology spread to areas of modern-day Korea in the 10th century, during the Koryŏ dynasty, 918-1392, and to Japan, by way of Korean craftsmen, in the first two decades of the seventeenth century (Cort 1986; Day 1992-94; Choo *et al.* 2002; Choo *et al.* 2004). The ability to produce "white gold" eluded Western craftsmen until the first decade of the eighteenth century, when Johann Friedrich Böttger succeeded in making hard-paste porcelain in Dresden (Schulle and Goder 1982; Kingery and Vandiver 1986b). Ehrenfried Walter von Tschirnhaus, whose experiments with high fired ceramics were funded by Augustus II the Strong, and who supervised Böttger before his death in 1708, also deserves significant credit for the discovery (Schönfeld 1998).

Augustus II founded the Meissen porcelain factory in 1710 and tried to maintain the factory's monopoly on porcelain production in Europe by keeping the facility under heavy guard and by curtailing the personal freedom of the workers. Despite these precautions production secrets made their way to Vienna in 1719 by way of a disgruntled specialist, Samuel Stöltzel, and contributed to the success of the Du Paquier factory as well as other European factories established later in the eighteenth century (Kingery and Vandiver 1986b; Busson 1994).

Recent research may lead to revisions in the history of early porcelain production in Europe, pushing back the date of the first hard-paste porcelain manufacture by about 25 years and locating successful production in England. The three small porcelain jars at the heart of this work are from Burghley House and were documented in a deed drafted in 1683 and proved in 1690. Non-destructive analyses of two of the jars using scanning electron microscopy (SEM) with energy dispersive X-ray spectroscopy (EDS) determined that the porcelain paste was highly vitrified and resembled hard-paste porcelain compositions with high clay content fluxed predominantly with potassium oxide (Spataro, Bimson, and Meeks 2008; Spataro *et al.* 2009). Interestingly, the thick glaze layer was found to be lead-silicate based, which is what was used on most types of soft-paste porcelain bodies, rather than being a high-fired alumino-silicate glaze, like that used on Chinese, Meissen and other European hard-paste porcelain objects.

This brief timeline of porcelain technology would not be complete without mentioning important milestones in the creation of porcelain-like wares, such as the formulation of white "stonepaste" bodies in Egypt during the Fatimid dynasty (909-1171). Such wares were made from a paste that contained a high proportion of finely crushed quartz mixed with smaller amounts of almost white clay and glass frit. The components of Medici porcelain bodies, produced for just over a decade in sixteenth-century Florence, were quite similar, including mostly crushed quartz sand, some white clay as well as additional crushed sand, sintered with salt and wine lees (Kingery and Vandiver 1984:141). Kingery and Smith (1984) considered stonepaste and Medici bodies antecedents of French soft-paste porcelain, which was first produced at the end of the seventeenth century at St. Cloud. These wares, along with eighteenth century French and English incarnations, also contained little clay in comparison with quartz sources, but differed from antecedents in that recipes included both lime and alkali metal sources, or lead in the case of some English factories (Kingery and Smith 1984; Tite and Bimson 1991; Tite 2008). The second half of the eighteenth century also saw the development of English bone-ash porcelain, bone china and soapstone porcelains, all of which are considered soft-paste porcelains (Kingery and Vandiver 1986b; Tite and Bimson 1991). Bone-ash porcelain recipes combined calcined animal bones with sand, ball clay and a source of calcium while bone china pastes contained a mixture of china clay and china stone in lieu of ball clay (Tite and Bimson 1991). In addition to soapstone, which is primarily composed of talc ($Mg_3Si_4O_{10}(OH)_2$), soapstone porcelain pastes also contained ground glass, frit, and a source of calcium in some cases (Tite and Bimson 1991).

The interpretation of qualitative and quantitative data on porcelain components

A consideration of the appropriateness of X-ray fluorescence spectroscopy for the analysis of porcelain

Kingery and Vandiver (1986b), Rice (1987), and Tite (1999, 2008) discuss the relevance of materials characterization and analysis methods in relation to informing researchers about raw materials, processing technologies, and the physical properties of the resulting ceramic objects. Importantly, these authors examine how the results of materials analysis may help to infer details about the cultural and socio-economic contexts of artifacts and technologies. The provenance and authenticity of objects is also highly relevant to the research and presentation of collections, although this can be challenging to establish in the case of works that were not scientifically excavated or which lack or have only discontinuous historic records (e.g. Yap 1992; Fuchs II and Mass 2009; Halfpenny and Mass 2009).

As mentioned earlier, glazed porcelain objects have a multi-layer, multi-phase structure. The body and overglaze enamel colors with discrete pigment particles are themselves composite materials. Most hard-paste porcelain bodies, for example, consist of mullite needle-shaped crystals in a glass matrix, though remains of quartz grains can also be present in K'ang Hsi Chinese porcelain (Tite, Freestone, and Bimson 1984; Kingery and Vandiver 1986b:153-155). Comprehensive characterization of porcelain objects thus needs to address their various layers individually and to identify their component phases. The examination of bodies at the micro-structural level is essential for deducing the nature of potential paste ingredients and learning about firing temperature and firing atmosphere based on the various phases present. A nuanced understanding of the visual impact of overglaze enamels is difficult to achieve without knowing whether transition metals are present as dissolved ions, as crystalline pigments or in the form of nanoparticles. Due to its ability to resolve features at the micron scale and to provide semi-quantitative or quantitative elemental analysis of phases, scanning electron microscopy coupled with energy or wavelength dispersive spectroscopy (SEM-EDS or SEM-WDS) has played a key role in our understanding of ceramics (for a discussion of early applications see Rice 1987:400-403). Despite its great utility, the use of this technique for the study of porcelains is limited to cases when fragments are available for study since it involves the examination of small samples that have been embedded, polished and often etched with hydrofluoric acid.

Instrument	Beam aperture/ collimator diameter	Other information
Thermo Scientific Niton XL3	3, 8 mm	Integrated camera
Innov-X Alpha Cam	1.5, 3.0, 6.0 mm	Integrated camera
Bruker Tracer III-V	2.41 mm (0.095 inches) – beam at surface is ≈ 4.3 x 3.4 mm	Elliptical opening on nosepiece is approx. 7 mm and 5.6 mm in diameter
Oxford Instruments X-MET 5000 and 5100	9 mm diameter at instrument nose	
Skyray Pocket-III, Quickshot XRF	6 mm	

Table 8.1:
X-ray beam diameter information for some handheld XRF systems.

X-ray fluorescence spectroscopy (XRF) gives us information on the atomic level. Because the lateral resolution (a measure of the ability to discriminate features in the x-y plane) of handheld instruments is on the order of millimeters as beam diameters range from 1.5 to 8 millimeters (Table 8.1) analysis of porcelain bodies and glazes using such units conflates the elemental composition of discrete phases and thus XRF can be considered a "bulk" characterization technique. Some commercially available open architecture portable XRF systems are equipped with polycapillary X-ray optics, which may offer beam diameters as small as about 70 μm, and bench-top micro-XRF systems can produce 10 μm diameter X-ray beams (e.g. HORIBA Scientific energy dispersive X-ray analytical microscopes with mono-capillary optics). However, lateral resolution on this finer scale is still unable to discriminate phases within bodies as large residual quartz grains are on the order of 10 μm, while transformed clay ingredients – such as mullite crystals and needles – are much smaller in size (Kingery and Vandiver 1986b:18-26).

A further limitation of handheld and portable, open architecture, XRF systems is that analysis takes place in the presence of air in the beam path, the argon content of which causes the attenuation of low energy characteristic X-ray lines from low atomic number (*Z*) elements. The bulk elemental components of porcelain bodies and clear glazes are predominantly low *Z*: oxygen, sodium, magnesium, aluminum, silicon, phosphorus, potassium, calcium, titanium (see Tables 8.2 and 8.3 for compositional ranges expressed as weight percent oxides). Portable vacuum pumps can be used to evacuate the beam path within handheld instruments to improve their sensitivity toward low *Z* elements, making it possible to detect aluminum ($Z = 13$) as long as the sample is in contact with, or very close to the aperture (Bruker AXS claims detection with their Tracer III-V, equipped with a rhodium X-ray tube and an Si-PIN detector, down to $Z = 12$, Mg). However, it is important to note that evacuation of the atmosphere within the instrument causes the flexible polymer window to form a concave surface, leaving a slight air gap between the instrument's aperture and a flat sample surface. The recent availability of silicon drift detectors (SDD) in handheld instruments (including Bruker AXS: S1 TURBO^{SD}; Thermo Fisher: Niton XL2 GOLDD; Oxford Instruments: X-MET5100; Innov-X: Delta Premium & Standard) promises to further improve the technique's sensitivity toward elements of *Z* less than 12 without the need for vacuum pumps. Portable, open architecture systems offer the ability to displace air from the beam path using helium during analysis and are also capable of detecting sodium ($Z = 11$) (though the experience of the authors is that the detection limit for Na is quite high, possibly on the order of 10 wt. % or greater).

Despite its limitations, X-ray fluorescence spectroscopy can provide a great deal of valuable information about the nature of the porcelain body and clear glaze as well as colored glazes and metallic decorations in a non-destructive and non-invasive manner. The discussion of porcelain components that follows emphasizes both qualitative and quantitative elemental analysis criteria for their characterization and provides guidelines for the interpretation of results with respect to the attribution of objects to particular factories and periods. The use of XRF and other non-destructive methods for materials characterization is preferred by the ethical guidelines of professional organizations in the field of conservation compared to techniques requiring the removal of samples. For example, Commentary 17 to the Guidelines for Practice of the American Institute for Conservation of Historic and Artistic Works states that "the conservation professional must use non-invasive analytical methods where possible" (American Institute for Conservation of Historic and Artistic Works). The microstructure of porcelain bodies serves as further motivation to avoid handling and sampling that may initiate cracks since the vitreous material poses little resistance to crack propagation and brittle fracture. Porcelain objects are most often curated as part of decorative arts collections or those highlighting the arts of Asia. Unless pursued as part of archaeometric studies that involve object fragments or production-related evidence, technical studies

Origin	Type	Na_2O	MgO	Al_2O_3	SiO_2	P_2O_5	K_2O	CaO	TiO_2	MnO	Fe_2O_3	SnO_2	PbO	Reference
Yuan Dynasty yingqing	Hard-paste	1,1	0,1	15,7	78,7	0,3	2,9	0,4			0,9			Tite *et al*. (1984)
Yuan Dynasty underglaze blue	Hard-paste	2,3		19,7	74,3	0,3	2,3	0,2			1			
Hop Lê (Vietnamese)	Hard-paste	0,24	0,57	15,52	78,59		1,76	1,49	0,67		1,16			Colomban *et al*. (2003)
Ch'ing Dynasty K'ang Hsi Famille Verte	Hard-paste	3		24,5	69,7		2,8	0,2						Kingery and Vandiver (1986)
Böttger (Saxon)	Hard-paste	0,2		33	61		0,1	4,8						Schulle and Ullrich (1982)
Meissen (Saxon)	Hard-paste	0,8		35	59		4	0,3						
Bristol (English)	Hard-paste	0,7	0,5	24	70,8		3	0,5			0,6			Tite and Bimson (1991)
Plymouth (English)	Hard-paste	0,7	0,3	23,1	72,3		3	0,3			0,4			
Terre de Lorraine (French)	Hard-paste	2,23	0,22	29,97	63,97		2,31	1,25	0,11	0,01	0,32			Maggetti *et al*. (2009)
	Calcareous SP	0,88	0,42	21,37	56,44	0,6	2,46	16,08	0,49	0,01	0,84			
St. Cloud (French)	Glassy SP	2,88	1,42	1,75	75,89		3,15	12,6	0,14		0,7	0,04	0,09	Kingery and Smith (1984)
Vincennes (French)	Glassy SP	2,66	0,83	2,11	72,84		4,32	15,69	0,1		0,3	0,06	0,2	
Sevres (French)	Glassy SP	2,15	0,78	2,23	74,25		0,3	15,68	0,07		0,3	0,05	0,18	
Chelsea (English)	Glassy SP	0,5	0,4	4,1	70,3		3,5	18,8	0,3	0,2			1,9	Tite and Bimson (1991)
Longton Hall (English)	Glassy SP	0,8	0,3	2,9	77,2		3,7	8,8					6,3	
Bow (English)	Bone-ash SP	0,8	0,6	8,7	45,6	18,6	1,1	23,6	0,5		0,56			
Worcester (English)	Soapstone SP	1	10,8	3,8	74,1		2,8	1,9	0,3	0,4			4,9	
Vauxhall (English)	Soapstone SP	1,5	10	4,4	78,2		3	2,1			0,6		0,3	
Liverpool (English)	Soapstone SP	2,2	11,9	4,3	70,3	1,6	2,9	4,8			0,8		1,3	
Buen Retiro (Spanish)	Soft-paste	0,49	12,42	4,65	79,81		2,08	0,19			0,36			Zucchiatti (2008)
Buen Retiro (Spanish)	Soft-paste	0,58	10,6	8,76	77,3	0,22	1,54	0,14	0,002	0,03	0,089			Aza *et al*. (2004)

Table 8.2:
Examples of bulk compositions of hard- and soft-paste porcelain bodies. Values reported as FeO were converted to Fe_2O_3 to facilitate comparison. SP = soft-paste.

Origin	Type	Na_2O	MgO	Al_2O_3	SiO_2	P_2O_5	K_2O	CaO	TiO_2	MnO	Fe_2O_3	SnO_2	PbO	Reference
Yuan Dynasty yingqing	Hard-paste	1,1	0,1	15,7	78,7	0,3	2,9	0,4			0,9			Tite *et al*. (1984)
Yuan Dynasty underglaze blue	Hard-paste	2,3		19,7	74,3	0,3	2,3	0,2			1			
Hop Lê (Vietnamese)	Hard-paste	0,24	0,57	15,52	78,59		1,76	1,49	0,67		1,16			Colomban *et al*. (2003)
Ch'ing Dynasty K'ang Hsi Famille Verte	Hard-paste	3		24,5	69,7		2,8	0,2						Kingery and Vandiver (1986)
Böttger (Saxon)	Hard-paste	0,2		33	61		0,1	4,8						Schulle and Ullrich (1982)
Meissen (Saxon)	Hard-paste	0,8		35	59		4	0,3						
Bristol (English)	Hard-paste	0,7	0,5	24	70,8		3	0,5			0,6			Tite and Bimson (1991)
Plymouth (English)	Hard-paste	0,7	0,3	23,1	72,3		3	0,3			0,4			
Terre de Lorraine (French)	Hard-paste	2,23	0,22	29,97	63,97		2,31	1,25	0,11	0,01	0,32			Maggetti *et al*. (2009)
	Calcareous SP	0,88	0,42	21,37	56,44	0,6	2,46	16,08	0,49	0,01	0,84			
St. Cloud (French)	Glassy SP	2,88	1,42	1,75	75,89		3,15	12,6	0,14		0,7	0,04	0,09	Kingery and Smith (1984)
Vincennes (French)	Glassy SP	2,66	0,83	2,11	72,84		4,32	15,69	0,1		0,3	0,06	0,2	
Sevres (French)	Glassy SP	2,15	0,78	2,23	74,25		0,3	15,68	0,07		0,3	0,05	0,18	
Chelsea (English)	Glassy SP	0,5	0,4	4,1	70,3		3,5	18,8	0,3	0,2			1,9	Tite and Bimson (1991)
Longton Hall (English)	Glassy SP	0,8	0,3	2,9	77,2		3,7	8,8					6,3	
Bow (English)	Bone-ash SP	0,8	0,6	8,7	45,6	18,6	1,1	23,6	0,5		0,56			
Worcester (English)	Soapstone SP	1	10,8	3,8	74,1		2,8	1,9	0,3	0,4			4,9	
Vauxhall (English)	Soapstone SP	1,5	10	4,4	78,2		3	2,1			0,6		0,3	
Liverpool (English)	Soapstone SP	2,2	11,9	4,3	70,3	1,6	2,9	4,8			0,8		1,3	
Buen Retiro (Spanish)	Soft-paste	0,49	12,42	4,65	79,81		2,08	0,19			0,36			Zucchiatti (2008)
Buen Retiro (Spanish)	Soft-paste	0,58	10,6	8,76	77,3	0,22	1,54	0,14	0,002	0,03	0,089			Aza *et al*. (2004)

Table 8.3:
Examples of bulk compositions of clear or white glaze layers on hard- and soft-paste porcelain bodies. Values reported as FeO were converted to Fe_2O_3 to facilitate comparison.

of porcelain objects most often involve complete objects or ones that have been restored to suggest completeness. Sample removal is thus discouraged due to a desire to avoid compromising the wholeness of the piece in addition to minimizing potential damage due to brittle fracture.

Porcelain bodies and clear or white glazes

Even without the ability to detect sodium and magnesium, the use of XRF to determine the bulk composition of porcelain bodies and clear glazes enables researchers to characterize these components based on the type and relative proportions of key ingredients. Specifically, ceramic pastes for porcelain are distinguished based on whether kaolin clay was used and the type of non-plastic inclusions present, which may include quartz, feldspars (alkali or plagioclase), mica, gypsum, calcium carbonate, bone ash, steatite (soapstone), sepiolite, fritted glass, as well as other minerals (Kingery and Smith 1984; Tite, Freestone, and Bimson 1984; Rice 1987:406-408; Tite and Bimson 1991; Aza *et al.* 2004; Tite 2008:220-222; Maggetti *et al.* 2009; Rapp 2009:189-191). XRF results are insufficient to draw conclusions about the specific nature of paste ingredients, but they allow some inferences to be made with the help of contemporary recipes and published studies on the correlation between composition, microstructure and possible paste components (see Table 8.2).

Hard-paste Chinese porcelain bodies were formulated with ground porcelain (or china) stone - a rock comprising quartz, mica and feldspar - to which kaolin was added if the stone did not already contain this clay component. No additional fluxes were needed since porcelain stone includes potassium- and sodium-containing minerals (Tite, Freestone, and Bimson 1984; Kingery and Vandiver 1986b). The bulk composition of Chinese and Vietnamese porcelain bodies (for examples see Table 8.2) is characterized by high silica (70-80%) and alumina content (15-25%) with the alkali oxides (soda and potassia) making up most of the remaining weight percent (for further examples see Leung *et al.* 1998; Leung, Stokes, *et al.* 2000; Wu *et al.* 2002). Hard-paste English porcelain bodies, produced from china clay, china stone and quartz sand, have remarkably similar compositions (Tite and Bimson 1991). The "aluminous-silicaceous" bodies of certain *Terre de Lorraine* objects, which were produced from a white clay in Northeastern France from the middle of the 18th century, also bear close resemblance to Chinese hard-paste compositions. Though no recipes survive, Maggetti *et al.* (2009) suggest that kaolinitic clay and kaolinite were mixed calcined flint (quartz) as well as glass frit. The so-called "feldspathic" porcelain bodies produced in Meissen from about 1730 were based on a mixture of alkali feldspar, "Schnorrsche Erde" (kaolin from Aue) and white Colditz clay (Mields 1960). Compared to Asian and English examples, the bulk composition of such bodies is a bit richer in alumina (30-35%) with correspondingly lower silica levels (55-60%) and includes primarily potassia rather than a more even mixture of alkali oxides (Schulle and Ullrich 1982). The paste of Böttger porcelain, produced in Meissen prior to 1730, was also prepared with kaolin from Aue and white Colditz clay, but with the use of calcined alabaster as a flux. This resulted in similar alumina and silica contents as encountered in the later Meissen bodies with about 5-8% calcia (Schulle and Ullrich 1982).

Spanish Buen Retiro porcelain bodies are classified as hard-paste, but differ significantly from other bodies in this category because their ingredients included magnesium-containing sepiolite ($Mg_4Si_6O_{15}[OH]_2{\bullet}6H_2O$) instead of kaolinite, along with quartz and feldspar (Aza *et al.* 2004; Zucchiatti *et al.* 2008). Handheld spectrometers equipped with silicon drift detectors promise the ability to detect magnesium, but users of instruments with Si-PIN detectors will likely not be alerted to the presence of this low Z element. Nonetheless, researchers may distinguish Buen Retiro porcelain pieces from other hard-paste types based on their high silica-to-alumina ratios and low calcia levels (see Table 8.2). The absence of lead in porcelain bodies and the use of potassium-rich feldspathic clear glazes help differentiate them from English soft-paste porcelain made with magnesium-containing soapstone (Pascual *et al.* 2006:140-141).

High calcia contents characterize a variety of soft-paste porcelain bodies. "Lime-rich" and "calcareous" compositions of *Terre de Lorraine* and products of other French manufactories as well as English glassy porcelain contain 10-20% calcia. Table 8.2 shows that the calcareous *Terre de Lorraine* bodies may be distinguished from other French soft-paste porcelains based on their high alumina and lower silica content, even though differences in soda content are not possible to measure with handheld XRF units (Maggetti *et al.* 2009). Though French soft-paste bodies can be considered "glassy porcelains" along with some English varieties, unlike the latter they were formulated without lead oxide, using quartz, alkali, clay and limestone (Kingery and Smith 1984; Tite and Bimson 1991). Again, the higher soda content of French soft-paste bodies will not be apparent, but the lead content of the English pastes can be detected easily with handheld instruments. Lead will also be detected in the clear glaze layers of both English and French soft-paste porcelains as the lower firing temperature of these bodies necessitated the use of glazes that could be fired at moderate temperatures. While high calcia content, 20-25%, is encountered in bone-ash porcelain, the use of calcined animal bones also contributes phosphorous, resulting in 15-20% diphosphorus pentoxide content (Tite and Bimson 1991).

The compositions of English soft-paste bodies made with soapstone (steatite, consisting primarily of talc, $Mg_3Si_4O_{10}[OH]_2$) and hard-paste Spanish Buen Retiro bodies prepared with sepiolite ($Mg_4Si_6O_{15}[OH]_2{\bullet}6H_2O$) share similarly high values for magnesia (about 10%) and silica (70-80%), but the latter tend to have higher alumina content (5-10% compared to 3-4%) (Aza *et al.* 2004; Zucchiatti *et al.* 2008). Tite and Bimson (1991:18) found that Worcester soapstone porcelain could be distinguished from other contemporaries based on its lead oxide content (5-7%), which is helpful since many Worcester pieces were unmarked (Reynolds 1989:8-9). As in the case of Buen Retiro porcelain, users of spectrometers with Si-PIN detectors may not detect magnesium in English soapstone wares. However, high silica-to-alumina ratios coupled with low potassia and calcia levels as well as the detection of lead in the body in pieces of English origin may alert researchers to the possibility that they are dealing with a magnesia-fluxed soft-paste porcelain body. The presence of high intensity lead peaks in spectra obtained from areas of clear glaze would further confirm the identification of such objects as soapstone porcelain (see Table 8.3).

Positioning the aperture of the handheld XRF on a relatively thick (about 150-200 μm) section of clear or white glaze will allow the analyst to collect spectra in which K emission lines for bulk glaze components (Mg, Al, Si, P, K, Ca, Ti, Mn, and Fe) will be restricted to the glaze layer, without interference from the porcelain body. This occurs because the relatively low-energy characteristic X-ray photons emitted by these elements in the porcelain body are absorbed by the clear glaze layer and thus cannot emerge and be detected by the instrument (for a detailed explanation see section – "Lateral and depth resolution issues in the analysis of porcelain objects"). Even in thinner regions of the glaze, around 75–100 μm, spectra will exclude contributions from the body for low *Z* elements: Mg, Al, Si, P, K, Ca, Ti. This means that glaze analysis may be at least semi-quantitative and allow the classification of the glaze layer, especially if spectra of the body are available for comparison.

Composition-based classification of glazes typically names the component or components that act as the main network modifier in the glaze formulation, which impacts the highest temperature at which the glaze and body are matured together (Rhodes 1973; Kingery and Vandiver 1986b; Rice 1987). Table 8.3 presents some examples for bulk glaze compositions on a variety of hard- and soft-paste porcelain bodies. Many Chinese porcelain glazes are considered lime-alkali type because the molar ratio of calcia to alkali oxides is greater than one ($CaO/R_2O>1$, where $R_2O = Na_2O + K_2O$). Kingery and Vandiver (1986b:262) cite 0.7 CaO to 0.3 R_2O or 2.33 as a typical molar ratio. Examples of such compositions are given by Tite *et al.* (1984) and Kingery and Vandiver (1986b), with 8-15% calcia,

		Weight %				Mole ratios			
Origin	Type	Na_2O	MgO	K_2O	CaO	CaO/K_2O	CaO/R_2O	RO/R_2O	Reference
Yuan Dynasty yingqing	Lime-alkali	1,4	0,5	2,1	14,6	5,6	1,78	1,86	Tite *et al.* (1984)
Yuan Dynasty underglaze blue	Lime-alkali	3,7	0,3	2,6	8,7	11,7	5,8	6,08	
Ch'ing Dynasty K'ang Hsi Famille Verte	Lime-alkali	2,1		3,3	10,1	5,1	2,61	2,61	Kingery and Vandiver (1986)
Ding, d1	Lime-alkali	0,69	1,81	3,11	3,9	2,1	1,58	2,59	Leung *et al.* (2000)
Ding, d2	Lime-alkali	0,55	2,77	3,16	5,44	2,9	2,29	3,91	
Ding, d3	Lime-alkali	0,66	1,81	2,47	3,41	2,3	1,65	2,87	
Ding, d4	Lime-alkali	0,73	2,1	2,23	4,28	3,2	2,15	3,62	
Cizhou, Guantai, stage 1	Lime-alkali	0,85	1,13	2,66	10,04	6,3	4,27	4,94	
Cizhou, Guantai, stage 2	Alkali-lime	0,79	0,94	5,06	2,61	0,9	0,7	1,05	
Cizhou, Guantai, stage 2	*Alkali-lime*	*0,63*	*0,53*	*4,25*	*2,86*	*1,1*	*0,92*	*1,16*	
Cizhou, Gauntai, stage 3	Lime-alkali	0,87	0,65	4,32	3,73	1,5	1,11	1,38	
Cizhou, Gauntai, stage 3	*Alkali-lime*	*0,8*	*0,58*	*4,52*	*3,28*	*1,2*	*0,96*	*1,2*	
Cizhou, Pangncheng	Lime-alkali	0,72	0,32	5,77	4,86	1,4	1,19	1,3	
Cizhou, Pangncheng	Alkali-lime	0,76	0,32	5,25	1,59	0,5	0,42	0,53	
Cizhou, Tianxizhen	Lime-alkali	0,6	0,66	4,92	11,73	4	3,38	3,64	
Böttger (Saxon HP)	Lime-alkali			1,3	13	16,8	16,8	16,8	Hornig-Suter (1985)
Bristol (English HP)	*Alkali-lime*	*1,5*	*2,3*	*4,2*	*3,6*	*1,4*	*0,93*	*1,76*	*Tite and Bimson (1991)*
Plymouth (English HP)	*Alkali-lime*	*1*	*1,8*	*4*	*2,7*	*1,1*	*0,82*	*1,58*	
Buen Retiro (Spanish SP)	Alkali	2,8	0,79	8,98	1,2	0,2	0,15	0,29	Zucchiatti (2008)

Table 8.4:
Alkali and alkaline earth oxide components of clear and white glazes and the impact of excluding soda and magnesia from the calculation of molar ratios that help classify glazes based on the dominant network modifier. $R_2O = Na_2O + K_2O$. RO = MgO + CaO. HP = hard-paste body. SP = soft-paste body.

2-4% potassia, and 1-4% soda (see Table 8.3). The glaze compositions published by Leung *et al.* (2000) for Ding and Cizhou porcelain from the tenth through fourteenth centuries are notable for their magnesia content (up to 2.77%), the detection of which requires handheld units with SDD detectors. The authors found a trend toward increasing alkali content in Cizhou porcelain produced at Guantai, which they based on the molar ratio of alkaline earth oxides (RO = MgO + CaO) and the sum of alkali and alkaline earth oxides (R_2O + RO). Indeed, some of the glaze compositions given by Leung *et al.* (2000) could be classified as alkali-lime type ($CaO/R_2O<1$), similar to those measured on English hard-paste porcelains by Tite and Bimson (1991), who recorded 3-4% calcia and 5-6% totals for potassia and soda. Zucchiatti and co-authors (2008) detected a much lower calcia-to-alkali oxide molar ratio in the glaze layer of a Buen Retiro porcelain and found that the glaze composition closely resembled that of an alkali feldspar to which a small amount of calcium had been added.

The inability of handheld instruments with Si-PIN detectors to detect either sodium or magnesium impacts researchers' ability to make fine distinctions between lime-alkali and alkali-lime type glazes. Table 8.4 demonstrates this by listing alkali and alkaline earth oxide values and their molar ratios for glazes on a variety of Chinese and European porcelains. Without the ability to quantify soda, researchers have to rely on the molar ratio of calcia to potassia instead of the molar ratio of calcia to alkali oxides. The highlighted rows in Table 8.4 show that this approximation leads to the incorrect classification of "borderline" glaze compositions with calcia and alkali oxide molar ratios close to one: relying on the calcia/potassia ratio would lead one to categorize the glazes on some Guantai wares, Bristol and Plymouth hard-paste porcelain as lime-alkali type, instead of alkali-lime. The relevance of such distinctions, and perhaps the need to use an instrumental approach capable of measuring and quantifying sodium and magnesium, depends on the goals of the particular research project.

The detection of lead, the key network modifier in the glaze layers of French and English soft-paste porcelains, is not a problem for handheld XRF units. Tite and Bimson (1991) subdivided these glazes as: high lead type (45-55% PbO), medium (30-40% PbO) or low lead type (about 20% PbO). Though, glassy and soapstone type English porcelain bodies also contain some lead, the concentrations are lower than those in the glaze. The ability to resolve lead emission lines contributed by the glaze, and to exclude photons generated in the body, depends on the lead content and the thickness of the glaze layer in the location of analysis. High lead type glazes have high total mass absorption coefficients that limit the detection of lead L lines to the glaze layer as long it exceeds about 80 microns in thickness (see Table 8.5, and section on "Lateral and depth resolution issues in the analysis of porcelain objects").

Analyte			Critical (infinite) thickness (μm)				Analysis depth (μm)			
	Absorption edge (keV)	Spectral line (keV)	Porcelain body	Clear glaze	Blue over-glaze enamel	High-lead clear glaze	Porcelain body	Clear glaze	Blue over-glaze enamel	High-lead clear glaze
Na (Kα)	1,08	1,04	4,3	3,8	2	1,7	1,4	1,3	0,7	0,6
Mg (Kα)	1,303	1,254	6	5,5	2,7	2,4	2	1,8	0,9	0,8
Al (Kα)	1,559	1,487	11,3	10,2	4,4	3,9	3,8	3,4	1,5	1,3
Si (Kα)	1,838	1,739	11,1	11,6	6,1	5,4	3,7	3,9	2	1,8
P (Kα)	2,142	2,01	8,3	8,3	5,8	5,4	2,8	2,8	1,9	1,8
K (Kα)	3,608	3,314	29,4	29,3	7,9	6,6	9,8	9,8	2,6	2,2
Ca (Kα)	4,038	3,692	39,7	35,1	9,9	8,1	13,2	11,7	3,3	2,7
Ti (Kα)	4,964	4,512	70,4	54,9	15,8	12,9	23,5	18,3	5,3	4,3
Cr (Kα)	5,988	5,415	120,3	92,7	25,4	20,7	40,1	30,9	8,5	6,9
Mn (Kα)	6,537	5,9	150,3	115,3	31	25,2	50,1	38,4	10,3	8,4
Fe (Kα)	7,111	6,405	183,2	140,1	36,9	30,1	61,1	46,7	12,3	10
Co (Kα)	7,709	6,931	230,1	175,3	45,2	36,9	76,7	58,4	15,1	12,3
Cu (Kα)	8,98	8,046	375,2	283,2	69,1	56,5	125,1	94,4	23	18,8
Rb (Kα)	15,201	13,396	1546,7	1148,3	105,4	84,2	515,6	382,8	35,1	28,1
Sr (Kα)	16,106	14,165	1838,9	1364,5	112,1	89,4	613	454,8	37,4	29,8
Y (Kα)	17,037	14,958	2298,5	1701,3	128,6	102,5	766,2	567,1	42,9	34,2
Zr (Kα)	17,998	15,775	2614,9	1935,1	127,4	101,3	871,6	645	42,5	33,8
Ba (Lα)	5,247	4,467	79,1	60,7	16,7	13,6	26,4	20,2	5,6	4,5
Ba (Kα)	37,441	32,068	17763,2							
Pb (Lα)	13,044	10,551	846,4	632,6	104,9	84,8	282,1	210,9	35	28,3

Table 8.5:
Values for the critical (infinite) thickness for analytes of interest in the porcelain body, clear glaze layer and an overglaze blue color. Any radiation excited deeper than the critical thickness will not be detected by the spectrometer. Analysis depth values for the same analytes represent the depth from which 90% of the characteristic X-rays originate. Angle of incidence: 50 degrees. Spectrometer takeoff angle: 65 degrees.

Those seeking to use major, minor and trace element data for provenance and authenticity research should be aware of some of the issues and challenges involved in relating composition-based groups to specific material sources (Budd *et al*. 1996; Adan-Bayewitz, Asaro, and Giauque 1999; Pernicka 1999; Tite 1999; Buxeda i Garrigós, Cau Ontiveros, and Kilikoglou 2003; Freestone 2005; Grave 2009). Most provenance studies of Chinese porcelain have not attempted to relate porcelain components to geological sources, but have focused on trying to establish compositional criteria for distinguishing porcelain made in different locations and time periods. Yap and co-authors initially grouped porcelain bodies based on barium and cesium content (Yap and Tang 1985b; Yap 1986) and then distinguished them using multivariate statistical analysis of data on twelve trace elements (Yap 1992). Leung and co-authors studied major and trace elements in porcelain bodies and clear glazes separately in order to differentiate wares produced at different historic Chinese sites (Leung, Peng, *et al.* 2000; Leung, Stokes, *et al.* 2000).

Stiegelschmitt and Tomandl (1985) analyzed about five hundred porcelain objects made at Meissen during the eighteenth century using XRF. Based on their work Hornig-Sutter (1984) reported typical and extreme concentration values for trace elements in the calcium-rich porcelain body formulations produced at Meissen during the factory's earliest period, prior to 1725, when alabaster was used as a flux. Zirconia (ZrO_2) content range was 200–300 ppm, titania (TiO_2) varied between 2500–4000 ppm, baria (BaO) and strontia (SrO) values were between 150 and 250 ppm. Due to the low calcium content of potassium-rich porcelain bodies made from about 1725, strontium levels were found to be lower than 100 ppm, titanium levels in did not exceed 2800 ppm (Stiegelschmitt and Tomandl 1985:977). Strontium cations have the same charge and are of similar size as calcium cations and thus they may substitute for the latter during the formation of gypsum as well as other calcium minerals, including calcite, dolomite, apatite and plagioclase feldspars (Rice 1987:420; Malainey 2011:42). During the transitional period of 1725–1730, during which the use of potassium feldspar flux in the porcelain body began, the zirconia content of bodies was below 20 ppm and rubidia (RbO_2) levels exceeded 150 ppm. Rubidium content is frequently correlated with potassium content, since the cations of these elements are of very similar size and rubidium could easily substitute for potassium during the formation of the potassium feldspar minerals (Rice 1987:420). Clear glazes in the same period had relatively high titania content, greater than 5000 ppm while zirconia levels exceeded 10 ppm (Hornig-Sutter 1984:3658; Stiegelschmitt and Tomandl 1985:977). In the following period, 1730–1765, the trace element content of porcelain bodies remained the same, however, Stiegelschmitt and Tomandl (1985) imply that the glaze layer contained lower levels of titania and zirconia. Porcelain bodies from the so-called Marcolini period, which starts in 1774 (Hornig-Sutter 1985:110), continue having zirconia levels below 20 ppm but are distinguished from earlier wares by low levels of rubidia, less than 140 ppm and frequently under 100 ppm. It is possible that the lower rubidium levels, compared to earlier feldspar-fluxed porcelain bodies, are the result of a shift in the source of raw materials. Rubidium isotope ^{87}Rb decays with time to form the strontium isotope ^{87}Sr, the resulting change in strontium isotope ratios ($^{87}Sr/^{86}Sr$), and Rb/Sr ratios, serves as the basis for the radiometric dating of mineral or rock deposits (Malainey 2011:188-189). Bodies made after about 1785 are differentiated from earlier periods by their higher zirconia content, which is approximately 100 ppm (Stiegelschmitt and Tomandl 1985).

Overglaze enamels, underglaze colors and gilding

Thanks to extensive research on the technology, composition and microstructure of colored glasses and glazes (examples of overviews include Kingery and Vandiver 1986b; Newton and Davison 1989; Tennent 1999; Weyl 1999; Eppler and Eppler 2000) the qualitative elemental information obtained by XRF, along with visual observations, can be used to infer details about the chemical state and form of colorants and opacifiers present.

Since X-ray fluorescence spectroscopy is an elemental analysis technique it can successfully identify colorants present as diffused or dissolved ions (for example the copper greens) and as discrete pigment particles (such as chromium oxide greens and copper oxide blacks). However ambiguities can arise in cases when an element can be present in a glaze in different forms and with different functions (e.g. as opacifier, flux or colorant). For example, the detection of lead, tin and antimony in a yellow overglaze enamel color could be related to the use of a tin oxide (SnO_2) opacifier with Naples yellow pigment (lead antimonate, $Pb_2Sb_2O_7$), or it may indicate a mixture of Naples yellow and lead tin yellow (type I: Pb_2SnO_4 or type II: $PbSn_{1-x}Si_xO_3$), or even use of a lead-tin-antimony triple oxide (Sandalinas *et al.* 2006). Therefore, caution must be exercised when interpreting data and, when possible, a complementary technique of analysis -such as *in situ* Raman microscopy or X-ray diffraction should be used to confirm hypotheses.

The detection of diagnostic elements may help establish when decoration was applied to a porcelain object based on the terminal dates of use for certain colorants and ancillary materials (e.g. zinc oxide, ZnO) for overglaze and underglaze enamels, as well as for fluxes used for gilding (e.g. lead silicate and bismuth) (see Table 8.7). For example Fuchs II and Mass (2009) used XRF analysis of overglaze colors to reinforce other evidence that a porcelain dish of Chinese origin, depicting the scene of the signing of the Declaration of Independence, is likely a 20th century work. It is important to note, however, that dates of usage are often slightly different for each of the European or Asian porcelain manufactories considered. For example chromium, an element that was discovered in 1797 by Vauquelin, was adopted as an overglaze enamel color in 1802 at Sèvres, but only documented as underglaze color at Meissen from 1817 and was officially listed as an actual component of Meissen overglaze enamels after 1837 (Hornig-Sutter 1985). Though European manufactories adopted the use of chromium-based greens at different dates, the detection of this element in green decorations would undoubtedly point to a nineteenth century application. Halfpenny and Mass (2009) used similar information to argue that a tureen, previously dated to around 1740, could not have been decorated at Meissen prior to the nineteenth century.

The analysis of both the porcelain body and colored decoration is important, especially when the provenance, authenticity or date of an object is in question. It was not unusual for factories such as Meissen, Du Paquier and Sèvres to have a number of clear-glazed but undecorated pieces in stock, waiting for a commission, so that there may be a time lapse between the formation of the body and its decoration, blurring the line between authentic and fake (e.g. Page and Chilton 2001:323). For instance, starting in 1848 undecorated original Sèvres pieces, which were either lost during the revolution or were part of a batch of lesser quality pieces and later sold by the manufactory personnel, started surfacing on the market after embellishment with faked overglaze enamel decoration (Craddock 2009:208).

The composition of the glaze matrix (e.g. alkali-lime silicate, lime-alkali silicate, lead-alkali silicate, etc.) and the firing atmosphere (oxidizing or reducing) can significantly affect the color imparted by the same ion (Weyl 1999). Since a comprehensive coverage of all possible cases is beyond the scope of this chapter, the following discussion of glaze colors focuses on lead-silicate glazes used on hard-paste porcelain. The colors purple, red, blue and black were used both for underglaze and overglaze enamel decoration. While their application technique and sequence differs, the colorants in overglaze and underglaze enamels are the same and thus will be treated together in the following overview.

Opacifiers ("Whites"). Oxides characterized by small particle size and high refractive indices are used to reflect and scatter light in order to confer opacity and "whiteness" to porcelain glazes. Traditionally tin oxide (SnO_2) and titanium dioxide (all polymorphs of TiO_2: brookite, rutile and anatase) have been used in Western production. In China, lead arsenate ($Pb_3[AsO_4]_2$) was used as a result of the eighteenth-century development of the famille rose palette (Kingery and Vandiver 1986a). In modern glazes, these oxides have been substituted by compounds based on zircon (Eppler and Eppler 2000).

Yellows. The main pigments traditionally used to obtain yellow overglaze enamels were Naples yellow ($Pb_2Sb_2O_7$) and lead-tin yellow (the orthorhombic lead-tin yellow I, Pb_2SnO_4, or the cubic lead–tin yellow II, $PbSn_{1-x}Si_xO_3$) in the East as in the West. Of the two forms of lead–tin yellow, type II is the more frequently detected phase. Either phase is only thermally stable up to about 900–950 °C, thus yellows based on lead antimonate, which is thermally stable to 1080°C, were preferred for higher firing glazes (Shaw 1962). In China, transparent yellow overglaze enamels based on iron were common before the rule of the Kanxi Emperor (1662-1722) and were followed by opaque overglaze yellow enamels based on lead and tin as well as lead and antimony, which have been documented in porcelain decorated with *Falangcai* (cloisonné enamel) and *Fencai* (famille rose) overglaze enamel (Miao, Yang, and Mu 2010). Miao and coworkers found that lead-tin yellow was used until the first decades of the twentieth century in famille rose decoration, after which it was substituted by lead antimonate. Chinese craftsmen influenced by western practices also used mixed lead, tin and antimony oxides for yellow decoration from the seventeenth century onwards. Many instances of such triple oxides have been documented in recent years by scientific investigations of European wares (Sandalinas *et al.* 2006). For example, Zucchiatti *et al.* (2008) reported on additions of both tin oxide (SnO_2) and iron oxide (Fe_2O_3) to lead antimonate yellows, leading to the formation of the pyrochlores $Pb_2Sb_{2-x}Sn_xO_{6.5}$ and $Pb_2Sb_{2-x}Fe_xO_{6.5}$, which they identified on eighteenth-century Spanish Buen Retiro hard-paste porcelain and which was also described in historical recipes. Additions of significant quantities of iron to yellow lead antimonate colors to obtain orange-yellow hues have been documented through analysis of a variety of European ceramic and porcelain glazes, confirming references to such practices in historical treatises (Piccolpasso, Lightbown, and Caiger-Smith 1980; Maggetti, Neururer, and Rosen 2009). For example, Tite (2009) cited levels of 3–13 weight percent iron oxide (FeO) dissolved in the glaze of Italian majolicas, while Sendova *et al.* (2007) found microscopic hematite particles in yellow della Robbia glazes colored with Naples yellow. Similar orange-yellow hues were also used on nineteenth-century French faience pieces and Meissen as well as Du Paquier porcelain (Mields and Lauschke 1965:81; Maggetti, Neururer, and Rosen 2009). Bezúr and Casadio (2009) reported the detection of zinc in Naples yellow, which suggests the presence of a lead-antimony-zinc triple oxide, on eighteenth-century Viennese Du Paquier and Venetian Vezzi porcelain and discussed the relation of this glaze formulation to "giallo dei Vasari" (pottery markers' yellow), which had not previously been detected on eighteenth-century porcelain but which was widely employed in Italian Renaissance majolicas (Dik *et al.* 2005; Rosi *et al.* 2011). In Meissen porcelain pieces, zinc has not been detected in yellow antimonate glaze layers prior to the nineteenth century. Starting around 1830, uranium based yellows became widely used in Europe to color glass, enamels, ceramics and porcelain overglaze enamels for decoration and restoration purposes, thanks to the ability of the salts to resist to temperatures up 1050 degrees Celcius

(°C). Towards the last quarter of the nineteenth century, the use of uranium glazes started to spread to Asia and began in the United States of America in the 1860s. The use of this radioactive element, which was discovered in 1789, tapered off after the 1940s and a little later in the East following rising concerns about its dangers to human health (Strahan 2001). Recent modern alternatives for colorants in yellow overglaze enamels include combinations of vanadium and tin oxide ($V:SnO_2$), with occasional additions of titanium dioxide (TiO_2) and iron oxide (Fe_2O_3), as well as praseodymium zircon [$(Zr,Pr)SiO_4$] (Colomban, Sagon, and Faurel 2001).

Reds and browns. Hematite (Fe_2O_3) has been used for both underglaze and overglaze red enamel decoration since ancient times, spanning numerous cultures and geographical regions. The skilled control of particle size and firing conditions allowed craftsmen to achieve subtle variations in tone and an extended range of chromatic effects. The presence of magnetite (Fe_3O_4) or the addition of manganese compounds results in brown hues (Wang *et al.* 2009). Zinc-iron-chromite compounds have also been occasionally used for browns (Colomban, Sagon, and Faurel 2001).

Blues. Cobalt oxides have been used throughout the centuries to impart a blue coloration to glasses, glazes and porcelain enamels. Occasionally cobalt can be found in underglaze or overglaze enamels as discrete pigment particles of cobalt aluminate ($CoAlO_4$) or cobalt silicates (Co_2SiO_4, with olivine structure), but predominantly cobalt is present as ions dissolved in the glassy matrix. Several transition metals can be added to the cobalt-rich frit to influence the final hue. Nickel and iron were added to darken and otherwise enhance the blue color; potassium was used to eliminate green undertones; and copper imparted a turquoise or greenish tint. Stiegelschmitt and Tomandle (1985) discussed the addition of zinc oxide to cobalt blues to adjust the final blue color of enamels at Meissen and dated such practice to 1765, the same date cited by Hornig-Sutter (1985). Fuchs II and Mass (2009) stated that only trace levels of zinc have been detected in Chinese export porcelain dating to the mid-nineteenth century and earlier.

Extensive literature deals with assigning the original cobalt ore sources of blue glazes to specific geological sources and probable time periods of exploitation based on particular types and levels of impurities. While this knowledge base is important, specific findings should be interpreted with an eye toward the possibility that nickel, copper and iron could also have been introduced as intentional additions in the making of the glaze colors. Gratuze *et al.* (1996) conducted an in-depth study aimed at identifying the characteristic fingerprints of metals associated with various cobalt sources. Specifically the authors described cobalt in association with antimony, copper and manganese for Near Eastern sources actively exploited until the twelfth century. Cobalt ores in the Freiberg area of the Erzgebirge mountains (located in present-day Germany), in use between the thirteenth and fifteenth centuries, contained zinc, lead, indium and iron, while cobalt sources in the Erzgebirge in use in the fifteenth and sixteenth centuries contained molybdenum, iron and generally high levels of nickel. Zucchiatti *et al.* (2006) reported measurements on several examples of cobalt blue Della Robbia glazes which contained significant levels of iron and nickel impurities with minimal amounts of arsenic until 1520. The authors linked this impurity pattern (or fingerprint) to cobalt oxide that was a byproduct of silver smelting, a high-temperature process that would have resulted in the loss of volatile arsenic. On the other hand, minerals rich in cobalt such as such as cobaltite (CoAsS) and erythrite ($Co_3(AsO_4)_2 \bullet 8H_2O$) were mined after 1520 in the Schneeberg (Erzgebirge) area and roasted in reverberating furnaces, leading to the presence of significant arsenic and bismuth in addition to nickel and iron impurities in the cobalt oxides produced, which were widely used in the western part of the Mediterranean until the eighteenth century. In agreement with such research Stiegelschmitt and Tomandle (1985) suggested that nickel, copper, arsenic and bismuth impurities detected in cobalt blue glazes of Meissen porcelain point to the use cobalt or cobalt/nickel sulfide or arsenide ores as raw material sources, such as the ores occurring at Schneeberg in Saxony and Neustadel.

A high cobalt-to-manganese ratio with virtually no nickel (Co/Mn≈0.3) is documented in fifteenth-century Vietnamese glazed ceramics, while a manganese-free cobalt ore is used for decoration of blue-and-white pieces today (Liem *et al.* 2000). Extensive analysis of blue-and-white Chinese porcelain, focusing on the cobalt oxide blue underglaze colorant, has uncovered comparable compositional clues on the ore sources used in various periods. Wen *et al.* (2007) reported synchrotron radiation (SR)-XRF data on iron-to-manganese ratios (Fe/Mn) during the Ming (1368-1644) period, one the most accomplished eras of production of blue and white wares in China, together with the Yuan (1271-1368) period. They were able to distinguish three different compositional fingerprints for various cobalt ores, which are specific to different time periods: Fe/Mn ratios over 11 prevalent between 1368-1424; Fe/Mn ratios mostly lower than 2.5 used between 1426-1505; and Fe/Mn ratios between 3 and 10 prevalent between 1506-1620. Their data confirmed and augmented previous research, which dated to ca. 1505 the change from foreign manganese-free arsenical cobalt ores, thought to have been imported from Persia, to local Chinese manganiferous ones. The Xuande period between 1426 and 1435 is to be considered a time of transition that saw the mixed use of imported and indigenous cobalt pigments. Similarly, the later period (1506-1620) may reflect use of a new technology or mixtures of imported and local cobalt ores. In examining manganese-to-iron (Mn/Fe) ratios of cobalt blue areas on Ming and Qing porcelain Yu and Miao (1999a) found that Ming wares of the Xuande period (1403–1435) have Mn/Fe ratios characteristic of manganese-poor and iron-rich imported ores, while those of the Qing period (Kanxi 1662–1722, Yongzheng 1723–1735, Qianlong 1736–1795) have ratios reflecting the use of manganese-rich indigenous ores. High levels of iron and manganese are also obtained after enrichment with high temperature sintering to obtain a dark blue pigment (Wen *et al.* 2007).

Although much more rare, the blue pigment prepared from the semi-precious stone lapis lazuli has also been detected in archaeological ceramics and eighteenth-century porcelain (Colomban 2003a). Colomban and Milande (2006) documented the presence of lapis lazuli used at the interface between the porcelain body and clear glaze in a 1715 sake bottle and a 1725 ewer stopper from Meissen. The blue colorant was purportedly used to increase the perceived whiteness of the glaze and of the overall ware devoid of painted decoration. Using micro-Raman spectroscopy Sendova and co-workers (2007) detected the presence of the chromophore S^{3-} in the lead silicate matrix of Della Robbia glazes colored with cobalt blue and called this a "lead ultramarine". Modern blue glazes are typically obtained with compounds of zirconium and vanadium (Colomban, Sagon, and Faurel 2001).

Greens. Translucent green overglaze enamels are typically based on copper oxide, which is thermally stable up to 1000 °C and imparts a green color through Cu^{2+} ions dissolved in the glassy matrix. Cassiterite (SnO_2) is sometimes added to bring about a paler color. To allow virtuoso shading and subtle gradation of color in painterly applications of green, Naples yellow and cobalt blue are also employed. Tite (2009) detected the use of copper green with lead antimonate particles in Della Robbia glazes, a combination that was referred to as "mixed copper" by Piccolpasso as opposed to "burnt copper", which meant copper by itself. Bezur and Casadio (2009) found many instances in which a combination of the three coloring agents was used to achieve a green hue in eighteenth-century Viennese Du Paquier porcelain. Stiegelschmitt and Tomandle (1985) discussed that in the eighteenth century the only known green colorant used in porcelain decoration was based on copper oxide and characterized Meissen green glazes as ranging from yellowish to blue hues, which were achieved with the addition of lead antimonate and cobalt oxide, respectively. Maggetti *et al.* (2009) also documented the use of cobalt blue in combination with a lead-tin-antimony triple oxide as well as iron and zinc (the latter present at levels of 0.1 weight % as determined by SEM/EDX) in opaque green glazes used in the faience workshop of Le Bois d'Épense, France, during the first decades of the nineteenth century. Following the 1797 discovery of the element chromium by Louis Nicolas Vauquelin in an ore from Siberia, chromium-based greens became available and their first documented use in porcelain glazes took place in Sèvres is in 1802 (Preaud and Ostergard 1997:58). Chromium-based green overglaze enamels containing some zinc oxide replaced copper greens in the Meissen color palette in the nineteenth century (Hornig-Sutter 1985:114).

The renowned aqua green tone of Vietnamese celadons is due to the dissolution of Fe^{2+}/Fe^{3+} (or Ti^{3+}) ions in the glaze, which was fired in a reducing atmosphere (Liem, Thanh, and Colomban 2002). Zinc oxide, which promotes higher color gloss, was not detected in authentic fifteenth-century Vietnamese celadons, but it is found extensively in colored enamels of modern Vietnamese, Chinese, Korean and Hungarian production (Colomban *et al.* 2004).

Blacks. Oxides of transition metals have been used typically to obtain a black coloration. Manganese oxide (MnO_2) and mixed iron and manganese oxides ($Fe_{2-x}Mn_xO_3$), such as those found on Buen Retiro hard-paste porcelain, are common blacks (Zucchiatti *et al.* 2008). The basic manganese oxide black was often modified with copper and cobalt oxides or more occasionally with discrete particles of black magnetite (Fe_3O_4). Manganese, copper and iron spinels ($A^{II}B^{III}{}_2O_4$), modified with different divalent (Co^{2+}, Mn^{2+}, Ni^{2+}, Fe^{2+}, Cu^{2+}, Mg^{2+}, Zn^{2+}) and trivalent (Fe^{3+}, Cr^{3+}, Mn^{3+}, Al^{3+}) cationic substitutions, have also been found to achieve various shades of black (Colomban, Sagon, and Faurel 2001). Bezur and Casadio (2009) identified the combined use of black overglaze enamels with purple of Cassius by porcelain painters to create three-dimensional modeling and hue modulation in figurative Du Paquier porcelain decoration that was executed in a painterly style. They also documented a black of complex composition, likely based on a black chromite-ferrite spinel ($MgFeAl[CrTi]_3O_4$) on an early Meissen turtle-shaped box in a private collection. Fuchs II and Mass (2009) discussed the absence of significant chromium in Chinese export porcelain black enamels predating the twentieth century.

Purples. The most widespread technology to obtain purple hues for porcelain used purple of Cassius, a preparation based on colloidal gold precipitated on stannous hydroxide discovered in the second half of the seventeenth century (Hunt 1981). This technology is still in use today. The color can be enhanced by additions of copper and cobalt to obtain deeper, bluish-purples. Different hues and color strengths can be obtained by varying the proportion of gold and tin in the recipe. Although the know-how to make purple of Cassius was available in China by the first quarter of the eighteenth century and the colorant was used extensively for the famille rose palette, Chinese wares were often decorated with ruby red glazes tinted with colloidal gold only, without the presence of tin. Ricciardi *et al.* (2008) found purples obtained with a combination of manganese-oxide and haematite on soft-paste Capodimonte porcelain in Italy (1743-1759). Manganese was also detected in purple underglaze colors of nineteenth-century Persian glazed tiles (Reiche *et al.* 2009).

Luster colors. Silver and copper metal nanoparticles concentrated in the topmost sub-micrometric layers of the glaze confer a glossy metallic decoration, which sometimes displays iridescent colors. Metallic luster was produced by applying a mixture of copper or silver compounds in a vinegar suspension to a glazed piece prior to firing it in a reducing atmosphere. The decoration resulted from the formation and diffusion of the metallic nanoparticles into the glaze. Depending on the composition, size and shape of the metal nanoparticles, luster decoration allowed the achievement of a wide range of colors: green, yellow, reddish, orange, purplish and brown. The surface often appears to have a metallic sheen when observed with reflected light and this is sometimes accompanied by iridescent or dichroic effects. Kingery and Vandiver (1986b) trace the beginning of painted overglaze luster decoration with elaborate designs, and its subsequent diffusion throughout the Mediterranean basin, to luster decoration on Islamic wares from the ninth through the twelfth centuries. Given the low concentration of metal particles in these metal-glass nano-composite layers and the thinness of these layers, most recent research has been conducted using synchrotron radiation or transmission electron microscopy, which also allowed for studies of oxidation states and nanostructure (e.g. Cotte *et al.* 2010; Gutierrez *et al.* 2010). Although present in very small concentrations, copper and silver can be detected with XRF and thus the types of luster may be differentiated. However, care must be taken in correlating copper and silver contents with luster color if relying on portable XRF measurements alone, as illustrated by Padovani *et al.* (2003) in their study of samples of Umbrian Renaissance luster pottery with a variety of techniques. The measurements revealed that gold and red colors can be attributed mainly to the silver and copper nanoparticles, respectively. Although copper is also present with silver in the gold luster, it was determined by extended X-ray absorption fine structure (EXAFS) to be mostly in the form of Cu^{+} and Cu^{2+} ions and thus not contributing to the perceived luster color, which is influenced only by the fraction of metal ions reduced to nanoparticles.

Gilding. Craftsmen at Sèvres and Meissen learned to "paint" in gold from silversmiths and goldsmiths, modifying their material choices to achieve a cooler or warmer metallic tone as desired. Fine particulate gold, precipitated from nitric and hydrochloric acid (aqua regia) with iron sulfate, was applied either by itself or with a flux to the clear glaze. The gold powder could also be applied on top of an iron oxide-rich red glaze layer, emulating the process of applying leaf gilding over a red bole layer in panel paintings and woodwork. The addition of silver to the gold powder created a cooler tone, which was termed "green gold" in the notebooks of the Sèvres Manufactury, while additions of copper created a warmer hue, so-called "red gold" (d'Albis 1998). Low levels of silver, however, should be considered an impurity of the gold. At Sèvres around 1781 a frit containing lead oxide, sand and calcined ("diaphoretic") antimony was used as a flux for gold powder in similar fashion to the practice at Meissen, where Höroldt employed a lead-rich silicate flux from 1731. Fluxes containing bismuth or lithium are indicative of later periods as bismuth oxide was introduced at Meissen after 1770 or 1775 (Hornig-Sutter 1985:113) and in 1780 at Sèvres (Stiegelschmitt and Tomandl 1985:978).

Silvering. Details in metallic silver were obtained by painting a mixture of finely ground silver chloride powder, turpentine oil and a flux on a glazed surface and firing the piece in a reducing atmosphere. The tendency of such details to tarnish makes it so that sulfur is often detected in association with the silver on blackened historic decorations.

Technical aspects of porcelain analysis using X-ray fluorescence spectroscopy

Instrumentation and operating parameters

Few examples of published porcelain research have been undertaken with handheld XRF instruments. While some of the instrumentation may have been portable, if not handheld, articles from the 1980s and 1990s do not disclose many details of instrument configurations. Most systems used radioisotope-based excitation and were equipped with energy dispersive spectrometers. Yap and co-authors relied on instruments with an americium radioisotope, ^{241}Am (Yap and Tang 1985a; Yap 1986, 1988). Cadmium radioisotope sources, ^{109}Cd, saw use by several researchers as well (e.g. Mazo-Gray and Alvarez 1992; Yap 1992; Page and Chilton 2001). Page and Chilton (2001) used an iron radioisotope, ^{55}Fe, source in addition to the cadmium radioisotope source. Exposure times of over 3 hours and sometimes over 13 hours were reported by Yap (1988) for instruments equipped with an annular americium radioisotope, ^{241}Am, source. Radioisotope-based spectrometers can be made portable with relative ease as the X-ray source is compact and does not require power.

Since these sources produce photons through specific decay mechanisms, their excitation spectrum is simpler than that of X-ray tubes and lacks the continuous bremsstrahlung component that can result in high backgrounds and thus lower sensitivities in X-ray tube-based instruments (Kneip and Laurer 1972). For the same reason, however, the range of characteristic X-ray lines that can be excited by a specific radioactive source is defined by its decay mechanism. ^{55}Fe sources produce manganese (Mn) Kα and Kβ X-rays at 5.9 and 6.5 keV, respectively, and thus are best suited for exciting the K lines of elements with low atomic numbers, sodium (Na) through titanium (Ti), that are present in the porcelain body and clear glaze (Bé *et al.* 2004a). These sources could also excite the L lines of rubidium (Rb) through zirconium (Zr), which are present as trace elements in the porcelain body and clear glaze, and the M lines of gold (Au), lead (Pb) and bismuth (Bi), which are relevant to the study of gilding. The primary decay products of ^{109}Cd sources are silver (Ag) Kα and Kβ X-rays at 22 and 25 keV, respectively, but a small amount of gamma-rays of 88.03 keV are also produced via the decay of the ^{109}Ag daughter nucleus (Bé *et al.* 2004a). Thus instruments with ^{109}Cd sources are most efficient at exciting the K lines of elements of medium atomic number, of which chromium (Cr) through arsenic (As) would be most relevant to studies of porcelain components. Since excitation of the K lines of gold (Au), mercury (Hg) and lead (Pb) lines is also possible, ^{109}Cd sources would allow the study of gilding as well. Americium, ^{241}Am, sources undergo complex decay processes that result primarily in the emission of gamma-rays of 59.5 keV, but also yield neptunium (Np) L lines between 11.89 and 22.2 keV, which are often removed by filters placed in front of the source (Bé *et al.* 2004b). The high energy of the gamma-rays makes ^{241}Am sources useful for the excitation of the K lines of cesium (Cs) and barium (Ba), trace levels of which have been used as indicators of provenance (e.g. Yap 1986).

X-ray tube-based excitation is more complex than that based on radioisotopes since it has a continuous, bremsstrahlung, component as well as characteristic X-ray lines of the anode material if the accelerating potential exceeds the excitation potential of those lines. Compared to radioisotope sources, X-ray tubes offer higher overall intensity excitation that is also "tunable": for a given anode material the energy distribution of the continuous component is controlled by the accelerating potential while the intensity is impacted by both the tube current and the accelerating potential (Jenkins 1999:3-5; Willis and Duncan 2008:3-3 to 3-15). The use of secondary targets and source filters offers further control (see Chapter 14 by Kaiser and Shugar). Since porcelain components contain a wide range of elements of interest, it is of significant advantage to be able to use several optimized sets of operating parameters for a single instrument.

Sample chamber-type energy-dispersive spectrometers with X-ray tube excitation have been used by a number of researchers. Yu and Miao (1996a, 1996c, 1999a) used a Philips DX-95 system equipped with a molybdenum-anode X-ray tube and a Si(Li) detector, which they operated at 35 kV accelerating potential and 50 μA tube current for 300 seconds live time in vacuum. The same anode and detector type was used by Leung and co-authors (1998; 2000) in an EDAX International DX-95 system that was operated with two sets of operating parameters: one for the excitation of elements from sodium through zinc (15 kV, 200 μA) and another set optimized for rubidium through barium (45 kV, 15 uA). X-ray tubes with rhodium anodes have also been used in sample chamber-type systems. Swerts *et al.* (1994) used a Tracor Spectrace 5000, operating it at 35 kV and 40 μA for 500 seconds. Although authors did not explicitly state this, it is likely that the sample chambers were placed under vacuum during analysis to increase their sensitivity to the K spectral lines of sodium, magnesium and aluminum. It must be noted that the cited studies involved fragmentary material (sherds) rather than entire porcelain objects since the instrument chambers have limited volumes and are optimized for the analysis of pellet-shaped samples or materials in sample cups.

Wavelength-dispersive XRF (WD-XRF) has also been used to study porcelain. The sample chambers of these systems pose the same limitations as discussed above for energy-dispersive instruments, though WD-XRF offers superior energy resolution. Both Maggetti *et al.* (2009) and Zucchiatti *et al.* (2008) used powdered samples, which the former used to create fused beads with lithium borate matrix. The grinding of porcelain body and glaze samples into powders and the preparation of fused disks does allow for improved matrix matching between samples and calibration standards, which is key for quantitative analysis. However, XRF as an analytical approach for porcelain objects is typically chosen due to the availability of instrumentation that allows *in situ* non-destructive studies as most routine studies in museum contexts involve whole or restored works, the sampling of which is preferably avoided. Should sample removal be feasible, the use of electron microscopy-based techniques offer certain advantages over XRF, such as the ability to visualize microstructure and to perform spatially-resolved elemental mapping and quantitative analysis at the sub-micron level (Schulle and Ullrich 1982; Tite, Freestone, and Bimson 1984; e.g. Kingery and Vandiver 1986b; Tite and Bimson 1991; Spataro *et al.* 2009).

Portable open-architecture XRF systems, which allow *in situ* analysis of complete and fragmented objects have been used to analyze Böttger stoneware (Miliani *et al.* 2009), which was developed prior to the successful production of hard-paste porcelain at Meissen, Chinese porcelain (Larsson 2008; Fuchs II and Mass 2009), as well as Meissen, Du Paquier, and early Italian-made hard-paste porcelain pieces (Bezúr and Casadio 2009; Halfpenny and Mass 2009). The custom-made portable instrument used by Miliani *et al.* (2009) was equipped with a tungsten-anode X-ray tube source, which was operated at 38 kV and 50 μA, and a Peltier-cooled silicon drift detector (SDD). Tungsten (W) anodes are not ideal for the study of porcelain as the presence of Rayleigh-scattered characteristic L lines at 8.39, 9.67 and 9.95 keV, in sample spectra may interfere with the detection of low-intensity copper (Cu Kα), gold (Au Lα) and mercury (Hg Lα) peaks, respectively (note that Miliani *et al.* mislabeled the W Lα line at 8.39 keV as Cu Kα in Fig. 2). Additionally, the silicon Kα peak at 1.74 keV may be impacted by the presence of the tungsten Mα line at 1.77 keV from the source.

The open-architecture Bruker Artax system used by Bezur and Casadio (2009) was equipped with a molybdenum-anode X-ray tube source, which is better suited for the analysis of porcelain because the zirconium K lines are the only spectral lines with which source-related peaks – the Rayleigh- and Compton-scattered Mo Kα - interfere. Bezur and Casadio (2009) used 50 kV accelerating potential and 800 μA anode current (the maximum power setting, 40 W, for the specific X-ray tube) for 360 seconds of live collection time on the porcelain body and for 180 seconds on the clear glaze, colored enamels, gilding and silvering. No source filters were used and helium was flushed through the conical beam restrictor of the X-ray source (0.65 mm diameter opening) as well as the detector nozzle at a rate of about 1.5-1.8 liters per minute. The authors explored the use of lower accelerating potentials (e.g. 20 kV) to shift the intensity distribution of the continuous portion of the excitation spectrum to longer wavelengths and to preferentially generate molybdenum L lines with the aim that these factors would improve the excitation of aluminum and silicon K lines in porcelain. The Artax system, however, does not allow users to achieve the maximum operating power by using high tube currents, such as 2000 μA, with low accelerating potentials, such as 20 kV. The intensity of the continuum (I_c) is proportional to the square of the accelerating potential [$I_c \propto (kV)^2$] and directly proportional to the current. Thus the use of a lower accelerating potential without the ability to significantly increase the current resulted in a lower overall intensity in the continuum source of excitation for elements with low atomic numbers and resulted in the need for very long collection times. The "tuning" of the excitation spectrum did not appear to enhance the sensitivity of the Artax system to aluminum and silicon, therefore the maximum power settings were used to throughout all measurements.

In their research on Du Paquier, Meissen and early Italian porcelain, Bezur and Casadio (2009) also employed a Bruker Tracer III-V handheld XRF, the X-ray tube of which was equipped with a rhodium (Rh) anode. Unlike the Artax system, the Tracer III-V does allow a significant increase in tube current for use in combinations with lower accelerating potentials, which helps compensate for the loss in the intensity of the continuum excitation at the lower voltage settings. Bezur and Casadio (2009) were able to optimize operating parameters for the quantitative analysis of the porcelain body's major elements (Al, Si, K, Ca, Ti,

Fe) by using 20 kV accelerating potential, 15 μA tube current, for 180 second live time measurements, with the instrument's interior under vacuum and without the use of a source filter. These conditions, however, were not well suited for the quantification of trace levels of rubidium (Rb) and strontium (Sr) in the porcelain body. The use of a 20 kV accelerating potential prevents the generation of rhodium Kα and Kβ X-ray lines from the anode and thus these photons are not available to excite the K spectral lines of rubidium and strontium. Additionally, the relatively weak K lines of rubidium and strontium that are excited by the continuum appear on top of the high background of the continuum radiation. If the quantification of trace levels of rubidium, strontium, yttrium and zirconium is sought and time permits, researchers would be well-served by taking a second set of measurement on the porcelain body using parameters similar to what Bezur and Casadio (2009) used to excite elements of medium and high atomic number in colored glazes and metallic decoration. These included: 40 kV accelerating potential, 1 μA tube current, for 180 second live time measurements, with the instrument's interior under vacuum and without the use of a source filter. Under these conditions the excitation spectrum includes both a continuum component, with a maximum intensity around 15-16 keV, as well as rhodium Kα and Kβ X-ray lines from the anode, all of which contribute to the excitation of the K spectral lines of metals and metalloids of interest in colored glazes and the L spectral lines of gold, lead and bismuth.

Lateral and depth resolution issues in the analysis of porcelain objects

Both lateral and depth resolution are relevant for assessing the appropriateness of handheld instruments for the characterization of porcelain artifacts. The lateral resolution of instruments depends on the diameter of the X-ray beam on the surface of the object. Since current handheld XRF systems are not equipped with mono- or polycapillary optics, the beam diameter at the surface is determined by pinhole-type divergence-limiting devices placed in the beam path, often referred to as "collimators". Not only is the diameter of the beam important with respect to the size of areas of interest, but so is the accuracy with which the aperture on the instrument's nose piece can be positioned by the analyst. Together, these factors determine whether the analyst is capable of discriminating details, such as areas of overglaze enamels of different color, or collecting meaningful data from small areas of the unglazed porcelain body. The layering of colors and the use of finely-rendered minute details on European and Chinese export porcelain can make the use of handheld XRF instruments particularly challenging. The integrated cameras of some handheld instruments aid users in aligning the instrument aperture with the desired analysis location (Table 8.1). If the instrument lacks a camera the operator is limited to using the relatively large aperture on the nosepiece of the instrument as a positioning aid even though the X-ray tube and its divergence limiting optics may produce a beam of smaller diameter (see Bruker Tracer III-V in Table 8.1.) Users of handheld instruments may be able to map the beam location and diameter with respect to the nosepiece aperture by exposing X-ray sensitive film that is registered to the nosepiece as they operate the instrument.

Since porcelain objects, other than biscuit ware, have a multilayer structure it is important to explore how depth resolution impacts analysis results. This requires that we understand both how deep the primary X-ray beam can penetrate (i.e. the penetration depth), from what depths fluorescent X-ray lines of interest can emerge (i.e. escape depth), and how these values relate to the thickness of various layers in a glazed porcelain object. When analyzing a surface with a clear glaze layer it may be possible for beam X-rays to excite atoms both in the clear glaze and the porcelain body. Though lower energy analytical lines coming from the body may be too attenuated by the clear glaze layer to be detected, some higher energy characteristic X-rays could emerge and contribute to the spectrum obtained. This scenario describes a case in which partial depth resolution is achieved: detection of elements with low energy lines is restricted to the clear glaze layer while higher atomic number elements are detected from both the glaze and the body.

The penetration depth of the primary X-ray radiation from the X-ray source depends on its energy (or wavelength) and the density of the sample. The escape depth of fluorescent X-rays is determined by the energy of the characteristic X-ray line of interest as well as the density (ϱ) of the sample. The concept of critical or infinite thickness allows us to connect the requirements that primary X-rays from the source must be able to travel into the sample in order to excite fluorescent X-rays and that the latter must be able to escape from the matrix in order to be detected by the instrument. Willis and Duncan (2008:9-10) define a specimen "to be 'infinitely thick' for a particular wavelength or energy when an increase in mass or thickness of the specimen results in no increase in emitted intensity of that wavelength or energy." This means that the emitted intensity of an infinitely thick specimen (I_∞) for an energy of interest is considered to be 1 (or 100%), while the emitted intensity thinner specimens (I_t) is between 0 and 1. Willis and Duncan (2008:9-11) further explain that any radiation from a given analyte element i that is excited deeper within the sample that the infinite (or critical) thickness, t, for that element "will never emerge from the sample and never be 'seen' by the spectrometer, i.e. will not be measured." To calculate the critical thickness, t, in centimeters the ratio I_t/I_∞ is considered to be 0.999, representing the case when

99.9% of the radiation for a given analyte element can emerge and be "seen" by the spectrometer (Equations 1-4). Note that the equations presented by Willis and Duncan (2008:9-11) can be derived from the formula for calculating transmitted beam intensity, as given by Jenkins (1999:7-8).

(1) $$log_e\left[1-\frac{I_t}{I_\infty}\right] = -\mu^*\varrho t$$

(2) $$log_e[1-0.999] = log_e[0.001] = -6.91$$

(3) $$-6.91 = -\mu^*\varrho t$$

(4) $$t_{0.999} = \frac{6.91}{\mu^*\varrho}$$

$t_{0.999}$ = critical or infinite thickness [cm]

μ^* = total effective mass attenuation coefficient [cm^2/g]

ϱ = density [g/cm^3]

(5) $$\mu^* = \mu(\lambda_0) \times \csc\Psi_1 + \mu(\lambda_f) \times \csc\Psi_2$$

$\mu(\lambda_0)$ = mass attenuation coefficient of the matrix for the wavelength of the incident radiation [cm^2/g]

$\mu(\lambda_f)$= mass attenuation coefficient of the matrix for the wavelength of the emitted radiation [cm^2/g]

Ψ_1= incident angle of radiation from X–ray source

Ψ_2= takeoff angle of spectrometer

(6) $$\mu(\lambda) = \mu_{z1}w_1 + \mu_{z2}w_2 + \ldots = \sum_{i=1}^{6} \mu_{z1}w_1$$

The calculation of the critical thickness for a specific element requires the calculation of the total effective mass attenuation coefficient μ^* of the sample matrix (e.g. the porcelain body, the clear glaze, overglaze enamels), which is the sum of mass attenuation coefficients at λ_0, the wavelength (or photon energy) of the primary radiation that is most likely to excite the emission of an X-ray line of an element, and at λ_f, which is the wavelength of the characteristic (analyte) radiation (Equation 5, Figure 8.1). The total effective mass attenuation coefficient takes into account the incident angle of the exciting radiation (Ψ_1) and the takeoff angle of the spectrometer (Ψ_2). If, for example, the element of interest is calcium (Ca), then the mass attenuation coefficient experienced by the exciting radiation, $\mu(\lambda_0)$, is calculated for incident X-rays with energy equal to the absorption edge of a calcium K-shell electrons, 4.038 keV. The mass attenuation coefficient experienced by the fluorescent calcium X-rays emerging from the sample, $\mu(\lambda_f)$, is calculated for the calcium K$\alpha_{1,2}$ spectral lines at 3.691 keV. At each wavelength or energy of interest, the mass attenuation coefficients μ of porcelain components are based on the individual coefficients μ_z of each component element with atomic number Z (Hubbell and Seltzer 2004) and their weight fraction w (Equation 6).

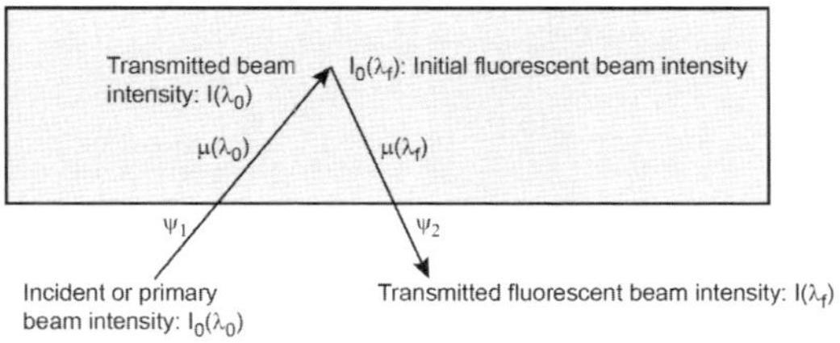

Figure 8.1:
Diagram illustrating the attenuation of the incident or primary X-ray beam at a given wavelength (such as the absorption edge of an element of interest) as it travels through the sample matrix as well as the attenuation of the fluorescent beam at the characteristic wavelength of an element of interest as it travels toward the detector.

Table 8.6 lists the weight percent oxide compositions and densities of porcelain components used in calculating the critical thickness values for a host of elements of interest in the analysis of porcelains, which are presented in Table 8.5. Additionally, Table 8.5 also lists the "analysis depths", representing the depths from which 90% of the characteristic X-rays originate ($I_t/I_\infty = 0.90$). Composition values for the porcelain body, clear glaze and blue overglaze enamel for a K'ang Hsi famille verte porcelain were reported by Kingery and Vandiver (1986b:153 and 161) and the composition of clear glaze with high lead content on Bow bone-ash soft-paste porcelain was taken from Tite and Bimson (1991:9). The density of a porcelain body was measured by Yu and Miao (1999b), while the densities of clear glazes and the blue overglaze enamel color were estimated using Fluegel's method (2007). The thickness of clear glaze layers on porcelain has been measured by a number of authors: Zucchiatti *et al.* (2008) found that clear glaze layers on Buen Retiro porcelain were 110–120 microns, while Yu and Miao (1996b) found that clear glazes on blue-and-white Chinese porcelain were thicker, 200–500 microns. Cross sections from Chinese porcelain presented by Larsson (2008:16-17) show clear glaze layers that are typically 250 microns thick, but may be as thin as 125 microns or as thick as 750 microns. Overglaze enamel layers tend to be thin compared to clear glaze layers, Zucchiatti *et al.* (2008) reported values under 20 microns for Buen Retiro porcelains.

	Porcelain body	Clear glaze	Blue overglaze enamel	High-lead clear glaze
	K'ang Hsi porcelain (Kingery and Vandiver 1986:153,161)			Bow porcelain (Tite and Bimson 1991)
Na_2O (%)	3	2,1	0,35	0,6
MgO (%)			0,35	
Al_2O_3 (%)	24,5	15,1	1	0,9
SiO_2 (%)	69,7	69,6	32,3	37,2
K_2O (%)	2,8	3,3	0,35	4,8
CaO (%)	0,2	10,1	0,35	0,7
Fe_2O_3 (%)			0,3	
CuO (%)			6	
PbO (%)			58,7	55,8
Density (g/cm^3)	2,337	2,45	3,74	4,127
Thickness (μm)		110-120†, 200-500‡	Less than 20†	

Table 8.6:
Weight percent oxide composition values used to calculate the critical (infinite) thickness and analysis depth values in Table 5. The density value for the porcelain body is from Yu and Miao (1999). Density of clear glazes and blue overglaze enamel was estimated using Fluegel's method (2007). Thickness values for the clear glaze and overglaze layers were measured by Zucchiatti *et al.* (2008) for Buen Retiro porcelain tiles† and by Yu and Miao (1996) for blue and white Chinese porcelain‡.

The critical thickness values in Table 8.5 highlight that porcelain bodies and most clear glazes are composed predominantly of low atomic number elements leading to a relatively low attenuation of both the primary X-ray beam and the characteristic X-ray lines of analytes, while lead-silicate clear glazes and overglaze enamels attenuate X-rays to a greater extent. The greatest depth from which the characteristic spectral line of a particular element can emerge is dependent on the element's atomic number and the energy of X-ray line. In the porcelain body, aluminum and silicon atoms in only the topmost 4.3 and 6.0 microns, respectively, contribute intensity to spectra collected, whereas energetic rubidium spectral lines can be excited and can emerge from as deep as 1546.7 microns. The fact that the critical thickness value for the zirconium Kα spectral line exceeds 2600 microns (2.6 mm) in a porcelain body matrix means that for accurate quantitative analyses of this trace element one must choose sections of exposed porcelain body that are at least 2.6 mm thick. Yap and Tang (1985b) chose to determine barium concentrations in porcelain bodies based on the Ba Kα line (32.07 keV). The critical thickness value in a porcelain body matrix for this analytical line exceeds 17 mm, which means that only measurements taken on sections that exceed 1.7 cm in thickness could be considered to have Ba Kα line intensities independent of thickness. Since this criterion is rarely fulfilled by porcelain objects, Yap and Tang (1985b) had to generate calibration curves for three different thickness values: 2, 3 and 4 mm. Had the authors chosen the Ba Lα spectral line (4.467 keV) they could have relied on a single calibration curve, since specimens and standards would have been considered infinitely thick with respect to the detection of barium at values of about 80 microns or greater.

The low critical depth values for the Kα lines of Al, Si, P, K, Ca, Ti, and the Lα line of Ba in a leaded glass matrix means that in spectra obtained from overglaze enamel surfaces the contribution of these elements is limited to the top overglaze color layer, if the thickness of the layer is around 20 microns (for layer thickness see Zucchiatti *et al.* 2008). Characteristic K lines for transition metal elements (e.g. Cr, Mn, Fe, Co, Cu), however, can be excited in, and emerge from other overglaze colors below the top color layer as well as from the clear glaze layer. If the clear glaze layer is relatively thin, ≤150 microns, then K lines for Rb, Sr, Y, Zr, and Pb may reach the detector from the porcelain body as well (for layer thickness, see Yu and Miao 1996b; Zucchiatti *et al.* 2008). Lead-containing clear glazes, especially those with high lead content, have similar mass attenuation coefficients as overglaze enamels, which results in similar critical thickness values for elements of interest. If the thickness of leaded clear glaze layers is similar to that of clear glaze layers without lead, around 100-500 microns (Yu and Miao 1996b; Zucchiatti *et al.* 2008), then the Kα spectral lines for copper and all lighter elements as well as the Lα spectral line for barium will be contributed only by the glaze layer. For glaze layers thinner than 150 microns, the intensity contribution from the Kα spectral lines of rubidium, strontium, yttrium and zirconium and the Lα of lead may also come from the porcelain body below the clear glaze layer.

The X-ray attenuation of unleaded clear glaze layers is similar to that of the body, which is reflected by critical thickness values in Table 8.5 that are quite close for these two matrices. Spectra obtained from clear glaze surfaces allow the resolution of the Al, Si, P, K, Ca, Ti, Cr and Ba content from the composition of the porcelain body if we assume that the thickness of the clear glaze layer is at the low end of the 100-500 micron range reported in the literature (Yu and Miao 1996b; Zucchiatti *et al.* 2008). If the thickness of the glaze is 200 microns or greater, then the Mn, and Fe content of the spectra will also be restricted to this layer. Given that the critical thickness for the Co Kα spectral line is about 175 microns, it is possible to detect cobalt - as well as Ni, Cu and As, which have higher-energy Kα lines – in underglaze blues as long as the clear glaze is not much thicker than 150 microns. In spectra of clear glaze layers, the Kα lines for Rb, Sr, Y, Zr, and Pb will reflect the contribution of both the glaze and the porcelain body. To summarize, if the analyst can position the instrument on a surface with a relatively

Elements detected	Coloring Agent	Possible additions	Color	Usage and terminal dates	Bibliographical references
Sn	SnO_2		White	Widespread in European porcelain	Piccolpasso *et al.*, 1980
Pb, As	$Pb_3(AsO_4)_2$		White	Used for Chinese wares, from the first quarter of the 18th C	Kingery and Vandiver, 1986a
Ti	TiO_2 (anatase, rutile and brookite polymorphs)		White	Antiquity	Colomban *et al.*, 2001
Zr, Si	$ZrSiO_4$ or ZrO_2		White	Modern times	Colomban *et al.*, 2001
Mn, Fe	MnO_2, Fe ions		Black		
Mn, Co	MnO_2, Co ions		Black		
Mn, Cu, Fe	spinels ($A^{II}B_2^{III}O_4$).	various divalent and trivalent cationic substitutions possible: Co^{2+}, Mn^{2+}, Ni^{2+}, Fe^{2+}, Cu^{2+}, Mg^{2+}, Zn^{2+} and three-valent ions: Fe^{3+}, Cr^{3+}, Mn^{3+}, Al^{3+}	black		Colomban *et al.*, 2001
Fe, Cr	$(Fe, Mg)(Cr, Al, Fe)_2O_4$ chromite		black		
Co	Co ions	(Zn additions at Meissen postdate 1760/65) Cu, Fe, Ni and K added to affect color (see text for suggested periods of use of ores with specific levels o f Mn, Fe, Ni ad As impurities)	Blue		Gratuze *et al.*, 1996
Na, Al, Si, S	S^{3-} ion		Blue	Lapis lazuli, used on Meissen and Italian wares (ancient ceramic pieces and Della Robbia glazes)	Colomban, 2003a
S	S^{3-} ion		Blue	S^{3-} ion in lead silicate matrix documented in Della Robbia glazes and termed "lead ultramarine"	Sandova *et al.*, 2007
Zr, V			Blue	Modern	Colomban *et al.*, 2001
Cu	Cu^{2+} ions	SnO_2 for lighter colors, Co blue and Naples yellow ($Pb_2Sb_2O_7$) for gradations of green	Green	Antiquity	
Cr	Cr_2O_3		Green	19th C	Hornig-Sutter, 1985
Pb, Sb	$Pb_2Sb_2O_7$	SnO_2 for lighter colors, Fe for yellow-orange; Zn for luminous yellows (19th C addition at Meissen, 18th C for Du Paquier; Mediaeval/Renaissance practice for Italian ceramics and glass)	Yellow	Used in the Near East in the early second millennium BC	Wainwright *et al.*, 1986

Pb, Sn, Sb	Pb-Sb-Sn triple oxide ($Pb_2Sn_2O_6$–$Pb_2Sb_2O_7$)		Yellow	16th C in Europe	
Pb, Sb, Zn	Pb-Sn- Zn triple oxide		Yellow	16th C Europe. At Meissen in the 19th C in Vienna (Du Paquier) in the 18th C. In Italian maiolicas and glass since the Renaissance ("giallo dei Vasari")	Dik *et al.*, 2005; Wainwright *et al.*, 1986
Pb, Sn	lead-tin yellow I (Pb_2SnO_4), orthorombic; lead–tin yellow II ($PbSn_{1-x}S_{ix}O_3$), cubic		Yellow	Middle ages	
Fe	Fe ions		Yellow	In transparent yellow glazesin China, mostly before the 17th C	Miao *et al.*, 2010
U	U^{6+} ions		Yellow	19th C until the 1940s	Strahan, 2001
V, Sn	V ions, SnO_2		Yellow	Modern	Colomban *et al.*, 2001
Zr, Pr, Si	(Zr, Pr) SiO_4		Yellow	Modern	Colomban *et al.*, 2001
Mn	Mn^{3+}		Purple	Identified in 19th c Persian glazed tiles	Reike *et al.*, 2009
Au, Sn	colloidal Au precipitated on Sn	Cu, Co for darker, bluish purples	Purple	2nd half of the 17th C	Hunt, 1981
Au	colloidal Au		Ruby red/ purple	China	Kingery and Vandiver, 1986a
Fe	Fe_2O_3 (haematite)		Red	Antiquity	
Fe Mn	Fe_2O_3 and Mn oxide		Brown	Antiquity	
Cr	$(CrO_4)^{2-}$ containing compounds		Brown	Documented in Capodimonte / Buen Ritiro 18th C porcelain	Ricciardi *et al.*, 2008
Zn, Fe, Cr	zinc-iron-chromite		Brown	18th C	Colomban *et al.*, 2001
Mn, Al	manganese alumina pink		Pink		Colomban *et al.*, 2001
Cr, Sn, Ca, Si	tin sphene structure with dissolved Chromium oxide		Pink		Colomban *et al.*, 2001

Table 8.7:
Elements used for the decoration of porcelain through the ages.

thick clear glaze coating without blue underglaze, then it is possible to resolve information regarding bulk glaze components (Al, Si, P, K, Ca, Ti, Mn, and Fe), but not with respect to typical trace-level components (Rb, Sr, Y, Zr) from the elemental composition of the porcelain body.

Yu and Miao (1999b) concluded the same based on their calculations of half-value thickness for major, minor and trace elements in clear glazes from which they inferred critical thickness values. They stated that during the analysis of typical glaze layers, which are 200–500 microns thick, only Rb, Sr, Y and Zr are contributed by the underlying porcelain body. They also found that analysis of clear glaze surfaces over blue underglaze decoration resulted in the contribution of Co and Mn to the spectrum by the underglaze layer. Based on these findings, they suggested that the analysis of glazed surfaces would result in obtaining the "signature" spectrum of a porcelain that comprised information from "the glaze, colorant and the body" (Yu and Miao 1999b:23) and that this "confirms the feasibility of the attribution of porcelains using EDXRF, which is based on the variation of the materials used for the glaze, colorant and body in different periods and localities." This approach, however, conflates trace element information (Rb, Sr, Y, Zr) from the clear glaze and underglaze blue decoration which could have been applied to an unglazed body at a different time and/or place. Additionally, Yap (1989) found deviations in Rb, Sr and Zr concentrations when analyzing glazed areas on the base versus on other areas of Qing dynasty porcelain (1644–1912), which he interpreted as evidence for the practice of using different glaze formulations for "base regions". Leung *et al.* took a conservative approach in their study of Song (960–1279) and Yuan dynasty (1271–1368) porcelain and excluded Sr, Y and Zr concentrations from their statistical analysis of clear glaze compositions "because some of the glaze layers can be as thin as 0.1–0.2 mm and for these heavier elements the interference from the underlying body cannot be totally avoided" (2000:132). The authors also collected separate measurements from unglazed areas of the porcelain body in order to track geographical and chronological differences in the composition of the body and clear glaze layers separately.

The impact of instrument apertures on the quantitative analysis of porcelain bodies

The difficulty of discriminating small features of interest on porcelain objects with handheld instruments has already been mentioned in relation to lateral resolution issues. While built-in cameras aid in the accurate positioning of the aperture on the front or "nose" of the instrument with respect to locations of interest, the ability of the analyst to carry out quantitative analysis of porcelain bodies is limited by the portion of the area of the aperture covered by the exposed porcelain body. If the aperture is not entirely covered by the exposed body the measurements cannot be considered quantitative. This is often the case for porcelain objects with narrow foot rings – such as smaller plates and bowls, teacups, teapots – and lids of containers with narrow unglazed rims.

To illustrate the effect that the incomplete coverage of an instrument's aperture has on measurements, binder-free kaolin pellets were prepared using a hydraulic press and a 13 mm die. The solid line in Figure 8.2 depicts a spectrum collected from the top circular surface of the pellet, which covered the aperture completely. The other spectrum was obtained by analyzing the edge of the pellet. The latter mimics the situation when the handheld spectrometer is used to analyze the unglazed porcelain body exposed on the bottom of a narrow foot ring. Not only are X-ray lines overall less intense in the spectrum of the pellet edge, the relative intensities of aluminum and silicon as well as potassium and calcium peaks also differ. Quantification of the spectrum would result in very low totals for the component oxides and the underestimation of alumina and potassia relative to silica and calcia, respectively. Awareness of the limitation of handheld instruments to obtain representative quantitative information from sample features that are smaller than their nose aperture is key for judging the appropriateness of this instrumental approach to specific objects.

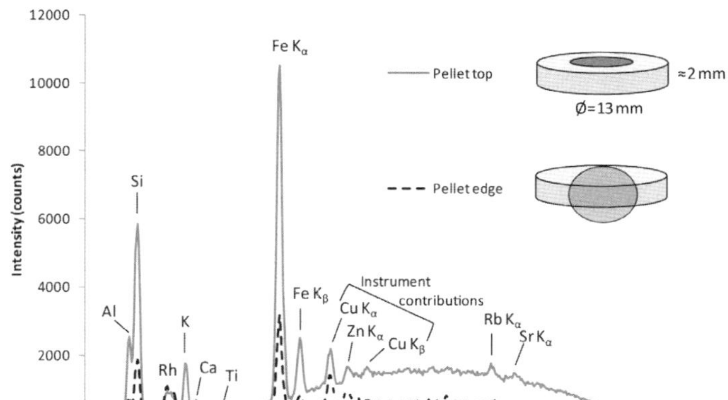

Figure 8.2:
Spectra obtained from the top surface and the edge of a pressed kaolin pellet using a Bruker Tracer III-V equipped with a Rh anode X-ray tube. Operating parameters: 20 kV, 15 μA, no filters, vacuum, 180s live time. The drawings illustrate the size of the elliptical opening on the instrument's nose relative to the pressed pellet.

Calibration standards

The quantitative analysis of porcelain bodies and clear glazes using XRF is challenging not only because most handheld and portable instruments lack the ability to detect sodium and magnesium, but also because calibration standards and analyte matrices should, ideally, have matching properties. Existing quantitative analysis of a wide range of soft- and hard-paste porcelain bodies (see Porcelain bodies and clear or white glazes), some of which was obtained using alternative analytical techniques, can be of help to researchers in determining the range of major, minor and trace element levels that should be bracketed by calibration standards if an empirical calibration method is employed. As cited earlier, the density of Chinese porcelain bodies has been determined empirically by Yu and Miao (1999b) to be 2.337 g/cm^3 (see Table 8.6). This figure is consistent with densities measured for fired kaolin clays (Baumann and Keller 1975), though one must keep in mind that hard-paste porcelain formulations also contain quartz, mica and feldspar. The density of clear glazes can be estimated based on their composition (Fluegel 2007), lime-alkali glazes being approximately 2.45 g/cm^3. Particle or phase size and distribution are matrix properties relevant to porcelain bodies but not clear glazes, unless they contain an abundance of gas or air bubbles.

Though fired porcelain tile calibration standards would most closely approximate actual porcelain matrices, their use has rarely been reported and they are not commercially available. Yu and Miao (1996c, 1996b) cut disks of 42 mm in diameter from actual historic porcelain shards to use as XRF calibration standards and determined peak intensities for thirteen trace elements (Ti, Mn, Fe, Co, Ni, Cu, Zn, Ga, Pb, Rb, Sr, Y, Zr) by analyzing the glazed surface of the disks. Researchers removed the top 1 mm – which included the clear glaze, blue underglaze and some of the porcelain body - from the remaining portion of the shards, then powdered this composite sample and determined the elemental composition by inductively coupled plasma optical emission spectroscopy (ICP-OES). Calibration curves were generated by plotting X-ray spectral line intensities against the concentrations determined by ICP-OES. This approach is problematic and impractical for a number of reasons, including that one must assume homogeneity of layer thicknesses and compositions throughout each shard that has been subdivided to be analyzed by XRF and ICP-OES. Additionally, the critical or infinite thickness values for Rb, Sr, Y and Zr in a clear glaze matrix most likely exceeded 1 mm (see Table 8.5). While the peak intensities of these elements in XRF spectra represented data from as deep as 1.9 mm, their concentration was only determined for the top 1 mm by ICP-OES.

While porcelain bodies are highly vitreous, they differ from glass standards due to the presence of crystalline phases in the matrix. The lack of fired certified reference materials, which has been noted by others investigating the suitability of XRF to the quantitative analysis of iron-containing ceramics (Papadopoulou *et al.* 2004), however, makes glass standards a reasonable choice for calibration purposes for both the porcelain body and the clear glaze layer. Glass certified reference materials (CRM) are available from a number of sources, including: National Institute of Standards and Technology (NIST), Breitländer Eichproben + Labormaterial GmbH, and the Society for Glass Technology. Certified values for elements contained in these glasses can be used for calibration purposes. However, since most glass CRMs are of the soda-lime type, their potassium content is often insufficient to allow empirical calibration for the analysis of potassium-rich porcelain bodies and glazes. It should be noted that Breitländer has CRMs with up to 8.4% potassia, though some contain lead oxide as well, and the company also markets synthetic glass standards for use in XRF calibration or in the validation of results with a good range of major, minor and trace element components (Reference Materials: BR A3, BR B2, BR C3, BR D3, BR E2, BR F2). A series of NIST certified reference materials is available to aid the determination of trace elements in glasses (NIST SRM 610, 611, 612, 613, 614, 615, 616, 617), which are best suited for creating calibration curves for rubidium and strontium in porcelain bodies and clear glazes. Stiegelschmitt and Tomandl (1985) prepared their own glass calibration standards was by adding K, Ca, Ti, Mn, Fe, Rb, Sr and Zr in some form to a base glass (73% SiO_2, 21.5% Na_2O and 5.5% Li_2O). The density of the base glass formulation is estimated at 2.43 g/cm^3 (using Fluegel 2007), which is close to the densities of porcelain bodies and unleaded clear glazes.

Pressed pellets prepared from clay and other mineral CRMs, alone or in combination with glass CRMs, have been used for calibration purposes with relative frequency for the analysis of archaeological ceramics (e.g. Papadopoulou *et al.* 2004, 2006; Papageorgiou and Liritzis 2007) and porcelain (Bezúr and Casadio 2009). Certified reference material powders of feldspar and clays are available from the National Institute of Standards and Technology (NIST), Breitländer Eichproben + Labormaterial GmbH, and Brammer Standard Company, Inc., among others. Papadopoulou *et al.* (2004, 2006) cite the use of SARM 69, which is derived from an iron-rich "pot shard" and is produced and distributed by MINTEK (South Africa) that may also be helpful in the quantitative analysis of porcelains, even though the latter contain much lower levels of iron. Practical information on the preparation and use of pressed pellets is available in published sources (e.g. Injuk *et al.* 2006) and is highly accessible through the sample preparation handbook in the knowledge base of SPEX SamplePrep (www.spexsampleprep.

com). Powder CRMs may need to be dried at elevated temperatures to eliminate water adsorbed onto particle surfaces. Thorough cleaning of pellet dies is crucial in order to avoid cross-contamination of the reference materials. The highly polished faces of steel pellets in die sets need to be inspected regularly to assess whether replacement is needed as the reference powders tend to be abrasive.

If XRF is used for *in situ* analysis of porcelains, then the "sample" is not powdered and pelletized in the same manner as the calibration standards and differences in matrix properties, such as particle size and density, will exist. Density differences may be overcome by applying sufficient pressure to the powdered CRM in the pellet die using a hydraulic press. Long *et al.* (1987) found that the density of pressed pellets of montmorillonite, a clay mineral, shows a logarithmic trend, measuring 2.18 g/cm^3 at 20,000 psi applied pressure, and "should converge to a density representative of the individual particles in the powder, which is 2–3 gm/cm^3 for montmorillonite". If one uses a 13 mm diameter pellet die and applies a 10 ton (US) load, this results in the exertion of 48.6 tons per square inch (tsi), which is equivalent to about 97,000 psi pressure. This value greatly exceeds the 20,000 psi applied by Long *et al.* (1987) to montmorillonite powders and suggests that pressed pellet densities similar to those of porcelain (around 2.337 g/cm^3) may be achieved.

Pressed pellets can be prepared from in-house mixtures of oxides or other compounds of the major, minor and trace components of porcelain bodies and glazes. Yap and Tang (1985b) created a series of standards for the quantification of barium based on addition by mixing a barium compound with ground-up historic porcelain and 9.1% starch binder (they did not state whether starch was measured as weight- or volume-percent). Yap (1988) used the same addition-based approach to create pellets from ground porcelain and known amounts of Mn, Fe, Ni, Cu, Zn, Ga, Ge and As. Leung and co-authors prepared pressed pellets containing known amounts of Na, Mg, Al, Si, K, Ca, Cr, Mn, Fe, Ni, Cu, Zn, Sr, Y, Zr and Ba in a matrix of alumina and silica, which they bonded with 5 weight percent polyvinylpyrrolidone (Leung *et al.* 1998; Leung, Peng, *et al.* 2000; Leung, Stokes, *et al.* 2000).

A case study: the composition of eighteenth century Viennese porcelain bodies

The success of hard-paste porcelain production in Vienna has long been tied to the 1719 elopement of Samuel Stöltzel, an "arcanist, paste maker, and kiln master" (Busson 1994), from Meissen to the Du Paquier factory. Possible transfer of technology in the other direction has been associated with the 1720 flight of the master painter Johann Gregorius Höroldt from Vienna to Meissen, who likely took with him knowledge regarding the color palette developed at the Du Paquier workshops (Busson 1994). In 2006 and 2007, the authors undertook research to explore whether material evidence bore out the above cited instances of transfer of knowledge and skill between the two factories due to the movement of key personnel (see Bezúr and Casadio 2009). The project was a joint venture among the Melinda and Paul Sullivan Foundation for the Decorative Arts, the Art Institute of Chicago, and the Metropolitan Museum of Art. The scope of the work included the study of a representative group of Du Paquier porcelain pieces in addition to objects made at Meissen as well as the Italian Vezzi (Venice) and Doccia (Florence) factories. A total of 92 objects or object sets, which comprised 136 individual pieces, were analyzed using X-ray fluorescence spectroscopy.

Compositional and statistical analysis methods

Objects in the collection of the Art Institute of Chicago, as well as a few privately-owned pieces, were analyzed using an open-architecture Bruker Artax system, which was equipped with a molybdenum-anode X-ray tube source. As discussed in the section "Instrumentation and operating parameters", 50 kV accelerating potential and 800 μA anode current were used and that helium was flushed through the 0.65 mm conical beam restrictor of the X-ray source and the detector nozzle at about 1.5–1.8 l/min. Live measurement times were 360 seconds on porcelain body surfaces and 180 seconds on the clear glaze, colored enamels, gilding and silvering. Measurements on porcelain pieces at the Metropolitan Museum of Art and on most privately-owned works were carried out using a Bruker Tracer III-V handheld XRF, the X-ray tube of which was equipped with a rhodium (Rh) anode. Operating parameters for the analysis of porcelain bodies and clear glazes were: 20 kV accelerating potential, 15 μA tube current, 180 second live time measurements, vacuum atmosphere in the instrument's interior, no filters. Clear glazes were also analyzed using the same set of conditions employed for colored glazes and metallic decoration: 40 kV accelerating potential, 1 μA tube current, second live time measurements, vacuum atmosphere in the instrument's interior, no filters.

Findings regarding underglaze and overglaze enamels as well as metallic decoration on Du Paquier objects have already been summarized in this chapter (see section – Overglaze enamels, underglaze colors and gilding), thus the following discussion focuses on the quantitative analysis of porcelain bodies. Though Schöde (1988:4) published data on the minor and trace element components of

Du Paquier porcelain bodies and clear glazes, the values and oxide formulas given appear to be incorrect and prevent direct comparison with compositional data on Meissen porcelain (e.g. Stiegelschmitt and Tomandl 1985). In order to ensure that the authors' method of quantitative analysis produced results comparable to extant literature, several early calcium-rich Böttger porcelain and later, so-called feldspathic, porcelain bodies were analyzed (see section – Porcelain bodies and clear or white glazes). The mean values, presented in Table 8.8 for oxide components show good agreement with data reported by Schulle and Ullrich (1982).

		Al_2O_3	SiO_2	Na_2O	K_2O	CaO	TiO_2	Fe_2O_3	Other	Rb_2O	SrO
		(wt %)	(wt %)	(wt %)	(wt %)	(wt %)	(wt %)	(wt %)	(wt %)	(ppm)	(ppm)
Böttger porcelain	Bezur and Casadio (n=14)	34,8	56		1,5	6,6	0,4	0,7		64	390
	Schulle and Ullrich	33	61	0,2	0,1	4,8			0,9		
Meissen porcelain	Bezur and Casadio (n=8)	33,3	61		4,2	0,5	0,1	0,5		642	80
	Schulle and Ullrich	35	59	0,8	4	0,3			0,9		

Table 8.8:
Mean values for Böttger and Meissen porcelain body composition determined by Bezur and Casadio (2009) and Schulle and Ullrich (1982, table 2, 45).

Certified reference materials supplied by the National Institute of Standards and Technology were used as calibration standards. Materials in powder form included: SRM 70 potash feldspar, SRM 76 burnt refractory, SRM 97 flint clay, SRM 98 plastic clay, SRM 99 soda feldspar, and SRM 198 silica brick. Pressed pellets were prepared from dried powders without the use of binder in a pellet die with 13 mm inner diameter with the aid of a Specac hydraulic press that was operated at 10 tons (US). NIST standard reference glasses 620, 610, 612, and 614 were also used – the last three served to calibrate the Artax instrument for the quantification of trace levels of strontium and rubidium. Aluminum, silicon, potassium, calcium, titanium, iron, rubidium, and strontium were quantified for measurements obtained with the Artax system. The elements quantified for Tracer III-V spectra were the same, excluding rubidium and strontium. Empirical calibration of both the Bruker Artax and Tracer III-V instruments was accomplished using the manufacturer's software. The "Compton range" for spectra obtained with the Tracer III-V system was set to 2.55–3.00 keV, which means that peak areas were normalized to this region of each spectrum.

Spectra of porcelain bodies were obtained from unglazed areas of objects that appeared free of discoloration and dirt. Measurements from at least three locations on each piece were collected and mean values were calculated for each constituent element. The mean weight percent elemental composition was converted into and reported as weight percent oxide (for SiO_2, Al_2O_3, K_2O, CaO, TiO_2, Fe_2O_3) and as parts-per-million oxide (for [Rb_2O, SrO]) data. Although the sum of weight percent oxide values was frequently less than 100 percent, due to experimental limitations and the presence of elements undetected by the instrument, normalized oxide weight percent values, which were calculated based on the sum of Al_2O_3, SiO_2, K_2O, CaO, TiO_2 and Fe_2O_3, were reported in order to facilitate comparison between values obtained using the different X-ray fluorescence spectrometers used as well as data reported by others, which is often normalized. A subset of the data on porcelain bodies obtained using the Tracer III-V handheld X-ray fluorescence spectrometer was found to be inaccurate due to the analysis of unglazed areas that were smaller in width than the aperture of the instrument (see "The impact of instrument apertures on the quantitative analysis of porcelain bodies"). The validity of measurements was assessed based on the ratio of un-normalized aluminum and silicon weight percent values, which was found to be below 0.5 for analyses of narrow unglazed areas that covered only a portion of the Tracer III-V instrument's aperture. Of the 136 individual porcelain pieces analyzed the body composition of only 79 was considered valid.

Normalized oxide weight percent values were not used in multivariate statistical analysis in order to avoid problems that can arise when studying correlations between variables that add up to 100 percent (see Aitchison *et al.* 2000; Thomas and Aitchison 2006). Instead, so-called "log-ratio" variables were calculated by taking the ten-base logarithms of the ratio of two compositional variables, i.e. un-normalized oxide weight percent. The three variable pairs – K_2O-CaO, TiO_2-CaO, and Fe_2O_3-CaO – were chosen because the equality of means hypothesis could not be rejected for data obtained by the Artax vs. the Tracer III-V spectrometers. In other words, these ratios allowed the pooling of compositional information collected by two different spectrometers. Systat 12 software was used to perform K-means cluster analysis based on Euclidean distances in order to help identify groups of Du Paquier objects with similar porcelain body compositions.

Highlights of findings on Du Paquier porcelain bodies

Unlike porcelain made at the Meissen factory, which can be clearly classified based whether a calcium- or a potassium-containing flux was used in the formulation of the body, the categorization of the majority of Du Paquier pieces, based on their potassia and calcia content, is more difficult owing to values that can be considered "intermediate" when compared to the extremes measured for Böttger and Meissen porcelain. Cluster analysis of the three log-ratio variables identified a group of ten Du Paquier objects with porcelain bodies that are very similar to those of calcium-rich Böttger porcelain (Table 8.9 and Figure 8.3). As in the case of Böttger porcelain bodies, the high proportion of CaO relative K_2O present suggests that the flux component of the paste in these Du Paquier pieces was primarily a calcium-containing material. The compositional similarities between calcia-fluxed Du Paquier and Böttger porcelain, however, cannot resolve whether the Du Paquier factory used alabaster, gypsum, limestone or perhaps white marble as a source of calcium.

	Al_2O_3 (wt %)	SiO_2 (wt %)	K_2O (wt %)	CaO (wt %)	TiO_2 (wt %)	Fe_2O_3 (wt %)	Rb_2O (ppm)	SrO (ppm)
Böttger porcelain (n=14)	34,8	56	1,5	6,6	0,4	0,7	64	390
Meissen (n=8)	33,3	61,5	4,2	0,5	0,1	0,5	642	80
Du Paquier CaO fluxed (n=10)	33,9	58,6	1,2	4,8	0,5	1	54	407
Du Paquier mixed flux, group A (n=9)	35,8	55,8	2,1	4,9	0,3	1,1	88	424
Du Paquier mixed flux, group B (n=6)	38,4	55,4	2	2,9	0,3	1	91	252
Du Paquier mixed flux, group C (n=22)	36,3	58,5	2,1	1,5	0,3	1,3	101	110

Table 8.9:
Mean oxide composition of porcelain paste groups identified by Bezur and Casadio (2009) using K-means cluster analysis.

Calcia-fluxed Du Paquier porcelain bodies differ slightly from Böttger porcelain in that they contain lower levels of alumina compared to silica. If one carries out rational analysis of the mean weight percent oxide composition of the two body types (Dinsdale 1986:23-25) - assuming that the starting minerals for both type of bodies included kaolinite ($Al_2O_3{\bullet}2SiO_2{\bullet}2H_2O$), alabaster ($Ca_2SO_4{\bullet}2H_2O$), quartz ($SiO_2$) and orthoclase ($K_2O{\bullet}Al_2O_3{\bullet}6SiO_2$) – then the results suggest that the recipe produced in Vienna contained a slightly lower proportion of kaolinite compared to quartz. The slightly higher mean iron oxide content of calcia-fluxed Du Paquier porcelain results in these bodies having a warmer white hue than that of Böttger objects. Rubidia-to-strontia and calcia-to-strontia ratios also differ between the calcia-rich bodies that were produced at the two factories. This, along with the higher iron content of Du Paquier bodies, may indicate the use of different raw materials sources for the calcium-containing flux and the clay. Figure 8.3 shows that all but one Du Paquier body studied could be differentiated from Böttger porcelain based on log-ratio variables. This finding underscores the importance of stylistic considerations when trying to distinguish these types of porcelain objects as well as the need for further compositional studies involving the quantification of trace elements.

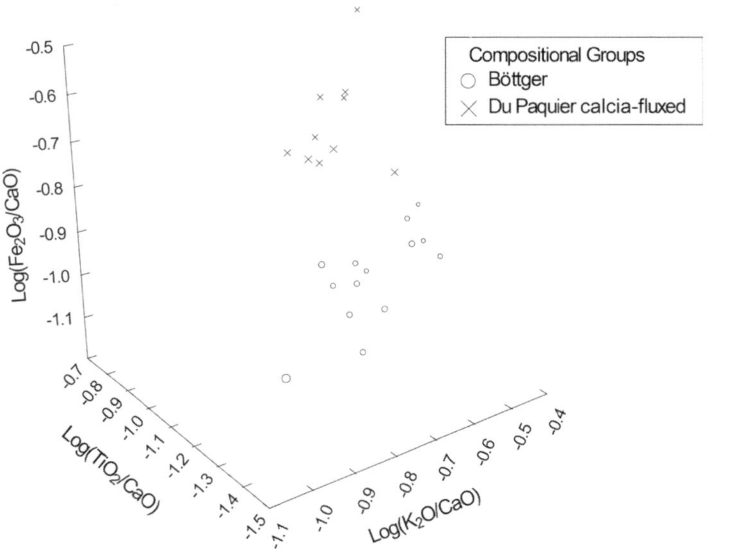

Figure 8.3:
The composition of calcium-rich Böttger and Du Paquier porcelain bodies represented as the logarithm of the ratio of three variable pairs.

The potassia and calcia levels of over two-thirds of Du Paquier objects is such that their ratio is intermediate compared to the extremes observed for Böttger, calcia-fluxed Du Paquier, and potassia-fluxed Meissen porcelain. The composition of this group of objects is highly variable, but it is possible to identify objects that form three relatively tight clusters with a mixture of these fluxes, which are identified in Table 8.9 as groups A, B and C. Because potassia is not the primary flux in Du Paquier porcelain bodies, they cannot, unlike the Meissen porcelain produced after about 1730, be described as feldspathic porcelain. Du Paquier porcelain bodies with a mixture of fluxes assigned to groups A and B are very similar, but members of group B contain lower amounts of calcia, making their potassia and calcia content almost equal and the total amount of these fluxes lower than values for members of group A. It is difficult to discern whether this difference reflects intentional changes in paste formulation, batch-to-batch variation, or natural variability in raw material sources. The analysis of a teapot and its cover at the Art Institute of Chicago resulted in their bodies being categorized, respectively, into group A and B. Such compositional differences between members of a matched set of objects are unlikely to have been intentional, and it is more plausible that the differences reflect the inherent variability of the manufacturing process. The within-group variability of body compositions in group C is greater than that of the other subgroups, however, objects in this cluster distinguish themselves by having slightly more potassia than calcia while the total amount of these oxides is lower than in any other Du Paquier compositional group. The iron oxide content within this cluster is higher than that measured in other groups, resulting in a creamy color.

Written evidence for a transfer of technology from Meissen to Vienna with the help of Samuel Stöltzel appears to be supported by the identification of Du Paquier objects with calcia-rich porcelain bodies that closely resemble the composition of Böttger porcelain. Scholars have attributed calcia-fluxed Du Paquier objects to either the first (c. 1720–25) or the second (c. 1725–35) period of the factory's production. Both these periods coincide with the production of calcia-fluxed porcelain – Böttger porcelain – at Meissen before about 1730. The exception is an armorial potpourri vase from the collection of Melinda and Paul Sullivan, which has been given a slightly later date, c. 1735. Interestingly, Du Paquier objects identified by scholars as having been produced during the factory's earliest period (c. 1720–25) were assigned to three different compositional subgroups. The hypothesis, that the grouping of porcelain bodies based on composition is independent of their period of manufacture, cannot be rejected based on the relatively small number of Du Paquier bodies analyzed. It is hoped that future research and analysis of more objects will allow more definitive conclusions regarding the correlation between body composition and object date or type (such as figure, tureen, vase, etc.).

Other non-destructive and minimally analysis methods applied to the study of porcelain

Raman spectroscopy

In recent years Raman microscopy has emerged as a powerful complementary tool to XRF for the non-invasive analysis of ceramics and porcelain artifacts (Colomban 2005c; Colomban 2005b, 2005a). Overglaze enamel colorants, clear glazes, and unglazed porcelain bodies can be studied by directing the laser excitation radiation to the surface of artifacts, while confocal microscope configurations extend the scope of this technique to the characterization of sub-surface layers such as underglaze colors based on cobalt blues (de Waal 2007; Kock and de Waal 2007).

Research in late 1990s began to demonstrate the ability of Raman spectroscopy to characterize the mineral pigments present on pottery shards or dispersed in the glassy matrix (Clark, Curri, and Laganara 1997; Clark and Curri 1998; Brooke, Edwards, and Tait 1999). Identification can be rapidly and unambiguously done by comparison with extensive reference pigment spectral databases developed for the analysis of paintings (Bell, Clark, and Gibbs 1997), illuminated manuscripts and other polychrome objects, as most of the colorants are the same (e.g. lead-tin yellow, Naples yellow, iron oxides, chromium oxide green, lapis lazuli, cobalt aluminate blue, etc.). A notable use of the technique involved detecting various polymorphs of titanium dioxide (TiO_2) used in white areas of a diverse range of objects, including ancient Chinese pottery shards (Zuo *et al.* 1999) and prehistoric Anasazi ceramics (Striova *et al.* 2006), long before the introduction of this material as an artist's pigment. One must note, however, that titanium dioxide is an accessory mineral in clays and thus the detection of titanium in porcelain bodies and clear glazes is to be expected prior to the use of titanium dioxide as a pigment. Other phases, such as cassiterite (SnO_2) opacifiers, can also be identified with Raman analysis, as can un-reacted feldspar fluxes, anatase (TiO_2) and quartz grains at the interface between glaze and ceramic bodies, offering clues on firing temperatures (Colomban and Milande 2006). Interestingly, occurrences of lapis lazuli at the glaze-body interface have been documented for early Meissen pieces, possibly employed as an optical brightener to enhance the whiteness of the porcelain (Colomban and Milande 2006). Instances of lazurite, which is not thermally stable above 1200 °C, also provide clues to the upper threshold of the glaze firing temperatures for the wares on which it is found.

A limitation of Raman microspectroscopy is its inability to characterize overglaze enamel colors that are obtained by the dissolution of transition metal ions (Cu, Co, etc.) in a glassy matrix. The presence of metal nanoparticles - gold particles in gold ruby glass, purple of Cassius, and silver and copper particles in lusterware - can only be inferred based on their modification of the Raman spectrum of the glassy matrix itself (Colomban and Truong 2004; Colomban and Schreiber 2005; Colomban and Slodczyk 2009). For these reasons, Raman spectroscopy is often combined with other techniques to fully characterize ceramic or porcelain objects (Colomban *et al.* 2004; Akyuz *et al.* 2007; Miao, Yang, and Mu 2010). A comprehensive selection of representative publications on overglaze and underglaze decoration of soft and hard-paste porcelain as well as pigments on pottery fragments, medieval glazed ceramics, Iznik pottery, fritwares, majolicas and terra sigillata wares studied with Raman micro-spectroscopy is described in a recent review by Vandenabeele *et al.* (2007). The analysis of the overglaze enamel color palette and the compilation of diagnostic Raman peaks for all the main traditional pigments and Raman-active coloring materials for wares produced at Sèvres (Colomban, Sagon, and Faurel 2001) and Meissen in the eighteenth and nineteenth centuries (Casadio 2004; Colomban and Milande 2006) deserves special mention in relation to the analysis of European porcelain discussed in this chapter.

The analysis of bodies and clear glazes is by far the most innovative application of Raman spectroscopy to the study of ancient ceramics and porcelain. Direct access to the ceramic body of glazed pieces is available due to the need to wipe the clear glaze off the foot, base or rim of objects prior to the second (glost) firing. Selection of laser excitation lines with high energies - in the ultraviolet or green wavelength range - is methodologically preferable, especially for analysis of the bodies, so as to take full advantage of the Raman effect, the intensity of which is inversely proportional to the excitation wavelength (laser excitation lines used in the cited literature are: 413.1, 457.9, 488, 514.32, 532, 632.8 6 and 647.1 nm). Additionally, high laser powers effect the thermally assisted removal of minute quantities of organic contaminants that may have accumulated in pores of the ceramic body, thereby minimizing fluorescence. When used with a microscope attachment the laser beam diameter is typically restricted to 1-2 microns, thus local heating due to laser irradiation of a small spot on the ceramic body is of limited concern. Overglaze enamels, on the other hand, may be more prone to thermal and photochemical damage, given the different thermal expansion coefficients of the body and glazes.

Spectroscopic markers for the differentiation of hard-paste and soft paste bodies have been identified for both ancient and modern porcelain, with the first comprehensive Raman assessment of the compositions used in major eighteenth-century European production centers having been published in 2001 (Colomban and Treppoz 2001). Characteristic spectroscopic signatures of β-wollastonite ($CaSiO_3$) and calcium phosphates (including tricalcium phosphate $Ca_3(PO_4)_2$) are linked to soft paste porcelain and bone china respectively. The spectra of hard-paste porcelain bodies, on the other hand, are dominated by two main broad bands that are related to Si–O bending and stretching vibrations and are centered at about 470-500 and 1000 cm^{-1}, which are characteristic of an amorphous silicate network. The band observed around 480 cm^{-1} represents the dominant mode in hard-paste porcelain, documenting structures composed of connected SiO_4 tetrahedra that are part of the glassy aluminosilicate networks, the formation of which is promoted by the use of feldspar as fluxing material. Additionally, hard-paste porcelain is identified by the presence of mullite phases ($3Al_2O_3 \cdot 2SiO_2$), which show especially intense bands in Meissen pieces. The Raman signature of calcite can be detected in Böttger porcelain, which is consistent with the use of alabaster ($CaSO_4 \cdot 2H_2O$) in the formulation of porcelain bodies and glazes before the introduction of feldspars for this purpose at Meissen (Colomban and Milande 2006).

Transferring knowledge from the field of modern glass and silicates analysis Colomban pioneered the use of the concept of a "polymerization index," which is derived from Raman spectra, to further our understanding of the structure of the aluminosilicates comprising porcelain and ceramic glazes and other iono-covalent materials (Colomban 2003b; Colomban and Paulsen 2005; Colomban and Slodczyk 2009). Because the presence of network modifiers - such as alkali and alkaline earths metal ions and lead cations - affects the $\nu 2$ bending vibration (centered at $\approx 460\ cm^{-1}$) of the bridging oxygen atoms between linked SiO_4 tetrahedra this mode has been used to characterize and quantify the degree of "polymerization" of a porcelain clear glaze (Liem *et al.* 2000; Colomban and Treppoz 2001).

Highly connected aluminosilicate structures will have a strong band at 500 cm^{-1}, while the Si-O-Si stretching combination band at approximately 1000 cm^{-1} becomes broader and more intense for glassy, amorphous aluminosilicates. Each broad band can be deconvoluted into components bands, representing vibrations of silicate structures with zero (SiO_4, identified as the Q^0 vibrational mode), one ($-SiO_3$ or Q^1), two ($=SiO_2$ or Q^2), three ($\equiv SiO$ or Q^3), and ultimately four ($\equiv Si-$ or Q^4) bridging/connecting oxygen atoms per tetrahedral structural unit of silica. The specific composition of the glassy or ceramic material - the relative proportions of network formers and modifiers, in particular - and the processing temperature will directly influence the ratios of the areas of the two bending and stretching bands A_{500}/A_{1000} (Colomban 2003b, 2004; Colomban, Tournie, and Bellot-Gurlet 2006; Colomban and Slodczyk 2009). After mathematical processing of the spectra of the ceramic glazes via multipoint baseline subtraction and curve-fitting, the weights of each Q^n spectral component can be calculated. Specifically, the integral area of each band component, bandwidth and center of gravity are calculated and an index of polymerization I_p is determined as A_{500}/A_{1000}. The value of I_p is closely related to the glass nanostructure and composition and because the specific composition of clear glazes was often varied for practical or aesthetic reasons – for example, to better match its thermo-mechanical properties to that of the body - the polymerization index serves a valuable parameter for provenance and authentication studies of porcelain and ceramic objects (Prinsloo *et al.* 2005; Colomban, Tournie, and Bellot-Gurlet 2006).

Ultimately, since different variables govern Raman and X-ray cross-sections, different information content can be unlocked when analyzing the same object with these complementary techniques. The intensity of X-ray powder diffraction patterns of most crystalline phases made of elements with similar atomic number and symmetry is proportional to the phase content, and the intensity of X-ray fluorescence spectral lines is proportional to the concentration of a specific element, given similar matrices. In contrast, the Raman spectrum will be dominated by the phases characterized by the most covalent, polarizable bonds. Thus not only compounds present in large abundance are identified, but also bands of secondary or minor phases may be enhanced in the Raman spectrum, a feature that could potentially aid in discriminating similar materials. This aspect, combined with the wide availability of a new generation of portable Raman instruments that allow the *in situ* analysis of museum objects, make Raman and XRF techniques an extremely powerful combination for the qualitative and quantitative characterization of ceramics, including porcelain.

Particle-induced X-ray emission spectroscopy

Particle-induced X-ray emission spectroscopy (PIXE) is another non-destructive technique that has been applied to the *in situ* characterization of ceramics, including porcelain. Most PIXE studies involve the use of a particle accelerator and thus comparisons with handheld XRF studies are inappropriate. Li *et al.* (2006) used accelerator-based PIXE to analyze Ru and Jun porcelain and assessed whether their body compositions could serve to distinguish these types. Swann and Nelson (2000) and Neelmeijer *et al.* (2003) quantified the composition of Böttger stoneware bodies in an attempt to distinguish authentic pieces from later reproductions. Zucchiatti *et al.* (Zucchiatti *et al.* 2006) and Gianoncelli *et al.* (2006) studied blue glazes on terracotta sculptures by Luca Della Robbia. While not handheld, the portable PIXE system employed by Pappalardo *et al.* (2004) for the characterization of blue glazes on Della Robbia sculptures, however, is comparable to portable XRF instruments in size and mobility. The authors used the two methods in a complimentary fashion, performing quantitative analysis of low Z elements with PIXE and focusing on the detection of medium and high atomic number (*Z*) elements with XRF.

Laser ablation inductively coupled plasma mass spectroscopy and laser-induced breakdown spectroscopy

The development of alternative or external cell designs has allowed for the *in-situ* application of laser ablation inductively coupled plasma mass spectroscopy (LA-ICP-MS) to the study of complete objects and fragments that exceed the dimensions of laser ablation sample chambers. The small size of sample craters – 20-100 microns in diameter - makes this technique minimally destructive. Repeated ablation of the same location allows depth profiling and makes *in situ* LA-ICP-MS well suited for the quantification of trace elements in silver-copper alloy objects that have silver-enriched surfaces (Devos, Moor, and Lienemann 1999; Sarah, Gratuze, and Barrandon 2007). The technique has been also used to study trace components of pigments in paintings (Smith *et al.* 2005).

Dussubieux and Cox *et al.* (2009) have applied *in situ* LA-ICP-MS to the quantitative analysis of ceramics and glass in addition to metals and performed cross-validation of the method using PIXE and instrumental neutron activation analysis (INAA). Though studies of porcelain objects using this technique have not yet been reported in the literature, they are likely forthcoming in the near future as LA-ICP-MS has become a key technique in research on ceramics and glass (e.g. Durrant and Ward 2005; Speakman and Neff 2005; Giussani, Monticelli, and Rampazzi 2009). The depth-profiling capacity of laser ablation would be of great value in "drilling" below the surface of unglazed porcelain areas to insure that quantitative compositional data is representative of the bulk material.

The application of laser-induced breakdown spectroscopy (LIBS) for the analysis of glasses, glazed ceramics and faience has been reported by several different research groups (e.g. Anzano *et al.* 2002; Melessanaki *et al.* 2005; Giakoumaki, Melessanaki, and Anglos 2007). Unlike LA-ICP-MS, LIBS does not require a closed cell with a carrier gas and thus can be performed *in situ* with greater ease, as long as appropriate eye protection measures are in place. The diameter of sample craters range between 50-200 microns, though micro-LIBS setups produce craters as little as 10 microns in diameter. Unlike LA-ICP-MS, LIBS can be performed with portable instrumentation and thus has the potential to offer minimally-destructive depth-resolved elemental analysis with similar convenience as portable and handheld XRF systems.

Conclusions

Porcelain objects represent a highpoint in the technical achievements of their time as their manufacture required careful material selection and processing, sophisticated pyrotechnology, and fine manual skills. A number of forming and decorating techniques may have been practiced within the same production unit at any given time and the output of manufacturers was often very stylistically diverse. While makers' marks can be helpful in relating a piece to a particular factory and period, the existence of authentic unmarked objects and the use of forged marks can complicate attribution. The potential for applying overglaze enamel and metallic decoration to glazed, but undecorated pieces at a later date leads to further challenges in establishing provenance. While archival records at some factories, such as Meissen and Sèvres, have allowed researchers to gain insight into once-secret production technologies and establish terminal dates for the use of particular materials and formulations, technical aspects of the production process at other facilities, such as the short-lived Du Paquier factory, must be inferred from the technical study of well-dated porcelain pieces themselves.

Portable and handheld X-ray fluorescence spectrometers are well suited to the *in situ* non-destructive analysis of porcelain objects, even if the technique and instrumentation cannot resolve questions about the nature and distribution of component phases in porcelain bodies and colored enamel layers. The latter may be approached with Raman spectroscopy, which has been shown to be a powerful complimentary technique to XRF. Despite the inability of many portable and handheld XRF systems to detect sodium and magnesium, researchers can still discriminate specific types of soft- and hard-paste wares based on the quantitative analysis of the porcelain body coupled with the qualitative analysis of the clear glaze layer (see Porcelain bodies and clear or white glazes). The right combination of operating parameters and quantification protocols allows the determination of trace-level components in porcelain bodies, of which rubidium and strontium may be particularly helpful in tracking changes in the source raw materials. In combination with careful visual assessment of color and transparency, the elemental composition of colored glazes provides a great deal of information about the colorants and opacifiers as well as auxiliary elements or compounds, such as zinc oxide, which are important to the preparation of colorants or the achievement of desired optical effects (see Overglaze enamels, underglaze colors and gilding).

Given the multi-layer structure of most porcelain objects, researchers must be aware of the depths from which fluorescent or secondary X-rays, generated by elements in the sample, can emerge to be detected and how these values relate to the thickness of various layers in a glazed porcelain object (see Lateral and depth resolution issues in the analysis of porcelain objects). It is equally important that users of the technique familiarize themselves with the dimensions of the X-ray beam at the point it exits the nosepiece of their handheld unit as this impacts the minimum dimensions of features from which representative spectra can be collected for quantitative analysis purposes (see The impact of instrument apertures on the quantitative analysis of porcelain bodies). It is also helpful to know the exact position of the X-ray beam relative to the opening on the nosepiece in order to help align the instrument relative to features of interest. In addition to proper instrument-sample geometry, the accurate quantitative analysis of porcelain bodies requires the use of suitable calibration and validation standards, as well as the correct use of empirical or fundamental parameters calibration protocols. Published studies have focused on the analysis of hard-paste porcelain objects, thus the selection of appropriate calibration standards for soft-paste porcelain types remains to be solved. Research into preparation of fired calibration tiles, using NIST-traceable components, is needed to address concerns regarding matched matrices between reference materials and samples.

Finally, the success of studies of porcelain objects using XRF depend not only on the soundness of the analytical methodology, but also on productive collaborative relations between scientists and experts in other fields: ceramic historians, archaeologists, art historians, geologists, archivists etc. Cross-disciplinary dialog is key to the definition of scientifically approachable problems as well as the interpretation and presentation of results in ways that enrich our understanding of past cultures, societies and their technical achievements.

Acknowledgments

The authors gratefully acknowledge support by the Andrew W. Mellon Foundation, the Melinda and Paul Sullivan Foundation for the Arts and the great generosity of the late Mrs. Eloise Martin. We also thank Jennifer Mass and Aaron Shugar for their helpful comments and suggestions, not to mention their patience with the slow pace of our writing process. We are grateful to our families, who endured the presence of laptops during weekends and vacations.

References

Adan-Bayewitz, D., F. Asaro, and R. D. Giauque. 1999. "Determining pottery provenance: Application of a new high-precision X-ray fluorescence method and comparison with instrumental neutron activation analysis." *Archaeometry* Vol. 41 (1):1-24.

Aitchison, J., C. Barceló-Vidal, J. A. Martín-Fernández, and V. Pawlowsky-Glahn. 2000. "Logratio analysis and compositional distance." *Mathermatical Geology* Vol. 32 (3):271-275.

Akyuz, S., T. Akyuz, S. Basaran, C. Bolcal, and A. Gulec. 2007. "FT-IR and micro-Raman spectroscopic study of decorated potteries from VI and VII century BC, excavated in ancient Ainos - Turkey." *Journal of Molecular Structure* Vol. 834:150-153.

American Institute for Conservation of Historic and Artistic Works. 2011. *Commentaries to the Guidelines for Practice* [cited 8/24/2011 2011]. Available from http://www.conservation-us.org/index.cfm?fuseaction=Page.viewPage&pageId=529&parentID=526.

Anzano, J. M., M. A. Villoria, I. B. Gornushkin, B. W. Smith, and J. D. Winefordner. 2002. "Laser-induced plasma spectroscopy for characterization of archaeological material." *Can J Anal Sci Spectrosc* Vol. 47:134-140.

Aza, Antonio H., Angeles G. Torre, Miguel A. G. Aranda, F. Jose Valle, and Salvador Aza. 2004. "Rietveld Quantitative Analysis of Buen Retiro Porcelains." *Journal of the American Ceramic Society* Vol. 87 (3):449-454.

Baumann, David, and W. D. Keller. 1975. "Bulk densities of selected dried natural and fired kaolin clays." *Clays and Clay Minerals* Vol. 23:424-427.

Bé, M.-M., V. Chisté, C. Dulieu, E. Browne, V. Chechev, N. Kuzmenko, R. Helmer, A. Nichols, E. Schönfeld, and R. Dersch. 2011. *Table of radionuclides (Vol. 1 - A = 1 to 151)*. Bureau International des Poids et Mesures 2004a [cited 11/26/2011 2011]. Available from http://www.bipm.org/utils/common/pdf/monographieRI/Monographie_BIPM-5_Tables_Vol1.pdf.

———. 2011. *Table of radionuclides (Vol. 2 - A = 151 to 242)*. Bureau International des Poids et Mesures 2004b [cited 11/26/2011 2011]. Available from http://www.bipm.org/utils/common/pdf/monographieRI/Monographie_BIPM-5_Tables_Vol2.pdf.

Bell, Ian M., Robin J. H. Clark, and Peter J. Gibbs. 1997. "Raman spectroscopic library of natural and synthetic pigments (pre- [approximate] 1850 AD)." *Spectrochimica Acta Part A: Molecular and Biomolecular Spectroscopy* Vol. 53 (12):2159-2179.

Bezúr, Aniko, and Francesca Casadio. 2009. "Artistic expression and technological mastery: A scientific evaluation of the materials." In *Fired by passion: Vienna Baroque porcelain of Claudius Innocentius Du Paquier*, edited by Meredith Chilton, Claudia Jobst and Katharina Hantschmann, 1165-1211. Stuttgart: Arnoldsche Art Publishers.

Brooke, C. J., H. G. M. Edwards, and J. K. F. Tait. 1999. "The Bottesford Blue mystery: a Raman spectroscopic study of post-mediaeval glazed tiles." *Journal of Raman Spectroscopy* Vol. 30 (6):429-434.

Budd, Paul D., R. Haggerty, A. M. Pollard, B. Scaife, and R. G. Thomas. 1996. "Rethinking the quest for provenance." *Antiquity* Vol. 70 (267):168-174.

Busson, Arnold. 1994. "The "indianische" porcelain from Rossau: The early history of Du Paquier porcelain and the relationship between the Meissen and Vienna manufactories." In *Claudius Innocentius du Paquier: Wiener Porzellan der Frühzeit 1718-1744*, edited by Elisabeth Sturm-Bednarczyk. Wien: Edition Christian Brandstätter.

Buxeda i Garrigós, J., M. A. Cau Ontiveros, and V. Kilikoglou. 2003. "Chemical Variability in Clays and Pottery from a Traditional Cooking Pot Production Village: Testing Assumptions in Pereruela." *Archaeometry* Vol. 45 (1):1-17.

Casadio, Francesca. 2004. "Decoration of Meissen porcelain: Raman microscopy as an aid for authentication and dating." In *Proceedings of the sixth Infrared and Raman Users Group conference (IRUG 6), Florence, March 29th - April 1st 2004*, edited by Marcello Picollo, 178-244. Padova: Il Prato.

Choo, C. K. Koh, K. H. Kim, Y. E. Lee, and J. S. Kim. 2002. "A scientific study of Choson white ware: early porcelain from a royal kiln at Kwangju Usanni." *Archaeometry* Vol. 44 (2):199-212.

Choo, C. K. Koh, Y. E. Lee, I. W. Shim, G. H. Kim, W. Y. Huh, S. C. Chun, and W. K. Choo. 2004. "Compositional and microstructural study of Koryŏ celadon and whiteware excavated from Sŏri Kiln in Kyŏnggi Province." *Archaeometry* Vol. 46 (2):247-265.

Clark, Robin J. H., and M. Lucia Curri. 1998. "The identification by Raman microscopy and X-ray diffraction of iron-oxide pigments and of the red pigments found on Italian pottery fragments." *Journal of Molecular Structure* Vol. 440 (1-3):105-111.

Clark, Robin J. H., M. Lucia Curri, and Caterina Laganara. 1997. "Raman microscopy: the identification of lapis lazuli on medieval pottery fragments from the south of Italy." *Spectrochimica Acta Part A: Molecular and Biomolecular Spectroscopy* Vol. 53 (4):597-603.

Colomban, P. 2005a. "Case study: glasses, glazes and ceramics - Recognition of ancient technology from the Raman spectra." In *Raman Spectroscopy in Archaeology and Art History*, edited by H. G. M. Edwards and J. M. Chalmers, 192-206. Cambridge: Royal Society of Chemistry.

———. 2005b. "Raman micro-spectrometry, a unique tool for on-site analysis and identification of ancient ceramics and glasses." In *Materials issues in art and archaeology VII: symposium held November 30-December 3, 2004, Boston, Massachusetts, U.S.A*, edited by P. B. Vandiver, J. L. Mass and A. Murray, 265-279. Warrendale: Materials Research Society.

Colomban, P. H., G. Sagon, L. Q. Huy, N. Q. Liem, and L. Mazerolles. 2004. "Vietnamese (15th Century) blue-and-white *Tam Thai* and lustre porcelains/stonewares: Glaze composition and decoration techniques." *Archaeometry* Vol. 46 (1):125-136.

Colomban, P., and O. Paulsen. 2005. "Non-destructive determination of the structure and composition of glazes by Raman spectroscopy." *Journal of the American Ceramic Society* Vol. 88 (2):390-395.

Colomban, Philippe. 2003a. "Lapis lazuli as unexpected blue pigment in Iranian Lâjvardina ceramics." *Journal of Raman Spectroscopy* Vol. 34 (6):420-423.

———. 2003b. "Polymerization degree and Raman identification of ancient glasses used for jewelry, ceramic enamels and mosaics." *Journal of Non-Crystalline Solids* Vol. 323 (1-3):180-187.

———. 2004. "Raman spectrometry, a unique tool to analyze and classify ancient ceramics and glasses." *Applied Physics A: Materials Science & Processing* Vol. 79:167-170.

———. 2005c. "Recent case studies in the Raman analysis of ancient ceramics: Glaze opacification in Abbasid Pottery, Medici and 18th century French porcelains, Iznic and Kûtayha Ottoman fritwares and an unexpected lapis lazuli pigment in Lajvardina wares." In *Materials issues in art and archaeology VII: symposium held November 30-December 3, 2004, Boston, Massachusetts, U.S.A*, edited by Pamela B. Vandiver, Jennifer L. Mass and Alison Murray, 153-160. Warrendale: Materials Research Society.

Colomban, Philippe, and Véronique Milande. 2006. "On-site Raman analysis of the earliest known Meissen porcelain and stoneware." *Journal of Raman Spectroscopy* Vol. 37 (5):606-613.

Colomban, Philippe, Gérard Sagon, and Xavier Faurel. 2001. "Differentiation of antique ceramics from the Raman spectra of their coloured glazes and paintings." *Journal of Raman Spectroscopy* Vol. 32 (5):351-360.

Colomban, Philippe, and Henry D. Schreiber. 2005. "Raman signature modification induced by copper nanoparticles in silicate glass." *Journal of Raman Spectroscopy* Vol. 36 (9):884-890.

Colomban, Philippe, and Aneta Slodczyk. 2009. "Raman intensity: An important tool to study the structure and phase transitions of amorphous/crystalline materials." *Optical Materials* Vol. 31 (12):1759-1763.

Colomban, Philippe, Aurélie Tournie, and Ludovic Bellot-Gurlet. 2006. "Raman identification of glassy silicates used in ceramics, glass and jewellery: a tentative differentiation guide." *Journal of Raman Spectroscopy* Vol. 37 (8):841-852.

Colomban, Philippe, and Françoise Treppoz. 2001. "Identification and differentiation of ancient and modern European porcelains by Raman macro- and micro-spectroscopy." *Journal of Raman Spectroscopy* Vol. 32 (2):93-102.

Colomban, Philippe, and Catherine Truong. 2004. "Non-destructive Raman study of the glazing technique in lustre potteries and faience (9-14th centuries): silver ions, nanoclusters, microstructure and processing." *Journal of Raman Spectroscopy* Vol. 35 (3):195-207.

Cort, Louise Allison. 1986. "Korean influences in Japanese ceramics: the impact of the Teabowl Wars of 1592-1598." In *Technology and style*, edited by W. D. Kingery and Esther Lense, 331-362. Columbus, Ohio: American Ceramic Society.

Cotte, Marine, Jean Susini, Joris Dik, and Koen Janssens. 2010. "Synchrotron-based X-ray absorption spectroscopy for art conservation: Looking back and looking forward." *Accounts of Chemical Research* Vol. 43 (6):705-714.

Craddock, P. T. 2009. *Scientific investigation of copies, fakes and forgeries*. 1st ed. Oxford: Elsevier / Butterworth-Heinemann.

d'Albis, Antoine. 1998. *The creation of hard-paste porcelain production at Sèvres*. Vol. 13. London: The French Porcelain Society.

Day, Judith H. 1992-94. "Influence and Innovation: Transplanted Korean Potters and Japanese Ceramics." *Bulletin of the Oriental Ceramic Society of Hong Kong* Vol. 10:54-58.

de Waal, D. 2007. "Raman analysis of underglaze pigments on porcelain art." *Journal of Raman Spectroscopy* Vol. 38 (7):956-957.

Devos, W., Ch. Moor, and P. Lienemann. 1999. "Determination of impurities in antique silver objects for authentication by laser ablation inductively coupled plasma mass spectrometry (LA-ICP-MS)." *Journal of Analytical Atomic Spectrometry* Vol. 14 (4):621-626.

Dik, J., E. Hermens, R. Peschar, and H. Schenk. 2005. "Early production recipes for lead antimonate yellow in Italian art." *Archaeometry* Vol. 47 (3):593-607.

Dinsdale, Allen. 1986. *Pottery science: materials, processes, and products, Ellis Horwood series in applied science and industrial technology*. Chichester: Ellis Horwood Limited.

Durrant, Steven F., and Neil I. Ward. 2005. "Recent biological and environmental applications of laser ablation inductively coupled plasma mass spectrometry (LA-ICP-MS)." *Journal of Analytical Atomic Spectrometry* Vol. 20 (9):821-829.

Dussubieux, Laure, Richard Cox, Mark Golitko, and Patrick Ryan Williams. 2009. ICP-MS with adaptable chamber laser: quantitative analysis of glass, metal and ceramic artifacts in the Museum environment. In *FACSS 2009*. Louisville, KY.

Eppler, Richard A., and Douglas R. Eppler. 2000. "Glazes and glass coatings." In. Westerville: American Ceramic Society. http://catdir.loc.gov/catdir/toc/fy033/99087054.html

Materials specified: Table of contents http://catdir.loc.gov/catdir/toc/fy033/99087054.html.

Finlay, Robert. 2010. *The pilgrim art: cultures of porcelain in world history, The California world history library*. Berkeley: University of California Press.

Fluegel, Alexander. 2007. "Global model for calculating room-temperature glass density from

the composition." *Journal of the American Ceramic Society* Vol. 90 (8):2622-2625.

Freestone, Ian C. 2005. "The provenance of ancient glass through compositional analysis." In *Materials issues in art and archaeology VII: symposium held November 30-December 3, 2004, Boston, Massachusetts, U.S.A*, edited by Pamela B. Vandiver, Jennifer L. Mass and Alison Murray, 195-208. Warrendale, Pa.: Materials Research Society.

Fuchs II, Ronald W., and Jennifer L. Mass. 2009. "Deciphering *The Declaration of Independence* on Chinese Export Porcelain." *American Ceramic Circle Journal* Vol. XV:169-187.

Giakoumaki, Anastasia, Kristalia Melessanaki, and Demetrios Anglos. 2007. "Laser-induced breakdown spectroscopy (LIBS) in archaeological science—applications and prospects." *Analytical and Bioanalytical Chemistry* Vol. 387 (3):749-760.

Gianoncelli, A., J. Castaing, A. Bouquillon, A. Polvorinos, and P. Walter. 2006. "Quantitative elemental analysis of Della Robbia glazes with a portable XRF spectrometer and its comparison to PIXE methods." *X-ray Spectrometry* Vol. 35 (6):365-369.

Giussani, Barbara, Damiano Monticelli, and Laura Rampazzi. 2009. "Role of laser ablation-inductively coupled plasma-mass spectrometry in cultural heritage research: A review." *Analytica Chimica Acta* Vol. 635 (1):6-21.

Gratuze, Bernard, Isabelle Soulier, Maryse Blet, and Lucy Vallauri. 1996. "De l'origine du cobalt: du verre à la céramique." *Revue d'archéométrie* Vol. 20:77-94.

Grave, Peter. 2009. "Melting moments: Modelling archaeological high temperature ceramic data." In *New directions in archaeological science*, edited by Andrew S. Fairbairn, Sue O'Connor and Benjamin Marwick, 215-232. Canberra: ANU E Press.

Gutierrez, P. Carolina, Trinitat Pradell, Judit Molera, Andrew D. Smith, Aurelio Climent-Font, and Michael S. Tite. 2010. "Color and golden shine of silver Islamic luster." *Journal of the American Ceramic Society* Vol. 93 (8):2320-2328.

Halfpenny, Patricia A., and Jennifer L. Mass. 2009. "Two converging perspectives on one Meissen tureen." *American Ceramic Circle Journal* Vol. XV:71-78.

Hornig-Sutter, Monika. 1984. "Meißener Porzellane des 18. Jahrhunderts im Visier von Kunstgeschichte und Naturwissenschaft. Teil I." *Weltkunst* Vol. 54 (23):3654-3658.

———. 1985. "Meißener Porzellane des 18. Jahrhunderts im Visier von Kunstgeschichte und Naturwissenschaft. Teil IV." *Weltkunst* Vol. 55 (2):110-115.

Hubbell, J. H., and S. M. Seltzer. 2010. *Tables of X-ray Mass Attenuation Coefficients and Mass Energy-Absorption Coefficients (version 1.4). Originally published as NISTIR 5632, National Institute of Standards and Technology, Gaithersburg, MD (1995)*. National Insntitute of Standards and Technology 2004 [cited August 26 2010]. Available from http://physics.nist.gov/xaamdi

Hunt, Leslie B. 1981. "Gold based glass and enamel colors." *Endeavour* Vol. 5 (2):61-67.

Injuk, J., R. van Grieken, A. Blank, L. Eksperiandova, and V. Buhrke. 2006. "Specimen preparation." In *Handbook of practical X-ray fluorescence analysis*, edited by Burkhard Beckhoff, Birgit Kanngiesser, Norbert Langhoff, Reiner Wedell and Helmut Wolff, 421-424. Berlin: Springer-Verlag.

Jenkins, Ron. 1999. *X-ray fluorescence spectrometry*. Edited by J. D. Winefordner, *Chemical analysis*. New York: John Wiley & Sons, Inc.

Jiazhi, Li. 1984. "The evolution of Chinese pottery and porcelain technology." In *Ancient technology to modern science*, edited by W. D. Kingery and Esther Lense, 135-162. Columbus, Ohio: American Ceramic Society.

Kingery, W. D., and D Smith. 1984. "The development of European soft-paste (frit) porcelain." In *Ancient technology to modern science*, edited by W. D. Kingery and Esther Lense, 273-292. Columbus, Ohio: American Ceramic Society.

Kingery, W. D., and Pamela Vandiver. 1986a. "The eighteenth-century change in technology and style from the famille-verte palette to the famille-rose palette." In *Technology and style*, edited by W. D. Kingery and Esther Lense, 363-381. Columbus, Ohio: American Ceramic Society.

Kingery, W. David, and Pamela B. Vandiver. 1984. "Medici Porcelain." *Faenza* Vol. 70 (5-6):441-460.

———. 1986b. *Ceramic masterpieces*. New York: The Free Press.

Kneip, Theo J., and Gerard R. Laurer. 1972. "Isotope excited X-ray fluorescence." *Analytical Chemistry* Vol. 44 (14):57A-68a.

Kock, L. D., and D. de Waal. 2007. "Raman studies of the underglaze blue pigment on ceramic artefacts of the Ming dynasty and of unknown origins." *Journal of Raman Spectroscopy* Vol. 38 (11):1480-1487.

Larsson, Mathilda A. G. 2008. *Estudo e caracterização de porcelanas orientais*, Departamento de Conservação e Restauro, Universidade Nova de Lisboa, Lisbon.

Leung, P. L., Z. C. Peng, M. J. Stokes, and Mike T. W. Li. 2000. "EDXRF studies of porcelains (800-1600 A.D.) from Fujian, China with chemical proxies and principal component analysis." *X-ray Spectrometry* Vol. 29 (3):253-259.

Leung, P. L., M. J. Stokes, Mike T. W. Li, Peng Zicheng, and Wu Shuicun. 1998. "EDXRF studies on the chemical composition of ancient porcelain bodies from Linjiang, Jiangxi, China." *X-ray Spectrometry* Vol. 27 (1):11-16.

Leung, P. L., M. J. Stokes, Chen Tiemei, and Qin Dashu. 2000. "A study of ancient Chinese porcelain wares of the Song-Yuan dynasties from Cizhou and Ding kilns with energy dispersive X-ray fluorescence." *Archaeometry* Vol. 42 (1):129-140.

Li, Guoxia, Weijuan Zhao, Rongwu Li, Hongwei Sun, Min Guo, Yanfang Wang, Hui Liu, Qingyun Zhao, Xinmin Sun, Wenjun Zhao, and Huansheng Cheng. 2006. "Proton induced X-ray emission (PIXE) analysis of sources of porcelain body of Ru Guan and Jun Guan porcelains." *Science in China Series G: Physics Mechanics and Astronomy* Vol. 49 (4):411-420.

Liem, N.Q., N. T. Thanh, and Philippe Colomban. 2002. "Reliability of Raman micro-spectroscopy in analysing ancient ceramics: the case of ancient Vietnamese porcelain and celadon glazes." *Journal of Raman Spectroscopy* Vol. 33 (4):287-294.

Liem, Nguyen Quang, Gerard Sagon, Vu Xuan Quang, Ha Van Tan, and Philippe Colomban. 2000. "Raman study of the microstructure, composition and processing of ancient Vietnamese (proto)porcelains and celadons (13-16th centuries)." *Journal of Raman Spectroscopy* Vol. 31 (10):933-942.

Long, Larry L., R. J. Bell, Jr R. W. Alexander, and Marvin R. Querry. 1987. "Preparation dependent properties of pressed pellets of montmorillonite in the far infrared." *Applied Optics* Vol. 26 (8):1372-1374.

Maggetti, M., C. H. Neururer, and J. Rosen. 2009. "Antimonate opaque glaze colours from the faience manufacture of Le Bois D'Épense (19th century, Northeastern France)." *Archaeometry* Vol. 51 (5):791-807.

Maggetti, M., J. Rosen, C. Neururer, and V. Serneels. 2009. "Paul-Louis Cyfflé's (1724-1806) *Terre de Lorraine*: A technological study." *Archaeometry* Vol. 2009.

Malainey, Mary E. 2011. *A consumer's guide to archaeological science. Analytical techniques, Manuals in archaeological method, theory and technique*. New York: Springer.

Mazo-Gray, V., and M. Alvarez. 1992. "X-ray fluorescence characterization of Ming-dynasty porcelain rescued from a Spanish shipwreck." *Archaeometry* Vol. 34 (1):37-42.

Melessanaki, K., A. Mastrogiannidou, S. Chlouveraki, S. Ferrence, P. Betancourt, and D. Anglos. 2005. "Analysis of Archaeological Objects with LMntla New Transportable LIBS Instrument." In *Lasers in the Conservation of Artworks*, 443-449.

Miao, J, B. Yang, and D. Mu. 2010. "Identification and differentiation of opaque Chinese overglaze yellow enamels by Raman spectroscopy and supporting techniques." *Archaeometry* Vol. 52 (1):146-155.

Mields, M. 1960. "Aus der Jugendzeit des europäischen Porzellans: Eine fachgeschichtliche Plauderei zum Böttger-Jahr (I. Fortsetzung)." *Sprechsaal für Keramik, Glas, Email* Vol. 93 (11):329-332.

Mields, Martin, and Rudolf Lauschke. 1965. *Praxis der Porzellan Malerei*. München: Porta Verlag.

Miliani, Costanza, Brenda Doherty, Alessia Daveri, Anette Loesch, Heike Ulbricht, Brunetto G. Brunetti, and Antonio Sgamellotti. 2009. "In situ non-invasive investigation on the painting techniques of early Meissen Stoneware." *Spectrochimica Acta Part A: Molecular and Biomolecular Spectroscopy* Vol. 73 (4):587-592.

Neelmeijer, Christian, M. Mäder, U. Pietsch, H. Ulbricht, and H.-M. Walcha. 2003. ""Böttger Stoneware" - Authentic or not?" In *Kolloquium aus Anlass des 350. Geburtstages von E. W.v. Tschirnhaus am 10. April 2001 in Dresden*, edited by Dagmar Hülsenberg, 29-33. Stuttgart: Steiner, Franz.

Newton, R. G., and Sandra Davison. 1989. *Conservation of glass, Butterworths series in conservation and museology*. London: Boston.

Padovani, S., C. Sada, P. Mazzoldi, B. Brunetti, I. Borgia, A. Sgamellotti, A. Giulivi, F. D'Acapito, and G. Battaglin. 2003. "Copper in glazes of Renaissance luster pottery: Nanoparticles, ions, and local environment." *Journal of Applied Physics* Vol. 93 (12):10058.

Page, John, and Meredith Chilton. 2001. "Appendix: Scientific analysis of Meissen commedia dell'arte sculpture in the Gardiner Museum collection." In *Harlequin unmasked: the commedia dell'arte and porcelain sculpture*, edited by Meredith Chilton, 323-327. New Haven: George R. Gardiner Museum of Ceramic Art with Yale University Press.

Papadopoulou, D. N., G. A. Zachariadis, A. N. Anthemidis, N. C. Tsirliganis, and J. A. Stratis. 2004. "Comparison of a portable micro-X-ray fluorescence spectrometry with inductively coupled plasma atomic emission spectrometry for the ancient ceramics analysis." *Spectrochimica Acta Part B: Atomic Spectroscopy* Vol. 59 (12):1877-1884.

———. 2006. "Development and optimisation of a portable micro-XRF method for in situ multi-element analysis of ancient ceramics." *Talanta* Vol. 68 (5):1692-1699.

Papageorgiou, Ioulia, and Ioannis Liritzis. 2007. "Multivariate mixture of normals with unknown number of components: An application to cluster neolithic ceramics from Aegean and Asia Minor using portable XRF." *Archaeometry* Vol. 49 (4):795-813.

Pappalardo, G., E. Costa, C. Marchetta, L. Pappalardo, F. P. Romano, A. Zucchiatti, P. Prati, P. A. Mando, A. Migliori, L. Palombo, and M. G. Vaccari. 2004. "Non-destructive characterization of Della Robbia sculptures at the Bargello museum in Florence by the combined use of PIXE and XRF portable systems." *Journal of Cultural Heritage* Vol. 5 (2):183-188.

Pascual, C., P. Recio, F. J. Valle, E. Criado, A. H. De Aza, R. Martínez, and S. De Aza. 2006. "The last period of "Buen Retiro" porcelain factory." In *Heritage, weathering and conservation: Proceedings of the International Conference on Heritage, Weathering and Conservation, HWC-2006, 21-24 June 2006, Madrid, Spain*, edited by R. Fort, 134-142. London: Taylor & Francis.

Pernicka, E. 1999. "Trace element fingerprinting of ancient copper: A guide to technology or provenance?" In *Metals in Antiquity (proceedings of the 1997 conference)*, edited by Suzanne M. M. Young, A. Mark Pollard, Paul Budd and Robert A. Ixer, 163-171. Cambridge, MA: Archaeopress.

Piccolpasso, Cipriano, R. W. Lightbown, and Alan Caiger-Smith. 1980. *The three books of the potter's art = I tre libri dell'arte del vasaio: a facsimile of the manuscript in the Victoria and Albert Museum, London*. 2 vols. London: Scolar Press.

Preaud, Tamara, and Derek E. Ostergard. 1997. *The Sèvres porcelain manufactory: Alexandre Brongniart and the triumph of art and industry, 1800-1847*. New Haven: Yale University Press.

Prinsloo, Linda C., Nigel Wood, Maggi Loubser, Sabine M. C. Verryn, and Sian Tiley. 2005. "Re-dating of Chinese celadon shards excavated on Mapungubwe Hill, a 13th century Iron Age site in South Africa, using Raman spectroscopy, XRF and XRD." *Journal of Raman Spectroscopy* Vol. 36 (8):806-816.

Rapp, George Robert. 2009. *Archaeomineralogy*. 2nd ed, *Natural science in archaeology,; Variation: Natural science in archaeology*. Berlin: London.

Reiche, Ina, Stefan Röhrs, Joseph Salomon, Birgit Kanngiesser, Yvonne Höhn, Wolfgang Malzer, and Friederike Voigt. 2009. "Development of a nondestructive method for underglaze painted tiles-demonstrated by the analysis of Persian objects from the nineteenth century." *Analytical & Bioanalytical Chemistry* Vol. 393 (3):1025-1041.

Reynolds, Dinah. 1989. *Worcester porcelain, 1751-1783: the Marshall collection, Ashmolean-Christie's handbooks; Variation: Ashmolean-Christie's handbooks*. Oxford: Phaidon-Christie's Ltd. in association with the Ashmolean Museum.

Rhodes, Daniel. 1973. *Clay and glazes for the potter*. Rev. ed. Philadelphia,: Chilton Book Co.

Ricciardi, Paola, Philippe Colomban, and Véronique Milande. 2008. "Non-destructive Raman characterization of Capodimonte and Buen Retiro porcelain." *Journal of Raman Spectroscopy* Vol. 39 (8):1113-1119.

Rice, Prudence M. 1987. *Pottery analysis*. Chicago: The University of Chicago Press.

Rosi, F., V. Manuali, T. Grygar, P. Bezdicka, B. G. Brunetti, A. Sgamellotti, L. Burgio, C. Seccaroni, and C. Miliani. 2011. "Raman scattering features of lead pyroantimonate compounds: implication for the non-invasive identification of yellow pigments on ancient ceramics. Part II. In situ characterisation of Renaissance plates by portable micro-Raman and XRF studies." *Journal of Raman Spectroscopy* Vol. 42 (3):407-414.

Sandalinas, C., S. Ruiz-Moreno, A. López-Gil, and J. Miralles. 2006. "Experimental confirmation by Raman spectroscopy of a Pb-Sn-Sb triple oxide yellow pigment in sixteenth-century Italian pottery." *Journal of Raman Spectroscopy* Vol. 37 (10):1146-1153.

Sendova, M., Zhelyaskov, V., Scalera, M. and Gulliford, C. 2007 "Micro-Raman Spectrocsocpy Characterization of Della Robbia Glazes." *Archaeometry* Vol 49 (4): 655–664

Sarah, Guillaume, Bernard Gratuze, and Jean-Noël Barrandon. 2007. "Application of laser ablation inductively coupled plasma mass spectrometry (LA-ICP-MS) for the investigation of ancient silver coins." *Journal of Analytical Atomic Spectrometry* Vol. 22 (9):1163-1167.

Schöde, Paul. 1988. "Der "Wiener' Schokoladenbecher von 1719: Eine Meissener Schöpfung!" *Keramos* (122):3-12.

Schönfeld, Martin. 1998. "Was there a Western inventor of porcelain?" *Technology and Culture* Vol. 39 (4):716-727.

Schulle, W., and W. Goder. 1982. "Die Erfindung des europäischen Porzellans durch Böttger - eine systematische schöpferische Entwicklung (The discovery of European porcelain by Böttger - a systematic creative development)." *Keramische Zeitschrift* Vol. 34 (10):598-600.

Schulle, Wolfgang, and Bernd Ullrich. 1982. "Ergebisse gefügeanalytischer Untersuchungen an Böttgerporzellan." *Silikattechnik* Vol. 33 (2):44-47.

Sendova, M., V. Zhelyaskov, M. Scalera, and C. Gulliford. 2007. "Micro-Raman spectroscopy characterization of Della Robbia glazes." *Archaeometry* Vol. 49 (4):655-664.

Shaw, Kenneth. 1962. *Ceramic colours and pottery decoration*. London: Maclaren.

Smith, Kari, Kristin Horton, R. John Watling, and Natalie Scoullar. 2005. "Detecting art forgeries using LA-ICP-MS incorporating the in situ application of laser-based collection technology." *Talanta* Vol. 67 (2):402-413.

Spataro, Michela, Mavis Bimson, and Nigel Meeks. 2008. "The earliest English porcelains? Part 3: A scientifi investigation of three pieces of porcelain presumed seventeenth century, from Burghley House." *English Ceramic Circle Transactions* Vol. 20 (1):187-194.

Spataro, Michela, Nigel Meeks, Mavis Bimson, Aileen Dawson, and Janet Ambers. 2009. "Early porcelain in seventeenth-century England: non-destructive examination of two jars from Burghley House." *The British Museum Technical Research Bulletin* Vol. 3:37-46.

Speakman, Robert J., and Hector Neff. 2005. *Laser ablation ICP-MS in archaeological research*. Albuquerque: University of New Mexico Press.

Stiegelschmitt, A., and G. Tomandl. 1985. "Untersuchung der Materialzusammensetzung alter Porzellane im Hinblick auf ihre kuntsthistorische Zuordnung." *Sprechsaal* Vol. 118 (10):974-978.

Strahan, Donna. 2001. "Uranium in glass, glazes and enamels: history, identification and handling." *Studies in Conservation* Vol. 46 (3):181-195.

Striova, J., C. Lofrumento, A. Zoppi, and E. M. Castellucci. 2006. "Prehistoric Anasazi ceramics studied by micro-Raman spectroscopy." *Journal of Raman Spectroscopy* Vol. 37 (10):1139-1145.

Swann, Charles P., and Christina H. Nelson. 2000. "Böttger stoneware from North America and Europe; are they authentic." *Nuclear Instruments and Methods in Physics Research Section B: Beam Interactions with Materials and Atoms* Vol. 161-163:694-698.

Swerts, J., A. Aerts, N. De Biscop, F. Adams, and P. Van Espen. 1994. "Age determination of Chinese porcelain by X-ray fluorescence and multivariate analysis." *Chemometrics and Intelligent Laboratory Systems* Vol. 22 (1):97-105.

Tennent, Norman H. 1999. *The conservation of glass and ceramics : research, practice, and training*. London: James & James.

Thomas, C. W., and J. Aitchison. 2006. "Log-ratios and geochemical discrimination of Scottish Dalradian limestones: a case study." In *Compositional data analysis in the geosciences: from theory to practice*, edited by Antonella Buccianti, G. Mateu-Figueras and Vera Pawlowsky-Glahn, 25-41. London: Geological Society.

Tite, M. S. 1999. "Pottery Production, Distribution, and Consumption: The Contribution of the Physical Sciences." *Journal of Archaeological Method and Theory* Vol. 6 (3):181-233.

———. 2008. "Ceramic Production, Provenance and Use – A Review." *Archaeometry* Vol. 50 (2):216-231.

———. 2009. "The production technology of Italian maiolica: a reassessment." *Journal of Archaeological Science* Vol. 36 (10):2065-2080.

Tite, M. S., and M. Bimson. 1991. "A technological study of English porcelains." *Archaeometry* Vol. 33 (1):3-27.

Tite, M. S., I. C. Freestone, and M. Bimson. 1984. "A technological study of Chinese porcelain of the Yuan dynasty." *Archaeometry* Vol. 26 (2):139-154.

Vandenabeele, Peter, Howell G. M. Edwards, and Luc Moens. 2007. "A decade of Raman spectroscopy in art and archaeology." *Chemical Reviews* Vol. 107 (3):675-686.

Vandiver, Pamela, and W. David Kingery. 1984. "Variations in the microstructure and microcomposition of pre-Song, Song, and Yuan dynasty ceramics." In *Ancient technology to modern science*, edited by W. D. Kingery and Esther Lense, 181-233. Columbus, Ohio: American Ceramic Society.

Wainwright, Ian N.M., John M. Taylor, and Rosamund D. Harley. 1986 ."Lead Antimonate Yellow." In *Artists' Pigments: A Handbook of their History and Characteristics*, vol. 1, edited by Robert L. Feller, pp. 219–54. Washington: National Gallery of Art 1986.

Wang, Lihua, Jian Zhu, Yan Yan, Yaning Xie, and Changsui Wang. 2009. "Micro-structural characterization of red decorations of red and green color porcelain (Honglvcai) in China." *Journal of Raman Spectroscopy* Vol. 40 (8):998-1003.

Wen, R., C. S. Wang, Z. W. Mao, Y. Y. Huang, and A. M. Pollard. 2007. "The chemical composition of blue pigment on chinese blue-and-white porcelain of the Yuan and Ming dynasties." *Archaeometry* Vol. 49 (1):101-115.

Weyl, Woldemar A. 1999. *Coloured glasses*. Sheffieild: Society of Glass Technology.

Willis, James P., and Andrew R. Duncan. 2008. *Understanding XRF spectrometry: A practical guide with worked examples*. Vol. 1. Almelo: PANalytical B.V.

Wu, J., J. P. L. Leung, Z. Li, and M. J. Stokes. 2002. "EDXRF studies on Chinese Yue ware." *X-ray Spectrometry* Vol. 31 (6):408-413.

Yap, C. T. 1986. "XRF Analysis of Nonya wares using an annular americium source." *Archaeometry* Vol. 28 (2):197-201.

———. 1988. "A quantitative spectrometric analysis of trace concentrations of manganese and cobalt in ceramics and the significance of As/Co and Mn/Co ratios." *Journal of Archaeological Science* Vol. 15 (2):173-177.

———. 1989. "EDXRF Studies on the variation in elemental concentrations of ceramic glazes." *X-ray Spectrometry* Vol. 18 (1):31-34.

———. 1992. "Multi-variate Analysis of Trace Elements from XRF Studies for Classification According to Origin." *Applied Spectroscopy* Vol. 46 (5):843-847.

Yap, C. T., and S. M. Tang. 1985a. "Energy-dispersive X-ray fluorescence analysis of Chinese porcelains using Am-241." *Archaeometry* Vol. 27 (1):61-63.

———. 1985b. "Quantitative XRF analysis of trace barium in porcelains by source excitation." *Applied Spectroscopy* Vol. 39:1040-1042.

Yu, K. N., and J. M. Miao. 1996a. "Non-destructive analysis of Jingdezhen blue and white porcelains." *Archaeometry* Vol. 38 (2):257-262.

———. 1996b. "Non-Destructive Analysis of Jingdezhen Blue and White Porcelains of the Ming Dynasty Using EDXRF." *X-ray Spectrometry* Vol. 25 (6):281-285.

———. 1996c. "Non-destructive and quantitative investigation of Jingdezhen blue and white porcelains using EDXRF." *Journal of X-ray Science and Technology* Vol. 6 (2):167-171.

———. 1999a. "Characterization of blue and white porcelains using the Mn/Fe ratio from EDXRF, with particular reference to porcelains of the Xuande period (1426 to 1435 A.D.)." *Applied Radiation and Isotopes* Vol. 51 (3):279-283.

———. 1999b. "Retrospective study of the feasibility of using EDXRF for the attribution of blue and white porcelains." *X-ray Spectrometry* Vol. 28 (1):19-23.

Zucchiatti, A., A. Bouquillon, I. Katona, and A. D'Alessandro. 2006. "The 'Della Robbia blue': A case study for the use of cobalt pigments in ceramics during the Italian Renaissance." *Archaeometry* Vol. 48 (1):131-152.

Zucchiatti, A., C. Pascual, M. D. Ynsa, L. Castelli, P. Recio, E. Criado, F. J. Valle, and A. Climent-Font. 2008. "Compositional analysis of XVIII century glazed, polychrome, layered porcelain by non-destructive micro alpha-PIXE." *Journal of the European Ceramic Society* Vol. 28 (4):757-762.

Zuo, J., C. Xu, C. Wang, and Z. Yushi. 1999. "Identification of the pigment in painted pottery from the Xishan site by Raman microscopy." *Journal of Raman Spectroscopy* Vol. 30 (12):1053-1055.

Chapter 9

Handheld XRF use in the identification of heavy metal pesticides in ethnographic collections[1]

Aaron N. Shugar and P. Jane Sirois

Introduction

A wide range of pest control techniques has been used on museum collections, in particular on natural history and ethnographic collections. These include the use of heavy metal compounds which are considered the most persistent forms of pesticide. The proliferation of handheld X-ray fluorescence spectrometers (XRF), with their non-destructive/ non-invasive nature of analysis, in the fields of art conservation and museum research has resulted in an increase in the investigation of ethnographic and natural history collections for residues of earlier pesticide treatments. The specific heavy metals that are of greatest concern are arsenic (As), mercury (Hg) and lead (Pb). Compounds containing these elements were applied to a wide range of artifacts of widely differing material compositions (e.g. wood, textiles, leathers, skins, feathers) in various ways including spraying in aerosol form, sprinkling with dust, painting or submerging. Contact with the contaminants may put individuals such as museum staff, visiting fellows and researchers, visitors using the collections, and native communities, to which many objects are being repatriated, at risk. Investigations into contaminated collections have been, and continue to be carried out and data are reported by a number of institutions worldwide (See Sirois *et al*. 2008 for more details).

The use of handheld X-ray fluorescence (XRF) to investigate heavy metal contamination in museum collections began around 1999 (Odegaard *et al.* 2006). Handheld XRF is now a standard method of analysis used to detect heavy metal contamination in museum collections. Several of the main concerns with analysis using handheld XRF are the lack of understanding of the materials being investigated, the physics of how X-rays interact with these materials, and the ability to assess the accuracy of the analytical results reported.

Factors that will affect XRF results, apart from the actual concentration of the elements of interest present, include: the methods of pesticide application, the density and thickness of the treated material, the overall composition of the material(s), the X-ray source used, the analytical parameters used to collect data, any additional adornments to the artifact (e.g. beads, paint layers), the surface geometry (i.e. non-flat or flat surfaces) and the availability of reference materials for calibrating instrumentation.

This chapter outlines some of the key issues that need to be considered when developing a sound methodology for the reproducible analysis of ethnographic collections by handheld XRF. Two of the key issues are understanding the construction and materials of the objects and choosing appropriate reference materials. The difficulties in obtaining reproducible quantitative results and their proper reporting will also be discussed.

Historical overview of inorganic pesticide use

Today's pesticide products, available as consumer products for home use, are properly labeled and designate the pesticide(s) being used, for example, pyrethrins and carbaryl (1- naphthyl methylcarbamate). Pesticides are also used by museums in certain situations such as to combat insect infestations when integrated pest management is not successful. The composition of currently used pesticides in museum contexts is known because of proper labeling, MSDS availability, and rigorous record keeping by conservation staff. This is not the case, however, with historic pesticides. Typically we do not know what pesticide was applied, how it was applied, when it was applied, and how many times it was applied to a collection artifact. There can be several reasons for this but two overriding issues prevail. First, record keeping was not as rigorous in the past as it is today, so complete documentation of potential pesticide treatments were not kept for each artifact. Second, the products that museums used for treatment may not have been labeled with their component ingredients. In fact, labels were not required on pesticides until the 1910 Insecticide Act in Canada, and in the USA until 1927 when the Caustic Poisons Act was enacted requiring that labels include warnings about dangerous chemicals present in the products as well as provide antidote information.

From extended research projects focusing on the history of pesticide treatments in museum collections it appears that the pesticides used were many and varied and were applied to ethnographic collections over long periods of time dating from as early as the early 1800s when arsenic was being used for natural history specimens (Pequinot *et al.* 2006) to the mid 1900s (Goldberg 1996, Conserve-O-Gram 2001, Hawks 2001, Odegaard, Sadongei and Boyer

2005, Sirois *et al.* 2008). It was even recommended for use in natural history specimens as recently as 1985 (Hangay and Dingley 1985). Although the range of pesticides used in museums is extensive they can typically be broken down into two overriding groups; inorganic and organic chemicals. Inorganic pesticides are the most persistent followed by organochlorines, carbamates, organophosphates (diazinon {O,O-Diethyl O-[2-isopropyl-6methylpyrimidin-4-yl] phosphorothioate} for example) and botanical compounds, which have a life span of up to several weeks. Organic compounds themselves can be broken down into two sub-groups; natural, which includes spices, alcohols, tobaccos and cedar, and synthetic which includes halogenated organics (aromatics including DDT {1,1,1-trichloro-2,2-bis(4-chlorophenyl)ethane (International Chemical Safety Card, Sept 9/11)} and paradichlorobenzene{1,4-dichlorobenzene}; and non-aromatics including carbon tetrachloride {tetrachloromethane}and methyl bromide {bromomethane}), organophophorous pesticides such as dichlorvos (2,2-dichlorovinyl dimethyl phosphate) and diazinon (O,O-diethyl O-2-isopropyl-6-methylpyrimidin-4-yl phosphorothioate), phenols such as pentachlorophenol (2,3,4,5,6-pentachlorophenol) and cresol (methylphenol) and carbamates such as bendiocarb (2,2-dimethyl-1,3-benzodioxol-4-yl methylcarbamate) and aldicarb (2-methyl-2-(methylthio)propionaldehyde O-methylcarbamoyloxime).

The most frequently encountered inorganic compounds used by museums include arsenical compounds (i.e. arsenic oxides), mercury compounds (i.e. mercuric chloride), and lead compounds (i.e. lead oxide(s) such as litharge and minium, the common white pigment lead carbonate hydroxide, and the pesticide lead arsenate). These inorganic compounds are persistent, having very long life spans, are highly toxic, and are difficult to dispose of or mitigate. Not only were many types of poisons used, but the application methods varied greatly. Objects may have been sprinkled with a powder, slathered with a paste, dipped or soaked in a solution, or sprayed with an atomized chemical. Several recipes and techniques for pesticide application have been reported in the literature. These include Hornaday's arsenical soaps used for taxidermy (Rowley, 1925; Wagstaffe and Fidler, 1968) in which powdered white arsenic (arsenic trioxide) was mixed with soap, subcarbonate of potash, camphor and alcohol. This resulting paste would then be applied to the inside of animal or bird skins. Similar hide poisons have been reported (Anderson, 1932) which combine arsenic trioxide, carbonate of soda (Na_2CO_3) and water. Arsenic trioxide can also be put into solution (boiled directly into water) prior to application. Mercuric chloride was applied to the surface of birds as well as to artifacts composed of feathers. In situations where multiple treatments occurred over time (possibly using different pesticides) the chemical compounds remaining today may well no longer be those originally applied.

Each of these application techniques has different implications for the location where the toxins reside in/on the object, their distribution on the surface, and their depth of penetration into the component materials. This, in turn, affects one's ability to detect and then remove these substances, making risk assessment problematic.

Health and safety considerations

Toxicity

The potential presence of these toxic heavy metal compounds in museum collections are a hazard[1] to those handling contaminated objects. Some of these compounds such as mercuric chloride, arsenic trioxide and lead arsenate have been identified in museum collections not only by XRF but by other analytical techniques such as X-ray diffraction (XRD) and Fourier transform infrared spectroscopy analysis (FTIR; Sirois 2001, Sirois *et al.* 2010). The physical effects from exposure could be severe depending on the handling situation, the frequency of exposure, and the method of exposure. Arsenic exposure can occur from direct handling. The trivalent form is the most toxic and was in primary use as a pesticide (as arsenic trioxide). As little as 3-4 mg of arsenic ingested daily can result in chronic toxicity/ health effects and 20 mg of arsenic in a child (as might occur with a repatriated object) may produce life-threatening toxicity (Agency for Toxic Substances and Disease Registry 2011). In Ontario, Canada, the time weighted average (TWA) for arsenic (airborne concentration of a biological or chemical agent to which a worker may be exposed in a work day or work week) is 0.01 mg/m^3 of arsenic in air (Ontario Regulation 490/09). Based on a 40 hour/wk exposure, the maximum allowable level would translate to 0.226 mg of arsenic dispersed in the air of a 10x10x8 ft room (22.6 m^3). Mercury exposure (as reported for metallic, inorganic compounds, and organic compounds), is even more likely to occur through direct handling than arsenic. 10 – 60 mg is cited for a fatal dose; (Agency for Toxic Substances and Disease Registry 1999). It primarily targets kidney functions and the nervous system (both acute and chronic exposures). Lead exposure can be via handling or ingestion. Exposure can cause anemia, vomiting, seizures and nervous dysfunction (Agency for Toxic Substances and Disease Registry 2007). With these potential symptoms in mind, one can readily see why museums are concerned about collection objects containing heavy metal pesticides and the potential exposure to them by their conservators, curators, researchers and guests. In addition, many artifacts are being repatriated as a result of the Native American Graves

Protection and Repatriation Act (NAGPRA) for example, and there is concern for the health and wellbeing of the native communities receiving these objects. Through a formal legal process in the United States and more informal methods in other countries, items that have been in museums for decades are returning to their communities of origin. Section 10.10(e) of the 1996 NAGPRA Final Regulations specifically requires museums and federal agencies to disclose "any presently known treatment of human remains, funerary objects, sacred objects or objects of cultural patrimony with pesticides, preservatives or other substances that present a potential health hazard to the objects or the persons handling the objects" (Native American Graves Protection Repatriation Act 1996).

Determining toxic compounds present

With the potential danger of pesticides present in collection objects, museum curators, collection managers, researchers, repatriation communities, potential visitors, and museum officials must be informed about whether collection materials are contaminated, the levels of contamination, and what artifacts are considered safe for handling.

Information from the literature can suggest what pesticides were in use during specific periods (see Goldberg 1996; Hawks 200; Odegaard *et al.* 2005 for a general overview; Péquignot *et al.* 2006) and in-house documentation may indicate what was specifically used on an artifact. Arsenical soap was one of the more traditional "preservatives" employed in the field of natural history since 1800 when the recipe was published (Péquignot *et al.* 2006). Similar techniques were said to be used on both natural history specimens and ethnographic objects made of organic material (Goldberg 1996; Hawks 2001). The Smithsonian Institution has done archival research to determine possible pesticides used in the National Museum of Natural History (NMNH) (Goldberg 1996). Some inorganic pesticides identified in that review, with associated dates if available, were: arsenic and corrosive sublimate (mercuric chloride) (1830s-60s); strychnine in arsenical solutions (reported in 1887) and Larvex (sodium aluminum fluorosilicate) (1930s). This may not include all the pesticide treatment(s) performed over the years as documentation is typically incomplete or sometimes excluded entirely. There may also be previous microchemical or instrumental analysis data that can provide information as to what might be present on an artifact. To identify the pesticide(s), determine their locations and quantify them, we rely on instrumental analysis. Ultimately both qualitative and quantitative results are required to accurately determine what pesticide(s) are present and in what concentration.

Determining Potential Exposure

Determining the potential exposure hazard and transfer of pesticides requires collecting information on:

- What pesticides are present?
- Where they are located?
- How much is present?
- How much can be transferred from the object to the person?

To determine potential exposure and answer these questions one must obtain information on how an artifact is handled and used to establish how transfer might occur. In some cases we know a great deal about handling protocols within the museum itself, but in other cases, such as with repatriated artifacts, we may not have a full picture of the object's use once it is returned to its proper native group. Occasionally, artifact use within native groups is shared with conservators and conservation scientists but for the most part, native groups prefer to keep the intended use of their artifacts private.

Ultimately, conservation scientists are not in a position to comment on issues related to exposure to pesticides and the potential health hazards arising from exposure. A trained medical toxicologist must make that determination. Conservation science can only address the first three questions listed above; however, information on an object's previous treatment(s), its handling and use, the pesticides present, and their concentrations can provide medical professionals with the necessary information for a proper toxicological assessment. Ideally a team of researchers (object user, custodian/guardian, scientist, toxicologist or other medical professional) should work together for a complete investigation of a collection. Including all the different members of this team in each phase of the project will promote success by ensuring the necessary information is obtained in a respectful manner, and that the information conveyed to the end user is accurate and useful. With that being said, the responsibility of performing the analysis and interpreting the results falls on the conservation scientist who potentially has a range of analytical techniques at their disposal.

Analytical techniques for pesticide identification

Numerous analytical techniques have been used for the identification of heavy metal pesticides. These include, but are not limited to; microchemical spot testing, scanning electron microscopy-energy dispersive X-ray spectroscopy (SEM/EDS), optical microscopy (OM), X-ray diffraction (XRD), atomic absorption spectroscopy (AAS), inductive coupled plasma spectroscopy (ICP), wet chemical reactions, Raman spectroscopy and X-ray fluorescence (XRF) (Found and Helwig 1995; Seifert *et al.* 2000; Palmer 2001; Sirois 2001; Sirios and Sansoucy 2001; Palmer *et al.* 2003; Odegaard 2005; Palmer *et al.* 2006; Sirois *et al.* 2008; Madden *et al.* 2010; Sirois *et al.* 2010). X-radiography has been used by some to successfully show the presence of heavy metal pesticides, in particular, in taxidermy specimens (Williams and Hawks 1987, Sirois 1989; Péquignot *et al.* 2006; Shugar, 2011). These imaging studies are typically used in conjunction with other analytical techniques to provide positive confirmation of the elements present. SEM/EDS, microscopy, AAS, ICP, wet chemical reactions and spot testing are considered destructive techniques as they require sampling. Raman does not require samples but requires potentially dangerous handling of an artifact and if improperly used, can damage the area being investigated. XRF, however, is non-destructive and non-invasive by nature.

The non-destructive nature of XRF becomes particularly important when considering that many of the artifacts currently being tested in museums in the USA fall under the NAGPRA act allowing for repatriation of artifacts to their rightful tribal groups. Very often these tribal groups request that no invasive testing be performed, but they do want to know if their artifacts pose a health hazard. With this in mind, XRF has rapidly become a favorite analytical instrument for the detection of elements indicating the use of heavy metal pesticides.

XRF analysis of inorganic pesticide residues

There is much promise for successful identification of pesticides using handheld XRF instruments, however, there are several issues that can make identification difficult and quantitative analysis of inorganic pesticide residues on ethnographic objects especially challenging. The challenges related to obtaining accurate quantification are related to both the instrumentation and the artifacts themselves. The following sections explain these issues in more detail and provide suggestions as to how to best calibrate instrumentation to acquire reproducible results for the identification and quantification of inorganic pesticides.

Instrument setup

The spectrometer should be configured for maximum performance for the specific application; configuring it to achieve the best peak resolution possible, the maximum peak to background signal and minimal interference in the energy region of the elements of interest are essential for obtaining good data. The most important instrument parameters that can be selected or changed to achieve this are: the type of X-ray tube and detector chosen at the time of purchase, the voltage and current used, the filters selected, the calibration method (or analysis mode) chosen and the reference materials selected.

Spectrometer hardware

The advancement in detector technology over the last few years has helped to clarify the spectral region of interest where many overlapping peaks for lead, mercury, and arsenic occur (Shugar 2009). Traditional Si(PIN) diode detectors used in handheld XRF instruments have a resolution of approximately 195 – 210 eV at manganese (Mn). This relatively low peak resolution does not allow for clean peak identification or peak separation. This problem has been improved slightly with newer handheld XRF instruments using silicon drift detectors (SDDs) which have a resolution of approximately 140 – 150 eV (Leutenegger *et al.* 2000; Bruker Corporation, 2008)

An X-ray tube should be selected that does not emit characteristic radiation in the energy range of the elements being studied. The energy range from 5 keV to 13 keV should be as free as possible form extraneous X-ray peaks (for example, Ag [Kα_1 22.16 kV], Rh [Kα_1 22.22 kV], or Mo [Kα_1 17.48 kV] targets are all acceptable but Re [Lα_1 8.65 kV] would not be). The X-ray tube must also generate X-rays of a sufficient energy to excite the inner shell electrons of the elements being analyzed. This is explained further in the instrument parameters section below.

Instrument parameters

In designing the best analytical protocol, the ability to adjust the voltage, current and filter in the spectrometer is necessary to optimize the experimental parameters and maximize the signal to noise obtained. Such flexibility is ideal for a highly specific application such as this one. Some XRF manufacturers pre-set these parameters with several choices available for example, preset conditions for bulk soil analysis may be 35 kV, 14 μA and an aluminum filter, while the conditions for alloy or filter analysis would differ. There is typically some ability to alter the voltage and current but generally not the filter, unless specific requests are made. With these instruments it will be important to select the correct mode – one that contains the best combination of parameters for your application. The manufacturers have chosen a set of effective parameters for specific common applications which may or may not be ideal for your analysis. This would be dependent on the object material, the geometry of the object relative to the X-ray beam, and how and where the pesticides were applied. For example, the conditions used for "soil" mode (as designated by the manufacturer) are designed to provide concentration data for infinitely thick samples with a low atomic number matrix, while "filter" mode is designed for samples on a minimally thick, low atomic number filter thereby allowing the sample matrix to be ignored in the quantification calculations. Some handheld instruments allow full control of their systems giving the operator the ability to change all three variables. Most manufacturers will allow you to specify the instrumentation setup and it is recommended that this be done. The voltage should be set high enough to excite the elements of interest, mainly lead, mercury, and arsenic but also bromine (Br), zinc (Zn) and chlorine (Cl) if possible. Unless an excitation voltage exceeding the binding energy of the element of interest is chosen, the elements will not be excited (The binding energies for Pb, Hg, and As are as follows; Pb $L_1$2s 15.816 KeV, Hg $L_1$2s 14.839 KeV, and As K1s 11.867 KeV). When looking for several elements, one should choose a voltage based on the highest binding energy. With this in mind, the parameters chosen should accentuate/effectively excite the region of interest in the resulting spectrum. For analysis of heavy metal pesticides the recommended voltage setting for one's instrument is 40 kV with the maximum current possible ensuring the dead time is limited to no more than 30% (typically scan times of approximately 60 seconds are considered acceptable).

The use of filters can greatly improve the background in specific regions of interest of the spectrum. By absorbing continuum energy above the mass absorption coefficient of the filter element, you can create a flatter/lower count rate region of the spectrum allowing for better determination of peak to background for individual element peaks that reside in that particular region. For example, Figure 9.1 shows the same sample analyzed using: no filter, a 25 μm Ti – 300 μm Al filter, a 25 μm Ti – 300 μm Al – 25 μm Cu filter and finally a 25 μm Ti – 300 μm Al – 75 μm Cu filter. It becomes evident that the filters containing Cu help reduce the background radiation in the spectrum between 10 – 15 KeV, directly where the characteristic arsenic, lead, mercury and bromine peaks are located. This results in a higher signal to background ratio and allows for easier interpretation of spectra. In addition, you will notice that the spectrum produced with no filter has additional peaks resulting from instrumental interferences and Bragg scattering (or diffraction). The downside of using thicker filters is that they reduce the count rate (counts/second). Ways of overcoming this would include increasing the current used, or extending the length of scan time. Since some instruments do not let you control the filters used, you should talk with your manufacturer and find out what has been installed. If possible, you should customize the filter. A 25 µm Ti – 300 µm Al – 75 µm Cu filter was found to be ideal for reducing the background in the spectral region of interest with scans of between 60 and 180 seconds, depending on the material being investigated.

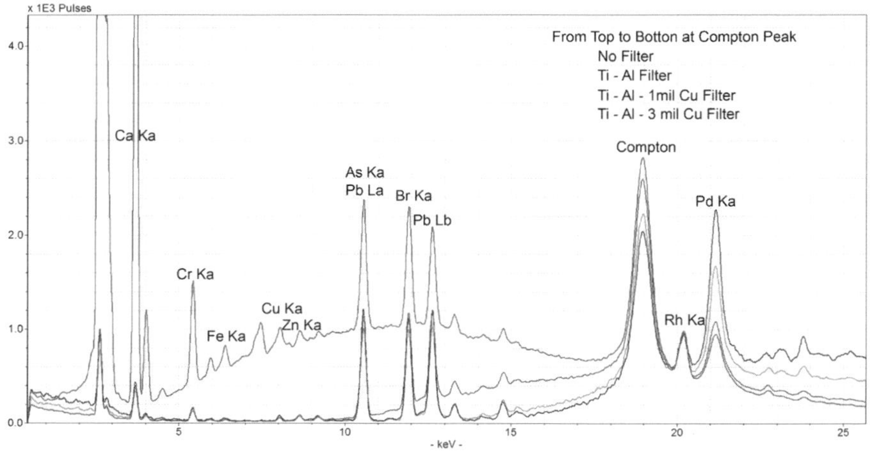

Figure 9.1:
Example of the effects on a spectrum when using various filters. Note that the continuum radiation is reduced in the region of interest with the use of an appropriate filter (all filters have a 25 µm Ti and 300 µm Al component. The copper component is either 25 µm or 75 µm).

Experimental setup

Before analysis is started, it is essential to consider what type of material(s) is being investigated, how many objects will be analyzed, what elements will be analyzed, how accurate the results need to be, is qualitative or quantitative data required and, what questions need to be answered. These issues will define how you design your experimental procedure and what type of data you collect. Depending on the answers to these questions, either qualitative or quantitative XRF analysis will be undertaken.

Qualitative XRF

A qualitative XRF survey involves detecting whether an element is present. In addition one can estimate its relative concentration (in comparison to other analyses), but not determine its absolute concentration (Bertin1970). This often is sufficient to determine whether there is a potential problem within a collection. It allows for a quick survey of a collection and can be a precursor to a more thorough investigation of specific objects later on.

A more comprehensive survey could be done by either determining what contaminants can be removed from the object (by analyzing a wipe drawn across the surface of the object if allowed by the client), or by doing a more quantitative analysis later on. Two topics relevant to qualitative analysis are the physical make-up of the object(s) and spectral interpretation.

Artifact composition

What is of greatest concern for the acquisition of data from ethnographic artifacts is the composition and structure of the artifacts themselves. The typical, although not exclusive types of materials that may have been treated with heavy metal pesticides include various types of wood, leathers, skins, feathers, quills, textiles (cotton, wool and other), stuffed animals and grasses. These different materials all have physical and chemical properties which can alter the X-ray spectrum obtained. The density, thickness, morphology, geometry and composition of the material all affect the X-ray signals being detected. An example of this is shown in Fig. 9.2, where spectra were acquired of different fur samples placed over a bulk arsenic-in-Chemplex reference material. The skin layer of these fur samples ranges from 1-2 mm however the fur thicknesses vary widely. The caribou fur, when compressed has a thickness of about 10 mm and when uncompressed is about 25 mm, the buffalo fur when compressed has a thickness of approximately 7 mm and when uncompressed is about 20 mm and lastly the beaver fur has a thickness of about 3 mm when compressed and is about 6 mm when not. The caribou fur is also the densest of the three types of fur. The typical elemental makeup for the substrate materials used in organic ethnographic artifacts includes H, C, O, S and N with Si, K, Ca, and Fe impurities usually present. The artifacts could also have varied methods of pesticide application, varied concentrations, and potential numerous treatment campaigns.

In addition, the objects may be adorned with decorative elements that have arsenic, lead and mercury or other potentially interfering elements in them. For example, moccasins may be covered with glass beads (lead and arsenic are common constituents, lead fluxes being used to lower the melting temperature of the glass and lead being a common component of glass colorants such as lead antimonate yellow ($Pb_2Sb_2O_6(O,OH)$). Arsenic is typically found in glasses as a fining agent or as an opacifying agent, or wooden masks may be painted with pigments derived from lead (lead white ($Pb_3(CO_3)_2(OH)_2$) or red lead (Pb_3O_4)), mercury (cinnabar or vermilion HgS) or arsenic (occasionally orpiment (As_2S_3) or realgar (As_2S_2)). These decorative elements have no relationship to a pesticide treatment, yet if present will affect XRF results. Some materials that have contributed to the X-ray signal from objects that are not due to pesticides are: paints and pigments, dyes, mordants, metal decorative elements or beads, and glass beads. It is important to document what areas you are analyzing and to check the reverse surface of the area analyzed to make sure that you are not analyzing additional layers or materials. An additional concern in the case of archaeological artifacts is that powdered metal corrosion can contaminate large sections of fabric or leather objects.

Spectral interpretation

Beyond the spectral interferences one normally has to deal with when interpreting XRF spectra [i.e. Rayleigh and Compton scatter, Bragg scatter, sum peaks, peaks due to the instrument (such as the X-ray tube housing) and escape peaks] the analysis of heavy metal pesticides complicates matters further. The spectra obtained from these analyses contain many spectral emission line overlaps in the region of interest. Pb Lα (10.55 KeV) and As Kα (10.54 KeV) emission lines overlap. At the As Kβ line (11.72 KeV) there is an emission line overlap with both Hg Lβ (11.82 KeV) and Br Kα (11.92 KeV). Both Pb Lβ (12.61 KeV) and Hg Lα (9.99 KeV) are fairly isolated and can be used for identification purposes, but with the overlap on both the As Kα and Kβ, it can be difficult to identify the presence of arsenic. Because of all these peak overlaps it is essential to examine the spectra themselves and not to rely solely on the concentrations provided by the spectrometer.

Estimating the concentration of an element identified in an XRF spectrum based solely on the peak height observed in the spectrum can be misleading. The principles of XRF rely on primary fluorescence from an element in a sample. Primary fluorescence is the fluorescence that results when an element has been excited by a beam of X-ray photons of higher excitation energy than the binding energy of the element. This results in the characteristic radiation for that element. Secondary fluorescence can occur when the primary characteristic radiation produced by one element interacts with an element with a lower binding energy than the characteristic radiation of the initially excited element, thereby producing characteristic radiation from the secondary element in the sample. This continues with tertiary fluorescence and so on. The result is that there can be a decrease in the reported intensities of higher energy elements present in the sample while simultaneously increasing the reported intensities of lower energy elements. This accounts for matrix effects within each artifact.

Another factor in estimating quantity from a qualitative analysis is the attenuation effects of elements that are not uniformly distributed throughout a material. A pesticide treatment may have been applied to only one side of an artifact with no documentation or visual markers to indicate this. An example of this phenomenon can be seen in Figure 9.2, which shows the XRF spectra obtained from the analysis of an arsenic reference material placed on the skin side of several types of skins with fur. The effect on the resulting spectra is quite revealing – demonstrating the high level of attenuation of the 10.54 keV As Kα X-rays through organic layers with thicknesses ranging from 3 to 10 mm of compressed fur attached to skin (i.e. holding the spectrometer closely and firmly to the surface of the fur). All the spectra were acquired with the spectrometer positioned on the fur side opposite from the arsenic reference. Arsenic was positively detected in all cases, however characteristic As emission lines have much lower counts than the actual (non-attenuated) value, and that discrepancy extends when examining thicker layers of fur and skin. The spectra in Figure 9.2 show that the thicker the layer of fur, the smaller the arsenic peak observed in the spectrum and the lower the ratio of the arsenic peak to the Compton peak. This simulation mimics natural history specimens which have been frequently reported to have been treated with arsenical soaps exclusively on the skin interior.

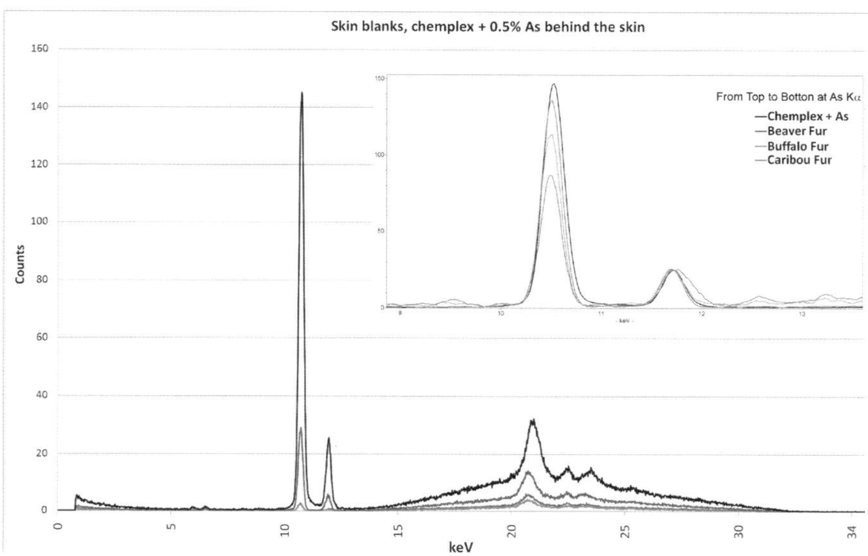

Figure 9.2:
Spectra showing the attenuation of As through various fur types. The inset shows the scans normalized on As Kβ (11.72 keV) clearly indicating the attenuation of the As Kα peak.

A benefit to this phenomenon is that the attenuation of the arsenic energy can be used to help determine the location of the treatment. The typical expected ratio of Kα to Kβ is approximately 5:1. The Kα peak attenuates more than the Kβ peak due to its lower energy resulting in a decrease of this ratio which is typically visually identifiable in the spectrum. Seeing the attenuation of the Kα peak would be an indication that the treatment was on the opposing side of the artifact. Ideally though, it is recommended that XRF spectra be collected from both sides of an artifact if possible to help address this issue.

Thickness and chemical makeup considerations of the sample matrix

Two material properties that affect the spectrum and the analytical mode chosen for data collection and interpretation are: its thickness and the chemical make-up of the object (its density and elemental composition). In addition, when considering what information is being collected by the XRF one has to consider the total area and volume of your sample that is being irradiated.

To determine the penetration of X-rays into the matrix material one must take the matrix composition, thickness (critical penetration depth), and density into account. The absorption of X-rays by the material being analyzed can be determined with the following equation:

$$I = I_o \exp^{[-(\mu/\rho)x]} \qquad (1)$$

where I is the intensity of the emerging radiation. I_o is the intensity of the incident radiation, a narrow beam of mono-energetic photons (this is not true in the case of the instruments we are typically using, but the minor variation in photon energies coming from the instruments is not significant, and therefore, the resulting values are close enough for our applications), that penetrates a layer of material with a mass thickness (x), defined as the mass per unit area, and density (ρ). The mass thickness is: the thickness t multiplied by the density ρ, therefore x=ρ t. The term μ/ρ, the mass attenuation coefficients, are basic quantities used in calculations to determine the penetration and the energy deposition in materials. This equation shows that the intensity of the emerging photons is exponentially affected by both density and thickness of the material being analyzed (Thomsen, Schatzlein and Mercuo 2005).

Another concept that is important in *in situ* XRF analysis of heterogeneous samples or samples with irregular surfaces or surface contamination where minimal or no sample preparation is involved, such as is the case with this application is the concept of the critical penetration depth. The critical depth perpetration of the incident X-ray beam (d_{crit}) is defined as the sample thickness from which 99% of a fluorescence signal originates. This is calculated with the equation:

(2) $$d_{crit} = m_{thick}/\varrho$$

where m_{thick} = saturation mass or the mass/unit area, ϱ = density of the sample material (Markowicz 2008). Therefore the extrapolation of results obtained for a relatively thin near-surface layer of thickness d_{crit} to estimate composition of bulk materials should be performed extremely carefully (Markowicz 2008). This is the case with many objects analyzed for inorganic pesticide residues.

Variable sample thickness is problematic when interpreting results from XRF data. Based on sample thickness alone, three models could be considered for XRF analysis of pesticide residues in various objects:

- The bulk model (infinite thickness). This is ideal but rarely encountered.
- The thin film model (where the layer investigated is of negligible thickness) and
- An intermediate thickness model where the layer being analyzed is somewhere between the other two models. For these samples the intensity of the characteristic X-rays of the element in question depends not only on the overall composition of the sample but also on its thickness.

If the object is essentially a thin film, the matrix (and therefore the composition of the object) does not contribute significant background to the XRF spectrum and so the composition of the object can essentially be ignored. For the bulk and intermediate thickness models, the matrix composition, the thickness (critical penetration depth), and density must be taken into account (see Figure 9.3). In the analysis of pesticide residues in anthropology and natural history collections the thin film and intermediate thickness models are most frequently encountered, for example with materials such as skins, fur, wood and thick textiles. The bulk model is rarely encountered. This leads to extreme difficulty in assessing spectra and assessment should be done extremely carefully as recommended by Markowicz (2008).

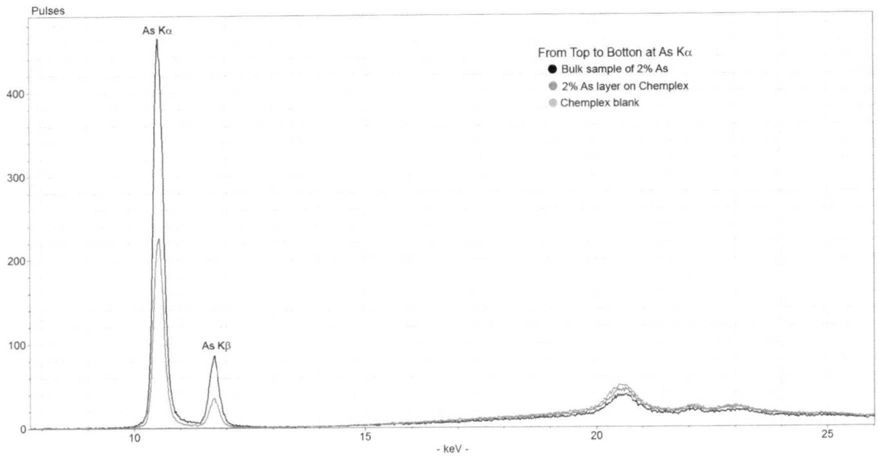

Figure 9.3:
The effect that the thickness of the contaminated layer has on the intensity of the arsenic X-ray peaks detected. The black spectrum shows the signal from an arsenic-containing sample considered infinitely thick, while the grey spectrum shows the signal from a sample with a thin arsenic-containing layer of the sample composition applied over a base of blank Chemplex™.

The formulae above also suggest the importance of knowing the chemical makeup of your object. As mentioned above, the typical elemental makeup for the substrate materials used in ethnographic artifacts are comprised mainly of low average atomic number composition, however, the density, porosity, shape, size and thickness will vary considerably which will have an exponential effect on the matrix effects.

An example of variable densities within even one material used in a variety of ethnographic artifacts is seen below with different types of wood (see Table 9.1). Not only can the thickness be variable between artifacts, but it can be variable within single artifacts. In addition, woods have densities ranging from $0.11–0.9x10^3$ kg/m^3. Since these two variables, density and thickness, have an exponential effect on the emerging energy from the sample, each wood type and each thickness has a great effect on the data produced.

Solid Density	(10^3 kg/m^3)
Ash, white	0,65 - 0,85
Aspen	0,42
Balsa	0,11 - 0,14
Basswood	0,3 - 0,6
Beech	0,7 - 0,9
Cedar, red	0,38
Douglas Fir	0,53
Elm, American	0,57
Maple	0,6 - 0,75

Solid Density	(10^3 kg/m^3)
Oak, American Red	0,74
Oak, American White	0,77
Oregon Pine	0,53
Pine, pitch	0,8 - 0,85
Pine, white	0,35 - 0,5
Pine, yellow	0,35 - 0,6
Redwood, American	0,45
Spruce	0,4 - 0,7

Table 9.1:
Range in densities of various woods from 0.11–0.9 10^3 kg/m^3 (The Engineering Toolbox 2011).

Surface geometry

In addition to an artifact's thickness and chemical makeup, its surface geometry can have an adverse effect on the resulting spectra. "Effects such as surface roughness, particle shape, particle size, and size distribution can all lead to no proportional relationships between spectral intensity and elemental composition" (Jenkins *et al.* 1995:271). This is often the case with ethnographic objects as their *in situ* analysis often presents rough surfaces, irregular shaped surfaces and variable geometries. This results in the object not being flush against the XRF window and these rough or irregularly shaped samples will affect the accuracy and precision of the analysis, altering spectral intensities.

All these factors can affect the resulting spectra collected from ethnographic samples and they exemplify the difficulty in achieving representative and reproducible data from the objects themselves. This can affect one's qualitative assessment of an object and its degree of pesticide residue, and may lead to misinterpretation of the spectra as well. Quantitative analysis also suffers from these effects but to a greater extent.

Quantitative analysis

When undertaking quantitative analysis, additional variables must be taken into account to obtain reliable results. The proper spectral correction procedures (or analysis modes) must be chosen or developed that will match the type of sample/object being analyzed. This will be influenced by the factors mentioned in the previous section (i.e. thickness and material composition of the sample being analyzed). Proper calibration of the spectrometer using suitable reference materials must be performed. Knowledge of the materials and construction of the objects must be known to prepare reference materials with matrices similar to the objects.

Initially, one should consider what materials will be analyzed and where the contaminants might be located. How the applied pesticide interacts with the material directly relates to how XRF can quantify a sample as can be seen in the example above discussing the attenuation of As through fur samples. XRF is an ideal analytical technique for quantification when dealing with homogenous materials but the analysis of heterogeneous materials is difficult at best (and really should be avoided) but may be overcome if one can match similar reference materials to the artifact in question.

As the application of pesticides, whether by brushing or dipping, does not create a homogenous distribution, artifacts will show uneven penetration in most cases. The interaction of the pesticide with the material can directly influence its location and chemical make-up, as can its initial concentration, and how evenly it was applied to the object. This can be seen as either a surface coating (where pesticides lie immediately on the surface of the artifact), as a surface enrichment (where there is some penetration into the artifact), as an uneven distribution throughout an artifact (where a gradient of concentration occurs with mobility of the pesticide through an artifact as is seen in taxidermy specimens), or as a more evenly distributed penetration on very thin samples or fine feathers. These different situations require different reference materials, calibrations and techniques of analysis for specific types of objects. Due to the varied composition of the substrate materials, the varied pesticides formulas and application methods, and the essentially inhomogeneous nature of these objects and their non-uniform geometry, high error quantification or qualitative assessment may be the best that can be achieved (Jenkins 1999; Fonicello 2007).

The protocol for creating proper reference samples for XRF analysis is well documented and generally involves preparing a homogenous, infinitely thick sample with a smooth, flat, regular surface. There are several techniques developed to accomplish this including preparing pressed pellets, creating borax glass beads as well as using a fusion technique that ensures a homogenous sample with a smooth surface.

Reference materials

The factors discussed above; incident radiation, sample thickness, material composition, and physical properties, show some of the information that needs to be considered, and is required when designing reference materials for a specific application. In addition, as stated earlier, it is important to know: what energy source will be used in the analysis, what elements are going to be identified, the density and thickness of the materials to be investigated, mass attenuation coefficients, and the geometry of the incident and take off angles of the X-rays. One ideally would like to match these characteristics to reduce any potential variations in resulting peak intensity between the reference materials and the unknowns. Several of these variables can be dealt with easily: the geometry of the systems are fixed; the voltage and filter set can be pre-assigned or determined; the remaining variables are much more difficult to deal with.

Typically we don't know exactly what compound has been used or where the pesticide resides on/in the artifact. In order to attempt to quantify the concentrations of these pesticides we first need to produce reference materials to calibrate our instruments. There are established methods for the production of reference materials (May *et al.* 2000) but with the variety and variation in material types encountered and their different physical properties, creating one set of reference materials is inadequate for proper calibration of ethnographic and natural history specimens.

There have been several attempts to create in-house reference materials used to both determine the instrument sensitivity and to fully calibrate instrumentation. Leathers have been impregnated with arsenical soaps or arsenical solutions (Bond, 2005). Mylar sheets with thin layer contaminants have been used to simulate thin layers of pesticide on the surface of artifacts (Sirois, in progress). Ink jet printing systems that can apply specific amounts of a known material on a substrate are also currently being investigated.

Chemplex^{tm} X-ray sample mix (48.7% C, 42.6% O, 8.1% H, 0.6% B) has been doped with arsenic, lead and mercury and pelleted in aluminum cups (by Sirois), left as loose powder, or pressed in X-ray sample cups (Shugar and Madden, forthcoming). Thin film samples have also been used as well as established standard reference materials (e.g. NIST lead paint films for portable XRF analyzers SRM 2579a).

These attempts to acquire or create reference materials are useful in helping to determine if artifacts were subjected to pesticide treatments, but lacked some specifics features required to ensure that accurate quantitative information could be achieved. There are several requirements when creating reference materials: 1. The concentration of the reference material should be known to a better degree of precision and accuracy that is required of the analyte, 2. Reference materials should have the same matrix as the analyte, and 3. The concentrations should vary independently when empirical absorption corrections are used.

With this in mind, four stages of reference material production have been established to characterize the materials and create sound calibration curves for analysis of heavy metal pesticides in museum collections [this work has been carried out in conjunction with the Smithsonian (MCI and NMAI) and NIST]. These stages are:

Stage 1: Basic single element reference materials need to be produced for mercury, arsenic, and lead to confirm the accuracy of reading for the XRF and to calibrate the instruments.

Stage 2: Compounds of potential common elements found on typical museum objects need to be produced to investigate matrix effects, beam attenuation and further calibrate the instruments.

Stage 3: A material-specific reference material needs to be produced that has similar properties (i.e. porosity, density, thickness) to the objects that we are studying (i.e. wood (various types), feathers, leather, skin, bone, etc…) and each type has to be prepared using specific pesticide treatments.

Stage 4: Cross section analysis of treated objects is needed to determine depth penetration of pesticide application on a micro-analytical scale and to better understand matrix effects with the media itself and the potential physics of X-ray excitation within the samples.

Several of these stages have been completed or in the process of being completed. For stage 1, there are currently fairly good single element reference materials for Pb (see NIST lead paint reference materials for handheld XRF SRM 2579a). Reference materials for arsenic have been produced to cover a wide range of arsenic concentrations in a pressed pellet of a similar density to red cedar ensuring homogenous sample preparation (Shugar and Madden, forthcoming). Initial calibration testing on handheld instruments using these reference materials shows good linear correlation for single element calibrations.

In developing more complete reference materials as described in stages two and three, the issues described earlier related to sample density, thickness, chemical makeup etc. need to be taken into consideration. One must determine the general density and chemical makeup of the objects in question as well as characterize the relationship between the Rayleigh and Compton peaks and the slope of the continuum radiation as it varies with matrix density. Less dense materials allow the incident X-ray beam to interact with a greater proportion of the sample, creating a greater concentration of continuum radiation as well as allowing for increased inelastic scattering, resulting in a shifted, increased, broader Compton peak (see Figure 9.4). As one can clearly imagine, changes in the slope of the spectrum as well as the altered intensity of the Compton peak make comparing two different spectra difficult. To create the best possible fit when making reference materials, one should attempt to match these variables to reduce error. Ensuring that the slope of the curve, the general peak height, and the Rayleigh and Compton ratio is similar is essential for creating quality reference materials.

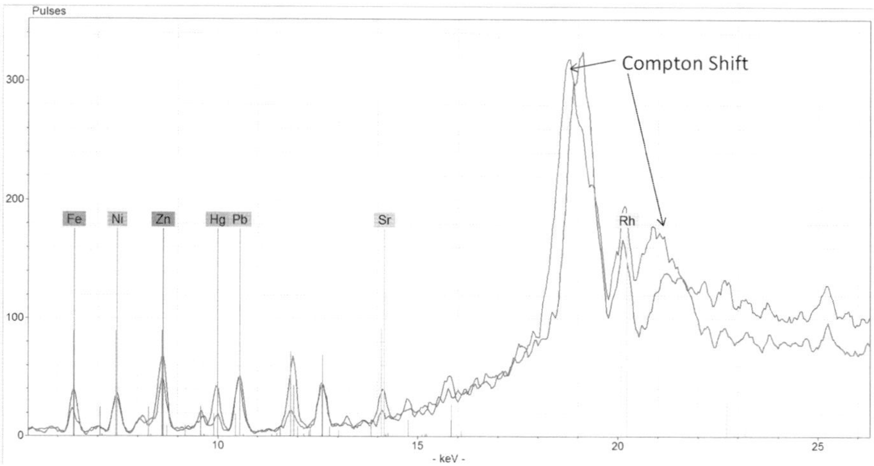

Figure 9.4:
Example of Compton shift and spread when a less dense sample is analyzed. In this case a denser flat feather has a Compton peak position at 19.10 kV with a peak spread of 1.16 kV. A less dense down feather has a peak position at 18.78 kV with a wider spread of 1.59 kV (measured at 120 pulses; Both samples run at 40kV and 7 μA for 60 seconds with a 25 μm Ti – 300 μm Al – 75 μm Cu filter).

To investigate the variation in densities and chemical compositions and their effects on XRF data, a collection of 69 ethnographic artifacts, representing a typical range of material types in museum collections, was analyzed. The artifacts investigated came from the Buffalo Museum of Science and comprise a range of materials including, ceramics, wood, horn, textiles and woven plant fibers. The spectra from these analyses created the baseline to begin addressing the various stages of reference material production. Reference materials for stage two instrument calibration have been produced using this dataset (see Table 9.2).

Based on the raw data collected, reference materials were chosen based on their densities matching those of the artifacts. Also, to help correct for matrix effects typical elements present in the artifact were added to the reference materials. Polyethylene (density of 0.910 g/cm^3) and polyvinyl chloride (density of 1.1 – 1.42 g/cm^3) pressed pellets and powdered samples were both prepared with varied amounts of lead, mercury, arsenic, bromine, iron, chromium and other elements that may have a direct matrix effect on elements of interest (see Table 9.2).

Ref.	Cl	Ca	Ti	Ba	V	Cr	Fe	Cu	Hg	As	Br	Pb	Cd
1						522	24		34	0	103	528	49
2						500			500	1250	502	503	50
3						500			503	1250	500	500	50
4						51			100	100	5	1004	25
5						749			50	100	1026	250	100
6						50			100		5	1022	25
7						756			78		1008	254	100
8						1254			529		257	50	125
9						1066			252		502	1260	75
10						651			813		1266	751	11
11						252			1000		751	100	5
12						505			38		101	500	50
13						100			1244		269	7022	150
14	810	637.29	1591	2200	48	114.6	23	117	25.3	31	808	107.6	140.8
15	93	307.23	767	257		17.7	23	16	4.5	4	98	13.7	21.7
16						0	27	3	0		0	0	0
17						50	23		100	0	5	1004	25
18						750	24	3	50	0	1001	253	100
19	66					1249	24		500		250	50	125
20						999	26		250		500	1253	75
21						650	26		750	0	1250	749	10
22						250	23	4	1001	0	750	100	5
23						500	24		15	0	100	500	50
24						100	26		1249	0	650	350	150
25						650	26		750	2239	1250	749	0
26						500			500	143	502	503	50
27						0			0		0	0	0
28						72			0.2		470	460	193
29						18			25		1100	98	104
30						43.5			83		10	1360	63
31						702			214.5		40	6	109
32						410.5			814.5		24	46	8.5
33						334			210		220	355	71.5
34						214.5			115.5		110	236	71
35						40.5			18		33	53	10
36	810	637.29	1591	2200	48	114.6	23	117	25.3	31	808	107.6	140.8

Table 9.2:
Selection of reference materials prepared from PE and PVC. Values are reported in PPM and are nominal concentrations that have not yet been confirmed by AAS, ICP-MS or any other bulk analysis method.

The results show a good match of spectral slope/curvature between the reference materials and the ethnographic objects (see Figure 9.5). This match is promising for the establishment of a calibration curve, but more work needs to be done to determine how successful this generalized calibration will be at determining pesticide concentrations in different ethnographic materials.

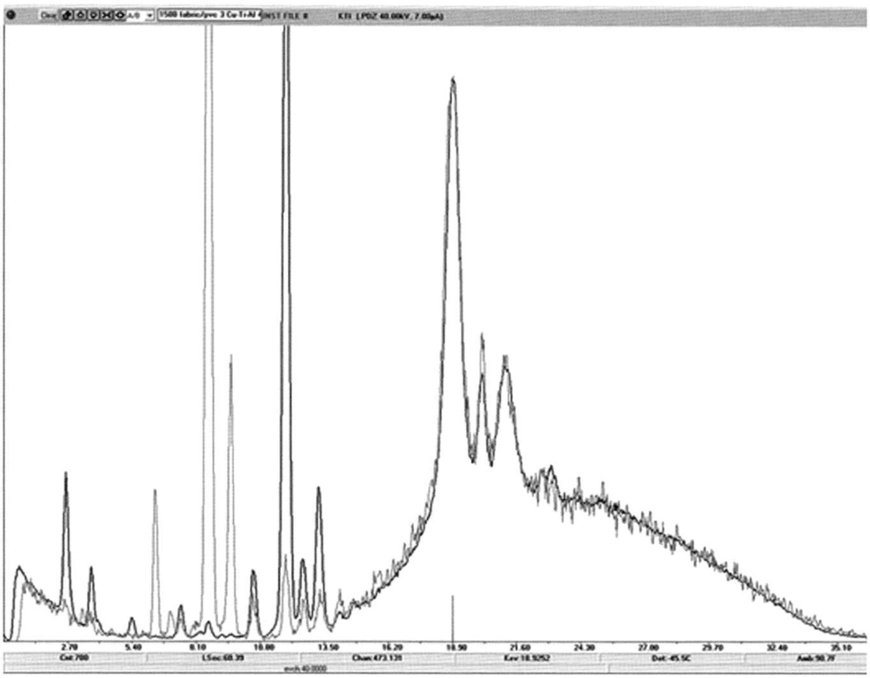

Figure 9.5:
Example of the shape of the continuous radiation and Compton peak height from an ethnographic artifact (grey) and the produced reference material (black: sample PVC 3) showing a similar curve and peak height.

Preparation of material-specific reference materials

Stages 3 and 4 in the model described above, where reference materials are created for specific applications, requires studying the materials of interest to understand their structure, the infiltration of pesticide treatments, and how those treatments changes over time prior to making appropriately constructed reference materials. Two experiments are presented here where we have started to investigate the relationship between an applied pesticide and the substrates wood and leather.

Various skin and fur samples

The spectra below show the difference in background obtained using the same X-ray parameters for different types of uncontaminated skins, some with different thicknesses. Two blank plastic samples, PVC and Chemplex are also included to show how the various different matrices compare. A matrix for the reference materials being constructed should be chosen which reflects the matrix of the objects. Figure 9.6 below shows the different spectra obtained from a variety of object materials.

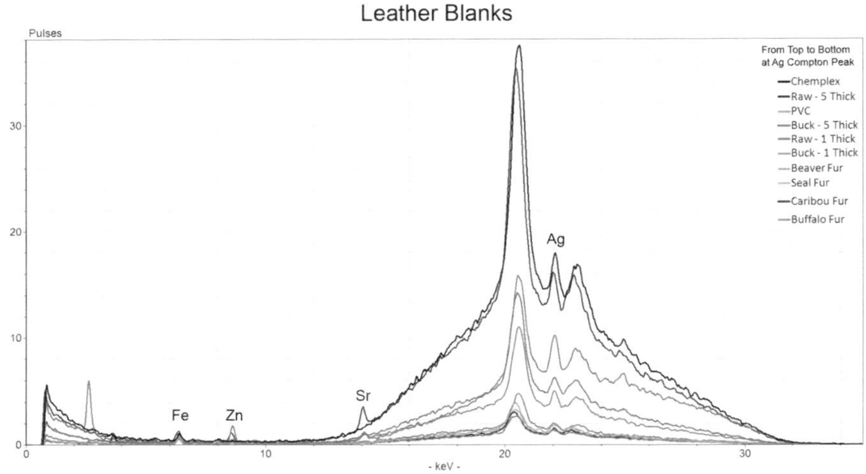

Figure 9.6:
Spectra of different types and thicknesses of uncontaminated skin and fur samples compared to two different plastic reference materials. (Spectrometer conditions: silver anode, 60 seconds acquisition time, 35 kV, 14 μA).

The different materials can generally be divided into three different groups, low medium and high backgrounds (continuum radiation, bremsstrahlung). The Chemplex blank had a high background which is similar in slope and peak height to the thick rawhide sample. The PVC had a medium background and matched closest to the slope of the thicker buckskin sample with slightly different peak heights. All of the fur samples, which had similar skin thicknesses with varying thicknesses of fur, had similar low backgrounds and peak heights and could possibly rely on a single set of reference materials for that particular object class. In general what is shown indicates that creating one set of reference materials expected to cover the diverse range of material types encountered with ethnographic materials is impractical and unlikely to work.

Arsenic-treated leather samples

Two types of leather, buckskin and rawhide, were treated with three different arsenic recipes, applied in two different ways, by spraying or by brushing (Bond 2005). The three different arsenic preparations applied were (see Table 9.3):

1 an arsenical soap solution comprised of: 1 kg of white bar soap, 1 kg of powdered white arsenic (As_2O_3), 190 g of subcarbonate of potash (K_2CO_3), 150g of camphor ($C_{10}H_{16}O$), 250 mL of ethyl alcohol (C_2H_5OH) (Bond, 2005). The soap used in the recipe was Nature Clean© All Natural Laundry Powder that contains: sodium chloride, sodium carbonate (between 15 – 50%), fatty polyglycoside, silicates, zeolite, sodium citrate, cellulose colloids (from cotton & wood; Frank T. Ross Ltd 2008)

2. a sodium arsenite solution comprised of: ¼ lb of arsenic trioxide (As_2O_3), ½ lb of carbonate of soda (Na_2CO_3), and one quart of water (ibid);
3. an arsenic solution comprised of: ¼ lb arsenic sesquioxide (As_2O_3), 2 gallons water, and an equal part of alcohol (C_2H_5OH) (ibid).

Skin	Application Method	Recipe 1 arsenical soap	Recipe 2 sodium arsenite sol'n	Recipe 3 arsenic sol'n	Penetration Depth (mm)
Buckskin	Brush	x			1,6
Buckskin	Brush		x		0,5
Buckskin	Brush			x	0,9
Buckskin	Spray	x			0,7
Buckskin	Spray		x		0,4
Buckskin	Spray			x	No arsenic detected
Rawhide	Brush	x			0,03
Rawhide	Brush		x		Surface particles
Rawhide	Brush			x	No arsenic detected
Rawhide	Spray	x			No arsenic detected
Rawhide	Spray		x		0,2
Rawhide	Spray			x	No arsenic detected

Table 9.3:
Sample composition (type of skin, type of arsenical pesticide and method of application) and detectable penetration depth of the treatment into the skin.

Visual examination of these samples showed that the arsenic formulations in several of the buckskin samples penetrated the full depth of the leather, as smaller amounts of residue and discoloration were found on the back side. The thickness of the buckskin samples ranged from 1.5 – 3 mm and the depth of penetration of the various pesticide formulations ranged from 0.4 to 1.6 mm as measured by

X-ray microanalysis and scanning electron microscopy. Most of the arsenic was present on the buckskin surface to a depth of about 1 mm after which the arsenic concentration tapered off but was still detectable all the way through the sample using energy dispersive X-ray microanalysis – both X-ray mapping and linescans.

The rawhide samples were prepared as cross-sections by embedding them in low viscosity epoxy resin (Epofix Cold-setting Embedding Resin by Electron Microscopy Sciences), microtoming them and then examining them by SEM/EDS using line scans and X-ray maps. The rawhide samples had very thin layers of arsenic on the surface (< 100 μm). Arsenic only penetrated into two of six rawhide samples, with the minimum depth being 0.03 mm and the maximum depth being 0.2 mm. Two of the three preparations penetrated the skins. Preparation 3, the one containing alcohol, did not penetrate the skins to nearly the degree of the other two.

Both the buckskin and the rawhide samples showed a higher concentration of arsenic at their surface, with a sharp drop in the arsenic levels when moving further into the cross-sections. The arsenic application to the hide samples resulted in a heterogeneous distribution of arsenic.

Examination of the different skins treated with different application methods of pesticide preparations showed that all of these factors play a role in the concentration of the arsenic present and its location. Based on these two experiments one can see how important it is to fully characterize the application technique in order to create a viable comparable reference material.

Model for skin reference materials

Using the data collected above, several mock ups were created to be used as potential reference materials. Several types of glue films in different thicknesses, expected to be similar in matrix to the skins were prepared. Based on comparative XRF spectra, hide glue was determined to be the best matrix to replicate the skins. It also provides a vehicle to evenly distribute the contaminant throughout the reference material in a homogenous manner during their preparation. In addition, the films can be encased in Mylar sleeves to make them safer to handle.

The two models proposed as reference materials include a thin film sample and an intermediate bulk thickness sample. The thin film of hide glue would contain an arsenic compound or other contaminant of interest and would simulate, for example, the thin film of arsenic present on the surface of a piece of rawhide. This could be placed over a rawhide blank to achieve the entire background of the contaminant film on rawhide. The second model is a thicker film of hide glue that would have the same background spectrum as buckskin with arsenic present throughout simulating arsenic present throughout a piece of buckskin. This second model matches the intermediate thickness model. Both could be placed in different locations on a sample to replicate the varying application possibilities that could have occurred in reality.

Arsenic-doped wooden samples

The investigation of arsenic doped red cedar samples was undertaken next. The depth of penetration, distribution and the X-ray attenuation of the arsenic doped samples was considered. A standard solution of 100 mg arsenic trioxide in 100 mL deionized water was made. The solution was heated until the arsenic was completely dissolved. Three wooden samples of thickness 2.5, 5, and 10 mm were prepared. The samples were then doped in three different ways. Dipped samples were prepared by submerging them, then immediately removing them for drying. Soaked samples were left in the solution for 5 minutes. Brushed samples were brushed on one side with a cotton swab. All samples were allowed to dry for 72 hours. The samples were then analyzed using handheld XRF. The results showed that the 2.5 mm wooden samples had the highest concentration of arsenic detected, while the 5 and 10 mm samples had similar levels at about half of the 2.5 mm samples (see Figure 9.7).

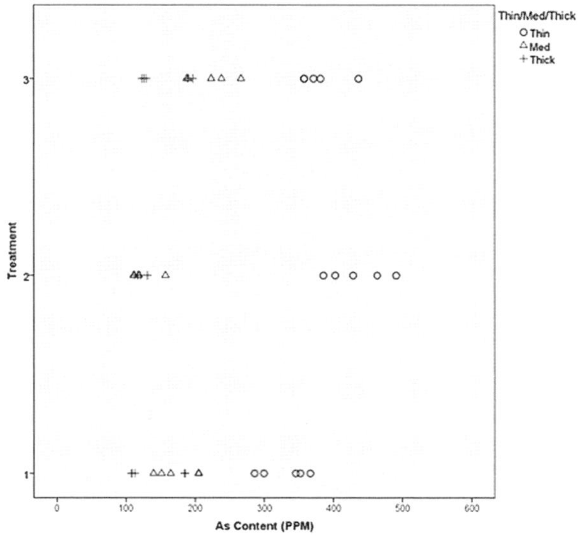

Figure 9.7:
Arsenic content of three doped treatments (1 – brushed, 2 – dipped, 3 – soaked) based on sample thickness (2.5, 5, and 10 mm). The variability in absorption is likely due to the saturation of the thinner samples as opposed to the thicker ones in combination with the X-ray emission depth. Samples all run at 40 kV, 1.8 μA with an Rh anode.

The results suggest that thicker samples showed limited saturation of the entire sample from each side, while the thinner sample was thin enough to be penetrated thoroughly by the solution from both sides. Based on LA-IPC (laser ablation inductively coupled plasma) analysis of a cross section of a 10 mm sample it was determined that the sample was fully saturated and showed a clear surface enrichment in arsenic with decreasing concentration towards the center of the block (Figure 9.8). While these initial investigations are useful for understanding the pesticide's interaction with an artifact, further work needs to be done to characterize the interaction of pesticides with wood, and other materials using different application techniques. The alarming, yet not surprising thing uncovered by these experiments is that the pesticide does not uniformly interact with/or penetrate into the wood substrate, accentuating the heterogeneous nature of these complex artifacts.

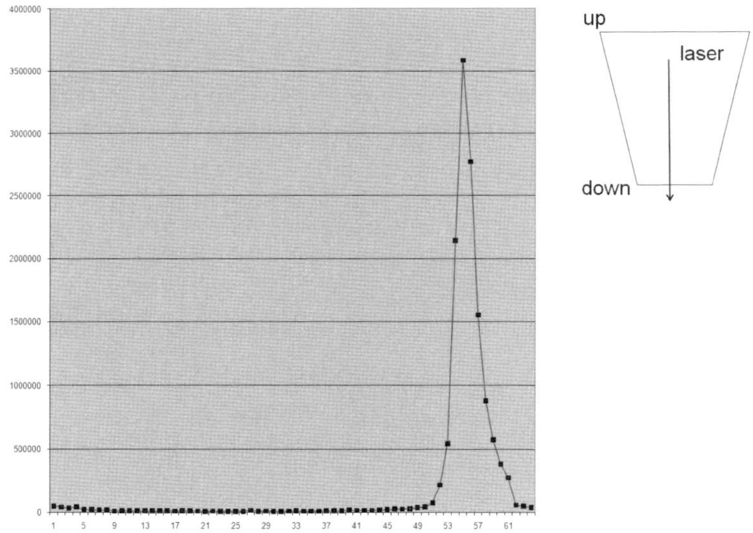

Figure 9.8:
LA-ICP analysis of doped wood samples showing concentration of arsenic at surface of wood (to the right). 'Up' refers to the top of the sample showing limited As while 'down' refers to the enriched As treated side. Overall the sample was 10 mm thick with approximately a 140 μm step size.

Issues related to creating reference materials

Standard methods used for the preparation of samples for analysis are designed to make uniform samples with a homogenous distribution of the elements of interest. These techniques include creating pressed pellets, putting samples into solution, or even fusing a borax bead from the sample. All unknown samples are recommended to be prepared in this manner. In fact, the quantitative measurement of heterogeneous materials is nearly impossible unless you know the makeup of your sample atom by atom, and you create reference materials that match this makeup perfectly. Jenkins (1999:141) states that "In practice, however, there are two major constraints that may prevent this ideal circumstance [quantitative analysis] from being achieved, these being sample size and samples heterogeneity." Herein lies our dilemma with ethnographic collections. As we discussed, they are variable in size, thickness, shape and are all heterogeneous in nature. Jenkins continues to state that samples for heterogeneous materials require significant pre-treatment prior to any type of quantitative analysis. What this ultimately indicates is that no matter what level of production we attain for reference materials, they can never fully be used for rigorous non-destructive quantitative analysis of the pesticides present in an artifact – only destructive analysis through ICP-MS, for example, will produce truly quantitative results, and even these results will not be readily reproducible given the inhomogeneity of the materials being studied. *It is critical to understand this when working with these materials!* While creating calibration curves can help reduce error, the resulting numbers provided cannot truly be fully quantitative, and should only be used as a guideline as to the presence of pesticides in broad categories of concentrations.

Reporting results

There are several schemes that are currently in use for reporting XRF analyses of pesticide residues. These include directly reporting quantitative results obtained from the instrument prior to specific calibration, or after a calibration has been performed. These values are typically derived by fundamental parameters computations or are empirically derived and provide values in ppm, wt.% or $\mu g/cm^2$.

Results may also be reported by their range of concentration along with recommendations as to the danger level; not detected (ND), not quantified (NQ), trace (<0.1%), moderate (0.1 to <1%), high (1 – 5%), very high (>5%). Results have also been reported as concentration ranges given color codes such as; green (safe, nothing or negligible contaminant detected), yellow (caution, some pesticide detected, handle with care, artifact may be used under certain conditions or with remediation), and red (higher levels of pesticides detected, artifact unsafe; Odegaard *et al*. 2006).

Currently the results of pesticide treatments are reported in both ppm and μg/cm^2. Toxicologists have requested that reporting be done based on total surface exposure. This was to help them make determinations as to the total potential exposure. It was thought that the pesticide application on artifacts acted like a thin coating rather than penetrating the artifact, so collecting bulk data was extraneous. More recent investigations have shown there are a variety of penetration depths and profiles associated with treatment (see for example the arsenic doped wood discussion above). In addition, whereas XRF is an ideal technique for identifying not only the composition of thin films, but also the thin film deposition thickness, this mostly relied on the ability to adjust the geometry of the instrument (see Piorek 2008). Unfortunately the geometry of handheld XRF instruments cannot be altered and although one can control the depth of penetration of the beam by adjusting the voltage, it is not practical when looking for higher Z elements additionally, the determining factors associated with thin film measurements are based on a clear layering of the film on the surface of the substrate which is not the case with ethnographic collections. For these reasons we recommend using bulk reporting in ppm unless there is a clear evidence of thin film application or the artifact itself can be considered a thin film (i.e. thin animal skin clothing).

Guidelines for analytical methods

A general outline is provided for a basic analytical procedure for a qualitative inorganic pesticide survey on a large collection of objects. An outline for quantitative analysis with large uncertainty ranges for very specific applications is also included. A quantitative analysis may be possible for specific materials that have been well-researched.

1. Method for qualitative analysis

1. Set up your instrument – know your X-ray tube beam (i.e. Ag anode, run at 40 kV and max μA).
2. Standardize the spectrometer as per the manufacturer's instructions to ensure proper spectral calibration (i.e. correct peak positioning).
3. Run a blank to ensure there is no contamination and that you know your system artifacts (can use SiO_2 blank, or Teflon or plexiglass).
4. Run reference materials to ensure your calibration is correct or to calibrate your system.

5. Determine what the lower limits of detection are for the elements of interest using relevant reference standards.
6. Examine your object closely to identify the materials and construction techniques and also any visible pesticide residues.
7. Choose your "mode" or operating conditions based on the construction of the object. If it is a thin item you will want to use a method comparable with that. If it is an object which will contribute significant background to the spectrum then a bulk mode should be used.
8. Choose the areas for analysis. Each different material/component should be analyzed. Document the area well both in writing and with photos if possible. Make special note of paints, pigment glass beads, metals, etc... as these materials may contain the elements you are interested in. For example lead is present in red lead and lead white paints which were frequently employed over a wide range of cultures and time periods. Vermilion, mercuric sulfide, is another red pigment which was frequently used. Beads often contain arsenic as a fining agent in the glass or as a colorant and lead as part of the glass former. If trying to detect pesticides, avoid painted and beaded areas.
9. Analyze the same place at least three times to ensure that the spectrometer is performing properly. This would also give you the error associated with the spectrometer and not the object.
10. Analyze each object several times in different places to determine whether the pesticide is heterogeneously applied. It likely was. This gives you the error associated with the object itself.
11. If it is possible to analyze both sides of the object do so, as it will assist with determining where the pesticide was applied as well as where it is currently located.
12. Always review the original spectra to confirm the peaks present rather than relying on the instrument's numerical data readout. Look at the ratios of the peaks to each other (check against expected X-ray peak intensity data, in particular when you have elements present with peak overlaps).

This type of analysis will give you results that are either positive or negative, identify the major and minor components of applied inorganic pesticides, and allow you to determine if inorganic pesticide contamination is high or low. It does not indicate specifically where the arsenic or other inorganic pesticide is located in the object, nor is it quantitative.

2. METHOD FOR QUANTITATIVE ANALYSIS

1. Examine the objects to determine how they are constructed and their materials of construction. This will determine what type of reference material(s) will be chosen for the analysis. Most of the objects analyzed in this application are either thin film samples or samples of intermediate thickness – very rarely do we have thick samples. The composition of the matrix, its thickness (critical penetration depth), and density must be taken into account when developing reference materials.
2. Determine whether you will use thin film model which is suitable for feathers, thin cotton, wool, gauze or filter samples such as wipes, or the intermediate thickness model which is the situation for many of our objects where the substrate contributes to the background radiation of the spectrum.
3. Choose or prepare proper reference materials.
4. Determine what instrument conditions or mode you will use. Some instruments have modes for the analysis of soils, thin films, plastics, dust wipes etc. Determine which one of these best suits your application.
4. Calibrate your system with the appropriate reference materials for the software application you will be using.
5. Run an instrument blank to make sure you know what your system artifacts are.
6. Determine the Lowest Detection Limits.
7. Determine error due to technique. Run your standard at least five times and do statistics on it to determine instrumental error.
8. Run unknown samples.
9. Examine spectra
10. Question your results and re-test.

Summary

The detection of inorganic pesticide residues on ethnographic, anthropological and natural history objects by portable or handheld XRF is a very quick and useful qualitative technique. This application is not particularly well suited for quantitative XRF analysis in a collection survey. The type of objects and range of materials encountered is too broad for a "one size fits all" solution where one or two sets of reference materials may be employed. For quantitative analysis to be successful, likely still with large inherent errors (due to the inhomogeneity of the substrate and also the pesticide application), a proper set of standards that match the object construction, in density, chemical composition, material thickness and location of contaminant must be prepared.

A good use of this technique may be to undertake a broad collection survey to determine whether or not inorganic pesticide residues are a problem and to use that data to determine how to more thoroughly characterize the objects. Two possible outcomes would be to set up more quantitative XRF analyses of specific types of objects or to analyze wipes for removable contaminant by XRF – a technique which is more easily quantifiable. One would also be able to recommend which artifacts might be best suited for a more invasive analytical technique to achieve quantification (Odegaard *et al.* 2005, Sirois and Sansoucy 2001).

Acknowledgements

We would like to thank Bruce Kaiser and Robert Shannon of Bruker-elemental for creating reference materials and providing prilimenary calibration filtering advice. Thanks also goes to Laure Dussubieux at the Field Museum for performing LA-ICP analysis, and to Kari Kipper who prepared the samples. We would like to acknowledge and thank Kathy Bond, Vancouver Art Gallery, for allowing us access to the arsenic-treated rawhide and buckskin samples she prepared, Kevin Bailey, a co-op student at the Canadian Conservation Institute for his work on preparing and analysing the prepared arsenic-impregnated leather samples to determine the penetration depth of the arsenic solutions, Melanie Raby, a co-op student at the Canadian Conservation Institute for her assistance with preparing the hide glue films for XRF analysis.

Notes

1. Hazard-describes a situation where there is a potential to cause harm, Risk- the chance of that harm occurring (West, 2008).

References

Agency for Toxic Substances and Disease Registry (ATSDR). 1999. Toxicological Profile for Mercury. http://www.atsdr.cdc.gov/toxprofiles/TP.asp?id=115&tid=24.

———. 2007. Toxicological Profile for Lead. http://www.atsdr.cdc.gov/toxprofiles/TP.asp?id=96&tid=22.

———. 2011. Medical Management Guidelines for Arsenic Trioxide (As_2O_3) http://www.atsdr.cdc.gov/MMG/MMG.asp?id=1200&tid=279.

Analytical Methods Committee. 1987. "Recommendations for the Definition, Estimation and Use of the Detection Limit in Analyst." (112 February 1987):199-204.

Anderson, Rudolph Martin. 1932. *Methods of collecting and preserving vertebrate animals, Bulletin / National Museum of Canada*. Ottawa: National Museum of Canada.

Bertin, E.P. 1970. *Principles and practice of X-ray spectrometric analysis*. New York: Plenum Press.

Bond, Kathleen. 2005. *Reliability of X-ray Fluorescence for the Quantitative Analysis of Arsenic in Contaminated Leather*. Research Project submitted to the for the Degree of Master of Art Conservation, Department of Art Queen's University, Kingston.

Bruker Corporation. 2006. Tracer III-V User's Manual.

———. 2008. Tracer III-SD User's Manual.

Burgess, C. 2000. *Valid analytical methods and procedures*. Cambridge: Royal Society of Chemistry.

Conserve O Gram. 2000. Arsenic health and safety update. 2/3: National Park Service.

———. 2001. Chronology of pesticide used on National Park Service collections. 2/16: National Park Service.

Fonicello, Nancy A. 2007. "Unique Problems with the Use of the Handheld XRF Spectrometer for Pesticide Surveys of Ethnographic Collections." *ICOM-CC Ethnographic Conservation Newsletter* (Number 28, February 2007):4-8.

Found, Christine, and Katie Helwig. 1995. "The Reliability of Spot Tests for the Detection of Arsenic and Mercury in Natural History Collections: A Case Study." *Collection Forum* Vol. 11 (1-2):6-15.

Frank T. Ross Ltd. 2011. *Nature Clean Laundry Powder MSDS* 2008 [cited September 9 2011]. Available from http://www.naturecleanliving.com/laundry_powder.

Goldberg, L. 1996. "A history of pest control measures in the Anthropology Collections, National Museum of Natural History, Smithsonian Institution." *Journal of the American Institute for Conservation* Vol. 35 (1):23-43.

Hangay, George, and Michael Dingley. 1985. *Biological museum methods*. 2 vols. Sydney: Academic Press.

Hawks, C. 2001. "Historical survey of the sources of contamination of ethnographic materials in museum collections." *Collection Forum* Vol. 16 (1-2):2-11.

Jenkins, Ron. 1999. *X-ray fluorescence spectrometry*. Edited by J. D. Winefordner, *Chemical analysis*. New York: John Wiley & Sons, Inc.

Jenkins, Ron, Dale Gedcke, and Robert William Gould. 1995. *Quantitative X-ray Spectrometry*. 2nd ed, *Practical spectroscopy*. New York: M. Dekker.

Leutenegger, P., A. Longoni, C. Fiorini, L. Strüder, J. Kemmer, P. Lechner, S. Sciuti, and R. Cesareo. 2000. "Works of art investigation with silicon drift detectors." *Nuclear Instruments and Methods in Physics Research Section A: Accelerators, Spectrometers, Detectors and Associated Equipment* Vol. 439 (2-3):458-470.

Markowicz, Andrzej A. 2008. "Quantification and Correction Procedures." In *Portable X-ray fluorescence spectrometry: capabilities for in situ analysis*, edited by P. J. Potts and Margaret West, 13-38. Cambridge, UK: Royal Society of Chemistry.

Native American Graves Protection Repatriation Act. 1996. Regulations, 43 CFR 10, Section 10.10 (e). http://ecfr.gpoaccess.gov/cgi/t/text/text-idx?type=simple;c=ecfr;cc=ecfr;sid=3af6a4dd63c0a64e172c5dfc5ac8c754;idno=43;region=DIV1;q1=NATIVE%20AMERICAN%20GRAVES%20PROTECTION%20AND%20REPATRIATION;rgn=div5;view=text;node=43%3A1.1.1.1.10.

Odegaard, Nancy, Scott Carroll, and Werner S. Zimmt. 2005. "Tests for Arseinc." In *Material characterization tests for objects of art and archaeology*, edited by Nancy Odegaard, Scott Carroll and Werner S. Zimmt, 40-43. London: Archetype Publications.

Odegaard, Nancy, Alyce Sadongei, and Leslie V. Boyer. 2005. *Old Poisons, New Problems: A Museum Resource for Managing Contaminated Cultural Materials*. Walnut Creek, CA: AltaMira Press.

Odegaard, Nancy, David R. Smith, Leslie V. Boyer, and Jae Anderson. 2006. "Use of Handheld XRF for the Study of Pesticide Residue on Museum Objects." *Collection Forum* Vol. 20 (1-2):42-48.

Ontario, Occupational Health and Safety Act. 2011. *Ontario Regulation 490/09, Designated Substances* 2011 [cited August 8 2011]. Available from http://www.e-laws.gov.on.ca/html/regs/english/elaws_regs_090490_e.htm#BK5.

Palmer, P.T. 2001. "A review of analytical methods for the determination of mercury, arsenic, and pesticide residues on museum objects." *Collection Forum* Vol. 16 (1-2):25-41.

Palmer, P.T., M. Martin, G. Wentworth, N Caldararo, L. Davis, S. Kane, and D. Hostler. 2003. "Analysis of pesticide residues on museum objects repatriated to the Hupa Tribe of California." *Environmental Science and Technology* Vol. 37:1083-1088.

Péquignot, Amandine, Fernando Marte, and David Von Endt. 2006. " L'arsenic dans les collections d'Histoire naturelle." *La lettre de l'OCIM* (105):4-10.

Piorek, S. 2008. "Coatings, Paint and Thin Film Deposits." In *Portable X-ray fluorescence spectrometry: capabilities for in situ analysis*, edited by P. J. Potts and Margaret West, 56-82. Cambridge, UK: Royal Society of Chemistry.

Rowley, John. 1925. *Taxidermy and Museum Exhibition*. New York: D. Appleton and Company.

Seifert, Steven A., Leslie V. Boyer, Nancy Odegaard, and David R. Smith. 2000. "Arsenic Contamination of Museum Artifacts Repatriated to a Native American Tribe." *The Journal of the American Medical Association* Vol. 283 (20):2658-2659.

Shugar, A. 2009. "Peaking Your Interest: an introductory explanation of how to interpret XRF data." *Western Association for Art Conservation Newsletter* Vol. 31 (3):8-10.

———. 2011. Personal Communication. Buffalo, NY, August 2011.

Sirois, P. Jane, Jessica S. Johnson, Aaron Shugar, Jennifer Poulin, and Odile Madden. 2008. "Pesticide Contamination: Working Together to Find a Common Solution." In *Preserving Aboriginal Heritage: Technical and Traditional Approaches. Proceedings of Symposium 2007*, edited by Carole Dignard, Kate Helwig, Janet Mason, Kathy Nanowin and Thomas Stone, 175-186. Ottawa: Canadian Conservation Institute.

Sirois, P.J. 2001. "The analysis of museum objects for the presence of arsenic and mercury: Non-destructive analysis and sample analysis." *Collection Forum* Vol. 16 (1-2):65-75.

Sirois, P.J., J. Poulin, and T. Stone. 2010. "Detecting Pesticide residues on museum objects in Candian collections – a summary of durveys spanning a twenty-year period." *Collection Forum* Vol. 24 (1-2):28-45.

Sirois, P.J., and G. Sansoucy. 2001. "Analysis of museum objects for hazardous pesticide residues: A guide to techniques." *Collection Forum* Vol. 17 (1-2):49-66.

Sirois, P.J., and J. Taylor. 1989. The Determination of Arsenic and Mercury in Natural History Specimens using Radioisotope Excited X-ray Energy Spectrometry and Scanning Electron Microscopy. In *Proceedings of the 14th Annual IIC-CG Conference. Toronto, May-1988*, edited by Johanne Wellheiser. Toronto: IIC-CG.

The Engineering ToolBox. 2011. *Wood Densities* [cited August 12 2011]. Available from http://www.engineeringtoolbox.com/wood-density-d_40.html.

Thomsen, Volker, Debbie Schatzlein, and David Mercuro. 2003. "Limits of Detection in Spectroscopy." *Spectroscopy* Vol. 18 (12):112-114.

———. 2005. "Tutorial: Attenuation of X-rays By Matter." *Spectroscopy* Vol. 20 (9):22-25.

Wagstaffe, R., and J. H . Fidler. 1968. *The Preservation of Natural History Specimens*. Vol. II. London: H. F. & G. Witherby Ltd.

West, Margaret. 2008. "Hazardous Substances in the Workplace." In *Portable X-ray fluorescence spectrometry: capabilities for in situ analysis*, edited by P. J. Potts and Margaret West, 83-97. Cambridge, UK: Royal Society of Chemistry.

Williams, S. L., and C. A. Hawks. 1987. "History of preparation materials used for recent mammal specimens." In *Mammal Collection Manangement*, edited by H.H.Genoways, C.Jones and O.L.Rossolimo, 21-49. Lubbock: Texas Tech University Press.

Woolson, E. A. 1975. *Arsenical pesticides: a symposium sponsored by the Division of Pesticide Chemistry at the 168th meeting of the American Chemical Society, Atlantic City, N.J., Sept. 9, 1974, ACS symposium series*. Washington: American Chemical Society.

Chapter 10

Using handheld XRF to aid in phasing, locus comparisons, and material homogeneity assessment at an archaeological excavation

Mary Kate Donais and David George

Introduction

The availability of a number of commercially produced, affordable, handheld X-ray fluorescence (XRF) spectrometers has meant that this instrument has become an important and valuable tool for field archaeologists. Characterization of various materials such as ceramics (Papadopoulou *et al.* 2006), glass (Ricci *et al.* 2004), metals (Karydas *et al.* 2004; Karydas 2007), and mortars (Donais *et al.* 2010) is achieved quickly and easily at site with no destruction of the artifacts. The advertised "point and shoot" design of these instruments does come with limitations, however. Care and caution are required to achieve meaningful results.

Some archaeological samples such as obsidian and pottery can exhibit sufficient homogeneity for very precise elemental measurements by handheld XRF spectrometry (Romano *et al.* 2006; Craig *et al.* 2007; Nazaroff, Prufer, and Drake 2010). Less homogeneous sample types such as mortars and cement floors instead necessitate multiple analyses distributed across the sample area to better assess and account for variations in elemental distributions (Donais *et al.* 2010). Sample type therefore plays a significant role and must be considered in deciding the number of repeat analyses to conduct across a number of sample locations. Instrument calibration can be a challenge due to the matrix dependence of XRF spectrometric signals. The factory calibrations provided with many handheld XRF spectrometers can be used to quickly process data and feed back almost instantaneous quantitative results. Caution must accompany this "point and shoot" approach, however, and these data are at best semi-quantitative in nature. Material-specific calibration is essential if accurate results are required. Not all research questions require a high level of accuracy, however, but instead dictate comparisons of data sets. Spectra or integrated peak areas may be sufficient to answer some research questions, therefore eliminating the need to quantify at all. This comparison-based approach to data evaluation can provide the archaeologists with immediate results to answer specific on-site questions, and thus aid in daily excavation decisions such as where to excavate next or the relationships among loci.

Prior to utilization of handheld XRF spectrometry at an excavation site, it is essential to evaluate what specific research questions might be better understood through the collection of elemental data. A common temptation is to analyze anything and everything available. This can lead to a significant waste of time with little of meaningful substance gained. Complex projects such as determining the provenance of materials have been successfully approached using XRF spectrometry (Popelka-Filcoff *et al.* 2008; Nazaroff, Prufer, and Drake 2010). More simple material comparisons that aid site archaeologists in evaluating with which features particular walls and structures are associated can also be achieved (Donais *et al.* 2010). Although a single analysis for either of these two situations can be performed in a few minutes, collection of enough data to draw valid conclusions will no doubt take much longer. Prioritization of samples to be analyzed and development of a research plan are therefore essential.

An important next step in utilizing handheld XRF spectrometry for field archaeology is to determine data collection protocols. Reproducibility and consistency in instrument operation is essential, especially if multiple researchers will be collecting data over the course of an excavation. Due to the heterogeneity of many archaeological samples, these protocols must include the collection of data *in situ* at systematic, multiple locations on the object with replicates collected at each *location*. Additional analyses of *ex situ* samples by a secondary method such as inductively coupled plasma atomic emission spectrometry, inductively coupled plasma mass spectrometry, or atomic absorption spectroscopy also can be performed to provide validation for XRF spectrometric data, necessitating the collection of these samples following established sample collection protocols. A number of protocols developed by our research group over multiple excavation seasons are discussed later in the chapter.

Lastly, the type of research question being asked will often dictate how data will be processed. For some situations, accurate quantitative data is necessary therefore requiring careful instrument calibration and use of primary or secondary reference standards. When material composition comparisons are being made (is sample “A” different from sample “B”), the qualitative comparisons of the spectra themselves or high error quantitative data may be sufficient to draw conclusions. As will be illustrated later in this chapter, our research utilizing handheld XRF spectrometry leans more toward material composition comparisons. A number of data processing approaches will be presented to illustrate various graphical and mathematical methods available to researchers. With proper planning and forethought, handheld XRF spectrometry can be a tremendously valuable tool for field archaeologists.

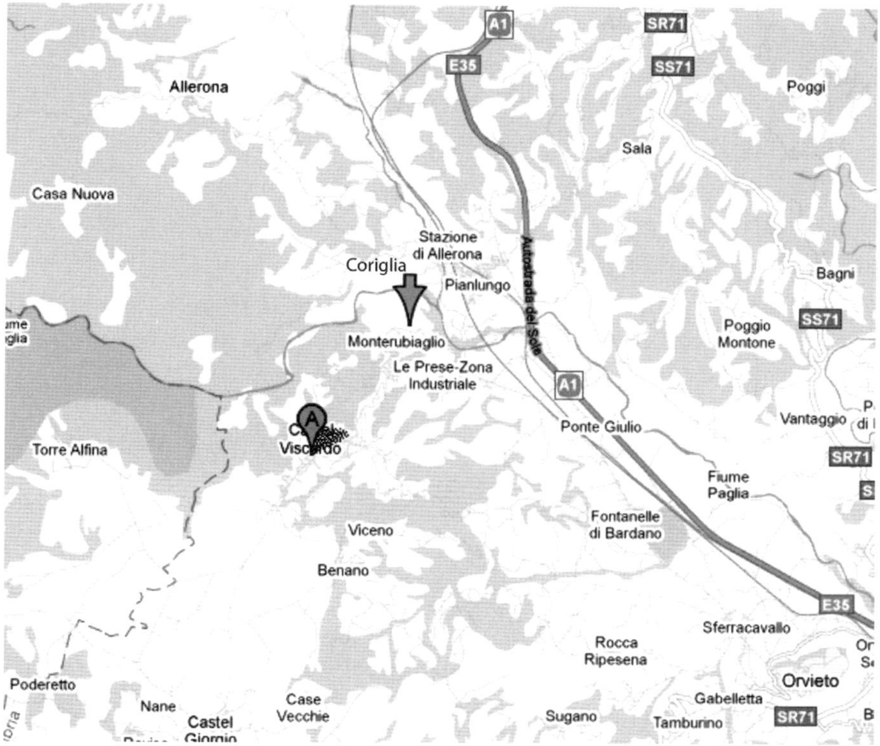

Figure 10.1:
Map showing location of Coriglia, Castel Viscardo within Orvieto region.

Saint Anselm College faculty and students have been excavating the site at Coriglia, Castel Viscardo, Italy since 2006, and have integrated portable chemical instrumentation into the research design since 2008 (Donais, Lebel, and Thibodeau 2008; Donais *et al.* 2009; Donais *et al.* 2010; Donais *et al.* 2011). The excavation at Coriglia, Castel Viscardo, located approximately 13 km northwest of Orvieto, Italy, is a multi-phased site beginning in the 9th century BCE as an Iron Age settlement and, with no interruptions, continues through the Late Antique and well into the Middle Ages and based on numismatic evidence into the 17th century CE. Excavations since 2006, directed by David George and Claudio Bizzarri under the auspices of the Soprintendenza per I Beni Archeologici dell'Umbria and the Parco Archeologico e Ambientale dell'Orvietano, have recovered a number of structures of Etruscan, Roman Republican, Imperial and late Antique dates as well as evidence for Medieval ecclesial structures. Considerable quantities of Etruscan and Roman pottery, fourth-style fresco, lead and terracotta pipes, and numismatic evidence from the middle of the second century BCE to the 17th century CE have been found. Products are both local as well as imported. The initial interpretation of the site was that it was an Etruscan settlement that, after the Roman conquest of Orvieto developed into a Roman villa. The finds, however, are inconsistent with this hypothesis. Our current working hypothesis is that Coriglia begins life as an Iron Age settlement and evolved into an Etruscan healing shrine around which a small town developed and grew. This town continued to grow into a somewhat larger village with a bath complex/shrine along a branch of the Via Cassia during the late Republic that remained in use until at least the 6th century CE. The location of the site within the Orvieto region is provided in Figure 10.1, and a map of the excavation site itself at the conclusion of the 2010 season is in Figure 10.2.

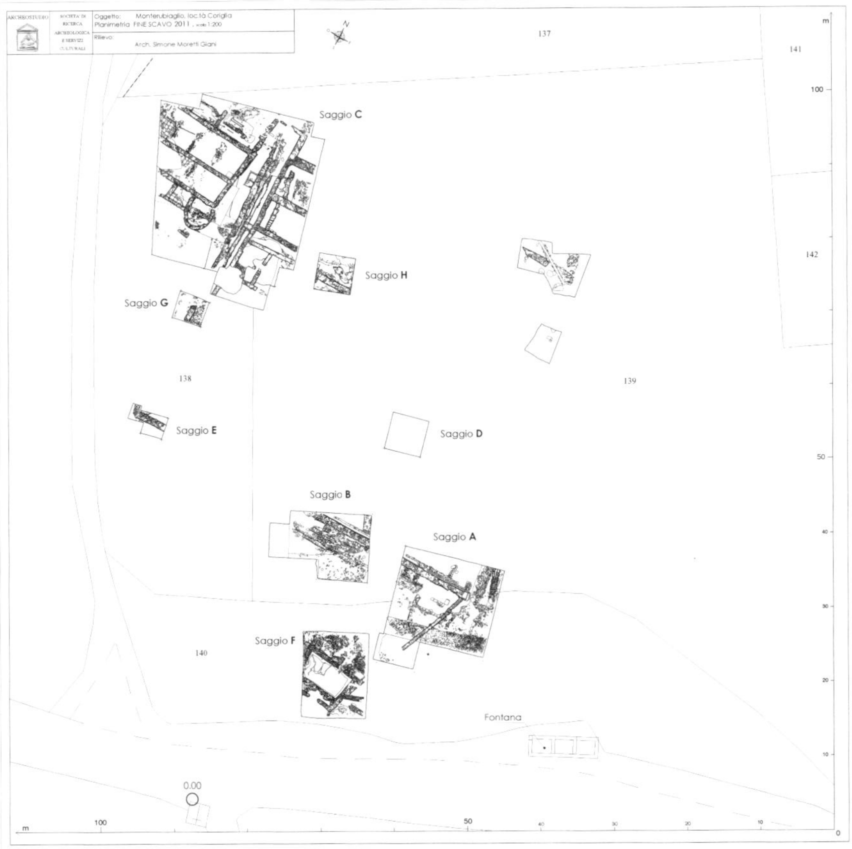

Figure 10.2:
Site map from the end of the 2011 excavation season.

During the last five years, a number of monumental structures have been uncovered for which handheld XRF spectroscopic data has been valuable to the site archaeologists in their preliminary determination of the relationships among

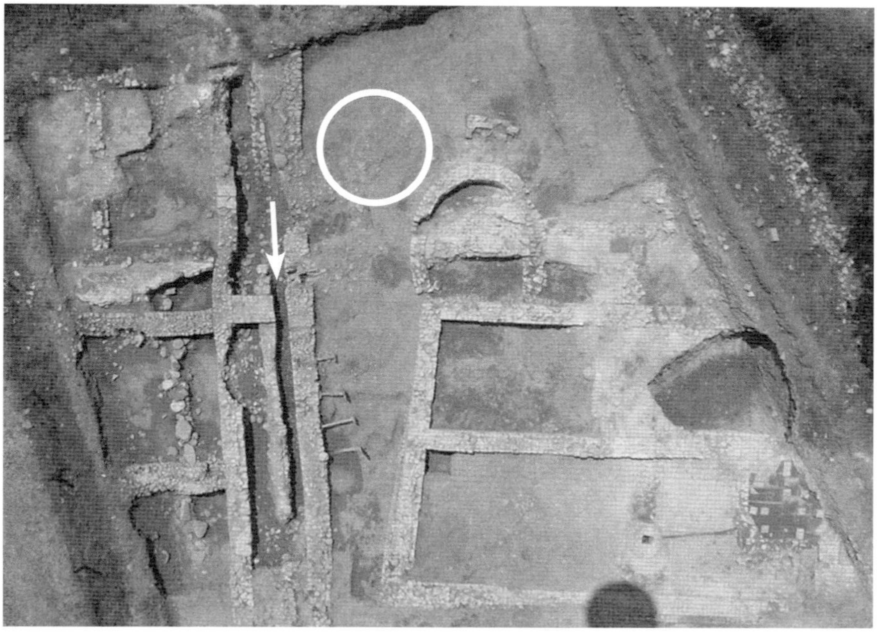

Figure 10.3:
Arial Photo of Trench C with exedra and hypocaust system. The circle indicates the location of the three connected lead pipes. The arrow indicates the suspected flow direction into the drainage system.

the various parts of the walls and structures to one another. In 2007 four lead pipes were found in the southwest corner of the site in Trench C, three of which were joined by five keys. The *in situ* pipes seem to be reused from earlier phases and appear to be associated with one of the last phases of the site as part of a water feature associated with a bath complex. There are a series of bathing complexes on the site that diachronically replace or repair earlier phases. During the 2006 and 2007 excavation seasons an apsidal structure, the inside of which was covered with hydraulic cement, was uncovered in Trench C and was interpreted as either being associated with a caldarium, the heating portion of a bath, or as part of a nymphaeum, a type of water feature. The discovery in 2010 of a lead fistula feature in the structure inclines the archaeologists toward the conclusion that the structure was most likely a nymphaeum. In 2008 additional rooms situated slightly north of the apsidal structure were excavated. None of these rooms were found adjacent to the existing structure, nor were they preserved as well. Thus, we do not know whether they were on the landscape sequentially or at the same time. In 2009 a hypocaust system, an under-floor heating system commonly used by the Romans, at the far northern end of the trench was uncovered but the association of the various components to each other was as yet unclear. Handheld XRF spectrometry has played a key role in better understanding this water system area and providing valuable information used by the site archaeologists for planning future excavation seasons. Figure 10.3 shows the excavation area containing the nymphaeum and hypocaust.

Considerable amounts of pottery, decorative terra cotta, and fresco also have been collected from the Coriglia site. The identification of the pigments, glazes, and slips used in producing these objects can contribute to a better understanding of the human activities and technologies that produced them (Lazic *et al*. 2003; Mazzocchin, Agnoli, and Colpo 2003; Gliozzo, Vivacqua, and Memmi 2008; Raskovska *et al*. 2010). The area is well connected via land and river routes, particularly the Paglia and Tiber, and thus participates in an exchange network that is tightly tied to central Italy and loosely to the Mediterranean exchange system. Also, production technologies tend to be regional and thus have the potential for determining the extent and range of trade partners. Handheld XRF spectrometry has been a valuable tool for the nondestructive characterization of these decorative archaeological finds.

As a means to illustrate various practical uses of handheld XRF spectrometry for archaeology, three separate scenarios will be discussed. Although these examples are obviously unique to the particular excavation and peoples being investigated by the authors, the types of research questions being asked can be generalized and easily applied to a variety to situations. Detail will be provided as to site preparation, the integration of global positioning system (GPS) data and the site architect's data into data collection protocols, and how *in situ* data collection is conducted to reduce error. Instrument calibration and accuracy as they pertain to the less than homogeneous samples archaeologists often face will be discussed. Various graphical and mathematical approaches for data interpretation using Excel® and Matlab® will be presented.

Handheld X-ray fluorescence spectrometer used for these studies

An Innov-X Systems Alpha Series handheld XRF spectrometer (now Olympus-Innov-X) was used for all data collection. The analyzer has a W anode X-ray tube excitation source and Si Pin diode detector. The spot size for the analyzer is approximately 170 square millimeters (although the actual spot size is oval). Control of the instrument and data storage is through a Hewlett Packard iPAQ personal digital assistant. The instrument was operated in "Soil" mode for all non-metallic samples such as mortars, water system deposit materials, and fresco; this mode uses a 40 kV excitation, a 9.5-10.0 μA tube current, a 2 mm aluminum filter, and a Compton Normalization algorithm (Fried 1967; Mantler *et al.* 2006). All data collected in this mode were reported in ppm levels except Fe and Ca which were reported in percent levels. Thirty second analysis times were used for all data collection. Supporting data by other instrumental techniques such as flame atomic absorption spectroscopy and Raman spectroscopy also will be presented as appropriate in the sections to follow.

Protocols used for all handheld XRF spectrometer data collection

Over time, protocols have been developed and refined for the successful collection of *in situ* data by handheld XRF spectroscopy at our excavation site. Sample types for which *in situ* data have shown utility thus far in our research efforts include mortars, floors, soils, and hydraulic cements. Other samples such as pottery sherds and fresco fragments are instead analyzed *ex situ* at the excavation laboratory. Simplified analysis protocols are used for *ex situ* samples with considerably less information being required for documentation purposes and less sample cleaning.

The 'toolbox' materials required to execute these protocols include the handheld XRF spectrometer, reference materials as closely matched as possible to the matrices to be analyzed (such as NIST SRM 1881a Portland Cement and NIST SRM 1886a Portland Cement for analyses of hydraulic cements), two hard-cover 3-ring binders (referred to as Binder A and Binder B), pens, a global positioning system (GPS), a digital camera, a tape measure, flags for marking sampling locations, trowels, and brushes. Four stages of information collection are required for each XRF data set: 1) scouting and marking of sampling locations followed by cleaning of the locations for data collection; 2) recording of the sampling location descriptions; 3) collection of handheld XRF data; and 4) recording of exact sampling locations by the site architect. We have found the assignment of one person to the stage one and stage two activities and a pair to the stage three activities provides the greatest efficiency. In stage four, the site architect collects data via total station and draws the excavated portions of the site. The points are related to the fixed point that is on the geospatial grid.

For the purposes of this discussion the term "data set" is used to describe all data collected for a particular study. The term "sampling location" refers to one location within the study area where an analysis is performed. A data set is therefore composed of the data collected for all sampling locations within the study area.

Stage 1: For each analysis, a sampling location is chosen which is as flat and smooth as possible and is larger than the analyzer spot size. For analyses along a wall or floor, a number of locations are chosen so as to best represent the overall material which is likely to exhibit significant inhomogeneity. As a general rule, analyses are conducted along a grid pattern every ten to fifteen centimeters if possible. For samples such as wall mortars, sampling locations on the top of the wall as well as along the side in between rocks are utilized. The sampling location is brushed to remove loose debris, scraped with a trowel until clean material is observed, and then cleaned with a brush again. The chosen sampling locations are each marked with a numbered flag; half of a bamboo skewer with folded-over colored lab tape works well and can be made as needed at the site. Figure 10.4 shows skewer flags placed in a wall.

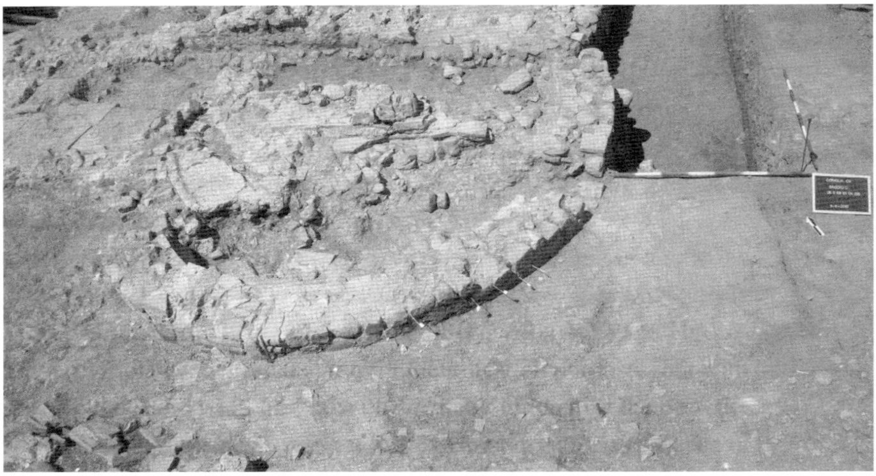

Figure 10.4:
Bamboo skewer flags on a wall.

Stage 2: With all the sampling locations chosen for the study area, details pertaining to each location are recorded on data sheets in Binder A. Basic information such as trench letter, locus number, a general description of the sample type (i.e. mortar on top of wall, drainage deposit material, etc.), and location number as indicated on the numbered flag are recorded first. For analyses along a wall or floor, a tape measure is used to assign a distance to each sampling location relative to a chosen reference point (locus end, trench bulk, etc.) with this value also being recorded in addition to the general description. A digital photograph of the sampling location is taken with an arrow positioned in the photograph area to indicate the specific location. An approximate analysis site location is also determined using a handheld GPS. The digital photograph number and GPS reading are also recorded on the data sheets. The filled-in data sheets from Binder

A are then transferred to Binder B for the data collection team. The location scout continues on to the next desired data collection area and starts at Stage 1 again. The data collection team proceeds on to Stage 3 based on the information now in Binder B and the sample location flags now in place.

Stage 3: The instrument window is cleaned with a cotton cloth and the instrument window positioned directly on the cleaned sampling location as indicated by the Location 1 flag. Fluorescence signal is then collected, holding the instrument with two hands to ensure its steadiness throughout data acquisition; any noticeable movements by the operator during data collection results in exclusion of that individual analysis from the study and a note indicating such in the field notebook. When possible, an instrument tripod or holder is used to improve stability during data collection. After the first analysis at a particular sampling location, replicates are then collected without moving the instrument from the analysis site. Our protocols require using two to three replicates per analysis site. The XRF data file names are recorded on the data sheet for Location 1. The data collection team then proceeds to the Location 2 flag and continues with the data collection process until all analyses in the data set are completed.

Stage 4: Along with the necessary locus shots and measurements that occur periodically at any archaeological excavation, we now also have the site architect collect data for the handheld XRF spectrometry analysis locations. These appear on the trench maps as shown in Figure 10.5.

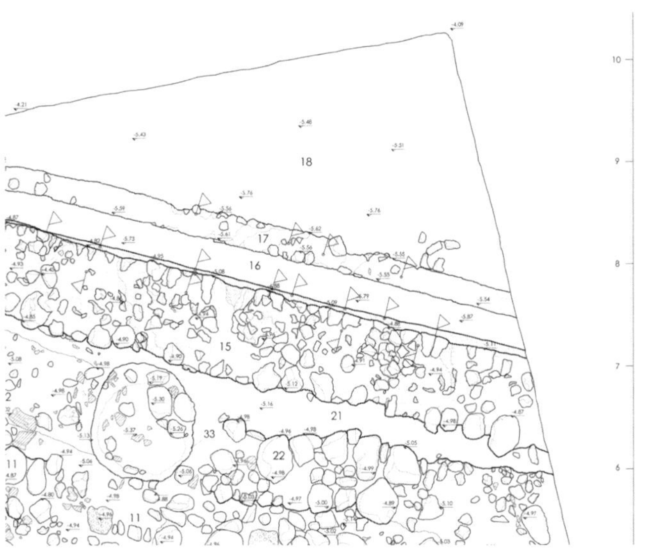

Figure 10.5:
Section of site map showing analysis locations as flags.

For each sample type investigated by handheld XRF spectroscopy, data also are collected from certified reference materials of a similar matrix. For mortars, tiles and cement floors, cement reference materials SRM 1881a and SRM 1886a from the National Institute of Standards and Technology (NIST) are used. The certified values for relevant elements in these materials are provided in Table 10.1. None of these materials were used by the instrument manufacturer to establish factory-set calibrations. Typically, analyses are conducted at the site on certified reference materials before and after collecting each data set plus after every ten to fifteen analysis locations. These replicate analyses are then utilized to check the instrument calibration and monitor for drift. By bracketing small sets of data, if at some point during a day of data collection the instrument is determined to be out of calibration, not all data is lost (Christian 2004).

Element	SRM 1881a	SRM 1886a
Fe	21610 ± 380 ppm	1063.1 ± 45.5 ppm
Ca	41.152 ± 0.243 %	48.51 ± 0.19 %
Ti	2195.4 ± 18.0 ppm	503.4 ± 53.9 ppm
Mn	725.2 ± 5.6 ppm	50.8 ± 1.4 ppm
Sr	304.4 ± 33.8 ppm	152.2 ± 50.7 ppm
Zn	392.9 ± 22.5 ppm	8.0 (no error provided)

Table 10.1:
NIST Standard Reference Materials used for calibration verification – certified values.

Most handheld XRF spectrometry instruments can be operated through either computer-control or by a personal digital assistant (PDA). Both operation modes have been used by our research team in the field with observed advantages and disadvantages to each. In general, computer control is more challenging in the field. Care must be taken in moving the laptop computer in and out of trenches while it is tethered to the instrument. Dirt can get on the keyboard and screen, and the screen can be difficult to see in the sun. Despite these disadvantages, computer-control provides much greater flexibility in how data are saved and organized. The full-sized keyboard allows for quick file name entry and data can be organized into folders as it is collected. Also, most importantly, spectra can be easily viewed and evaluated as they are obtained.

The main advantages to PDA operation are throughput and ease of use. The quickest data collection setting for PDA operation is to save data as numerical files in the order they are collected without having to name each file; the file numbers are recorded in Binder B together with the sampling location information so they can be matched up later. By using numerical file names, considerable time is saved between analyses which ultimately translate into an increased sample throughput. File names can be input individually on the PDA using the stylus and small screen although this can be a slow process. The PDA also can be set to provide "quantitative" results for each sampling site as it is analyzed based on pre-established calibration data; these can be viewed as each sample is analyzed for on-the-spot data evaluation. It is important to note, however, that accurate quantitative analyses by XRF spectrometry require matrix-matched standards and homogeneous samples. No archaeological artifacts can be considered homogeneous and thus, quantification, even semi-quantification should be considered suspect. That being said, comparing the resulting data between similar artifacts that have similar layered structures, in a large survey format can indicate if clear chemical differences are present. Pre-established calibration data can be utilized to obtain these 'semi-quantitative' results in the field. Disadvantages to PDA operation include the post-analysis time required to update all the numerical data file names with the actual sample descriptions plus any additional file organization within folders that is required. Also, the small PDA screen makes viewing and evaluating spectra more difficult compared to a full-sized computer screen.

For many analysis locations, samples are also physically collected and transported to the Saint Anselm College chemistry laboratories for additional and more extensive analyses. A digital photograph is taken of each sample *in situ*. The digital photograph number, GPS location, trench letter, locus number, and a general description of the sample type are recorded in a sampling notebook. This notebook is separate from the three-ring binders discussed in the preceding paragraphs. A second copy of this data is recorded on a sample card. The sample is then removed from its location and sealed in a sampling bag. This bag and the sample card are then sealed in a second sample bag. All samples returned to the college are catalogued and sorted by excavation year and sample type in plastic bins.

Much of the data discussed in this paper were collected by undergraduate students across several disciplines; there are no experiential, coursework or academic major requirements for participation in the excavation. Many students choose to participate in the excavation for multiple years to build upon their experiences with each additional season. In addition, chemistry majors often attend the excavation to complete a portion of their required senior research and consequently spend more time using the XRF spectrometer than other students. To this end, protocol and instruction sheets for instrument operation, data collection, and data transfer were created to ensure proper data collection. Multiple copies of these sheets were laminated and brought to the excavation site for daily use by the students. Students were provided with X-ray safety and instrument training at the beginning of each excavation season with second and third-year excavators requiring only a review of its operation. Particular care is taken regarding instrument handling and data collection as they relate to radiation safety with an emphasis on the radiation path and the instrument "active" light during data collection. Also, no digging by other excavation participants is allowed within a four meter perimeter surrounding the XRF spectrometer data collection area. Typically, an experienced multi-year user is with the instrument at all times.

Case study 1: tracking of lead within a water system

Trench C is the largest trench within the excavation area at the Coriglia site, covering approximately 500 square meters in a roughly twenty by twenty-five meter rectangular area. Of particular importance to this section of the site is its access to water. A natural spring exists just uphill to the south and is currently utilized extensively by the local community. A number of lead pipe sections and connective keys in relative proximity to this water supply were uncovered in 2006, 2007, and 2008 (indicated by the circle in Figure 10.3 with a close up view in Figure 10.6) and are suspected to have once been fed by both a fresh water spring as well as a thermal source located in antiquity on the site but having migrated 0.75 kilometers to the north in the early modern period. The fresh water spring still rises in the proximity of the site. What appears to be an associated drainage system (indicated by the arrow in Figure 10.3 and an axial view in Figure 10.7) dating between 1 BCE to 3rd century CE was previously discovered at the site. The system is comprised of 20 meters of fistulae, 17 meters of drainage channel, and five other drainage channels and drainage basins. To the south in Trench F there are at least three vascae (pools that are part of water features) and likely a series of fountains. When the lead pipes were uncovered they were isolated and no longer connected to the structures. Intact lead pipes are a rare archaeological find (Paparazzo 1994; Boni *et al.* 2000; Ortloff and Crouch 2001; Durali-Mueller *et al.* 2006). The evidence suggests that this pipe system was removed and repurposed several times.

A fundamental research question for the site archaeologists was to determine the general structure of this water system. Of particular importance was to find evidence that these lead pipes connected into the larger water system. A logical hypothesis was that the pipes fed water from the uphill spring source into the narrow drainage trench cutting through the center of Trench C. For a water system utilized over many years this would likely mean that, in repeated cycles, lead from the pipes would dissolve in small amounts into the water, flow into the drainage system, evaporate, and leave layered mineral deposits containing lead on the drainage system walls. Analysis of the deposit material confirming the presence of lead would be a first step toward providing chemical support for this hypothesis.

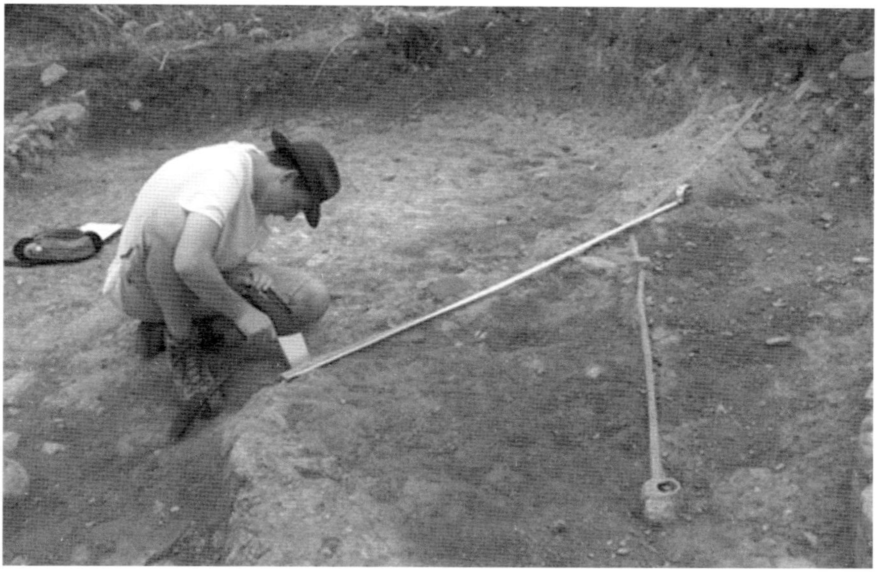

Figure 10.6:
Lead pipes in Trench C.

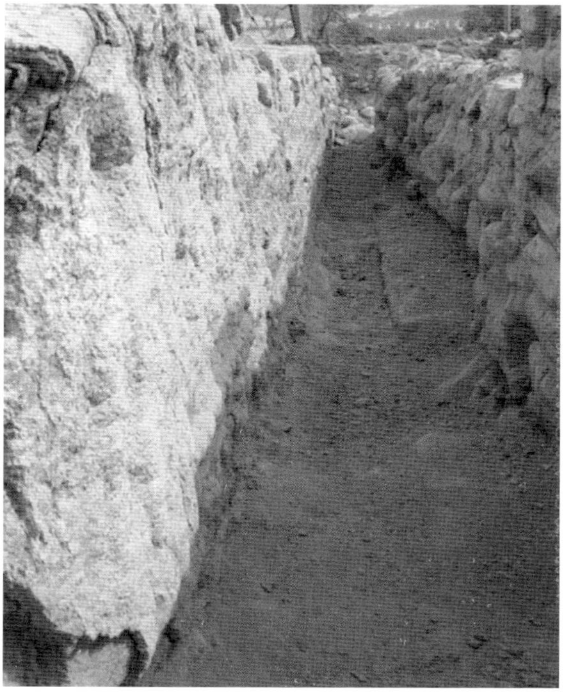

Figure 10.7:
Axial view of drainage system.

The XRF spectrometer was calibrated by the instrument manufacturer to identify lead over the range 18.9 ppm to 5532 ppm using the following materials: National Institute of Standards and Technology (NIST) SRM 2709 San Joaquin Soil, Resource Technology Corporation RTC-408, NIST SRM 2711 Montana Soil, NIST SRM 2710 Montana Soil, and two custom-mixed standards. These soil reference materials are not specific to the deposit material matrix, however; quantitative results presented in this section of the chapter should therefore be considered semi-quantitative. Analyses were conducted on the west wall of the drainage system where a thick layer of deposit material was found along most of the locus. Considering the lead distribution would likely not be uniform along the wall, short data collection times of thirty seconds each were used with five analyses at each location. Data were collected along both vertical and horizontal directions on the wall. As this deposit material was unique to the site with no reference material available to study method accuracy, multiple samples were collected and removed from the site for additional analyses by a second confirmatory method. A sample XRF spectrum showing the Pb in the deposit material is provided in Figure 10.8.

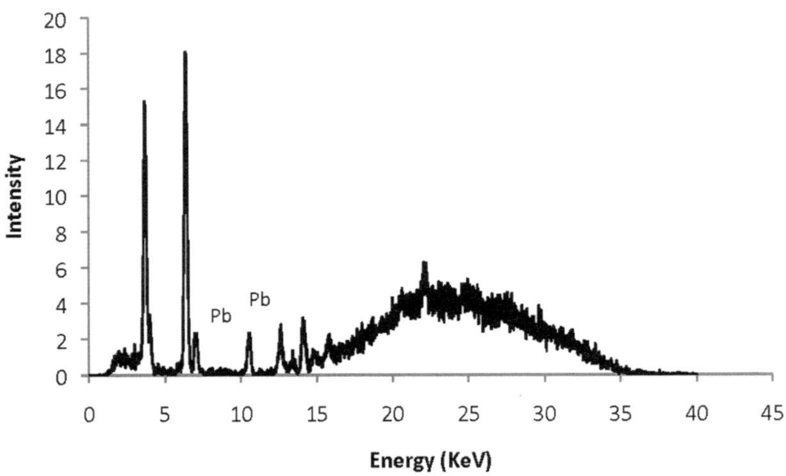

Figure 10.8:
XRF spectrum showing Pb in deposit material.

Precision for the XRF spectrometric data ranged from 5.7 to 7.4 percent relative standard deviation. For lead measurements at twelve locations along the horizontal distance of the drainage system, no data trend was observed but instead exhibited a wide concentration range from 134 ppm to 569 ppm with a 43.0% relative standard deviation and an average value of 373 ppm. Concentrations from the floor to top showed a maximum at about 30 cm from the floor, possibly indicating the fill height to which the drainage system was most often exposed. The lead concentration at the top of the wall was approximately 25% of that at the maximum concentration showing that the drainage system most likely did not fill completely at all times.

The accuracy of the XRF spectrometric findings was evaluated by additional analyses using flame atomic absorption spectroscopic (FAAS) analyses of the material at the Saint Anselm College chemistry laboratories. This secondary, independent method was used to aid in data evaluation because matrix-matched standards were not utilized for the *in situ* data collection by XRF spectrometry. Five 0.1 g subsamples of the deposit material were heated in 0.5 M hydrochloric acid at 90°C for thirty minutes in a programmable digestion block. The 217.0 nm atomic absorption wavelength for lead was used for data collection with a 0.5 nm bandpass. The flame was air/acetylene with a 1.1 L/min fuel flow and 7.0 mm burner height. The instrument was calibrated from 0.5 ppm to 4.0 ppm with acid-matched aqueous standards. The FAAS results ranged from 99 to 245 ppm with an average of 161 ppm. This range is lower than those found by XRF spectrometry (134 to 569 ppm), although the two sets of results certainly overlap in the 134 to 245 ppm range. Also, the difference could result from the FAAS results being based on five samples from one large section of the deposit material and not along the horizontal distance of the drainage system used for the *in situ* XRF spectrometric measurements. The conclusion can certainly be made that lead at significant levels is present in the deposit material. The next step to this study is focused on lead isotope ratio measurements by inductively coupled plasma mass spectrometry (ICP-MS). The lead isotope ratios for the deposit material are being compared to those of the lead pipes and the surrounding soil. Agreement between the two data sets would provide further support for the lead pipes being the lead source in the deposit material.

This simple application of handheld XRF spectrometry measurements illustrates how the instrument can be utilized to quickly determine "detectable" or "not detectable" levels for a particular element of importance to the research at hand. With no lead found in the samples, the research direction would have been entirely different to that currently being followed. This example also provides one approach for evaluating accuracy through confirmatory analyses by a second independent method. Other researchers also have used secondary methods to evaluate and compliment handheld XRF spectrometric results, including the use of laboratory-based XRF spectrometry (Desnica and Schreiner 2006; Nazaroff, Prufer, and Drake 2010), micro-particle induced X-ray emission spectroscopy (Desnica *et al*. 2008), laser ablation inductively coupled plasma mass spectrometry (Scheid *et al*. 2009), and instrumental neutron activation analysis (Scheid *et al*. 2009; Nazaroff, Prufer, and Drake 2010). Many of these techniques are laboratory-based and/or are destructive, making them less suited for field archaeology. Also, sample composition and homogeneity can have significant effects on method accuracy, often necessitating close examination of the instrument calibration (Shugar 2010).

Case Study 2: differentiation of tiles in a hypocaust

During the 2009 excavation season a hypocaust, the heating system for a bath, was uncovered in the northwestern corner of Trench C. An intact hypocaust would consist of a lower tile floor, tile pillars resting on that floor, and an upper floor supported by the pillars on which the bath water would sit. The space between the two floors would be heated by a separate fire box which in turn heats the water for the bath. As shown in Figure 10.9, two sections of floor tiles (Locus 173 and Locus 212) and tiles used to form a floor and pillars (Locus 205) were uncovered. Aspects related to a water system had been discovered in this area of the site to that point, but no substantial bathhouse was excavated. The discovery of a hypocaust establishes the site at Coriglia, Castel Viscardo as having regional importance

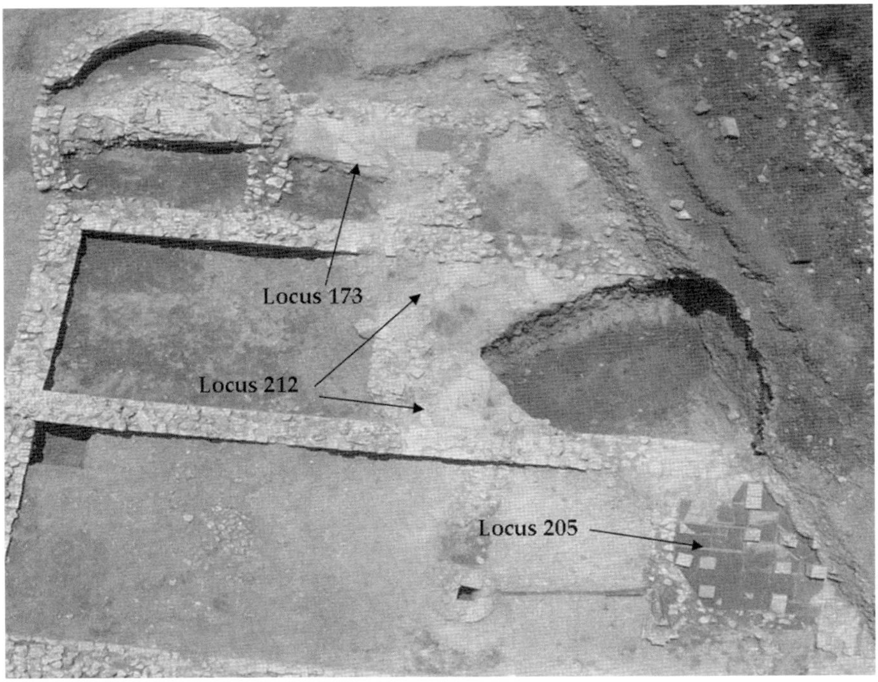

Figure 10.9:
Trench C showing Loci 173, 212, and 205.

since bathhouses were only found in cities with some significance. The elemental compositions, in conjunction with establishing a stylistic and dimensional similarity of these hypocaust tiles may provide valuable information for the site archaeologists regarding the tiles' manufacturing location and fabrication technologies. If all the tiles were made at the same time, in the same factory, using similar raw materials and formulation, their compositions should be similar. Based on visual inspection of the tiles, no colorant appears to have been added to the clay and no glaze was applied; they retain their post-fired reddish gray color.

Three replicate measurements were taken for each locus at five randomly chosen locations on the floors and tiles for a total of fifteen replicates for each locus. Sample preparation was identical to the method discussed above. Two analyses of NIST SRM 1881a Portland Cement were performed prior to the hypocaust samples and then every fifteen analyses for a total of eight replicates interspersed throughout the data set. The results for NIST SRM 1881a were used post-data collection to adjust the instrument calibration factor as necessary to improve accuracy. Typically, little drift was observed and allowed for one minor adjustment per element to be made to the entire data set. The XRF instrument was configured to measure K, Ca, Ti, V, Cr, Mn, Fe, Co, Ni, Cu, Zn, As, Se, Rb, Sr, Zr, Mo, Ag, Cd, Sn, Sb, Ba, Hg, Pb, and Bi using a factory calibration based on measurements of NIST SRM 2709 San Joaquin Soil, NIST SRM 2711 Montana Soil, and NIST SRM 2710 Montana Soil. Of these elements, K, Ca, Ti, Cr, Mn, Fe, Co, Zn, Rb, Sr, Zr, Ba, and Pb were found at measureable levels. LEAP mode, optimized for the detection of light elements such as K and Ca, was not used for these measurements. Elevated levels of K and Ca can be measured adequately using the high kV settings of Soil mode, although these are not the ideal detection parameters for these elements. Some elements such as Co, Zn, and Zr varied significantly both within each locus and among all three loci. Other elements such as K, Ca, Fe, Mn, and Sr exhibited good precision within each locus so therefore warranted closer study. The factory linear calibration ranges for these five elements were as follows: 2.03% to 2.45% for K, 1.25% to 2.88% for Ca, 2.89% to 3.5% for Fe, 540 ppm to 1.01% for Mn, and 230 ppm to 330 ppm for Sr. As stated earlier, attempting to acquire quantitative data from non-uniform materials is not possible, but in direct comparison of similar artifacts, such as hypocaust tiles, the results are valuable and can be used for clarification and classification of materials.

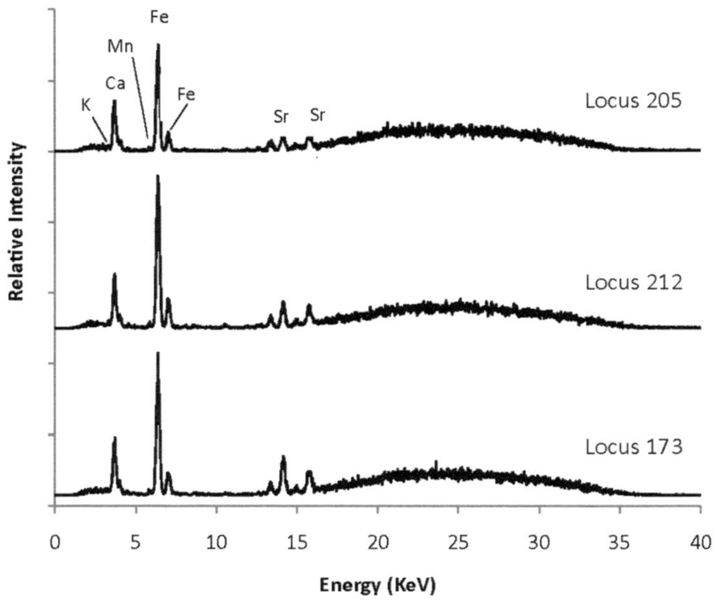

Figure 10.10:
XRF spectra of Loci 173, 212, and 205.

The purpose of the SRM 1881a analyses was to verify method accuracy and confirm that the resulting data was not affected by instrument fluctuations. This material was chosen for its similarity to the ancient mortars, hydraulic cements, and tiles which are of great interest to these researchers. The Innov X software allows for post-data collection adjustment of results. After the data sets for all three loci were collected, the XRF spectrometric results obtained for SRM 1881a were compared to the material's certified values. Knowing the expected result for the NIST material allowed for empirical changes in the factory-set "Element Rates" for each element within the software. Increasing the "Element Rate" value for an individual element increases the concentrations reported for that element. The results for each element in the NIST material were individually adjusted until all replicate analyses for that element were considered valid, ideally within 15% of the certified value. Typically, one adjustment was made for each element and then applied to an entire data set; this adjustment was usually minor compared to the calibration factor used just prior to the most recent analyses. This process was then repeated for all certified elements in the NIST material detected by the XRF spectrometer. These adjustments were made with no effect on the original raw data. The process described here for data validation is certainly not as rigorous as the utilization of matrix-matched standards for generation of our own calibrations and would not be suggested if you were attempting to obtain low error quantitative data values. For the purposes of material comparison, however, this approach was deemed sufficient. Although the quantitative results are interesting in showing what elements are at a trace levels versus a percent level, the relative amounts of each element are far more important in allowing for the differentiation of one material from another. The means and 95% confidence intervals for the hypocaust data are provided in Table 10.2 below. XRF spectra for the three loci are provided in Figure 10.10.

Locus	K (%)	Ca (%)	Fe (%)	Mn (ppm)	Sr (ppm)
173	0.984 ± 0.043	19.6 ± 1.3	4.51 ± 0.24	1900 ± 130	458 ± 51
205	0.693 ± 0.035	12.6 ± 0.9	4.08 ± 0.26	1510 ± 160	285 ± 21
212	1.106 ± 0.044	18.9 ± 0.9	5.08 ± 0.26	1800 ± 130	479 ± 93

Table 10.2:
Summary of XRF data for hypocaust tiles.

Although the means and confidence intervals can be compared to determine if differences exist among the analyzed materials, a graphical examination of all the data points in the form of a 3-D scatter plot can better show clustering. The data sets, means, standard deviations and 95% confidence intervals for K, Ca, Fe, Mn, and Sr in the three loci were organized and calculated using Excel. These data were then imported into Matlab as separate variables. The "scatter3" and "plot3" functions in Matlab were used to produce the graphs in Figure 10.11 and Figure 10.12. The mean for each data set and the 95% confidence interval error bars are indicated in black.

As can be seen by the two figures, Loci 173 and 212 are clustered together while Locus 205 is separate. Two graphs were produced to verify that this trend existed over multiple elements which were confirmed. Also note that the same trends were observed prior to the post-data collection adjustments. For quick data analysis on site the unadjusted data can be used as comparators, noting that the absolute concentration values may change upon evaluation of NIST standard used for quality control data. The cause for the apparent similarity between Loci 173 and 212 and their dissimilarity to Locus 205 is not definitively known. One possibility is that the pillar tiles of Locus 205 were produced using a clay material of different composition or production process compared to the floor tiles of Loci 173 and 212.

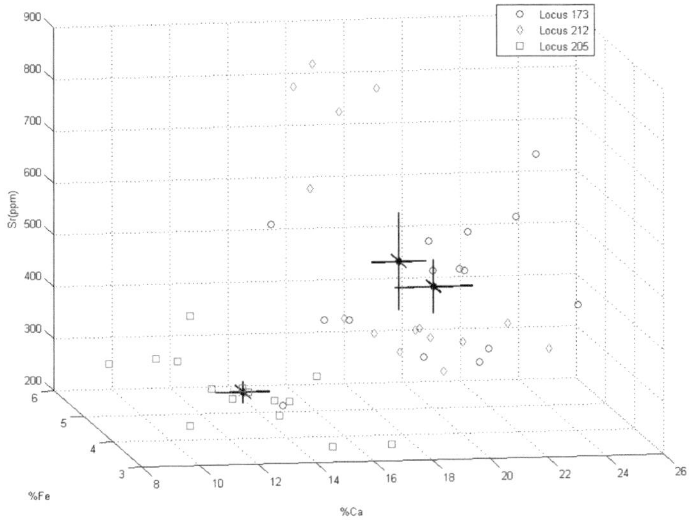

Figure 10.11:
Scatterplots of elemental data for Loci 173, 212, and 205.

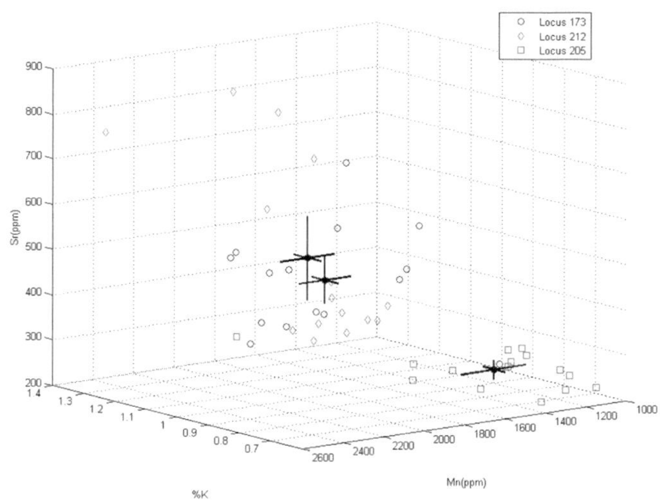

Figure 10.12:
Scatterplots of elemental data for Loci 173, 212, and 205.

A two-way analysis of variance (ANOVA) also was performed using Excel on the Ca, Fe, Sr, K, Mn, and Sr data for the three loci. All elemental concentrations were higher in Loci 173 and 212 compared to those found in Locus 205, suggesting the possibility that these elements are associated with the same tile component and that this component was used in a higher percentage in the preparation of the floor tiles compared to the pillar tiles. The tiles were likely made by simply adding water to raw materials and scoring the resulting clay into the desired tile sizes followed by firing. The association of Ca and Sr together is expected as they likely come from the same raw material source.

As illustrated by this example, XRF spectrometry can be exceptionally useful as a comparator between large groups of similar material types – is material "A" different than material "B"? This comparison can be done in the more simple format shown here with graphed elemental concentrations, or via comparisons among entire spectra using advanced statistical techniques such as principal components analysis (PCA). In addition, the two-way ANOVA provided further support for the visual comparison in the 3-D scatter plot. These data required one day of data collection time followed by one day of data evaluation time. This provides the site archaeologists with close to immediate results, thus guiding them during the current excavation. Also, where this example used the produced concentrations for each element, one could do a similar comparison using the raw total counts under peak for each element of interest.

Case Study 3: Identification of pigments using handheld XRF and portable raman

For definitive pigment identification and characterization, the elemental data provided by XRF spectroscopy often requires support via a complimentary technique such as Raman spectroscopy (Comodi *et al.* 2004; Ricci *et al.* 2004; Desnica *et al.* 2008). Where XRF provides elemental data it cannot provide compound identification or identity of elements lighter than Al. Raman spectroscopy has undergone similar changes in technology to those of XRF spectroscopy, and many portable instruments are now available (Bruni and Guglielmi 2007; de Waal 2009). The portable designs of both XRF and Raman permit easy and quick data collection at archaeological sites and on museum pieces. Identification of pigments on artifacts can help in identifying the age of finds and offer potential information related to trade partners based on the availability of the minerals used to produce the pigments (Bell and Clark 1997; Smith *et al.* 2005; Vandenabeele, Edwards, and Moens 2007). The complementary nature of Raman spectroscopy as a molecular technique and XRF spectroscopy as an elemental technique is particularly valuable for pigment studies such as those discussed here. As well, the portable and non-destructive nature of both these instruments is ideal in that alteration of the artifacts was avoided.

Our protocols for the treatment of fresco fragments, most of which can be stylistically identified as Pompeian fourth style dating to the third quarter of the 1st century CE, are to leave the fragments in the condition they are when taken *ex situ*. This allows time for study before a conservator works on them either for long term storage or display at the end of the excavation. Thus elements of the soil matrix sometimes remain attached. With this in mind, the XRF data for fresco samples are not provided as quantitative information but instead are reported as elements found at "significant" and "trace" levels. This is due to the XRF spectrometer providing data for the combined surface pigments and underlying fabric based on the X-rays' depth penetration into the sample; analyses of the fabric from the side and/or back of the fragments allows differentiation between it and the pigments and thus determination of elements only in the pigments. This lack of a discrete pigment layer must be taken into account when analyzing fresco samples by XRF spectrometry. Care is taken when analyzing the fresco fragments to gently brush away soil matrix from the analysis locations. Also, pure pigment samples from a commercial source are analyzed together with the artifacts to ensure method accuracy.

Raman spectra were measured with an Enwave Optronics (Irvine, CA) EZRaman-I-A2 high sensitivity field portable Raman system. The laser source was an Enwave Optronics frequency stabilized 785 nm diode laser system with a variable output up to 300 mW. The optical system is an Enwave Optronics high sensitivity Peltier cooled CCD spectrograph with a spectral range of 250 to 2100 cm^{-1} with optical resolution of 4.5 cm^{-1}. The laser light and the return signal are carried over a 1 meter armored fiber optic from the main unit to a hand held probe, which contains focusing optics and a OD >8 filter for Rayleigh scattering rejection. The probe tip is interchangeable and for most samples a contact probe was used to permit direct contact with the samples. The system runs on AC power or with a rechargeable lithium battery. Data acquisition and instrument operation was managed with a built in Comact PC using Enwave Optics EZRaman Reader software. For comparison of spectra, CrystalSleuth software (Downs 2006) was used.

Numerous fresco pieces have been recovered from Castel Viscardo Trench C with XRF and Raman spectroscopic data being collected on most to determine the identities of the applied pigments. Pigments found in significant amounts that were studied using both Raman and XRF spectroscopies were a purple pigment, two red pigments, a pink pigment, an orange pigment, and various yellow-brown pigments.

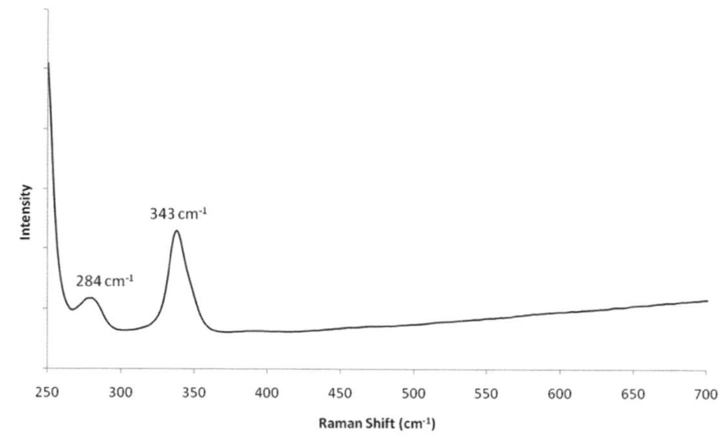

Figure 10.13:
Raman spectrum of fresco fragment with vermillion as pigment.

The XRF data for about half of the thirty red fresco pieces indicated significant levels of Hg. A red pigment derived from the mineral cinnabar (HgS) has been documented as commonly used by the Romans in this region (Baraldi *et al.* 2007; Desnica *et al.* 2008) and is therefore the likely identity of this pigment. The Raman spectrum for a fresco fragment with this red pigment is shown in Figure 10.13. Note the distinctive Raman signature for vermillion in the 250 to 400 cm^{-1} wavenumber shift region (Bergio and Clark 2001). The broad peak from 450 to 2140 cm^{-1} wavenumber shift is due to fluorescence background. Vermillion is clearly identifiable by using a combination of Raman and XRF spectroscopies.

A second red pigment, darker in hue than vermillion and referred to as brick-red for the remainder of this discussion, also was observed on a number of fresco pieces. The XRF data for this pigment indicated significant levels of Fe and trace levels of Pb. The Raman spectra for the brick-red pigment exhibited a weak signal for hematite, Fe_2O_3, and supports the XRF results. Hematite has been previously identified on Roman wall paintings (Mazzocchin *et al.* 2003) and would therefore be an expected pigment used at Castel Viscardo. The trace lead indicated in the XRF spectra may be caused by PbO or Pb_3O_4, both red pigments derived from natural minerals, or may be from contamination. No Raman signal was observed for lead oxides in these fresco fragments, however, so the form of lead could not be determined (Bell and Clark 1997); this is likely due to the trace level of Pb and the low sensitivity and high fluorescence background of the portable Raman spectrometer used for these studies. A number of other fresco fragments with pink, orange, and yellow pigments also produced similar Raman and XRF spectroscopy results – weak hematite shifts, significant Fe, and trace Pb. Lastly, fresco pieces with purple pigment produced a strong hematite Raman signal. XRF and Raman spectra of all five pigments are shown in figures 10.14 and 10.15, respectively. Based on analyses from the back and sides of various fresco fragments, the Ca, Sr, and a portion of the Fe measured in the fresco fragments result from the underlying background matrix onto which the pigments were applied. The characteristic Raman peaks at 284 cm^{-1}, 404 cm^{-1}, 494 cm^{-1}, and 607 cm^{-1} are indicative of the pigment hematite (Bergio and Clark 2001; Baraldi *et al.* 2007).

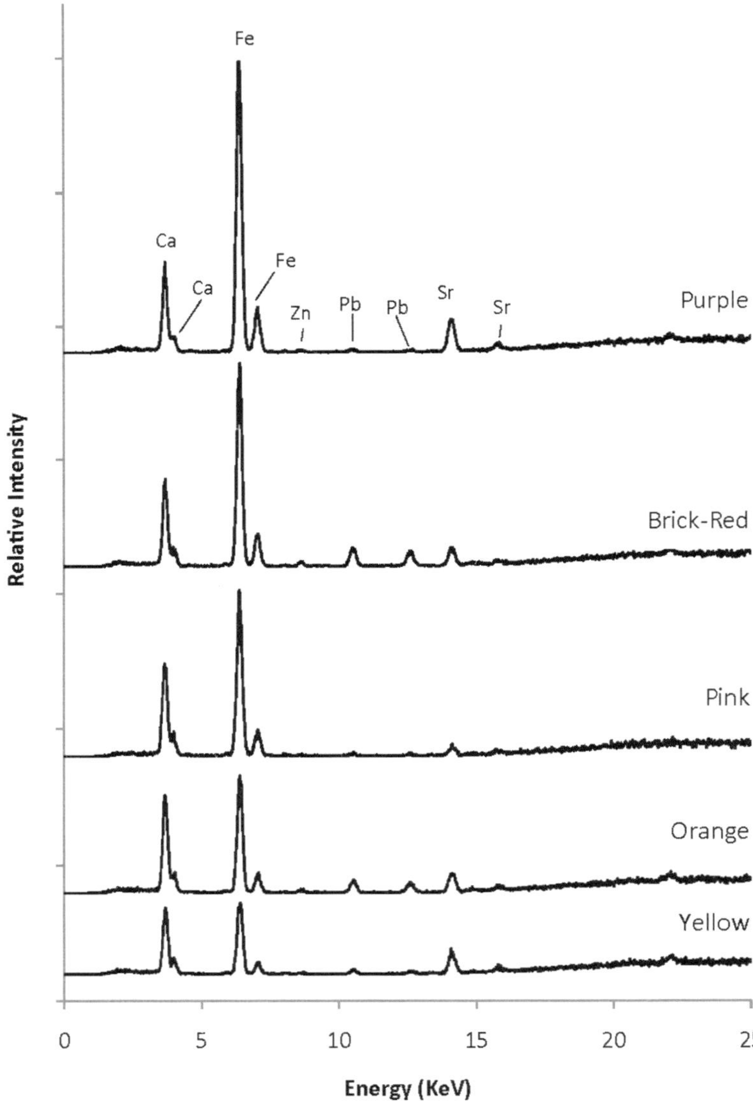

Figure 10.14:
XRF spectra of hematite-based fresco pigments.

The chemical distinction among these five pigments is based on the purity of the Fe_2O_3. The purple pigment is hydrated Fe_2O_3 with no clay or silica mixed into it (Mazzocchin, Angoli *et al.* 2003). Logically, this pigment produced the strongest Raman signal and the XRF spectrum with the highest iron content. The other four pigments are various ochres, Fe_2O_3 mixed with clay and silica at different hydration levels. All these pigments are naturally occurring, so environmental conditions and particle size largely control the color produced by the material (Damiani *et al.* 2003; Popelka-Filcoff *et al.* 2008; Akyuz *et al.* 2009). Yellow ochre (goethite) is the hydrated form whereas red ochre is the non-hydrated form. While the Ca and Sr signals remain relatively constant when comparing the XRF spectra, the Fe levels decrease from purple, brick-red, pink, orange, and finally with yellow at the lowest amount. The pink pigment is likely the result of red ochre being mixed with a white pigment such as $CaCO_3$ as suggested by Mazzocchin and Agnoli (Mazzocchin *et al.* 2003).

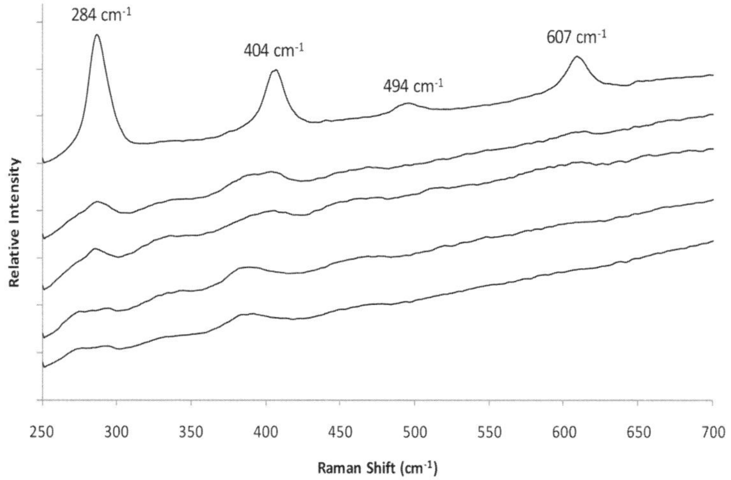

Figure 10.15:
Raman spectra of hematite-based fresco pigments.

This example illustrates how XRF spectra can be evaluated qualitatively for identification of natural minerals such as those used as pigments. While XRF spectrometry alone will not allow for definitive material identification, this technique performed in conjunction with another complimentary technique, such as Raman spectroscopy, is a powerful combination. Instrument calibration was not a concern using this qualitative approach which therefore simplified the experimental design. Analyses of pure pigments obtained from Natural Pigments

LLC were performed to determine major elements for the XRF spectrometric work and the potential shifts for the Raman spectroscopy work.

Conclusion

After three seasons integrating handheld XRF spectroscopy into the Coriglia excavation design, the applications of this versatile technique have grown and will most likely continue to grow for years to come. A variety of scenarios have been explored by our research team in which XRF data have played a key role including the following: 1) "detectable" versus "not detectable" for a particular element; 2) comparisons among materials of similar matrices; and 3) mineral identification. Taking into account the limitations and cautions that must accompany handheld XRF spectrometry analysis, among which are sample homogeneity, sample thickness, and calibration limitations, the successful use of this technique for field archaeology has been demonstrated. A number of statistical and graphical approaches also were illustrated to show how XRF data can be evaluated and subsequently utilized to provide support for common archaeological questions. Lastly, the protocols that the Coriglia research team has developed are detailed as a guide for others considering including handheld XRF spectrometry into their archaeological research designs.

Acknowledgements

The authors would like to thank Innov X Systems for the loan of the handheld XRF spectrometer in the summers of 2008 and 2009 through their Academic Relations Grant Program. We would also like to thank Symphotic Tii for the loan of the portable Raman spectrometer for the 2009 and 2010 excavation seasons. This research would not have been possible without the tremendous support and encouragement of the scientists at these two instrument companies. The following undergraduate students collected and evaluated a portion of the data presented in this work: Cindy Lebel, Brad Duncan, Anna Daigle, and Sara Wojtas.

References

Akyuz, S., T. Akyuz, S. Basaran, I. Kocabas, A. Gulec, H. Cesmeli, and B. Ucar. 2009. "FT-IR and EDXRF analysis of wall paintings of ancient Ainos Hagia Sophia Church." *Journal of Molecular Structure* Vol. 924-926:400-403.

Baraldi, P., C. Baraldi, R. Curina, L. Tassi, and P. Zannini. 2007. "A micro-Raman archaeometric approach to Roman wall paintings." *Vibrational Spectroscopy* Vol. 43 (2):420-426.

Bell, I.M., and Robin J. H. Clark. 1997. "Raman spectroscopic library of natural and synthetic pigments (pre- ~ 1850 AD)." *Spectrochimica Acta, Part A Molecular and Biomolecular Spectroscopy* Vol. 53 (12):2159-2179.

Bergio, L., and Robin J. H. Clark. 2001. "Library of FT-Raman spectra of pigments, minerals, pigment media and varnishes, and supplement to existing library of Raman spectra of pigments with visible excitation." *Spectrochimica Acta, Part A Molecular and Biomolecular Spectroscopy* Vol. 57:1491-1521.

Boni, M., G. Di Maio, R. Frei, and I.M. Villa. 2000. "Lead Isotopic Evidence for a Mixed Provenance for Roman Water Pipes from Pompeii." *Archaeometry* Vol. 42 (1):201-208.

Bruni, S., and V. Guglielmi. 2007. "Applications of a compact portable Raman spectrometer for the field analysis of pigments in works of art." *Springer Proceedings in Physics* Vol. 116 (Lasers in the Conservation of Artworks):407-414.

Christian, Gary D. 2004. "Good Laboratory Practice: Quality Assurance of Analytical Measurements." In *Analytical Chemistry*, 124-140. John Wiley & Sons, Inc.

Comodi, P., M. Bernardi, A. Bentivoglio, G. D. Gatta, and P. F. Zanazzi. 2004. "The production and technology of glazed ceramics from the middle ages, found in the Saepinum territory (Italy): a multimethodic approach." *Archaeometry* Vol. 46 (3):405-419.

Craig, N., R. J. Speakman, R.S. Popelka-Filcoff, M.D. Glascock, J. D. Robertson, M.S. Shakley, and M.S. Aldenderfer. 2007. "Comparison of XRF and PXRF for analysis of archaeological obsidian from southern Peru." *Journal of Archaeological Science* Vol. 34:2012-2024.

Damiani, D., E. Gliozzo, I. Memmi Turbanti, and J. E. Spangenberg. 2003. "Pigments and plasters discovered in the House of Diana (Cosa, Grosseto, Italy): an integrated study between art history, archaeology and scientific analyses." *Archaeometry* Vol. 45 (2):341-354.

de Waal, D. 2009. "Micro-Raman and portable Raman spectroscopic investigation of blue pigments in selected Delft plates (17-20th Century)." *Journal of Raman Spectroscopy* Vol. 40 (12):2162-2170.

Desnica, V., K. Skaric, D. Jembrih-Simbuerger, S. Fazinic, M. Jaksic, D. Mudronja, M. Pavlicic, I. Peranic, and M. Schreiner. 2008. "Portable XRF as a valuable device for preliminary in situ pigment investigation of wooden inventory in the Trski Vrh church in Croatia." *Applied Physics A Materials Science & Processing* Vol. 92 (1):19-23.

Desnica, Vladan, and Manfred Schreiner. 2006. "A LabVIEW-controlled portable x-ray fluorescence spectrometer for the analysis of art objects." *X-Ray Spectrometry* Vol. 35 (5):280-286.

Donais, M.K., B. Duncan, D. George, and C. Bizzarri. 2010. "Comparisons of Ancient Mortars and Hydraulic Cements through In-Situ Analyses by Portable X-Ray Fluorescence Spectrometry." *X-Ray Spectrometry* Vol. 39:146-153.

Donais, M.K., D. George, B. Duncan, S. Wojtas, and A. Daigle. 2011. "Evaluation of Data Processing and Analysis Approaches for Fresco Pigments by Portable X-Ray Fluorescence Spectrometry and Portable Raman Spectroscopy." *Analytical Methods* Vol. 3 (5):1061-1071.

Donais, M.K., D. George, S. Roberts, J. Roberts, and E. Wu. 2009. Analyses of Ancient Roman Pigements by Portable X-Ray Fluorescence and Raman Spectroscopies. In *Conference of the Federation of Analytical Chemistry and Spectroscopy Societies*. Louisville, Kentucky.

Donais, M.K., C. Lebel, and K. Thibodeau. 2008. Investigations of metals in drainage system deposit material and lead pipe in Castel Viscardo. In *Conference of the Federation of Analytical Chemistry and Spectroscopy Societies*. Reno, Nevada.

Downs, R.T. 2006. The RRUFF Project: an integrated study of the chemistry, crystallography, Raman, and infrared spectroscopy of minerals. In *19th General Meeting of the International Mineralogical Association*. Kobe, Japan.

Durali-Mueller, S., G.P. Brey, D. Wigg-Wolf, and Y. Lahaye. 2006. "Roman lead mining in Germany: its origin and development through time deduced from lead isotope provenance studies." *Journal of Archaeological Science* Vol. 34:1555-1567.

Fried, M.H. 1967. *The Evolution of Political Society*. New York: Random House.

Gliozzo, Elisabetta, Paola Vivacqua, and Isabella Turbanti Memmi. 2008. "Integrating archaeology, archaeometry and geology: local production technology and imports at Paola (Cosenza, Southern Italy)." *Journal of Archaeological Science* Vol. 35 (4):1074-1089.

Karydas, A. G. 2007. "Application of a portable XRF spectrometer for the non-invasive analysis of museum metal artefacts." *Annali di Chimica (Rome, Italy)* Vol. 97 (7):419-432.

Karydas, A. G., D. Kotzamani, R. Bernard, J. N. Barrandon, and Ch Zarkadas. 2004. "A compositional study of a museum jewelery collection (7th-1st BC) by means of a portable XRF spectrometer." *Nuclear Instruments & Methods in Physics Research, Section B Beam Interactions with Materials and Atoms* Vol. 226 (1-2):15-28.

Lazic, Violeta, Francesco Colao, Roberta Fantoni, Antonio Palucci, Valeria Spizzichino, Ilaria Borgia, Brunetto G. Brunetti, and Antonio Sgamellotti. 2003. "Characterisation of lustre and pigment composition in ancient pottery by laser induced fluorescence and breakdown spectroscopy." *Journal of Cultural Heritage* Vol. 4 (Supplement 1):303-308.

Mantler, M, JP Willis, GR Lachance, BAR Vrebos, K-E Mauser, R Kawahara, RM Rousseau, and PN Brouwer. 2006. "Quantitative Analysis." In *Handbook of Practical X-Ray Fluorescence Analysis*, edited by B Beckhoff, B Kanngießer, N Langhoff, R Wedell and H Wolff, 309-410. Berlin/Heidelberg/New York: Springer.

Mazzocchin, G. A., F. Agnoli, and I. Colpo. 2003. "Investigation of Roman age pigments found in pottery fragments." *Analytica Chimica Acta* Vol. 478 (1):147-161.

Mazzocchin, G. A., F. Agnoli, S. Mazzocchin, and I. Colpo. 2003. "Analysis of pigments from Roman wall paintings found in Vicenza." *Talanta* Vol. 61 (4):565-572.

Nazaroff, A.J., K.M. Prufer, and B.L. Drake. 2010. "Assessing the applicability of portable X-ray fluorescence spectrometry for obsidian provenance research in the Maya lowlands." *Journal of Archaeological Science* Vol. 37 (4):885-895.

Ortloff, C.R., and D.P. Crouch. 2001. "The Urban Water Supply and Distribution System of the Ionian City of Ephesos in the Roman Imperial Period." *Journal of Archaeological Science* Vol. 28:843-860.

Papadopoulou, D. N., G. A. Zachariadis, A. N. Anthemidis, N. C. Tsirliganis, and J. A. Stratis. 2006. "Development and optimisation of a portable micro-XRF method for in situ multi-element analysis of ancient ceramics." *Talanta* Vol. 68 (5):1692-1699.

Paparazzo, E. 1994. "Surface and interface analysis of a Roman lead pipe "fistual": microchemistry of the soldering at the join, as seen by scanning Auger microscopy and X-ray photoelectron microscopy." *Applied Surface Science* Vol. 74:61-72.

Popelka-Filcoff, R.S., E.J. Miksa, J.D. Robertson, M.D. Glascock, and H. Wallace. 2008. "Elemental analysis and characterization of ochre sources from Southern Arizona." *Journal of Archaeological Science* Vol. 35 (3):752-762.

Raskovska, A., B. Minceva-Sukarova, O. Grupce, and Ph. Colomban. 2010. "Characterization of pottery from Republic of Macedonia II. Raman and infrared analyses of glazed pottery finds from Skopsko Kale " *Journal of Raman Spectroscopy* Vol. 41:431-439.

Ricci, C., I. Borgia, B. G. Brunetti, C. Miliani, A. Sgamellotti, C. Seccaroni, and P. Passalacqua. 2004. "Perugino's palette: Integration of an extended in situ XRF study by Raman spectroscopy." *Journal of Raman Spectroscopy* Vol. 35 (8/9):616-621.

Romano, F. P., G. Pappalardo, L. Pappalardo, S. Garraffo, R. Gigli, and A. Pautasso. 2006. "Quantitative non-destructive determination of trace elements in archaeological pottery using a portable beam stability-controlled XRF spectrometer." *X-Ray Spectrometry* Vol. 35 (1):1-7.

Scheid, Nicole, Stefan Becker, Marc Dücking, Gabriele Hampel, Jens Volker Kratz, Peter Watzke, Peter Weis, and Stephan Zauner. 2009. "Forensic investigation of brick stones using instrumental neutron activation analysis (INAA), laser ablation-inductively coupled plasma-mass spectrometry (LA-ICP-MS) and X-ray fluorescence analysis (XRF)." *Applied Radiation and Isotopes* Vol. 67 (12):2128-2132.

Shugar, Aaron. 2010. Variation in data reported from handheld XRFanalysis resulting from minor alterations to machine calibration. In *38th International Symposium on Archaeometry*. Tampa, Florida.

Smith, Kari, Kristin Horton, R. John Watling, and Natalie Scoullar. 2005. "Detecting art forgeries using LA-ICP-MS incorporating the in situ application of laser-based collection technology." *Talanta* Vol. 67 (2):402-413.

Vandenabeele, Peter, Howell G. M. Edwards, and Luc Moens. 2007. "A Decade of Raman Spectroscopy in Art and Archaeology." *Chemical Reviews (Washington, DC, United States)* Vol. 107 (3):675-686.

Chapter 11

Handheld XRF elemental analysis of archaeological sediments: some examples from Mesoamerica

Hector Neff, Barbara Voorhies and Federico Paredes Umaña

Introduction

The highly portable, durable handheld XRF instruments that have appeared on the market over the past few years hold tremendous potential for archaeology. The instruments can easily be transported to remote field stations, where they can either be set up and used in a field lab or used on-site, to analyze artifacts or sediments *in situ*. The present paper focuses on sediment analysis, which is useful for activity-area identification (e.g., Hutson and Terry 2006; Dahlin *et al.* 2007; Wells *et al.* 2007; Wells and Terry 2007) and in stratigraphic interpretation.

Traditionally, archaeologists have described sediments according to color, texture, moisture content, and other visual clues, usually in qualitative terms, and with some attention to soil scientists' classification systems. Some measure of objectivity is introduced by keys, such as the Munsell soil color chart, which is as common in archaeological field toolkits as line levels and plumb-bobs. Qualities such as color, however, vary depending on composition, sediment history, organic content, moisture content, and other conditions, not to mention the eyesight and expertise of the observer. The advent not only of handheld XRF but of other kinds of portable analytical instruments, such as FTIR (Weiner 2010), promises to take much of the subjectivity and guesswork out of archaeologists' interpretations of ancient sediments, as the examples to be discussed here illustrate.

The two examples to be discussed here are from excavations in southern Mesoamerica. In the first example, XRF is used to analyze samples from an extensive but shallow, fine-textured layer entombed within Tlacuachero, an archaeological shell midden (a midden is basically a heap of mollusk shells and other debris discarded by people in the past) dated to the Late Archaic period (~2000 - 3500 BC; see Voorhies 2004). Since its discovery in the mid-1970s (Voorhies 1976), this stratigraphic feature has been identified as a "clay floor." The goal in this case is to test hypotheses about what activities might have been carried out on the "floor" and to explore whether there is spatial variation across the floor in the elemental indices of different activities. In addition, a surprising twist on the accepted characterization of the feature as a "clay floor" emerged from the XRF analysis.

The second example discussed here is from the Formative through Classic-period archaeological site of Ataco, in the highlands of western El Salvador. In this case, the goal is to identify the source of a volcanic ash layer found at approximately two meters depth at most locations across the site and to understand the depositional processes that mixed the volcanic ash up or down within the sediment profile. Ultimately, this information bears on the stratigraphic integrity of deposits at Ataco.

Some honest advertising is in order here: Neither of the two studies reported here was actually undertaken in the field. The Archaic floor analyses were done in the IIRMES lab in Long Beach, in conjunction with LA-ICP-MS analysis of the same sediment samples. The Ataco study was undertaken in the field laboratory in El Salvador. Both studies *could have* been conducted *in situ*, with the handheld XRF instrument either mounted on a tripod or held by hand, but it turned out in both cases to be more convenient to extract samples in the field and to analyze them later in a more comfortable laboratory setting. Sampling for the Tlacuachero study was done on a roughly uniform grid across the "floor," whereas the Ataco study emphasized uniform vertical sampling in stratigraphic test excavations.

Methods

Slightly different sample-preparation procedures were used in the two studies. For the Tlacuachero floor samples, which were intended for LA-ICP-MS as well as XRF, preparation involved homogenization with mortar and pestle, infusion with an internal standard (indium), and consolidation with an XRF binder (a waxy material combined in a ratio of approximately 4:6 with the powder before pressing). In the Ataco study, sediments were dried, crushed, and poured into a sample holder supplied by the XRF manufacturer. A thin-film bottom was changed prior to every analysis, The second study thus quite closely approximates what would be obtained by handheld XRF of *in situ* sediments.

Analytical procedures used in the two studies took advantage of the flexibility of the Bruker Tracer III-V handheld XRF instrument with rhodium anode. This instrument allows variable X-ray tube settings and beam filtering in order to optimize signal-to-noise ratios for elements of interest. In the studies reported below, elements from magnesium to niobium were determined from Kα lines, with some Lα lines in this spectral region measured as well. Each sample was analyzed twice. Kα lines from magnesium to iron were determined with the high voltage set at 15 kV and the anode current at 15 μA, with no filter, and with the detector chamber evacuated to less than 20 Torr. Kα lines from iron to zirconium were determined using a filter consisting of thin foils of .006" copper, .001" titanium, and .012" aluminum, 40 kV high voltage, and 15 μA anode current. The Lα line for barium was measured on the first analysis, and the Lα lines for lead and thorium were measured on the second analysis.

Peak intensities were calibrated to elemental concentrations using a set of obsidians assembled by Mike Glascock at University of Missouri Research Reactor Center. The obsidians had been analyzed by neutron activation analysis, XRF, and ICP-MS. Net counts on which the calibrations were based were saved separately and are used in the first of the two following case studies.

Food processing at Tlacuachero, Chiapas, Mexico

Tlacuachero is one of several shell-midden sites in the mangrove estuary zone of Pacific coastal southern Chiapas, Mexico. The shell deposits, which date from the Middle and Late Archaic periods (approximately 3500 BCE – 2000 BCE) consist of bedded remains dominated by a single species of marsh clam, *Polymesoda radiata* (Voorhies 1976, 2004; Kennett and Voorhies 1996). Several prepared floors encountered within the shell deposits have remained enigmatic due to the near-total absence of macroscopic artifacts. As mentioned in the introduction, these features have been identified as "clay floors" since they were first excavated in the 1970s (Voorhies 1976).

In an effort to understand what activities might have been carried out on the floors, Voorhies recently exposed approximately 15% of the total estimated area of the uppermost floor and a smaller percentage of the underlying floor. Features visible on both floors include post molds, depressions thought to have been locations where rocks had been placed, and finger holes (Figure 11.1). A total of only two artifacts were found on the top floor, suggesting that it was cleaned off before new deposits of shell began to accumulate above it. The excavation recovered small bones and sediment samples from both floors, and ongoing analyses are focusing on plant microfossils (phytoliths), faunal identification, and elemental and isotopic characterization.

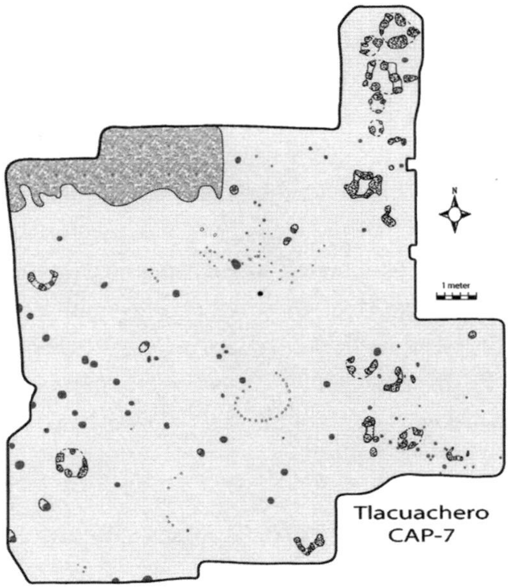

Figure 11.1:
Map of features on the excavated portion of the Tlacuachero floor. Small circles are finger molds; larger filled circles are post molds; irregular shapes are rock depressions.

XRF measurements on 103 samples from Floor 1 provide a basis for inferring what activities might have been carried out on the floor and how they might have varied in relation to the visible features. In particular, the hypothesis that plants and/or animals were processed for consumption on the floor can be tested by mapping variability in the concentrations of biologically important elements, such as phosphorus and trace metals (e.g., zinc, copper, iron, manganese). Phosphorus is a component of crucial biological molecules, including DNA, and the metals and other trace elements are involved in a variety of metabolic pathways and therefore are found in tissues of virtually all organisms. Phosphate analysis using wet-chemical techniques has long been used in archaeology for identifying anthropic soils (e.g., Sjoberg 1976), and rapid approaches phosphate survey have recently become developed (Rypkema *et al.* 2007); what handheld XRF adds is seven greater speed of analysis and the ability simultaneously to determine a range of biologically important elements in addition to phosphorus.

In order to evaluate the impact that particular biological tissues might have on the geochemistry of the floor, dried samples of fish of various sizes and shrimp were included in the analysis along with samples of shell and pozol (corn meal). A sample of mud from the mangrove forest floor adjacent to Tlacuachero was also included as a background reference sample.

The bulk chemistry of the floor sediments is dominated by calcium and phosphorus, both of which are far higher than the surrounding mangrove mud sample. Enriched calcium is expected considering that nearly pure calcite in the form of burned and unburned shell overlie and underlie the floors. Phosphorus is far higher than expected in shell, which provides a preliminary indication that the floor was, in fact, used to process plant or animal tissues for consumption. Overall, the floor-sediment bulk chemistry, as judged from the low-energy spectra collected on this project, resembles the bulk chemistry of bone, which is mostly hydroxylapatite (Figure 11.2). In light of previous characterizations of this feature as a "clay floor," this is a surprising result. Silicon and aluminum are so low in the floor that a significant clay-mineral component can be ruled out. It is highly unlikely that the feature is made of ground bone, however. Currently, the best interpretation is that it represents *in situ* formation of hydroxyapatite from the interaction of calcite from the shell with organic-rich materials, which released their phosphates.

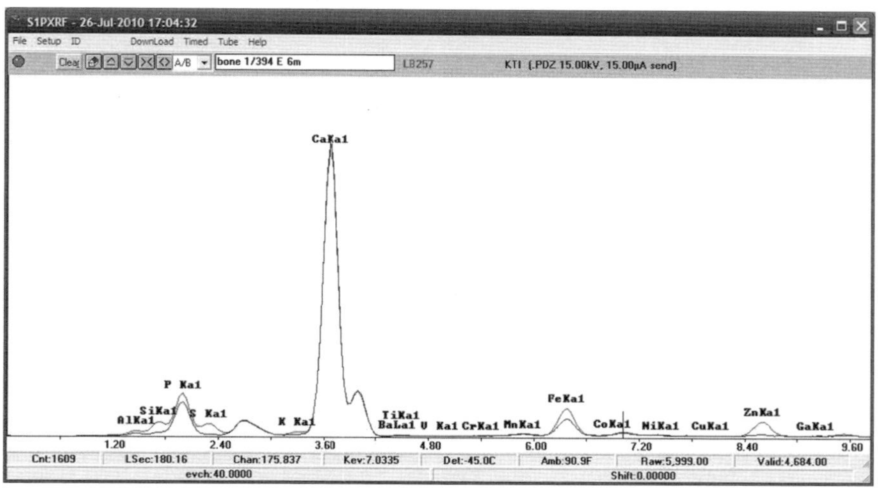

Figure 11.2:
XRF spectrum from bone overlain on one of the Tlacuachero floor samples. Spectrum was taken with accelerating voltage of 15 kV and 15 μA beam current, with no filter and with the detector chamber evacuated.

In many traditional analytical methods of soil analysis one would expect to acquire quantifiable concentrations of the elements present. This requires a sound calibration to be undertaken, but in many cases, the question being asked can be altered slightly so that raw data can be used and calibration is not necessary. Because a calibration suitable for calcium- and phosphorus-dominated matrices has not yet been developed, statistical analysis of the floor data was carried out on the net intensities data. The floor layer as a whole, as noted, is a relatively uniform matrix of diagenetically formed hydroxyapatite-like mineral, and thus there should be little matrix-related variation in senstivity for phosphorus. Future plans call for the development of calibrations for phosphorus and other low-energy elements using a series of clay reference materials mixed with bone ash in varying proportions.

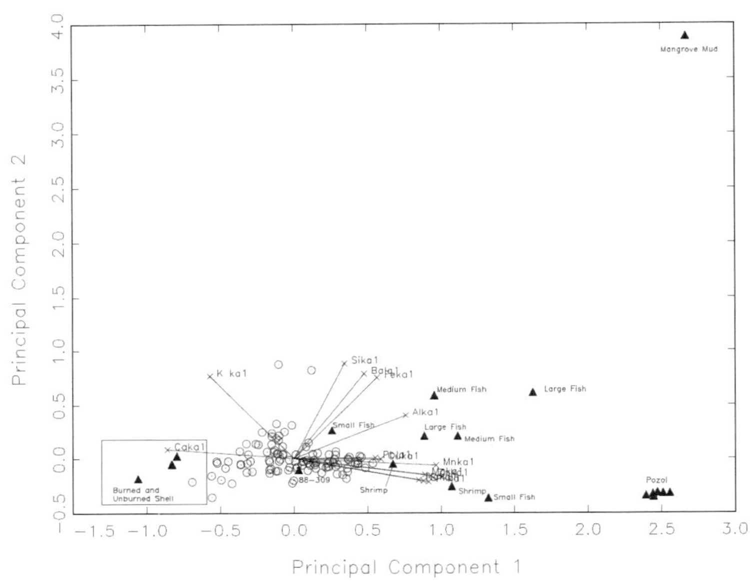

Figure 11.3:
Biplot derived from principal components analysis of the correlation matrix of net intensities data for Tlacuachero floor samples. Analytical data for samples of fish, shrimp, pozol, shell, and mangrove mud are projected into the PCA space. Vectors connect the origin with coordinates of the variables in order to illustrate patterns of elemental correlation and determinants of the object point scatter (Neff 1994).

Principal components analysis (PCA) was used to identify the major axes of elemental variation in the data. The PCA program was written by the senior author in the Gauss programming language (Aptech Systems) and is used routinely in artifact sourcing and other archaeometric studies (e.g., Neff 2002). As shown in Figure 11.3, Principal Component 1 subsumes a negative correlation between calcium on one hand and phosphorus, zinc, manganese, magnesium, and sodium on the other hand. PC2, meanwhile, expresses correlated enrichment of potassium, iron, aluminum, barium, and silicon. Projection of the shell, mangrove mud, and food items into the PCA space indicates that shrimp, fish, and pozol would all contribute to higher scores on PC1, whereas shell (near pure calcite or aragonite) contributes to lower scores. The mangrove mud, being much higher in silicon and aluminum differs dramatically from all floor samples and from the analyzed foods.

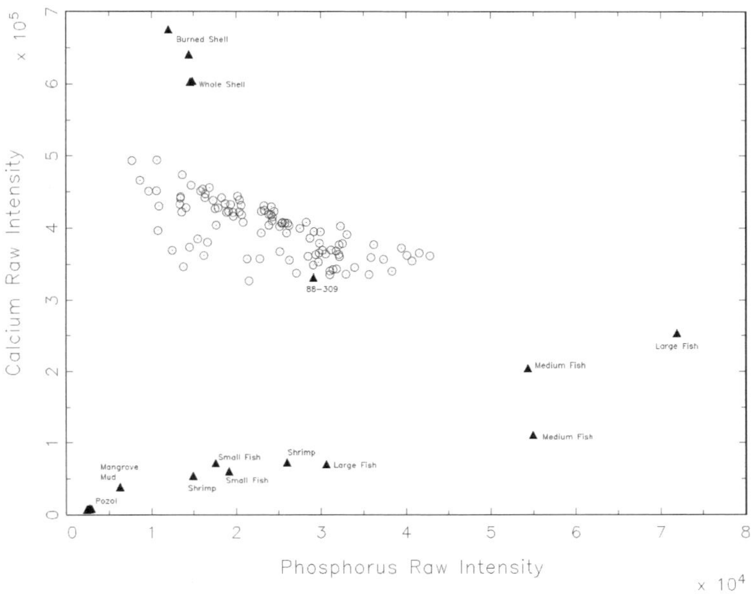

Figure 11.4:
Bivariate plot of phosphorus and calcium net intensities in Tlacuachero floor samples and samples of fish, shrimp, shell, pozol, and mangrove mud also analyzed by XRF.

Bivariate plots of the net intensities data further clarify the patterns of enrichment and dilution in the data. As detected by the PCA, calcium is negatively correlated with phosphorus and several other elements (Figure 11.4). This perspective also illustrates how addition of fish and shrimp tissues to the shell would create intermediate compositional profiles within the range of variation of the floor compositions. The pozol, however, is too low in phosphorus to be the sole food contributor to the floor's composition. Note too how low the mangrove mud samples are in both phosphorus and calcium.

Zinc levels in the floor are enriched relative to the shell matrix in which the floor is embedded (Figure 11.5). Moreover, this enrichment is plausibly due to processing of shrimp and/or fish on the floor, as the relative concentrations in the foods indicate. While processing maize (pozol) would enrich zinc, as noted before, the low phosphorus levels in maize exclude it from consideration as the sole source of elemental modification.

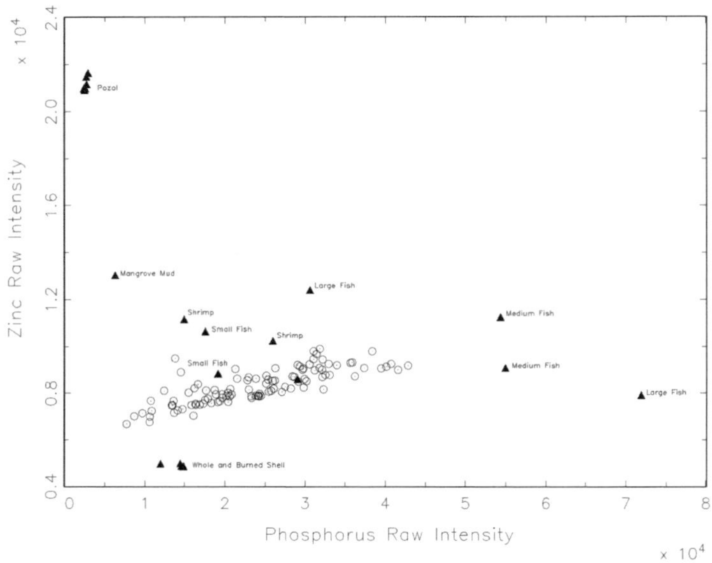

Figure 11.5:
Bivariate plot of phosphorus and zinc net intensities in Tlacuachero floor samples and samples of fish, shrimp, shell, pozol, and mangrove mud also analyzed by XRF.

In summary, PCA of the data from the Tlacuachero floor indicates that elemental variation is plausibly attributed to processing of food, especially seafood tissue. Processing of fish, shrimp, and probably other marine and estuarine organisms left behind biologically important elements like phosphorus and zinc, which combined with the calcite-rich material of the floor, probably forming a diagenetic hydroxyapatite with varying trace zinc levels. To reiterate, the overall floor composition (e.g., Figure 11.2) approximates bone, but is probably a hydroxyapatite formed *in situ* from the reaction of shell with organic materials contributed by food processing or other activities. While the possibility that plants, such as maize, were processed cannot be ruled out (and may help account for enriched zinc, for example), maize was certainly not the only food that contributed to the composition of the floor. Future analysis of phytoliths may shed additional light on what plants might have been in the environment during the period that the floor was in use.

Variation in elemental enrichment across the floor suggests systematic variation in the intensity of food processing activities. Scores on Principal Component 1, the food-processing dimension, are highest in the immediate vicinity of two clusters of finger holes and are also high in the area of rock emplacements (Figure 11.6; compare with Figure 11.1). This indicates enrichment of phosphorus (Figure 11.7), zinc (Figure 11.8), and other food-residue elements relative to calcium.

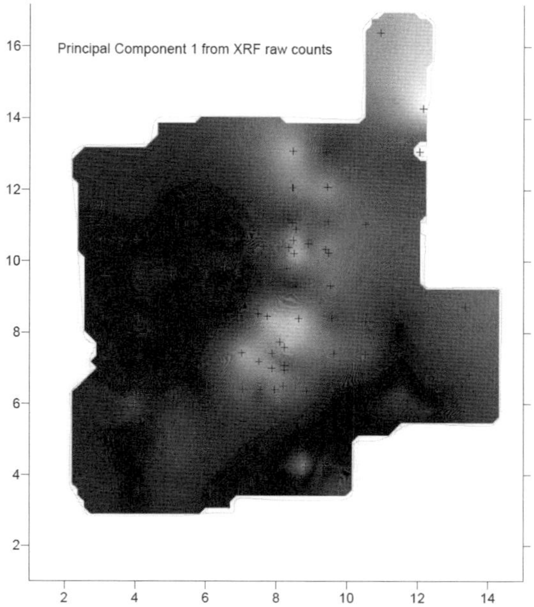

Figure 11.6:
Map of scores on Principal Component 1 across the Tlacuachero floor. Whites are high values; blacks are low values. Plusses indicate sampling locations.

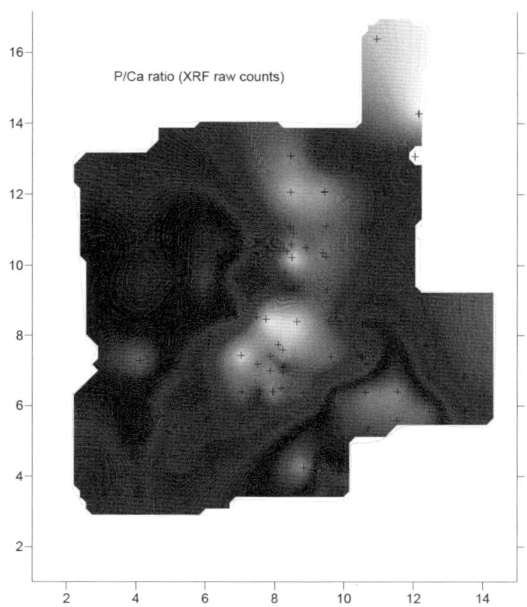

Figure 11.7:
Map of phosphorus/calcium ratios across the Tlacuachero floor. Whites are high values; blacks are low values. Plusses indicate sampling locations.

Future analysis of the Tlacuachero floor samples will compare the distribution of elemental indicators of food processing with the distribution of plant microfossil and faunal (bone) indicators. At this point, it is clear that the floor carries an elemental signature of food processing in the form of phosphorus, zinc, and other elemental concentrations that could not be derived from any conceivable mixture of shell, local mud, or plant residues. Thus, the geochemical data accord with the hypothesis that the Tlacuachero floor was devoted to processing tissues derived from marine and estuarine organisms.

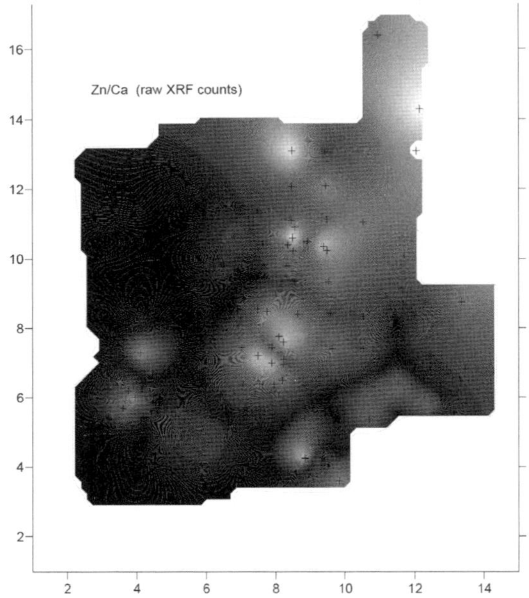

Figure 11.8:
Map of zinc/calcium ratios across the Tlacuachero floor. Whites are high values; blacks are low values. Plusses indicate sampling locations.

Geochemical evidence for sedimentary processes at Ataco, El Salvador

A major event in the recent geological and cultural history of El Salvador was the explosive eruption of Ilopango Volcano, an event now dated to approximately 429 CE (Dull *et al.* 2001). The eruption caused severe social disruption in the immediate vicinity of the volcano and deposited a layer of white, rhyolitic air-fall tephra across much of El Salvador and westward as far as the eastern Pacific coast of Guatemala (Sheets 2008). This layer is now called the *Tierra Blanca Joven* (TBJ) tephra.

At Ataco, in the western highlands of El Salvador, a white volcanic tuff layer thought to the TBJ tephra was encountered in different states of preservation at most locations excavated recently by Paredes. Questions about this ash layer that can be addressed via handheld XRF are (1) is it, in fact, the TBJ ash and (2) what natural or cultural processes account for the different states of preservation of the ash layer?

Sediment samples from Ataco were analyzed in the field lab in Concepción de Ataco by handheld XRF during March - April, 2010. A total of 77 samples of soil and volcanic ash samples was analyzed. The samples included intact samples of the volcanic ash layer at Ataco and, for comparison, samples of known TBJ tephra from a roadcut near San Salvador. Other samples of sediment included samples from underneath and above the tephra layer. Two series of samples were from short, bamboo cores driven through the ash layer and into the underlying sediments within excavation units. Two longer stratigraphic sequences from the sidewalls of excavation units (Units 4 and 30) were analyzed as well, and these proved particularly useful for stratigraphic interpretation.

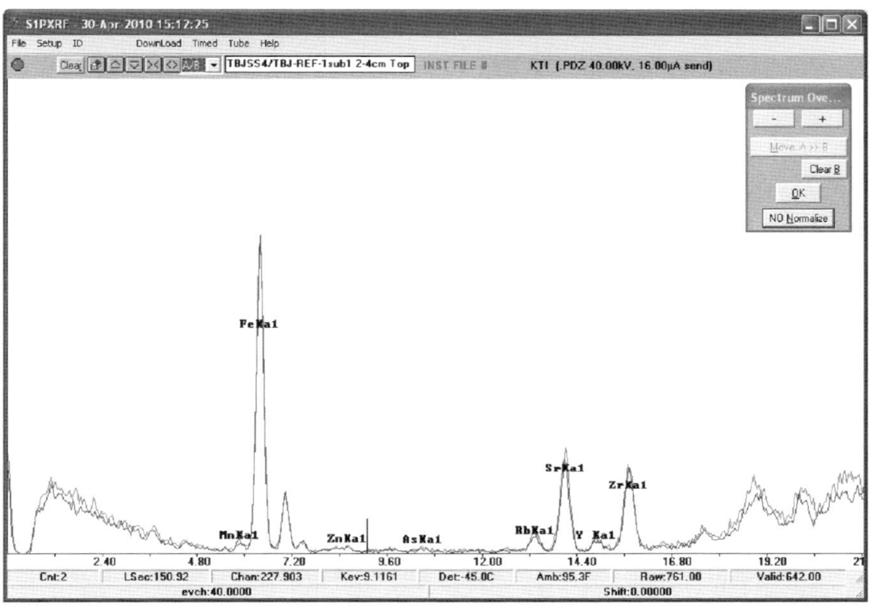

Figure 11.9:
Spectum from a sample of TBJ ash from a road cut near San Salvador overlain on spectrum of ash from tephra layer at Ataco. Spectra were taken with 40 kV accelerating voltage and 15 microamps beam current, with beam filtered by thin foils of copper (150 μm), aluminum (50 μm) and titanium (300 μm), and no vacuum.

The most intact volcanic ash samples from Ataco match the elemental profile of TBJ ash. This is clear from a comparison of a spectrum from TBJ to a spectrum from intact ash layers at Ataco (Figure 11.9). Bivariate plots of the elemental concentrations data also demonstrate that intact ash samples from Ataco fall within the range of variation of TBJ ash (Figure 11.10). It is worth noting, however, that even many intact Ataco ash samples diverge chemically from the pure ash, showing elevated titanium (Figure 11.10) and elevated aluminum together with lower silicon, characteristics that are indicative of incipient formation of clay within the ash layer. In sum, however, there is little doubt that the ash layer encountered in excavations at Ataco originated with the eruption of Ilopango volcano around 430 CE.

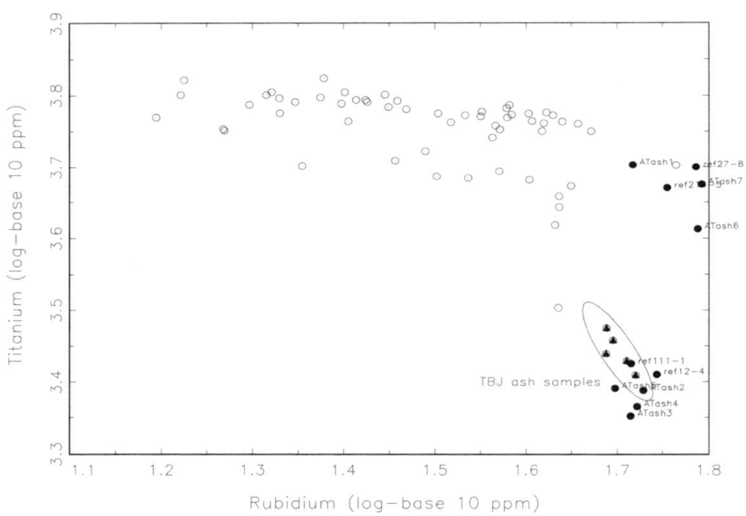

Figure 11.10:
Bivariate plot of rubidium and titanium log concentrations in all sediment samples from Ataco together with TBJ volcanic ash samples from near San Salvador. The ellipse represents 90% confidence level for membership in the TBJ ash group. Labeled samples are those identified in the field relatively unweathered samples of the Ataco ash layer.

In PCA, the first principal component is the longest axis through the data, the second principal component is orthogonal to the first and represents the second longest axis in the data, and so on. The principal components may be more readily interpretable than the raw elemental data, since the redundancy in raw elemental data is concentrated on single axes. Both the objects (ash and sediment samples) and the variables (elements) can be represented in the same principal component space, so that the correlated effects of specific elements on the overall configuration of the data points can be inferred (Neff 2002). In the Ataco sediment data, the first two principal components derived from a PCA of the correlation matrix of the sediment XRF data subsume almost 73% of total variance, while the first three components subsume over 81% of the total variance. Much of the variation in sediment chemistry thus can be summarized within a space of greatly reduced dimensionality (Figures 11.11 and 11.12). The first three axes also have fairly clear interpretations, as discussed in the following paragraphs.

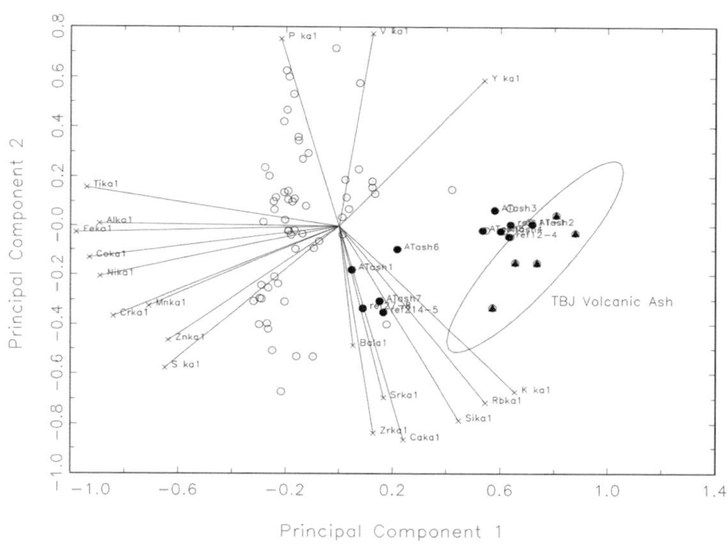

Figure 11.11:
Biplot derived from a principal components analysis of the correlation matrix of logged XRF data for TBJ ash samples and sediments from Ataco. Labeled samples are relatively intact ash samples from Ataco. Ellipse represents 90% confidence level for the TBJ volcanic ash group.

Figure 11.11 shows the ash and sediment samples plotted along with the variable (element) coordinates on the first two principal components of the data. The plot indicates that the first principal component is determined largely by correlated enrichment-dilution of iron, titanium, manganese, and other transition metals. Samples with high concentrations of transition metals have low scores on PC1, while those with relatively low concentrations have high scores. In effect, PC1 indicates the extent to which the TBJ ash compositional signature affects the sediments elemental profile: as the plot shows, the TBJ ash samples along with the intact ash samples from Ataco have extremely high values on PC1. The Ataco ash samples that do not cluster with the TBJ samples apparently are weathered and/or mixed with other sediment components. PC2, meanwhile, indicates enrichment of phosphorus and vanadium and dilution of several elements, including calcium, strontium, and zinc.

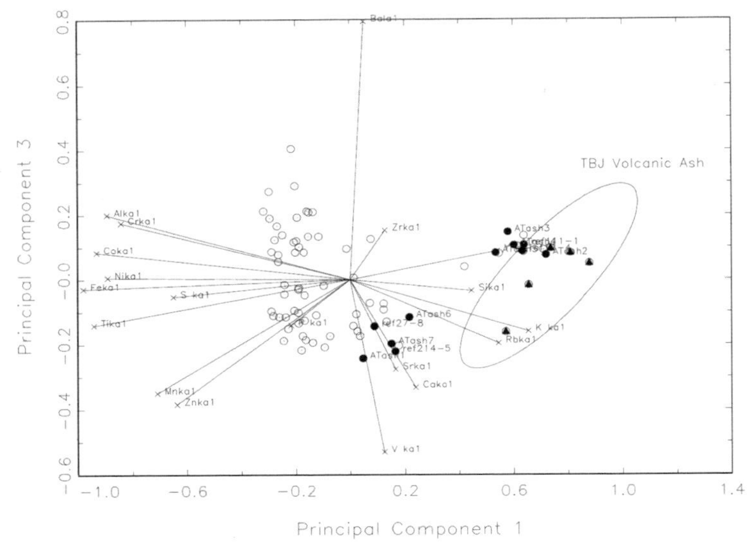

Figure 11.12:
Biplot of PC1 and PC3 derived from a principal components analysis of the correlation matrix of logged XRF data for TBJ ash samples and sediments from Ataco. Labeled samples are relatively intact ash samples from Ataco. Ellipse represents 90% confidence level for the TBJ volcanic ash group.

Figure 11.12 shows the object and variable coordinates on Principal Components 1 and 3. PC3 expresses enrichment of barium together with dilution of vanadium, calcium, and several other elements. It is worth noting too that the Ataco ash samples that do not cluster with the TBJ samples on PC1 also plot relatively low on PC3, indicating that these more-weathered ash samples are depleted in barium relative to the unweathered ash. This observation suggests that barium concentrations in sediments are determined in part by presence and weathering state of TBJ ash in the sediments. This geochemical effect of the TBJ ash, however, is independent of the main effect, which is expressed on PC1. The sampled sediments include two sequences taken at regular intervals within excavation units. These sequences provide a means to examine stratigraphic patterning in the XRF elemental data.

Figure 11.13 shows PC1 scores (x axis) plotted against depth below the present ground surface in Units 4 and 30. PC1 monitors contribution of TBJ ash to the sediment signature. These plots show the extent to which TBJ ash is incorporated into sediments at different depths. Sediments laid down before the eruption of Ilopango in 430 CE are, predictably, very low on PC1. This is clearest in the case of Unit 4, where the ash layer lies at a depth of about two meters, but the excavation penetrated deposits up to 4.5 meters in depth. The peak PC1 score at 2.15 meters depth coincides with the actual TBJ tephra layer in this unit (this is the unlabeled circle that lies close to the TBJ ellipse in Figures 11.11 and 11.12). Low scores at

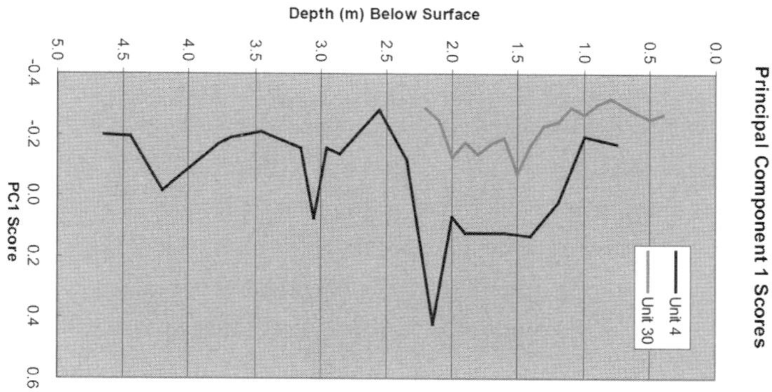

Figure 11.13:
PC1 scores of sediment samples from Units 4 and 30 plotted against depth from the ground surface.

the bottom of this unit show that, predictably, TBJ ash had no effect on sediment chemistry before the 430 CE eruption and that there was little downward mixing following the eruption. Above the ash layer, however, PC1 scores remain high for at least 0.5 meters, indicating that sediments above the TBJ tephra are affected by underlying ash deposits. Impact of the TBJ elemental signature on overlying sediments could be a result of erosion of ash from upslope, human disturbance, or other processes that would have mixed ash into soil and sediment layers that were developing after 430 CE. In Unit 30, the ash layer is not intact, but traces of it appear near the bottom of the unit; consistent with these observations, PC1 scores are high between about 2.0 and 1.5 meters depth, indicating some effect of TBJ ash on the sediments that accumulated after 430 CE. In more recent times, toward the top of Units 4 and 30, the TBJ ash has made a smaller contribution to sediment chemistry, as indicated by the low values of PC1 near the top of each excavation.

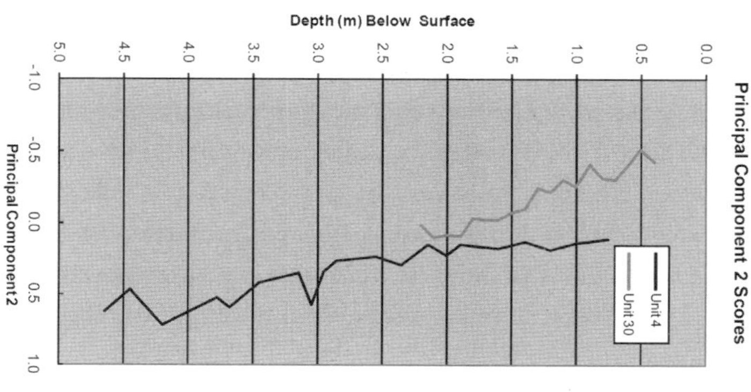

Figure 11.14:
PC2 scores of sediment samples from Units 4 and 30 plotted against depth from the ground surface.

PC2, which expresses enrichment of phosphorus, shows a very regular increase with depth in both Unit 4 and Unit 30 (Figure 11.14). Figure 11.15 confirms that the PC2 gradient is, in fact, a phosphorus gradient, with phosphorus increasing in concentration regularly with depth beneath the surface. A viable interpretation of these observations is that phosphorus, an important plant nutrient, is being taken up by the coffee trees and other plants at the surface, but that this impact on soil chemistry declines with depth, so that phosphorus concentration increases. The spike in phosphorus at around two meters in Unit 30 is approximately at the level where large incensario fragments were found, and it is possible that the enriched phosphorus at this location originates in organic remains discarded along with the incensario. Similarly, the spike in phosphorus to almost 1900 ppm at 3.05 m in Unit 4 may represent a stratum in which organic-rich garbage accumulated.

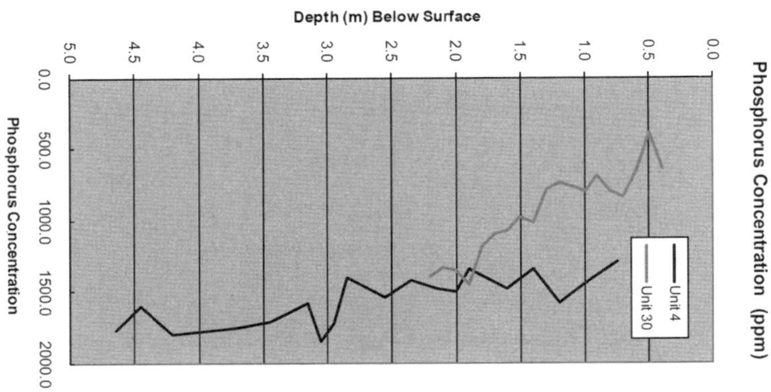

Figure 11.15:
Phosphorus concentrations in sediment samples from Units 4 and 30 plotted against depth from the ground surface.

Interestingly the sequence of samples from one of the bamboo cores driven through the volcanic ash layer shows a similar increase in phosphorus content with depth (Figure 11.16). This confirms that the surface immediately underlying the TBJ ash was stable and had plants growing on it, depleting soil phosphorus, at the time of the Ilopango eruption. Just as interestingly, the other bamboo core is even more depleted in phosphorus below the TBJ layer, with most measurements below detection. This may indicate the extent to which plant growth had depleted soil phosphorus immediately before the TBJ eruption. Alternatively, since the samples from below the ash at this location are highly enriched in aluminum and iron, they may represent clay-rich material quarried and brought in for construction or surfacing of a public space.

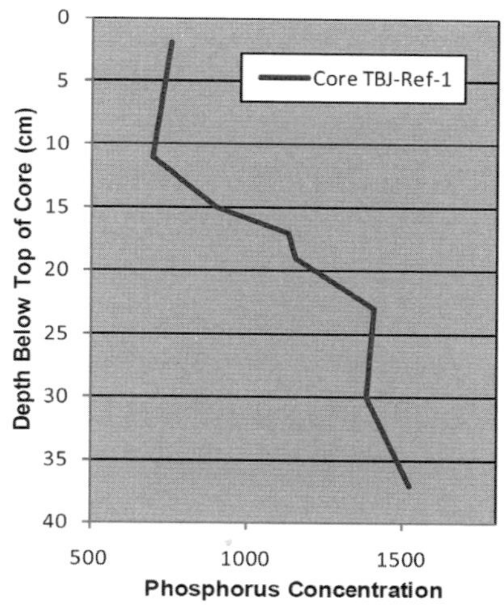

Figure 11.16:
Phosphorus concentration vs depth below top of core in TBJ-Ref-1. Uppermost 10 cm are relatively intact TBJ ash.

Finally, PC3 scores (Figure 11.17) show a pattern of low scores in levels above the TBJ ash. In Unit 30, PC3 scores are highest above about 1.3 m, which is where the effect of TBJ ash (based on PC1 scores) begins to disappear from the sediment elemental signature. Similarly, PC3 scores are low above 2 m in Unit 4, but they increase below about 2 m, i.e., below the TBJ ash layer, before dropping slightly and then leveling out at about 3 m. Figure 11.18 confirms that the observed trends in PC3 are in large part due to barium concentration in the sediments. Based on Figure 11.18, barium enrichment appears to extend from just below the TBJ ash (about 2 m) down approximately 0.8 m to 1.0 m, to approximately 3.0 m. It is also worth noting that barium increases at the very bottom of Unit 30, below the ash, to concentrations very close to those observed below the TBJ ash in Unit 4.

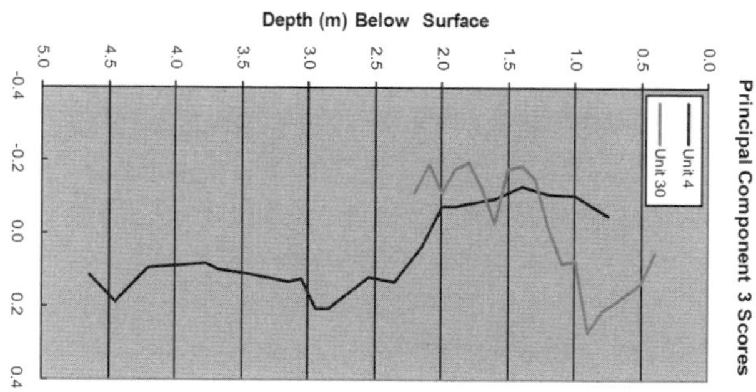

Figure 11.17:
PC3 scores in sediment samples from Units 4 and 30 plotted against depth from the ground surface.

Barium enrichment in the layers below the TBJ ash appears to be due to mobilization and leaching of barium from the volcanic glass and its uptake in underlying sediments. In accord with this interpretation, the more weathered (less intact) ash samples from Ataco are depleted in barium relative to the TBJ samples from near San Salvador and relative to the most intact Ataco ash samples (Figure 11.12), which indicates that the weathering process involves leaching of barium.

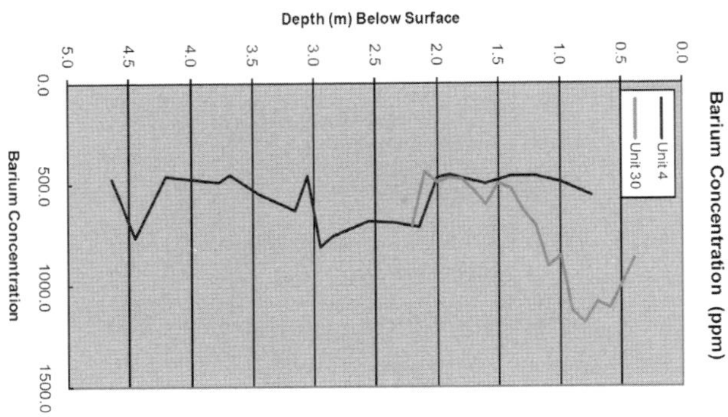

Figure 11.18:
Barium concentrations in Units 4 and 30 plotted against depth below the ground surface.

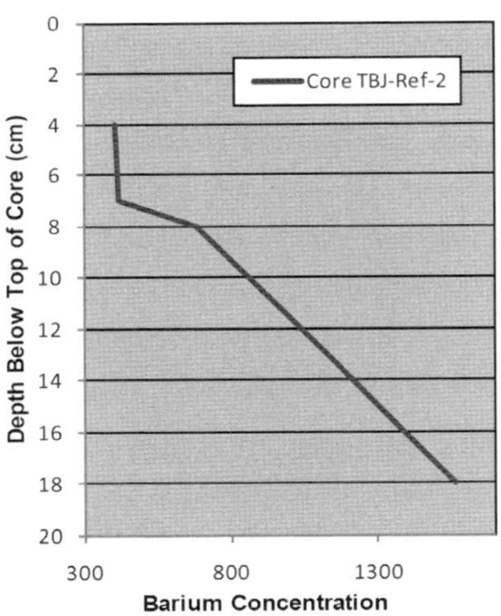

Figure 11.19:
Barium concentrations in samples from bamboo core #2 (TBJ-Ref-2) plotted against depth from the top of the core.

Bamboo core #2 (TBJ-Ref-2) provides direct evidence that mobilization of barium affects the composition of TBJ ash and adjacent sediments within the soil environment at Ataco (Figure 11.19). The topmost sample in this core, which is otherwise quite similar to TBJ ash, has a barium concentration of only 403 ppm (versus an average of 747 +/- 47 in the five TBJ ash samples). Barium then increases with depth below the intact ash, to almost 1600 ppm at 18 – 20 cm.

In summary, analysis of TBJ ash samples together with relatively intact ash, weathered ash, and sediments from Ataco indicates that the ash layer at Ataco is almost certainly a remnant layer produced originally by the eruption of Ilopango Volcano in 430 CE. The ash fell on a relatively stable surface that in some places supported growing plants that were depleting soil phosphorus near the surface, just as do the modern coffee trees; in some places, this 430 CE surface had so little phosphorus as to suggest that it may have been an imported construction material nearly devoid of organic matter. After the eruption, natural processes, such as erosion of ash from mounds and animal and human disturbance, mixed TBJ ash into the sediments that were accumulating on top of the TBJ ash. Little downward mixing of TBJ ash appears to have occurred, but weathering of the ash mobilized barium, which was fixed by clays and other components in the sediments just below the ash, such that an enriched-barium layer underlies the TBJ layer in some locations.

Conclusion

Many characteristics of the archaeological record, often the most informative ones, are invisible to the naked eye and can only be perceived or measured with instrumental assistance (Weiner 2010). In the past, the need for instrumental measurement has limited investigation of sub-visible portions of the record to laboratory settings, where the requisite analytical instrumentation could be kept in favorable conditions and maintained. Thin-section microscopy, elemental analysis by a variety of techniques, pollen and phytolith analysis, residue analysis, and many other kinds of laboratory-based studies have made crucial contributions to our understanding of the micro-record, but micro-properties of the largest portion of archaeological deposits – the sediments in which artifacts and ecofacts are embedded – have often been ignored. On most projects, this "back dirt" is tossed aside after a few cursory observations of color and texture. Only when specialists have been brought in to sample for later laboratory analysis have archaeological sediments yielded up some of the microscopic-level information they contain (e.g., Hutson and Terry 2006; Berna *et al.* 2007; Dahlin *et al.* 2007; Wells *et al.* 2007; Wells and Terry 2007; Goldberg and Berna 2010).

Portable analytical instruments, such as handheld XRF spectrometers, reduce the cost of analysis by eliminating most sample preparation, eliminating the need for packaging, cataloguing, and transporting samples, and in principle even eliminating the need to *take* samples (if analyses are made *in situ*). With reduced analytical costs, it becomes feasible to subject the sedimentary matrix to the kinds of analyses usually reserved for fancy pottery, obsidian, metals, and other artifacts encountered during excavation. The examples discussed here are intended to illustrate the kinds of information that such fieldanalytical efforts may yield.

The Tlacuachero floor, a feature that for 35 years had been interpreted as an intentionally constructed clay floor, now appears, as a result of analysis, to be powdered and packed hydroxyapatite formed by the interaction of calcite from shell and phosphates liberated from food residues and other waste products deposited on the surface. Mapping of biologically important elements across the floor documents variation in the intensity of these food-processing and waste-deposition activities.

In the Ataco example, elemental analysis confirmed that a volcanic ash layer encountered during excavation was the TBJ tephra produced by the major eruption of Ilopango Volcano during the first millennium CE. Additionally, the analysis suggested new interpretations of the surfaces on which the ash fell as well as some of the post-eruption processes that modified the vertical and horizontal distribution of the ash and elemental components of the ash.

Although neither example discussed here actually involved analysis at the site of excavation, the instrument used – a Bruker Tracer III-V – is fully compatible with use in the field. It is hand held and battery powered, and could have been used to analyze the Tlacuachero floor on a grid placed over the floor without removing samples. Similarly, at Ataco, the instrument could have been used to characterize sediments down the profiles without sample removal. In-field analysis would have been faster and would have generated many more analyses, but at the cost of some additional noise introduced by uneven surfaces, particle size variation (due to elimination of grinding), variable moisture content, and other conditions. Such tradeoffs will have to be addressed on a case-by-case basis, but one promising compromise model is one in which a well-equipped field lab is set up close to the site of fieldwork, so that samples can be removed and analyzed in "real time" and the results fed back immediately into the design of excavation work (Weiner 2010: Chapter 11). Such a lab could house not only handheld XRF but also portable FTIR, facilities for preparing micromorphology samples (e.g., Goldberg and Berna 2010), and other portable analytical instruments and facilities. But whether portable analytical instruments are deployed on site or near the site, the spread of analytical technology out of the laboratory certainly represents an exciting trend in archaeology, one that promises to have profound effects both on archaeological methods and on understanding of the past over the coming decades.

References

Berna, F., B. Behar, R. Shahack-Gross, J. Berg, E. Boaretto, A. Gilboa, H. Sharon, S. Shalev, S. Shilstein, N. Yahalom-Mack, J. Zorn, and S. Weiner. 2007. Sediments exposed to high temperatures: reconstructing pyrotechnological processes in Late Bronze and Iron Age strata at Tel Dor (Israel). *Journal of Archaeological Science* 34:358-373.

Dahlin, Bruce H., Christopher T. Jensen, Richard E. Terry, David R. Wright, and Timothy Beach. 2007. In search of an ancient Maya market. *Latin American Antiquity* 18:363-384.

Dull, Robert A., John R. Southon, and Payson Sheets. 2001. Volcanism, ecology and culture: A reassessment of the Volcan Ilopango TBJ eruption of the southern Maya realm. *Latin American Antiquity* 12 (1):25-44.

Goldberg, P., and F. Berna. 2010. Micromorphology and context. *Quaternary International* 214:56-62.

Hutson, Scott R., and Richard E. Terry. 2006. Recovering social and cultural dynamics from plaster floors: chemical analyses of ancient Chunchucmil, Yucatan, Mexico. *Journal of Archaeological Science* 33 (3):391-404.

Kennett, D. J., and B. Voorhies. 1996. Oxygen Isotopic Analysis of Archaeological Shells to Detect Seasonal Use of Wetlands on the Southern Pacific Coast of Mexico. *Journal of Archaeological Science* 23:689-704.

Neff, Hector. 2002. Quantitative techniques for analyzing ceramic compositional data. In *Ceramic Production and Circulation in the Greater Southwest: Source Determination by INAA and Complementary Mineralogical Investigations*, edited by D. M. Glowacki and H. Neff. Monograph 44. Los Angeles: The Cotsen Institute of Archaeology at UCLA.

Rypkema, Heather A., Wayne E. Lee, Michael L. Galaty, and Jonathan Haws. 2007. Rapid, in-stride soil phosphate measurement in archaeological survey: a new method tested in Loudown County, Virginia. *Journal of Archaeological Science* 34:1859-1867.

Sjoberg, A. 1976. Phosphate analysis of anthropic soils. *Journal of Field Archaeology* 3:447-454.

Voorhies, Barbara. 1976. *The Chantuto People: An Archaic Period Society of the Chiapas Littoral, Mexico*. Papers of the New World Archaeological Foundation no. 41. Provo: New World Archaeological Foundation.

———. 2004. *Coastal Collectors in the Holocene: The Chantuto People of Southwest Mexico*. Gainesville, FL: University Press of Florida.

Weiner, S. 2010. *Microarchaeology: Beyond the Visible Archaeological Record*. Cambridge: Cambridge University Press.

Wells, E. C., C. Novotny, and J. R. Hawken. 2007. Quantitative modeling of soil chemical data from inductively coupled plasma—optical emission spectroscopy reveals evidence for cooking and eating in ancient Mesoamerican plazas. In *Archaeological Chemistry: Analytical Techniques and Archaeological Interpretation*, edited by M. D. Glascock, R. J. Speakman and R. S. Popelka-Filcoff. Washington, DC: American Chemical Society.

Wells, E. C., and R. E. Terry. 2007. Introduction to the special issue: Advances in geoarchaeological approaches to anthrosol chemistry, Part II: Activity area analysis. *Geoarchaeology: An International Journal* 22 (4):387-390.

Chapter 12

X-Ray fluorescence of obsidian: approaches to calibration and the analysis of small samples

Jeffrey R. Ferguson

Introduction

For decades, archaeologists have used a number of techniques to determine the compositions of geological sources of chemically homogenous materials and then attempted to match artifacts to sources to understand trade and exchange of material objects. Such provenance research is common with glassy rhyolites (obsidian) that were used wherever available as a source of raw material for flaked stone artifacts such as projectile points and cutting tools. Handheld X-ray fluorescence, of all of the analytical techniques available for obsidian compositional analysis, has the potential to make the greatest impact by combining non-destructive analysis with rapid results, relatively low equipment and analysis cost, and the option of in-field analysis. While handheld XRF is a powerful research tool, its successful use in obsidian provenance research requires an understanding of X-ray physics, igneous petrology, the calibration process, and the ability to test a sufficient variety of homogeneous and well-characterized reference materials suitable for developing a valid calibration curve. The potential for "point and shoot" handheld XRF is hampered in part by the all of these factors.

This chapter addresses precision requirements of a matrix-specific calibration, the choice and analysis of reference standards, spectral normalization to account for differences in sample geometry and size, and the use of secondary targets to fine tune the spectra to maximize sensitivity to the elements of interest. While the focus here is on obsidian provenance research, many of the same basic principles apply to almost all applications of XRF.

Compositional studies of artifacts have been an integral part of archaeological scientific investigations for decades, providing insight into many aspects of past human behavior, and obsidian sourcing studies are often presented as the "poster child" of adherence to the provenance postulate (Weigand, *et al*. 1977; Neff 1998). The requirements for materials to permit successful provenance studies as stated in the provenance postulate are that measurable intersource variability must exceed intrasource variability. Obsidian is a nearly ideal material for provenance studies because it occurs in a relatively limited number of geologic contexts and typically exhibits uniform chemistry within particular outcrops. There are numerous problematic cases of sources with complex and even overlapping chemistry (Braswell and Glascock 1998; Glascock *et al.* 1999; Ambroz *et al.* 2001), but in general obsidian studies are extremely successful at linking artifacts with geologic sources wherever it was used in the past.

Obsidian is a glassy rhyolitic rock, meaning that it has high silica content and a disordered atomic structure (Shackley 2005). The disordered atomic structure allows obsidian to fracture predictably and with extremely sharp edges. Obsidians are formed only in specific volcanic events when magma of the right chemistry is cooled under specific conditions making it a remarkably rare material. The fracturing ability and limited natural outcrops made obsidian a valuable resource that in many cases was carried or exchanged across great distances.

Provenance studies are particularly appealing to archaeologists because they can document migrations, logistical movements, exchange relationships, and tool production, use, and discard in a spatial context (Duff 1999; Shackley 2002; Carter *et al.* 2007; Eerkens *et al.* 2007; Ferguson *et al.* 2010). Obsidian has the additional benefit of potentially providing direct temporal information through the measurement of water absorption; a process known as obsidian hydration (Friedman and Trembour 1978, 1983; Origer 1989). When coupled with temporal information acquired either through obsidian hydration analysis or by other means, such temporally-sensitive artifact types, stratigraphic sequences, or direct association with dated materials, it is possible to examine changes in obsidian procurement patterns over time spans ranging from decades to hundreds of thousands of years. No doubt the reasonable cost of obsidian sourcing relative to the valuable cultural information obtained has entrenched obsidian provenance in the world of contract archaeology, driving further research in the development of analytical techniques, geologic characterization, and the cultural interpretation of sourcing results.

This chapter is aimed at providing a technical background for the analysis of obsidian using XRF, with a particular focus on proper calibration techniques and the analysis of small artifacts. This chapter is included in a volume focusing on handheld XRF because handheld XRF users tend to be most likely to underestimate the complexity of proper analysis. The analytical issues presented here are common to both handheld and lab-based XRF. This paper is not intended to discuss the geologic context of obsidian in detail, proper methods of source sampling and characterization, or the anthropological implications of obsidian provenance studies. There are a number of excellent publications on these topics, starting with volumes by Shackley (1998a, 2005) and edited volumes by Glascock and others (Taylor 1976; Hughes 1984, 1989; Glascock 2002; Glascock *et al.*

2007) and these volumes contain references to many of the hundreds of relevant publications.

Analytical techniques for obsidian

Unlike many other geologically-derived materials typically recovered from archaeological sites, obsidian sources tend to follow the requirements of the provenance postulate for major, minor, and trace elements, permitting the successful application of a variety of analytical techniques. Specific details of the analytical techniques are readily found in existing publications (i.e. Shackley 1998b) so the discussion here is limited to the differences between the techniques and their advantages and limitations as they relate to obsidian analysis.

Visual sourcing is occasionally put forth as a more cost-effective method for examining large assemblages in contexts where all geologic sources are well known (Weisler and Clague 1998), however the reliability of visual sourcing is highly questionable (Bettinger *et al.* 1984; Braswell *et al.* 2000). On more than one occasion I have seen archaeologists working in the American Southwest get excited after finding an artifact with a green translucent color, a feature characteristic of the Pachuca source in Central Mexico. In one case it turned out to be from Pachuca, comprising one of the handful of Mexican obsidian artifacts recovered from north of the Mexican border (except for a small cluster in the southern tip of Texas [Hester *et al.* 1991; Ferguson and Skinner 2006]). In the second, the artifact was from the Antelope Wells source in southwestern New Mexico that is somewhat similar to the Pachuca source as it also very high in zirconium and has a green translucent color. While there may be a few cases of material so visually unique as to allow some correct visual source assignments, there is an error rate most researchers are not comfortable accepting.

Neutron activation analysis (NAA) was one of the first techniques used for the chemical characterization of archaeological obsidian (Gordus *et al.* 1968; Glascock *et al.* 1994; 1998). NAA allows for the precise determination of a wide range of elements that can vary depending on the nuclear properties of the elements, irradiation conditions, and timing of the gamma count(s), but it is an expensive technique and typically requires the destructive analysis of at least some portion of the artifacts (although, as described later, there are options for non-destructive NAA of small samples). NAA also requires considerable sample preparation time and cost in the form of irradiation and the use of high-purity sample containers.

Another common technique is inductively coupled plasma mass spectrometry (ICP-MS), often incorporating a laser (LA-ICP-MS) to ablate, or vaporize, an extremely small portion of the artifact for analysis in the ICP-MS. LA-ICP-MS can provide data on a wide range of elemental compositions, comparable to NAA, but can suffer from short and long term difficulties with analytical accuracy due to difficulties with consistent calibration (Speakman *et al.* 2007). LA-ICP-MS is often touted as a non-destructive analytical method, but almost every instrument in use today requires the removal of a portion of a sample in order to fit inside the laser ablation chamber.

Techniques with limited application to the compositional analysis of obsidian include proton induced X-ray emission (PIXE; Pollard *et al.* 2007; Bellot-Gurlet *et al.* 2008) and electron microprobe analysis (EPMA; Merrick and Brown 1984; Pollard *et al.* 2007). These techniques often involve the same basic methods as XRF, but they vary in the means of removing the electron from the inner shell of the atom in order to create the emitted X-ray as a result of outer shell electrons dropping down to fill the void. Many of the same issues discussed here for XRF apply to these techniques as well. Additional analytical techniques include, but are not limited to: natural radioactivity (Leach *et al.* 1978), obsidian density (Reeves and Armitage 1973), fission track analysis (Bigazzi *et al.* 1998), and numerous chemical methods (Tykot 2004).

XRF offers the potential for totally non-destructive analysis, portable and handheld operation, minimal or no sample preparation, rapid results, widely available analytical instrumentation, low cost, and quantitative analysis of some of the most important elements used for discriminating obsidian sources and providing provenance data. All of the other techniques described above require large stationary equipment – ranging from the size of a chest freezer to a nuclear reactor – and require the sample to fit the space requirements of an instrument sample chamber rather than introducing a small instrument to the object of interest as is possible with handheld XRF. Most of the commercial obsidian provenance laboratories employ non-portable energy-dispersive (ED-XRF) as their main instruments, yet portable instruments can provide high quality data in the lab while also allowing in-field analysis including obsidian artifacts in countries or curation facilities that will not allow their removal of artifacts. While the decreasing cost, increasing availability, and portability have all contributed to making XRF a more attractive analytical technique to a broader spectrum of researchers across many disciplines, the fundamental physical principles behind the technique and the need to understand them have not changed. The following section describes some of the most important issues in handheld XRF analysis of obsidian, with particular attention to some of the issues encountered in the research at the Archaeometry Laboratory at the University of Missouri Research Reactor (MURR).

XRF methodology

One of the first considerations in XRF analysis is whether or not quantitative analysis is necessary, or even possible, with the sample set under investigation. While qualitative comparative spectral overlay can be used to assign individual obsidian artifacts to particular sources, it is not the most efficient means of processing large numbers of samples and can lead to difficulties in discriminating geologic sources with similar elemental ratios (i.e. the shape of the spectra are similar but the magnitudes differ). Without the ability to extract quantitative concentrations for certain elements, it is not possible to compare new analyses with previously published compositional data, limiting comparisons to those samples previously analyzed using the same equipment. Thus, quantitative analysis is necessary in obsidian studies. In contrast, qualitative analysis can be highly useful in, for example, attempting to determine if some large pit features were used to smelt iron or to fire limestone to produce lime. In this case qualitative spectral overlay comparing areas inside the features to areas of the surrounding natural soil showed significantly higher calcium and strontium levels in deposits at the bottom of the features compared to outside the features and no difference in the amount of iron. Additional tests may be warranted, but it seems likely that the features were used as lime kilns strictly based on spectral comparisons (Ferguson 2010).

The main reason for not performing quantitative analysis on the pit features described above is I have not yet developed a comprehensive calibration for the analysis of soils, and the difficulties in properly quantifying such a heterogeneous material in the field make the process difficult at best. The "point and shoot" claims by some instrument manufacturers have led to misleading data being published without a proper matrix-specific calibration (i.e. Morgenstein and Redmount 2005; Goren *et al.* 2011). A single calibration is simply incapable of such a broad range of different matrices, and matrix-specific calibrations are required for accurate quantitative analysis due to a number of factors, including absorption properties of the matrix, secondary fluorescence, and different concentration ranges for specific elements.

Developing accurate quantitative approaches for obsidian analysis allows the rapid (although not fool-proof) assignment of geologic sources to artifacts using a variety of multivariate statistical methods, something not possible with qualitative spectral overlay. Proper calibration is the key to efficient and accurate source assignment. MURR has a long history of obsidian analysis by NAA and consequently a massive source library including almost every known obsidian source in the world. The MURR handheld XRF generally uses a calibration based on compatibility with NAA data in order to maximize the potential to find sources not anticipated in a particular study, and thus not yet fully characterized by XRF. The most common deviations between MURR XRF data and that produced at other labs using only XRF are in the iron and zirconium concentrations. Unfortunately NAA is not as capable of producing highly accurate zirconium concentrations due to the production of both ^{95}Zr and ^{97}Zr during ^{235}U fission. It is possible to correct for the uranium interference, but the correction has a large uncertainty (Glascock *et al.* 1986).

The differences between the iron concentrations reported by the Berkeley XRF Laboratory and MURR are more complicated to explain, but Mike Glascock (pers. com. 2011) provides a possible explanation. The published values for iron in many USGS standards used by some XRF labs for calibration are generally listed as total iron. USGS standards, as a historical convention resulting from the use of gravometric techniques, report concentrations for multiple iron oxides. It is possible to convert the weight percent into ppm concentrations, but there is often the false assumption that all of the iron exists as the single oxide reported. This can result in inaccurate iron concentrations entered into the calibration standards. NAA iron concentrations are based on a single NIST standard (fly ash) and the concentration in the standard is not reported as an oxide, but as total iron. NAA calibration is linear and only requires a single standard for iron. The MURR XRF calibration is based on the data derived from NAA measurements of the 40 calibration standards and thus is not susceptible to the oxide weight percent conversion problem.

Obsidian XRF calibration

Proper calibration is not a simple process. It is important to have a broad range of concentrations in one's standards that cover the full range of concentrations anticipated in the unknown samples. This was made clear with the initial analysis of obsidians from the Central Rift Valley of Kenya. The initial calibration included samples from 40 well-known obsidian sources in the Americas[1]. For these obsidians, iron rarely exceeds 6-7 percent, zirconium is rarely above 1,000 ppm, and niobium is rarely over 200 ppm. It was clear from the spectral overlay that many of the Kenyan samples had abnormally high concentrations of these three elements, yet the concentrations calculated by the calibration produced lower concentrations as the peak areas increased. Bruce Kaiser (personal communication 2009) has termed this phenomenon the "wet noodle effect" in which peak areas for particular elements that exceed the range of the standards used in the calibration can trail off in extreme directions, sometimes leading to inverse relationships between peak areas and concentrations.

The data from the 2009 Kenya source samples (n=141, collected by Stanley Ambrose) serve as an example of the need to include the complete range of variability for each element. Figure 12.1 is a plot of the NAA-determined concentrations for iron (X-axis) and the XRF-determined concentrations for iron (Y-axis) for both the original and extended calibrations. As noted above, the original calibration produced decreasing iron concentration for larger peaks once the concentrations exceeded the range in the calibration. The problem was corrected by including a couple Kenya obsidian samples with high iron, zirconium, and niobium concentrations, previously analyzed by NAA. The corrected calibration produces XRF results comparable to NAA.

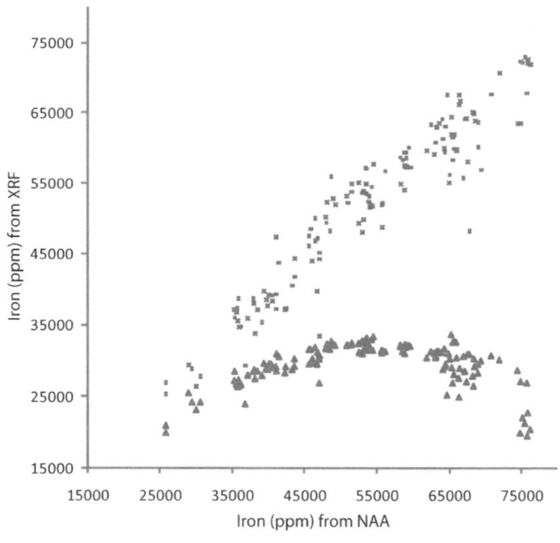

Figure 12.1:
Comparison of the initial and extended MURR XRF calibrations for iron. The lower curve (triangles) are the initial concentrations reported for a group of 141 source samples from Kenya. The upper linear spread ("x" symbols) are the same analyses recalculated using the extended calibration. The Y-axis values are the XRF concentrations for iron and the X-axis values are concentrations determined by NAA.

The current MURR calibration uses either 40 or 50 different obsidian secondary standards previously characterized by NAA at MURR (the main calibration uses only 40 standards, and the calibration for high iron, zirconium, and/or niobium samples uses an additional 10 standards, this is explained in greater detail below). The calibration produces concentrations for 19 elements (Na, Al, K, Ba, Ti, Mn,

Fe, Co, Ni, Cu, Zn, Ga, Pb, Th, Rb, Sr, Y, Zr, and Nb). Although not all these elements are relevant to the overall composition of obsidian, they are included to improve the overall calibration. Only six elements are routinely used in obsidian provenance studies. These six elements and their concentration ranges are listed in Table 12.1. It is necessary to not only bracket the expected range of concentrations, but to have uncorrelated variation between the elements in the suite of standards. This multi-elemental variability helps the calibration algorithms account for peak interferences and secondary fluorescence (especially for the elements Rb-Nb where the spacing of the peaks allows for the Kβ peaks from a particular element to interfere with the Kα peaks of the element two units greater in atomic number). It is also important to test calibrations using samples of known composition that are not included in the calibration (Mauser 2006).

	Minimum (ppm)	Maximum (ppm)
Fe	3554	69319
Rb	11	481
Sr	0	488
Y	12	662
Zr	61	3455
Nb	4	372

Table 12.1:
Minimum and maximum concentrations for standards used in extended MURR XRF calibration.

Unfortunately, concentrations for yttrium and niobium are not available using NAA, and NAA has a relatively high detection limit (approximately 50ppm) for strontium. Standard values for yttrium, niobium, and low strontium concentrations were averaged from the ED-XRF analysis of the same standards at two other obsidian studies laboratories: Northwest Research Obsidian Studies Laboratory in Corvallis Oregon, and the Berkeley Archaeological XRF Lab at the University of California, Berkeley. Methods for analysis and calibration used by these laboratories are available through their respective web sites (Northwest Research Obsidian Studies Laboratory 2011; Shackley 2011a).

Shackley (2010) recently commented on the need to use certified international standards in calibration. This is certainly an ideal goal that would help to increase inter-laboratory compatibility, but there are a number of problems with this approach. First of all, most certified standards are available only as powders. While the powders can be pressed into pellets there is the potential for analytical differences when analyzing solid glass samples using a calibration based on pressed powders. Empirical tests of this powder versus solid sample difference have shown minimal, if any, effect. Shackley (2011b) has analyzed both solid and pressed pellets from the single boulder used by USGS to create the RGM-1 standard and found very little difference. A recent test at MURR examined a polished solid piece of obsidian from the Whitewater Ridge source in Oregon and compared the spectra to loose powder and a pressed pellet from the same rock. There was a reduction in the backscatter for the powder and pellet samples, but there was no apparent difference in the peaks of the main elements of interest (Rb-Nb). Others have observed more significant differences as a function of grain size of powdered samples. Liritzis and Zacharias (2011) observed differences between solid and powdered obsidians ranging from up to 5% for K, Ca, Sr, Zr, and Rb, to 20 and 25% for Fe and Ti respectively. The different matrix effects of different types of geologic materials (i.e. basalt versus rhyolite) may be more problematic but are not yet fully studied.

Second, there are very few certified standards for obsidian or compositionally similar materials. The inter-elemental interactions in XRF are incredibly complex, and it is desirable to have a large number of calibration standards that not only bracket the range of concentrations in the samples, but also include a large mix of these elements. The goal of the MURR calibration is to use a large number of secondary standards in the actual working range for obsidian rather than include a number of certified standards with concentrations irrelevant for obsidian analysis. It is critical to create a matrix-specific calibration. Finally, it is unlikely given the compositional variations of obsidians that simply using a series of common standards will allow one to match source and artifact data collected at different laboratories and using different instruments. Proper calibration using appropriate standards can increase the probability of identifying possible sources for an unassigned artifact in published data using other laboratories or techniques, but assigning an artifact to a particular source unless the sample and source material are analyzed on the same instruments using the same settings and calibration is unwise. Some other analytical techniques, such as NAA, are much more stable due, in part, to linear standardizations and allow for calibration using a small number of standards. These techniques are not susceptible to matrix effects, and thus specific procedures to allow for inter-laboratory exchange of data are warranted (Graham *et al.* 1982; Glascock *et al.* 1998), but XRF is based on different excitation and absorption phenomena and requires a more conservative approach. A recent inter-laboratory comparison of quantitative XRF of copper alloys found that the accuracy of the data was a function of variables such as the type of calibration used (fundamental parameters with and without standards and empirical calibrations), the number of standards used in the calibration, but other variables such as detector type and valid count rates, had little apparent impact on the accuracy (Heginbotham 2011).

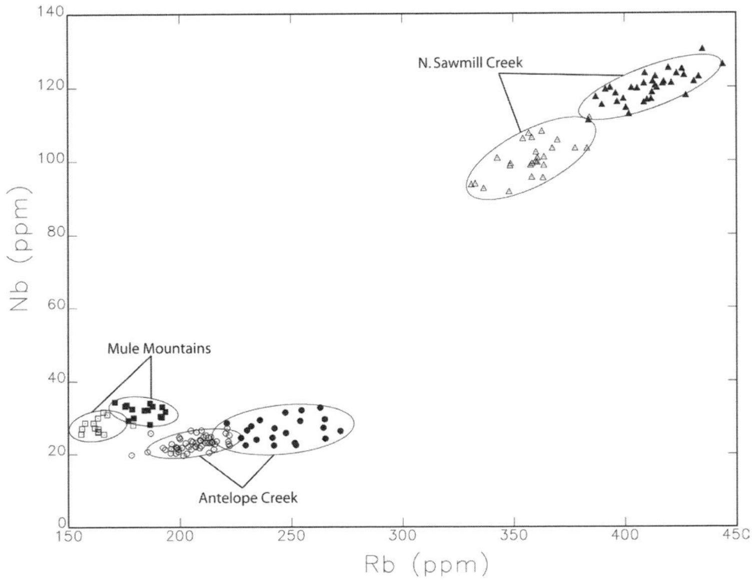

Figure 12.2:
Bivariate plot of rubidium and niobium for source samples from Mule Creek. The open symbols are source samples analyzed at MURR and the solid symbols are source samples analyzed at the Berkeley XRF Lab. The specific samples are not the same for each lab. Ellipses represent 90 percent confidence intervals for membership in the group.

The goal of archaeological obsidian provenance studies is to match obsidian artifacts to specific geologic sources of obsidian. While every effort should be made to determine accurate elemental concentrations by using appropriate standards in a robust calibration, consistent separation of sources is the ultimate goal. Shackley (2005:7) correctly described the balance between precision and the ultimate goals of obsidian provenance research: "[s]ome physicists and chemists in the past have made what I consider the mistake of focusing on precision versus the archaeological accuracy we seek in source provenance studies, and archaeologists have been guilty of the opposite – ignoring precision and accuracy in provenance studies by wholly trusting the analyst because they don't understand the process." In the case of XRF at MURR, the calibration is more valuable the closer it matches the NAA data to facilitate the identification of artifacts from sources not yet analyzed by XRF. The differences between the MURR calibration and other major labs are relatively minor. For example, Table 12.2 provides a comparison between the Berkeley Lab and MURR for some of the major sources in New Mexico. While there are sufficient differences to complicate inter-laboratory sharing of data, both labs produce quantitative elemental concentration data capable of discriminating most sources, including two compositionally-similar sources from Mule Creek (Figure 12.2).

		n	Fe mean	s.d.	Rb mean	s.d.	Sr mean	s.d.	Y mean	s.d.	Zr mean	s.d.	Nb mean	s.d.
El Rechuelos Rhyolite	Berkeley	15	6676	296	152	6	9	3	23	1	77	3	47	2
	MURR	11	3803	114	138	7	11	1	22	2	71	4	39	4
Obsidian Ridge	Berkeley	20*	9735	666	207	11	5	3	63	3	183	7	98	4
	MURR	10	7289	185	186	5	6	1	54	4	168	7	89	3
Cerro del Medio	Berkeley	2*	10315	1817	160	9	10	1	43	1	172	4	54	1
	MURR	22	7053	396	143,6	7	10	1	41	3	167	9	50	3
Bear Springs Peak	Berkeley	24*	6593	431	116	5	43	4	21	2	108	4	53	5
	MURR	7	4877	206	110	4	51	5	20	1	108	5	47	3
Grants Ridge	Berkeley	10	8253	474	564	34	4	1	76	2	118	4	196	7
	MURR	10	5016	256	471	22	11	1	67	5	101	5	171	10
Horace Mesa	Berkeley	12	9568	321	530	17	4	1	87	3	142	4	236	7
	MURR	13	5945	412	422	21	13	1	73	5	117	5	205	12
Mule Creek (Antelope Creek)	Berkeley	21	9619	385	246	15	16	3	42	4	122	7	27	3
	MURR	54	6441	380	206	10	21	3	37	4	106	5	23	2
Mule Creek (N. Sawmill)	Berkeley	34	8423	564	411	15	6	2	71	2	118	7	120	4
	MURR	26	5007	300	358	13	12	2	63	4	98	7	101	5
Mule Creek (Mule Mountains)	Berkeley	15	7936	297	184	7	10	2	25	2	120	5	32	2
	MURR	13	4861	221	163	6	13	2	24	2	116	5	28	2
Gwynn Canyon	Berkeley	20	8512	539	226	13	19	2	31	2	156	7	22	3
	MURR	12	6079	266	205	5	21	2	30	2	156	12	22	2
Antelope Wells	Berkeley	25	22521	1914	348	22	7	1	128	6	1255	47	97	5
	MURR	3	18178	900	271	8	8	1	115	6	1069	50	82	9

*Fewer samples with iron concentrations.

Table 12.2:
Mean and standard deviations (in ppm by XRF) for some of the major sources in New Mexico for both the Berkeley Lab and MURR. Berkeley Lab concentrations taken from the laboratory website (http://www.swxrflab.net/swobsrcs.htm).

Maximizing the potential of XRF for obsidian

The goal of XRF analysis is to maximize the fluorescence of, and thus sensitivity to, elements best able to discriminate geologic sources of obsidian. The most useful group of elements that occur in sufficient concentrations for accurate quantification by ED-XRF are the elements Rb through Nb (atomic numbers 37-41). These elements generally occur in concentrations of tens to hundreds of parts per million and the $K\alpha_1$ fluorescence energies for each of these elements are roughly in the middle of the spectrum of energies produced by most handheld XRF instruments operating at 40kV.

The probability of fluorescing a particular atom is related to the excitation energy of the atom relative to the energy of the incident X-ray. The probability of fluorescence is greatest when using incident X-rays with energies just above the excitation energy of the atom in the sample. For example, iron has a $K\alpha_1$ of 6.4 keV, and there is no chance of K-shell electron fluorescence from incoming X-rays with energies less than or equal to 6.4 keV, while X-rays slightly higher than 6.4 have the highest probability of fluorescence. The probability of fluorescence decreases exponentially with higher energy. Thus, the goal should be to produce an abundance of X-rays with energies just above the absorption edge of the elements of greatest interest. The elements Rb-Nb have $K\alpha_1$ energies from 13.40 - 16.62 keV. An ideal incoming array would contain X-rays primarily above 17 keV and tapering off at much higher energies.

Bruce Kaiser (with Bruker Scientific) and R. Jeffrey Speakman (with the Museum Conservation Institute at the Smithsonian) have developed a filter that acts like a secondary target to absorb the incoming X-rays below about 18 keV in order to both reduce the background of scattering X-rays at the same energy as Rb-Nb and to remove the X-rays of low energy that do not significantly contribute to the fluorescence of Rb-Nb. The filter consists of thin foils of copper (150 μm), aluminum (50 μm) and titanium (300 μm). The copper absorbs the X-rays below 18 keV, but the fluorescence of copper produces a large spike of X-rays at approximately 8-9 keV. The titanium layer absorbs the 8.05 keV X-rays, but produces another peak between 4.5 and 5 keV as a result of titanium fluorescence. The aluminum layer absorbs the titanium peak, and the resulting aluminum fluorescence does not have sufficient energy to impact sample analysis. Figure 12.3 shows the difference in background and increased sensitivity for the same obsidian sample analyzed under the same conditions with and without the filter.

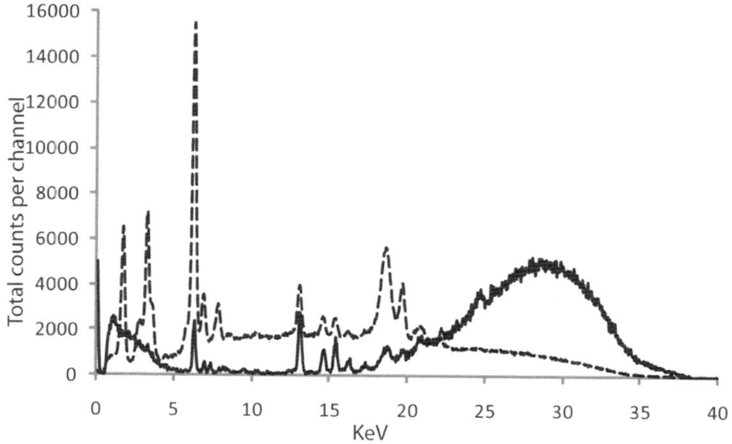

Figure 12.3:
Spectra of the same large flake of obsidian showing the difference in background when using the copper-titanium-aluminum filter (solid line) versus without the filter (dashed line). The filter reduces the total number of X-rays hitting the sample so the filtered data have been multiplied eight times. Note the minimal background under the peaks between 13 and 17 keV that correspond to the elements rubidium to niobium.

Normalization and small/thin sample analysis

Samples selected for XRF analysis should ideally be infinitely thick for the elements of interest. Infinite thickness is defined as the thickness at which additional sample thickness does not result in additional fluorescent X-rays. Infinite thickness is different for each element in direct correlation with the excitation energy and varies between matrices, although sample matrix differences are not addressed here because obsidians are relatively similar in their overall high-silica matrix. For example, 99 percent of the iron fluorescent X-rays (6.4 keV) are from up to 0.1mm from the surface of a pure silica matrix (obsidian is close to a pure silica matrix). The X-rays with the highest probability of exciting iron have relatively low energy and thus are not capable of penetrating deep into a sample, and even if they did fluoresce an iron atom deep in the sample, it is unlikely for the low energy fluorescent X-ray to make it back through a thick sample and into the detector. While iron peaks in a spectrum result primarily from fluorescence at the surface of the sample, under the same conditions, zirconium Kα X-rays (15.8 keV) escape from up to 2.2 mm deep, and barium Kα X-rays (32.2 keV) from up to 17 mm deep.

The differences in infinite thicknesses are critical to an understanding of the difficulties in analyzing thin samples (I will expand on the analysis of thin samples in the following section), but normalization can help account for some of the issues related to sample thickness variability. Normalization was initially developed as a means to neutralize the effects of small fluctuations in the output of the X-ray sources. The method involves selecting a region of the spectrum that does not have peaks derived from elements of interest in the sample and that directly reflect the amount and energy of X-rays hitting the sample. The Bruker Tracer III-V PXRF in use at MURR uses a rhodium source and thus produces a characteristic peak of X-rays for Rh Kα at approximately 20.2 keV due to Raleigh scatter of the source X-rays on the sample. This interaction of Rh X-rays with the sample produces both elastic and inelastic scatter resulting in both the Raleigh (elastic scatter at ~ 20.2 keV as mentioned above) and Compton (inelastic scatter at approximately 18.5-19.5 keV) peaks being produced in the spectrum based on infinitely thick samples. In order to get quantitative concentrations for a batch of samples it is necessary to normalize the peak areas in the rhodium Compton peak and apply the same correction to the other peaks in the spectra. This minimizes the differences in the incoming X-ray profile.

While the Compton peak serves well for normalization in most obsidians, there are certain sources with elevated niobium concentrations that have a sufficiently large niobium Kβ peak to interfere with the rhodium Compton peak area. In these cases, the calibration interprets the additional peak area from niobium as increased X-ray output requiring comparable reduction in all peaks in the spectra. Obsidians with such elevated niobium concentrations are rare worldwide, but they are common for the sources in the Rift Valley of Kenya. For Kenyan obsidian a slightly modified version of the calibration that normalizes to a portion of the backscatter in a region above obvious elemental interferences is used. This avoids the niobium effects on normalization, but it moves the normalization region even farther from the peaks for the main elements of interest (Rb-Nb) making correction for sample size even more difficult. XRF calibration requires a series of compromises to arrive at the best possible data. There is no obvious one-size-fits-all answer.

The same normalization procedure used to account for tube fluctuations can also be used, with somewhat less success, to account for variability in sample size, particularly thickness. Many applications of XRF involve thin samples, such as the analysis of thin films and filters. This type of thin sample is relatively easy to correct for if the samples have consistent thicknesses and it is possible to create comparable calibration standards (Markowicz 2008; Piorek 2008). Although it is possible to create thin slices of obsidian, artifacts have complex morphology and thicknesses are not only inconsistent between samples, but across a single sample. By understanding the physics behind the fluorescence of thin samples as well as the potential errors inherent in normalization it is possible to confidently determine the source of obsidian artifacts that are only a couple millimeters in diameter and less than 0.5mm thick. Figure 12.4 shows two overlaid spectra of a large/thick flake and a small/thin flake from the same obsidian cobble.

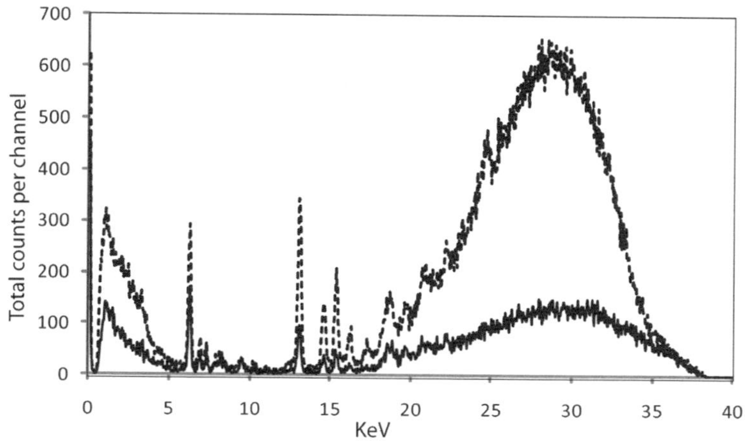

Figure 12.4:
Comparison of large (dashed line) and small (solid line) flakes from the same cobble. The large flake covered the entire beam with a thickness of 7mm, and the small flake measured 4 by 3mm and 0.4mm thick.

The current solution to small sample analysis at MURR involves a number of strategies. First of all, we use an instrument with a small beam cross-section. Figure 12.5 is a briefly exposed dental X-ray of the Bruker Tracer III-V PXRF compared to that of the benchtop ElvaX. The Bruker beam focuses on an area of approximately 2 x 3 mm. Although a narrow beam is more susceptible to variation in heterogenous samples, obsidian is usually homogenous enough to not create any problems. The beam in the Tracer III-V is large compared to that of a micro-XRF and small beam use has great potential for the analysis of obsidian microdebitage. The small beam improves the analysis of samples with a small diameter by simply covering more of the beam. If samples were infinitely thick, the small sample diameter would easily be accounted for by the typical normalization process and no data shift should occur.

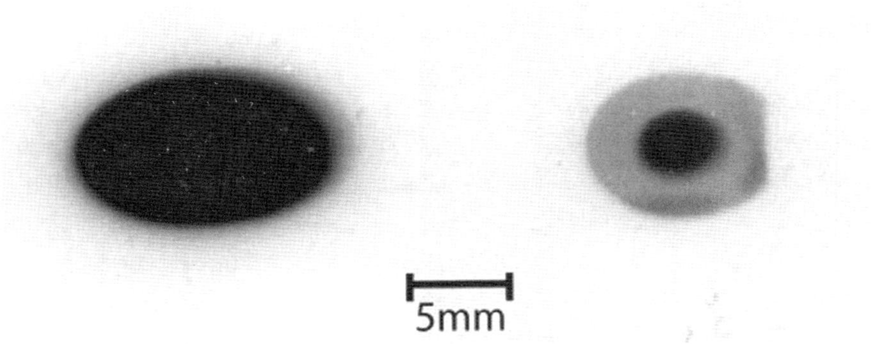

Figure 12.5:
Dental X-ray film briefly exposed to the Elva-X (left) and Tracer III-V (right) beams. Note the much smaller beam area of the Tracer.

The real problems arise with the analysis of samples of less than infinite thickness. For the purposes of this discussion, consider an example in which we are only concerned with obtaining concentration data for iron and niobium. The niobium $K\alpha_1$ peak is between 16.3 and 16.9 keV, and this is relatively close to the rhodium Compton peak used for normalization that is between 18.5 and 19.5 keV. The closer the two peaks are in energy, the closer they are to having the same infinite thickness value, thus the two peaks are absorbed at similar rates with increases in sample thickness. When the calibration normalizes to the rhodium Compton peak, and the same correction factor is applied to the niobium peak the quantitative results are reasonable. But, iron has an infinite thickness in obsidian measured in microns, and thus the size of the iron peak does not change as a function of sample thickness until the samples are extremely thin. This may not seem like a problem, but the normalization applies the same correction to all peak areas, whether it is needed or not. Therefore, in a sample thin enough to generate only half the rhodium Compton peak area, the niobium peak area (and thus the calculated concentration) will be correctly increased by a factor of two, but the iron peak area will be significantly over-corrected. The primary elements of interest in obsidian characterization, Rb-Nb, are all relatively close to the rhodium Compton peak, and the normalization does a good job of correcting for sample size although there is an increasing overcorrection moving down in energy. This overcorrection is extreme for low-energy elements like iron. When analyzing assemblages that include small samples, it is best to discount data from all elements with fluorescence energies below that of rubidium.

It should now be clear why the calibration discussed previously that accounts for the elevated niobium levels in some East African obsidians is an example of a difficult compromise. The East African calibration sacrifices accuracy of small sample analysis because of the higher energy of the normalization region in order to avoid the negative effects of high niobium concentrations. In other areas of the world, namely the western United States, the niobium levels are low enough that sample size is more important and one should use the calibration that normalizes to the rhodium Compton region.

The combination of the variably erroneous normalization correction in thin samples and the overall reduced counts for small/thin samples causes a slight shift in the data the correlates with sample size. It can make source assignment by whatever means (i.e. bivariate plots, cluster analysis, Mahalanobis distance projection) more difficult. Bivariate plots can show the correlation in the scattered data that should co-vary with the correct source data across all plots for elements Rb-Nb. It is very important to confirm the source assignment with direct spectral overlay, keeping in mind the expected shifts resulting from small samples.

A number of recent projects on obsidian from a number of sites in central and southern California has recently been conducted at MURR to demonstrate that XRF is effective for small sample source assignment. Of the more than 1,000 artifacts analyzed by XRF, 40 of the smallest and most difficult to assign samples were selected for additional analysis by short-irradiation NAA. The advantages of the short NAA are that the samples become only slightly radioactive and can be safely handled after only a brief decay period, the technique is a bulk technique that is not subjected to variations in sample morphology because the sample rotated, and it can produce precise data for about six elements on samples of only a few milligrams. Of the forty samples, XRF source assignment was confirmed for 38 of them. One sample was not assigned to any source by XRF but was assigned by NAA, and one sample was incorrectly assigned by XRF to a source with similar composition. This error rate of five percent is higher than one would like, but it occurs on the smallest and most challenging samples. The error rate for the entire study, including the larger artifacts, would likely be much lower. Similar tests have been done with smaller assemblages from the American Southwest with similarly encouraging results.

The current approach to small sample analysis by XRF is not yet an ideal solution. Attempts have been made to create an element-specific correction factor that uses the rhodium Compton peak as an indicator of sample mass. One possibility is to use a two-element standard behind the artifact that provides a good approximation of the thickness of the sample in the beam. The standard would contain sufficient concentrations of two elements, one with a peak below the elements of interest and one above. In some preliminary tests using a pressed pellet containing gallium and indium in which the lower energy element (in this case gallium) would have a peak area reduced in relation to the higher energy element (indium) in a ratio reflecting sample mass. This method has some promise, but it is time consuming to keep the standard in proper position and this would also greatly increase the time required to interpret the final data. It is important to verify source assignments, particularly with small samples, using other analytical techniques such as NAA or LA-ICP-MS, but the ultimate solution to small artifact analysis may involve micro-XRF.

Conclusions

XRF has held a prominent position in the provenance study of archaeological obsidian for decades, and all indications are that it will continue to be a major analytical technique. Handheld XRF instruments provide some logistical advantages over lab-based instruments, but aside from a few, relatively minor, limitations the two groups of instruments pose the same analytical potential as well as limitations. Handheld XRF has been the unfair subject of recent skepticism in the literature (i.e. Shackley 2010). The problems arise not from the portability of a particular instrument, but from the lower cost of most portable instruments that places the technology in the hands of researchers with insufficient understanding of the physics, calibration methods, and analytical limitations. This chapter begins to address some of these problems.

Notes

1. A completely new set of calibration standards is in preparation that will cover the known concentration ranges in obsidian for the elements iron, rubidium, strontium, yttrium, zirconium, niobium and barium. This new calibration set will include 40 obsidian samples from around the world. The concentrations will be derived from a combination NAA, microwave digestion ICP-MS, and XRF analysis of the individual rocks. A publication detailing the concentrations of the original set of calibration standards in currently in preparation, and considering it will soon be obsolete upon the completion of the new set of calibration standards the values are not reported here.

References

Ambroz, Jessica A., Michael D. Glascock, and Craig E. Skinner. 2001. "Chemical Differentiation of Obsidian within the Glass Buttes Complex, Oregon." *Journal of Archaeological Science* Vol. 28 (7):741-746.

Bellot-Gurlet, Ludovic, Olivier Dorighel, and Gérard Poupeau. 2008. "Obsidian provenance studies in Colombia and Ecuador: obsidian sources revisited." *Journal of Archaeological Science* Vol. 35 (2):272-289.

Bigazzi, G., G. Poupeau, Z. Yeğingil, and L. Bellot-Gurlet. 1998. "Provenance Studies of Obsidian Artefacts in Anatolia Using Fission-Track Dating Method: An Overview." In *L'obsidienne au Proche et Moyen Orient*, edited by M.-C. Cauvin, A. Gourgaud, B. Gratuze, N. Arnaud, G. Poupeau, J.-L. Poidevin and C. Chataigner, 70-89. Oxford: BAR International Series.

Braswell, Geoffrey E., John E. Clark, Kazuo Aoyama, Heather McKillop, and Michael D. Glascock. 2000. "Determining the Geochemical Provenance of Obsidian Artifacts from the Maya Region: A test of the Efficacy of Visual Sourcing." *Latin American Antiquity* Vol. 11 (3):269-282.

Braswell, Geoffrey E., and Michael D. Glascock. 1998. "Interpreting Intrasource Variation in the Composition of Obsidian: The Geoarchaeology of San Martin Jilotepeque, Guatemala." *Latin American Antiquity* Vol. 9 (4):353-369.

Carter, Tristan, Gérard Poupeau, Céline Bressy, and Nicholas J. G. Pearce. 2006. "A New Programme of Obsidian Characterization at Çatalhöyük, Turkey." *Journal of Archaeological Science* Vol. 33 (7):893-909.

Duff, Andrew Ian Lyman. 1999. *Regional Interaction and the Transformation of Western Pueblo Identities, A.D. 1275-1400*. PhD, Department of Anthropology, Arizona State University, Phoenix.

Eerkens, Jelmer W., Jeffrey R. Ferguson, Michael D. Glascock, Craig E. Skinner, and Sharon A. Waechter. 2007. "Reduction Strategies and Geochemical Characterization of Lithic Assemblages: A Comparison of Three Case Studies from Western North America." *American Antiquity* Vol. 72 (3):585-597.

Ferguson, Jeffrey R. 2010. In-Field Qualitative X-Ray Fluorescence of Kiln Features at LA155669, Capitan Mountains, New Mexico. Report on file at MURR.

Ferguson, Jeffrey R., Stephen H. Lekson, Margaret Nelson, Toni S. Laumbach, and Karl W. Laumbach. 2010. Characterization of Carbon-Painted Ceramics from Southwestern and South Central New Mexico. Paper presented at the Society for American Archaeology Meeting, at St. Louis, Missouri.

Ferguson, Jeffrey R., and Craig E. Skinner. 2006. Geochemical Evidence of Mesoamerican Prismatic Blades in the United States. In *Society for American Archaeology Meeting*. San Juan, Puerto Rico.

Friedman, I. and F. W. Trembour. 1978. "Obsidian: The Dating Stone." *American Scientist* Vol. 66 (1):44-45.

———. 1983. "Obsidian Hydration Dating Update." *American Antiquity* Vol. 48 (3):544-547.

Glascock, M. D., H. Neff, K. S. Stryker, and T. N. Johnson. 1994. "Sourcing Archaeological Obsidian by an Abbreviated NAA Procedure." *Journal of Radioanalytical and Nuclear Chemistry* Vol. 180 (1):29-35.

Glascock, Michael D. 2002. *Geochemical Evidence for Long Distance Exchange*. Westport, Connecticut: Bergin and Garvey.

Glascock, Michael D., Geoffrey E. Braswell, and Robert H. Cobean. 1998. "A Systematic Approach to Obsidian Source Characterization." In *Archaeological Obsidian Studies: Method and Theory*, edited by M. Steven Shackley, 15-65. New York: Plenum Press.

Glascock, Michael D., Raymond Knuselman, and Daniel Wolfman. 1999. "Intrasource Chemical Differentiation of Obsidian in the Jemez Mountains and Taos Plateau, New Mexico." *Journal of Archaeological Science* Vol. 26 (8):861-868.

Glascock, Michael D., Robert J. Speakman, and Rachel S. Popelka-Filcoff. 2007. *Archaeological Chemistry: Analytical Techniques and Archaeological Interpretation*. Vol. 968. Washington, DC: American Chemical Society.

Gordus, Adon A., Gary A. Wright, and James B. Griffin. 1968. "Obsidian Characterized by Neutron-Activation Analysis." *Science* Vol. 161 (3839):382-384.

Goren, Yuval, Hans Mommsen, and Jörg Klinger. 2011. "Non-Destructive Provenance Study of Cuneiform Tablets Using Portable X-Ray Fluorescence." *Journal of Archaeological Science* Vol. 38 (3):684-696.

Graham, Christoper C., Michael D. Glascock, James J. Carni, James R. Vogt, and Thomas G. Spalding. 1982. "Determination of Elements in National Bureau of Standards' Geological standard Reference Materials by Neutron Activation Analysis." *Analytical Chemistry* Vol. 54 (9):1623-1627.

Heginbotham, Arlen, A. Bezur, M. Bouchard, J. M. Davis, K. Eremin, J. H. Frantz, L. Glinsman, L. Hayek, D. Hook, V. Kantarelou, A. G. Karydas, L. Lee, J. Mass, C. Matsen, B. McCarthy, M. McGath, A. Shugar, Sirois J, D. Smith, and R. J. Speakman. 2011. "An Evaluation of Inter-Laboratory Reproducibility for Quantitative XRF of Historic Copper Alloys." In *Metal 2010: International Conference on Metal Conservation*, edited by P. Mardikian, C. Chemello, C. Watters and P. Hull, 244-255. Clemson, South Carolina: Clemson University.

Hester, T. R., F. Asaro, F. Stross, H. Michael, A. C. Kerr, and P. Headrick. 1991. "An Overview of the Results of the Texas Obsidian Project." *La Tierra* Vol. 18 (1):4-7.

Hughes, Richard E. 1984. *Obsidian Studies in the Great Basin*. Contributions of the University of California Archaeological Research Facility No. 45. Berkeley: Archaeological Research Facility, University of California.

———. 1989. *Current Directions in California Obsidian Studies*. Contributions of the University of California Archaeological Research Facility No. 48. Berkeley: Archaeological Research Facility, University of California.

Leach, B. Foss, Stanley Warren, and Barry Fankhauser. 1978. "Obsidian from the Far North of New Zealand: A Method of Sourcing Based on Natural Radioactive Emission." *New Zealand Journal of Science* Vol. 21:123-128.

Liritzis, Ioannis, and Nikolaos Zacharias. 2011. "Portable XRF aof Archaeological Artifacts: Current Research, Potentials and Limitations." In *X-Ray Fluorescence Spectrometry (XRF) in Geoarchaeology*, edited by M. Steven Shackley, 109-142. New York: Springer.

Mauser, Karl-Eugen. 2006. "Standardless Methods. In Handbook of Practical X-Ray Fluorescence Analysis." In, edited by B. Beckhoff, B. Kanngießer, N. Langhoff, R. Wedell and H. Wolff, 384-393. Berlin: Springer.

Merrick, Harry V., and Francis H. Brown. 1984. "Rapid Chemical Characterization of Obsidian Artifacts by Electron Microprobe Analysis." *Archaeometry* Vol. 26 (2):230-236.

Morgenstein, Maury, and Carol A. Redmount. 2005. "Using Portable Energy Dispersive X-Ray Fluorescence (EDXRF) Analysis for On-Site Study of Ceramic Sherds at El Hibeh, Egypt." *Journal of Archaeological Science* Vol. 32 (11):1613-1623.

Neff, Hector. 1998. "Units in Chemistry-Based Ceramic Provenance Investigations." In *Unit Issues in Archaeology*, edited by A. F. Ramenofsky and A. Steffan, 115-127. Salt Lake City: University of Utah Press.

Northwest Research Obsidian Studies Laboratory. 2011. *Introduction to Obsidian Characterization Studies* 2011 [cited September 2011]. Available from http://www.obsidianlab.com/info_xrf.html.

Origer, Thomas M. 1989. "Hydration Analysis of Obsidian Flakes Produced by Ishi During the Historic Period." In *Current Directions in California Obsidian Studies*, edited by Richard E. Hughes, 69-77. Contributions of the University of California Archaeological Research Facility No. 48. Berkeley: University of California.

Pollard, Mark, Catherine Batt, Ben Stern, and Suzanne M. M. Young. 2007. *Analytical Chemistry in Archaeology*. Cambridge Manuals in Archaeology. Cambridge: Cambridge University Press.

Reeves, R. D., and Gaye C. Armitage. 1973. "Density Measurements and Chemical Analysis in the Identification of New Zealand Archaeological Obsidians." *New Zealand Journal of Science* Vol. 16:561-572.

Shackley, M. Steven. 1998a. *Archaeological Obsidian Studies: Method and Theory*. New York: Plenum Press.

———. 1998b. "Current Issues and Future Directions in Archaeological Volcanic Glass Studies: An Introduction." In *Obsidian Geology and Archaeology in the North American Southwest*, edited by M. Steven Shackley, 1-14. Tucson: University of Arizona Press.

———. 2002. "More than Exchange: Pre-Ceramic through Ceramic Period Obsidian Studies in the Greater North American Southwest." In *Geochemical Evidence for Long-Distance Exchange*, edited by Michael D. Glascock, 53-87. Westport, Connecticut: Bergin and Garvey.

———. 2005. *Obsidian: Geology and Archaeology in the North American Southwest*. Tucson: University of Arizona Press.

———. 2010. "Is There Reliability and Validity in Portale X-Ray Fluorescence Spectrometry (PXRF)?" *SAA Archaeological Record* Vol. 10 (5):17-20,44.

———. 2011a. *ThermoFisher Scientific Quant'X Analsyis and Instrumentation* 2011 [cited September 2011]. Available from http://www.swxrflab.net//anlysis.htm.

———. 2011b. "An Introduction to X-Ray Fluorescence (XRF) Analysis in Archaeology." In *X-Ray Fluorescence Spectrometry (XRF) in Geoarchaeology*, edited by M. Steven Shackley, 7-44. New York: Springer.

Speakman, Robert J, Michael D. Glascock, Robert H. Tykot, Christophe Descantes, Jennifer J. Thatcher, Craig E. Skinner, and Kyra M. Lienhop. 2007. "Selected Applications of Laser Ablation Inductively Coupled Plasma-Mass Spectrometry to Archaeological Research." In *Archaeological Chemistry: Analytical Techniques and Archaeological Interpretation*, edited by Michael D. Glascock, Robert J. Speakman and Rachel S. Popelka-Filcoff, 275-296. Washington, DC: American Chemical Society.

Taylor, R. E. 1976. *Advances in Obsidian Glass Studies: Archaeological and Geochemical Perspectives*. Park Ridge, New Jersey: Noyes Press.

Tykot, R. H. 2004. "Scientific Methods and Applications to Archaeological Provenance Studies." In *Proceedings of the International School of Physics "Enrico Fermi" Course CLIV*, edited by M. Martini, M. Milazzo and M. Piacentini, 407-432. Amsterdam: IOS Press.

Weigand, Phil C., Garman Harbottle, and Edward V. Sayre. 1977. "Turquoise Sources and Source Analysis: Mesoamerica and the Southwestern U.S.A." In *Exchange Systems in Prehistory*, edited by T. K. Earle and J. E. Ericson, 15-34. New York: Academic Press.

Weisler, Marshall I., and David A. Clague. 1998. "Characterization of Archaeological Volcanic Glass from Oceania: the Utility of Three Techniques." In *Archaeological Obsidian Studies: Method and Theory*, edited by M. Steven Shackley, 123-128. New York: Plenum Press.

Chapter 13

Handheld XRF analysis of Maya ceramics: a pilot study presenting issues related to quantification and calibration

Jim J. Aimers, Dori J. Farthing and Aaron N. Shugar

Introduction

The investigation of archaeological ceramics has a long and varied history with regard to the analytical instrumentation used (for general examples, see Peacock 1970; Bishop *et al.* 1982; Rice 1987; Pollard *et al.* 2007). In recent years newer applications have been used for the analysis of ceramic materials as well, including ICP-MS (Fermo *et al.* 2008; Mannino and Orecchio 2011) and INAA (Glascock 1992; Neff 2000). In most cases the motivation to obtain chemical concentrations from archaeological ceramics has been to establish the source of the clay matrix. This has proven possible using instrumentation with low detection limits (i.e. trace element analysis techniques such as NAA, ICP, AAS, and WD-XRF). Handheld X-ray fluorescence spectrometry was developed in the early 1960's (Piorek 1997) but did not enter the world of archaeology, outside of isolated research, until the early to mid 2000's when the instrumentation became more affordable (e.g. Uda *et al.* 2000; Cesareo *et al.* 2004; Ida and Kawai 2005; Newman and Loendorf 2005). With the flourishing use of handheld XRF by non-trained scientists and other researchers who may not be trained in the basic (and advanced) theories of X-ray fluorescence, its misuse and the misinterpretation of results is prevalent (see chapter 1 of this volume for more detail).

Several papers have recently been published dealing with the provenancing of ceramics using handheld XRF with varied success (e.g., Morgenstein and Redmount 2005; Tagle and Gross 2010; Barone *et al.* 2011; Goren *et al.* 2011; Speakman *et al.* 2011). Unfortunately, the 'boxed' calibrations that come with these instruments are not designed to deal with the complex nature of archaeological ceramics. Ceramics are by nature heterogeneous with numerous components (such as temper) all having variable particle size. They can have surface alterations and coatings, and over time, the chemistry of the surface can alter as well. In addition, archaeological ceramics often have altered chemical surfaces related to their burial environment. Manufacturer calibrations are more geared to modern applications and modern materials that are uniform in makeup, and expecting calibrations designed for these purposes to be effective with archaeological ceramics is unreasonable. The desire for quantification, whether it be for provenancing studies or characterization studies, requires the user to create material specific calibrations (see Hein *et al.* 2002).

This paper discusses current investigations of Maya ceramics from Belize. The focus of this study is not to determine the source/provenance of the clay bodies, but to investigate the potential for establishing handheld XRF as an on-site analytical tool for the characterization and potential classification of ceramics based on their chemical signatures. The development of an empirical calibration is presented including the process involved in creating reference materials for that calibration.

Overview of Maya chronology and pottery

The date of the arrival of people in the Maya lowlands is currently a matter of debate (see Lohse 2010), but lies somewhere in the Archaic period (8000-2000 B.C.) with maize pollen indicating farming by about 3000 B.C. (Pohl *et al.* 1996). The Preclassic period dates from roughly 2000 B.C. to A.D. 250 with the earliest well-documented Maya pottery about 1100-900 B.C. in the Cunil ceramic complex of the Belize Valley (Sullivan and Awe 2012). By the Late Preclassic period (often dated 250 B.C. to A.D. 250) Maya pottery was very well made and styles were widely spread across the entire Maya lowlands. Although most of the significant cultural aspects of Maya civilization were in place by the Late Preclassic, the subsequent Classic period (A.D. 250-800) is generally considered the height of Maya development. The Classic Maya lived in a literate, highly stratified society which produced monumental art and architecture and elaborate polychrome pictorial pottery. In the Terminal Classic period many sites were abandoned and profound changes swept across the Maya world (Aimers 2007b). Dates vary because different sites were abandoned or transformed at different times from about A.D. 750-1050 but the Terminal Classic has traditionally been dated to about A.D. 800-900. Pottery of the Terminal Classic varies across the lowlands but it was still well-made with an emphasis on elaborate modeling and incising over polychromy. The Postclassic period follows the Terminal Classic and ends at the arrival of the Spanish in the Maya area at about A.D. 1540. Postclassic pottery also emphasizes incising and modeling and is typically well-made. The pottery of the Postclassic period is the focus here.

The Postclassic period and its pottery

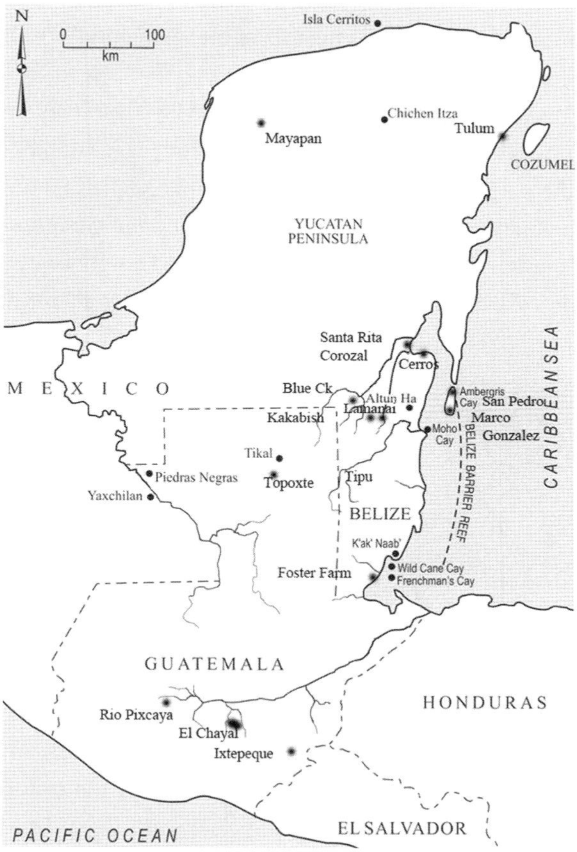

Figure 13.1:
Map of the Maya region showing key Postclassic and trading sites relevant to this study (After McKillop 2005).

In the early years of Maya archaeology the Postclassic period was neglected as a period of cultural decline following the Classic "collapse", but more recent research at Postclassic sites has revealed population movement, innovative political strategies, increased exchange and commercialization, iconographic innovation, and intense stylistic interaction (Smith and Berdan 2000, 2003). A key characteristic of the Maya Postclassic period is evidence of extensive trade, especially among sites along the Caribbean coasts and on rivers, and involving sites in the northern half of the peninsula such as Mayapán and Chichén Itza (Figure 13.1). A better understanding of the economic and political milieu of the Postclassic would be greatly aided by more detailed documentation of the nature and degree of interaction among Postclassic sites, and one of our most informative artifact classes is pottery.

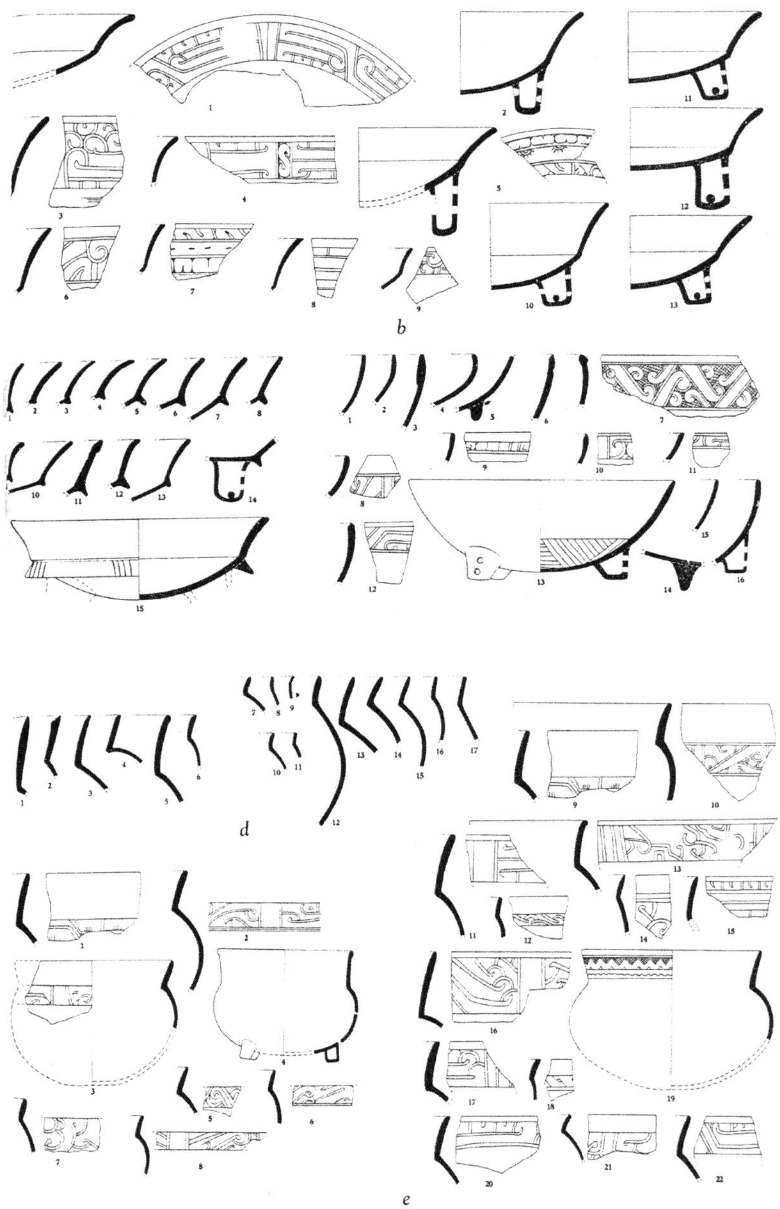

Figure 13.2:
Red Payil Group Ceramics. Payil Red is the plain type, Palmul Incised is the incised version (after Sanders 1960: Fig 4, 5).

Aimers has been investigating the Postclassic period since his dissertation research on the Maya collapse and its aftermath (Aimers 2003, 2004c, 2004a, 2007b) and particularly with research on the pottery of two sites that were not abandoned in the Terminal Classic: Tipu (Aimers 2004a; Aimers and Graham 2012a) and Lamanai (Aimers 2008, 2009, 2010; Aimers and Graham 2012b). In the summer of 2011 Aimers began a pilot study to investigate the chemical variability of a plain red type (Payil Red) and a related incised type (Palmul Incised) (see Figures 13.2 and 13.3) using XRF with samples from the inland site of Tipu, and the site of San Pedro on the island of Ambergris Caye. These types are not particularly common but they are widely distributed in the Late Postclassic period, and they are much more common at coastal sites and are thought to have been produced along the coast of the Mexican state of Quintana Roo (e.g., the sites of Ichpaatun, Tancah and Tulum, see Figure 13.1; (Sanders 1960; Aimers 2009). As we discuss later, the original goal of the research was to identify compositional groupings within these types which might help in addressing trade and exchange in the Postclassic period. We do not expect to link the pots to their production location except in rare cases (see comments below), but we hope that chemical characterization can help us map the distribution of pottery types better than surface style and form alone. These distributions can help in the construction of inferences about Maya pottery production and trade. A larger study is planned to follow the pilot study with more stylistic types and samples from more sites.

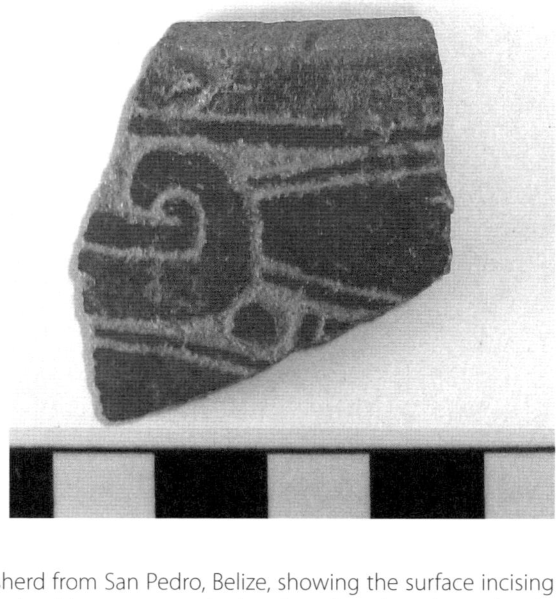

Figure 13.3:
Palmul Incised sherd from San Pedro, Belize, showing the surface incising and the red slip. This example also has blue pigment which is thought to have been distributed from the site of Mayapán (Mexico).

Major analytical techniques used in maya pottery studies

In the Maya area, a stylistic classification system known as type-variety has dominated the study of pottery since its introduction from American Southwestern archaeology in the 1950's and 1960's (Smith, Willey, and Gifford 1960). Type-variety organizes pottery hierarchically into *wares* (based on broad characteristics of paste and/or surface), *groups*, which are clusters of *types* (defined by a set of attributes such as color and decorative treatment), and *varieties* which are often based on single attributes. So, Payil Red and Palmul Incised are types within the Red Payil Group of Tulum Red Ware. Each of these types only has a single variety because these types are macroscopically quite homogenous in paste and surface treatment—this is one reason they were chosen for the XRF study (see Cecil 2012 for more detail on the pastes, NAA data and petrography).

Type-variety has been used widely because it is a rapid and inexpensive "low-tech" way to organize the thousands (sometimes millions) of sherds that are produced by excavations at sites in the Maya area (Aimers and Graham 2012b). Aimers' research to date has involved assessing interaction among sites and regions using type-variety analysis of pottery from hands-on examination of collections from across the Maya lowlands (see e.g., Aimers 2004a, 2004b, 2007a, 2008, 2009, 2010). Type-variety provides a common language for archaeologists and has facilitated the comparison of pottery across sites and regions in addressing issues as varied as chronology, function, trade/exchange, and cultural meaning. Nevertheless, type-variety has been subject to a number of important criticisms. One of the most important issues is the characterization of fabrics (which Mayanists generally call pastes) at the ware level (Rice 1976). Paste variation has been used by some archaeologists as a key discriminating attribute and thus used to make distinction at the highest (ware) level (e.g., in Rice's work cited in this chapter), but it has been considered by others to be a minor factor and occurs randomly in, for example, type or variety descriptions or to create varieties (Gifford 1976). Attention paid to paste has tended to vary with research questions. Those interested in manufacture and production, have tended to privilege paste variation for the insight it can provide into these issues. Those interested in consumer choice, stylistic analysis and comparison, or meaning, have often considered paste variation irrelevant.

Thus, type-variety classification is problematized by inconsistencies in the treatment of paste variation that are not weaknesses in the system itself, but result from the fact that like all classifications, type-variety methods vary according to the research questions addressed (Aimers 2012a). Still, it is reasonable for Mayanists to seek greater accuracy, consistency, and comparability in the characterization of paste variation, and the oldest established technique for the close examination of paste variation is petrography (Jones 1986, 1991). Maya petrographic studies have been surprisingly rare in comparison to work in the Old World and to the amount of research on pottery in the Maya area. This is probably because Maya pottery is stylistically varied, exceptionally elaborate and often well-preserved, so macroscopic characterizations have been adequate for chronology and broad comparative studies. Petrography is of course time-consuming and destructive which poses a problem with large or complex sample sets. Petrographic studies of pottery have tended to focus on issues related to manufacture, production, and distribution (e.g., Rice 1977, 1991, 1996; Cecil 2001b, 2001a; Cecil and Pugh 2004; Howie 2005; Cecil 2009). One of the challenges for the petrographic study of Maya ceramics is that the geology of the Maya area is relatively poorly mapped (see extensive comments about these issues in Howie 2005:120-161 for Belize) so until more sampling of geology and clays is done it can be very difficult to tie pottery to its clay sources. Successful petrographic studies have tended to be focused on a fairly local level (e.g., Cecil and Rice's work in the Petén Lakes; Howie 2005) where the geology is well known or distinctive, and/or where clay sampling has been undertaken.

Materials science approaches to the study of archaeological ceramics are advancing rapidly. Petrography is of course well established, and recent studies of archaeological pottery have used XRF (Bakraji *et al.* 2010; Bakraji *et al.* 2006; Hall 2001; Thomas *et al.* 1992), portable XRF (PXRF) (Papadopoulou *et al.* 2007; Papadopoulou *et al.* 2006; Papageorgiou and Lizritis 2007), mineralogical analysis using XRD (McCaffery *et al.* 2007; Mitchell and Hart 1989; Rasmussen *et al.* 2009; Stanjek and Häusler 2004; Zhu *et al.* 2004), trace chemistry determination by NAA (many, e.g., Glascock 1992; Glascock *et al.* 2004; Hancock *et al.* 1989; Lopez-del-Rio *et al.* 2009; Olin and Blackman 1989), structural and microstructural characterization techniques such as SEM (Ownby *et al.* 2004; Palanivel and Meyvel 2010) or combinations of various techniques (Marghussian *et al.* 2009; Padilla *et al.* 2005; Speakman *et al.* 2011). The best overview on the use of all of these techniques in the analysis of archaeological pottery was done by Rice (1987).

Of the elemental analysis techniques, Mayanists have considered NAA to be the most valuable because of its sensitivity, accuracy, few matrix effects, and the range of trace elements that can be identified, including rare earth elements (Neff 1992:2). The disadvantage of NAA is its cost and the fact that it can only be conducted in facilities with research reactors. In addition, the sample size required for NAA is quite small, typically a small drilling is all that is required. For this reason sample heterogeneity could have an adverse effect on the resulting chemistry obtained. This issue is recognized by researchers and now larger samples are taken and powdered for analysis (see Speakman *et al.* 2011 for example). Many of the other elemental analysis techniques (e.g., PIXE) are also expensive and require equipment that is not easy to acquire. This has led to continued interest in petrography and the use of XRF and XRD because many universities and museums have access to these instrumental methods. Although XRD analysis does not provide elemental analysis data insights, it compliments other techniques because it provides information on the mineral makeup of analyzed samples. For example, Tenorio *et al.* (2010) used NAA, XRD, and SEM in a study of pottery from Lagartero, Chiapas, Mexico.

Current trends in Maya ceramic analysis

The introduction of a new and readily accessible analytical technique typically results in optimism about its utility for the investigation of archaeological problems. For example, Culbert and Schwalbe (1987) published an early study of the application of standard XRF to Maya ceramics from Tikal (see also Schwalbe and Culbert 1988). This study and others were criticized concerning issues of precision and especially comparability of results to other studies (Bishop *et al.* 1990:543; Neff 1992:4). Recently, portable and handheld XRF technology created a similar wave of optimism but critical evaluations did not lag far behind. Shackley (2010) provides the most straightforward critique of handheld XRF on issues of reliability and validity as well as the "near religious fervor" with which the technology has been embraced by people who are not adequately familiar with the methodological and interpretive issues involved (Shugar 2009: also addresses these issues). This is certainly the case in Maya studies. The Mayanist here (Aimers) learned of handheld XRF relatively recently and was excited by what appeared to be a fast way to acquire "hard" compositional data in the field. Like many others he had no background in XRF methodologies and no awareness of the challenges of sensitivity, precision, accuracy, and comparability of results using this new technology. This chapter brings together the differing experience of the three authors in an investigation of these issues in relation to archaeological pottery.

In the study of Maya pottery new analytical techniques, after a period of enthusiastic experimentation, are typically absorbed into research projects which combine them with more established techniques. In particular, petrography combined with quantitative chemical analysis broadens the scope of an investigation to include both the paste makeup (e.g, the characteristics of the clay body, natural inclusions, and temper which are more indicative of the *Chaîne opératoire* or specific manufacturing process), and its particular chemistry (as stated above – potentially to source clays). Type-variety, despite its problems, is also still a very useful organizational and descriptive structure for Maya pottery. There is broad agreement that results from multiple techniques of analysis are always more useful than any one alone (Culbert and Rands 2007). In a discussion of the challenges of characterizing Aegean pottery Day *et al.* (1999:1034) reached a similar conclusion:

> ... different sources of chemical variation emphasize the need for the integration of other information; mineralogical, technological and stylistic; which enables the researcher to attribute differences to provenance or aspects of clay paste technology. The complex interplay of these natural and human sources of variation means that such analyses cannot take place in isolation, in a "black-box" approach. On the contrary, it is imperative for mineralogical and elemental analyses, at least in the Aegean, to be conducted in an integrated programme which exploits complementary types of archaeological and analytical information.

Pottery variability and the potential of XRF and handheld XRF

Inter-observer inconsistency is always an issue in type-variety, especially for rare types (Aimers 2012b) but many types, including the ones discussed in this chapter, are recognized by experienced specialists with little if no debate. So, why is there a need for XRF and other means of compositional characterization? In the case of Red Payil Group sherds, the problem is their macroscopic consistency. We know that these types are widely distributed and we assume that like most widespread types, they were made by multiple producers and probably at multiple locations. Pool and Bey (2007:36) note that the "vast majority of [Maya] pottery was made and consumed locally" (see also Arnold *et al.* 1991). This has been found repeatedly for Maya ceramics, most famously with the Preclassic Sierra Group types which are very stylistically consistent across the entire Maya lowlands. This pilot project was designed to see if XRF could detect intra-type compositional groups that could be investigated and hopefully confirmed by other techniques such as petrography, SEM, and XRD. The longer-term thinking was that if standard XRF would reveal compositional groups in this otherwise homogenous pottery, handheld XRF would have the potential to do the same. The ability to use handheld XRF on large numbers of samples in the field could allow the establishment of what are essentially technological varieties of Payil Red and Palmul Incised. In some ways, this new portable and handheld technology could solve what Aimers (2007a) has called "The Curse of the Ware"—the inconsistent treatment of paste in type-variety (see Rice 1979 for an extended discussion of this issue).

Another reason for the interest in handheld XRF is that the ability to export large number of sherds from Belize, or any country is difficult at best, making traditional analysis difficult, and analysis of large sample groups even more problematic. Being able to transport the XRF to the field would allow for on-site analysis of the sherds and could help archaeologist direct their research questions *in situ*.

Benchtop XRF sample preparation and analysis

To investigate the viability of the XRF as a "discriminating" tool, a selection of samples representing the Payil Red and Palmul Incised types were analyzed for their major and trace element compositions in the Department of Geological Sciences ar SUNY Geneseo. All samples were analyzed with a PANalytical AXIOS Sequential WD-XRF Spectrometer. The XRF uses a 4 kW Rh-anode X-ray source and both a flow and a scintillation detector. The flow detector is ideally suited for the analysis of transition elements and the scintillation detector is ideal for the analysis of heavy elements. A set of internal curved crystals (including the following options: LiF 200, LiF 220, PE 002, and GE 111), are also used in every analysis to disperse the X-rays emitted from the sample according to their different wavelengths using diffraction. The crystals are connected to a turret that rotates to insert one crystal at a time into the beam path. The crystal that is selected depends upon the element that is being analyzed (Table 13.1). The XRF is operated at voltages that range from 10 kV to 60 kV and currents that range from 10 mA to 125 mA. Typically flow detectors and scintillation detectors have resolutions below 1000 eV, but when used in unison with a crystal spectrometer, that resolution is greatly improved to range from approximately 12 eV (LiF 220) to 31 eV (LiF 200) (Jenkins 1999:100).

All pottery samples were cleaned with water and an ordinary nylon-bristle toothbrush to remove soil and particulate matter that was loosely adhered to the pottery and not considered original to the pottery body. The samples were then crushed using a SPEX SamplePrep Mixer/Mill and a hardened steel grinding canister equipped with 2 grinding balls. The grinding process produced a powder that passed through an 80-mesh sieve size. This was achieved by milling

~10 grams of sherd material for 3 minutes. If, after milling, large pieces of sample remained, the sample was milled for an additional 1 to 3 minutes depending on the abundance of large fragments. Between each sample, the ball mill was cleaned by milling quartz sandbox sand for 3 minutes. The cleaning sand was disposed of and the canister was then blown out with compressed air to remove any additional sand. Small sized particles are easier to fuse into glass beads/disks because they have a greater amount of surface area and dissolve easier in the fusing process. Small and even particle sizes are also essential for preparing compacted pellets because the small and even-sized particles are easier to homogenize and will compact into a more coherent flat-faced pellet with no major nicks and divots in the sample surface. Both fused beads and well-made pressed pellets are essential for obtaining the best possible XRF analysis because they minimize matrix effects, which can skew the data and not accurately represent the overall chemistry of the sherd. The powdered materials are also in the ideal form for X-ray diffraction (XRD) analysis and samples can be analyzed first with XRD and then the powder can be re-used in the preparation of XRF samples.

Crystal name	PANalytical's suggestions for use
LiF 200 crystal	Used for routine analysis for elements ranging from K to U
LiF 220 crystal	Used for routine analysis for elements between V and U. It is not as reflective as the LiF 200, but has a higher dispersion effect.
Upgrade to PE (002) curved crystal	Used for elements between Al and Cl
Ge (111) curved crystal	Used for P, S and Cl
PX1 synthetic multilayer crystal	Used for elements between O and Mg

Table 13.1:
XRF analyzing crystals and their suggested uses during analysis.

Major element chemistries were measured on samples prepared as fused glass beads (see Burke *et al.* (1998) for more information on this technique). The creation of a glass bead entails melting a flux (in our case, a lithium borate-based flux), which dissolves the powdered sample at high temperatures. The molten mixture is cooled to produce an extremely homogenous glass bead that is exceptional for avoiding matrix issues associated with particle size and mineralogy. 0.5 ± 0.001 grams of powdered sample were mixed with 6 ± 0.001 grams of fusion flux. The fusion flux was comprised of 33% lithium metaborate, 66% lithium tetraborate and 1% lithium bromide, which is a non-wetting agent that deters sticking of the glass bead to the platinum mold. Once the sample and the flux were mixed together, the mixture was poured into a platinum crucible. Using an automated fusing machine, the crucible was evenly heated to ~1150°C, mixed well (while avoiding the creation of bubbles), and then the molten mixture was poured into a platinum mold. When cool, the resultant glass bead was analyzed with a PANalytical AXIOS XRF. The analytical program, IQ+ used internal standardization to quantify the element concentrations. The initial standardization was based upon fundamental parameters that were improved by analyses of laboratory-generated standards containing known amounts of major elements (the IQ+ suite). The initial standardization is regularly checked by weekly analyses of two glass monitor standards (BGSMON and AUSMON-F) as well as the occasional analysis of an in-house set of geological samples which allow us to monitor for drift, background levels and quantitative accuracy. The two glass monitor samples came with the XRF. AUSMON-F is a drift monitor standard that was specifically chosen to coordinate with measurements of silicate materials and is available through multiple vendors including Analytical Reference Materials International and Brammer Standard. Our current accuracy for major elements is ~ ±1 wt.% for high concentration major elements. After every analysis we manually inspected each spectrum to make sure that the "search-and-match" function of the analytical program identified all the peaks. We also force the analytical program to strips Br from the results. Br is a constituent of the flux and is never considered as a major element. Br must be stripped from the analysis because it interferes with the aluminum peaks; the position of the bromine L-lines at 1,480.4 eV overlap with the aluminum K-lines located at 1,486.3 and 1,486.7 eV. Since the quantity of Br in the sample was known, stripping it from the spectrum was trivial and did not have a detrimental effect on the Al analyses. Glass beads are ideal materials for major element analyses because they are extremely homogenous and will not generate analytical errors due to grain sizes or preferential grain orientation as is the case when mica or clay is present [see Jenkins *et al.* (1995) for a discussion of grain-size related errors]. In general, the concentration of any element as it relates to the intensity of the X-ray peak is mathematically described as:

$$C = K \bullet I \bullet M \tag{1}$$

C represents the concentration of a specific element, K is the calibration constant determined from the analysis of standards, I is the intensity of the peak and M is a correction factor that accounts for matrix effects. The M value accounts and corrects for a variety of parameters including particle size, particle size distribution, crystallographic nature, and grain orientation (see Rousseau 2006 for insight on the calculation for M). In this simple equation, M is potentially the greatest source of error because of the variety of parameters it incorporates. Vitrifying a sample simplifies the calculation of M, thus strengthening the certainty associated with the resulting concentration measurement.

Even though glass beads minimize matrix effects, they were not used for trace element analysis. The preparation of glass beads is a dilution process, making them not ideal for trace element analysis due to the low concentrations of the elements being studied. Trace element data were obtained on pressed pellets instead of glass beads. Pressed pellets were prepared by combining 6 ± 0.001 grams crushed sherd (using the same crushing techniques described above) with 1 ± 0.001 grams cellulose binder (PrepAid Cellulose Binder $(C_6H_{10}O_5)_n$ from SPEX CertiPrep). The mixture was homogenized and then pressed into a pellet using a steel die and plunger set. The mixture was pressed with a hydraulic press at 25 tons for 15 minutes and then the pressure was slowly released over a 1-minute timeframe to avoid cracking the disk. Pressed pellets provide a concentrated amount of material that is well homogenized and flat, which is important for obtaining a correct analysis. Changes in sample height will shift the location of peaks on the energy spectrum. Once the data is collected, it is quantified by comparing the analysis to blank monitor standards (to account for drift), synthetic calibration standards (which create the empirical foundation for the analysis), and a set of geological standards (which help improve the initial standardization). The primary foundation for the trace analysis is a set of synthetic standards that work in conjunction with a software system that 1) defines the background values around the analytical peak locations, 2) accounts for peak interactions (as was the case for Br and Al in the major element analysis) and 3) accounts for matrix parameters in the standards. Detection limits vary by element but are reported to range between 0.4 ppm to 1.3 ppm for many of the elements of interest (Sr and Y at 0.4, Rb at 0.6, Zr at 1.1 and Nb at 1.3; Jenkins 1999:119) (see Table 13.2 for trace element results).

The geological standards were prepared in the SUNY Geneseo XRF laboratory using the same techniques described above. These standards are widely used by the geological community and supplied by the United States Geological Survey (see Wolf and Wilson (2007) for a brief overview of the USGS Standards program). The set of standards used at SUNY Geneseo was chosen because they best represent the variety of samples analyzed by their XRF facilities. Should this pilot project become more substantial, the potsherd analyses would benefit and be improved by using ceramic standards that might better represent the nature (in terms of mineralogy and matrix) of the Maya pottery and our initial analysis can be re-calculated based on the improved standardization.

Sample #	Sc	V	Cr	Mn	Co	Ni	Cu	Zn	Ga	As	Br	Rb	Sr	Y	Zr	Nb
T1	32	39	60	79	3	13	5	53	8	0	3	4	134	23	157	7
T2	29	74	55	255	6	23	13	55	13	4	2	84	67	25	155	13
T4	32	48	65	172	3	14	4	52	9	5	3	6	139	23	224	10
T5	30	41	57	58	2	15	4	82	9	3	3	4	120	30	160	8
T7	27	62	52	315	5	25	11	81	10	3	7	51	134	31	212	16
T8	27	44	69	217	3	16	5	100	10	0,47	3	4	114	19	225	9
T9	23	62	55	216	5	27	14	99	13	3	1	54	108	30	195	15
T10	43	31	29	464	5	18	5	29	4	0	1	6	69	13	70	2
T12	34	45	61	354	3	17	5	67	9	9	2	6	124	21	193	8
T13	28	46	75	153	2	16	6	56	10	3	5	12	220	25	249	11
T14	26	50	91	324	3	24	9	71	11	4	5	12	252	35	294	12
T15	27	43	65	156	2	18	7	110	10	4	3	8	85	15	206	9
T16	26	72	58	725	8	26	13	93	11	2	4	61	136	31	202	14
T17A	0,05	11	0	58	4	7	0	51	11	6	2	55	59	17	121	9
T17B	0,05	11	0,92	63	4	7	0	52	11	2	1	57	62	16	124	9
T18	0,22	2	0	53	2	6	0	60	9	2	3	53	97	18	113	8
SP1	0	11	0,1	14	1	7	0	26	7	6	5	5	456	13	148	6
SP2	0,47	0	0	11	0	7	0	30	8	7	7	7	450	19	136	7
SP3	0,37	2	0	11	5	11	0	41	10	1	9	5	449	18	166	9
SP4	0,05	7	2	17	2	5	0	28	8	7	5	8	592	18	185	7
SP5	0,05	6	0,89	17	2	4	0	28	6	5	4	8	453	20	137	6
SP6	0,32	6	0	11	1	5	0	41	8	8	11	7	330	20	156	7
SP8	0,05	1	0	17	0	7	0	27	8	5	5	6	362	15	148	7
SP9	0,05	3	0,32	14	1	4	0	31	7	6	10	8	394	16	184	7
SP10	0,05	3	0	0	0	4	0	36	10	5	4	8	255	21	184	9
SP11	0,05	0	0	10	0	5	0	39	8	8	13	7	388	17	185	8
SP12	0,05	5	0	39	3	9	0	33	9	9	3	11	387	21	236	9
SP13	0,05	3	0	13	1	5	0	45	10	6	8	6	363	17	213	9
SP14	0,04	0	0	30	13	7	0	37	10	5	9	10	348	17	212	9
SP15	0,05	1	0	3	1	3	0	40	10	4	14	8	479	19	165	9
SP16	0,1	0	0	11	1	9	0	150	10	5	6	8	504	14	182	9
SP17	0,14	0	0	57	2	17	0	50	14	8	6	11	400	36	331	15
SP18	0,12	0	0	12	3	8	0	29	8	4	6	6	469	16	141	7
SP19	0,03	0	0	80	4	9	0	46	12	9	9	14	499	32	304	13
SP20	0,05	4	0	74	2	11	0	51	14	7	6	14	329	28	335	14

Mo	Cd	Sn	Sb	I	Cs	Ba	La	Ce	Nd	Sm	Yb	Hf	Ta	W	Pb	Th	U
2	8	9	3	11	3	220	48	114	47	7	0	3	2	8	14	9	3
0,25	5	11	3	9	10	362	24	57	27	3	0	3	3	12	16	11	3
2	6	11	3	10	2	279	37	84	34	5	0	4	2	10	17	11	2
1	7	9	2	9	0,08	235	75	156	67	12	0	4	2	8	17	9	2
0,34	8	10	3	7	6	377	33	71	32	6	0	7	2	10	22	11	3
2	1	9	3	7	2	276	31	72	28	5	0	6	3	10	17	10	2
0,18	2	11	2	5	3	369	32	65	30	8	0	4	3	13	19	12	2
0,44	5	9	6	12	5	213	18	47	10	9	0	3	2	4	9	5	3
2	4	8	2	7	3	202	47	96	38	8	0	5	2	8	16	9	3
3	3	9	3	10	6	253	47	101	44	9	0	6	2	13	19	13	3
2	2	8	3	11	2	265	42	74	33	6	0	6	2	12	21	15	3
3	5	8	1	7	0,95	175	21	51	22	3	0	4	1	10	16	9	2
0,67	5	9	0,73	4	6	360	38	70	32	7	0	5	2	10	25	11	3
0,39	4	11	5	6	0	0	0,07	4	3	7	0	5	2	6	14	10	3
0,53	6	11	4	9	0	0	0,08	9	2	0	0	4	2	4	14	10	3
0,27	6	11	3	7	0,22	0	0,7	10	2	17	0	3	3	1	13	10	2
2	2	9	2	8	1	0,9	0	19	0	0	0	3	3	4	11	10	5
2	6	9	2	11	17	0	0	25	6	0	0	3	6	5	15	10	4
3	1	7	2	6	0	11	0	10	0	0	25	5	0	0	16	12	5
3	5	10	2	11	0	0	0,09	4	1	6	0	5	2	2	13	12	8
2	7	9	4	12	0	0	0,08	5	2	3	0	3	2	2	11	10	4
3	5	8	0,64	7	1	0,06	0	8	4	0	0	3	2	2	15	10	5
2	5	9	0	8	0	0	0	8	1	5	0	4	0	8	13	9	5
1	4	10	3	7	0	0	0	8	3	3	0	5	4	3	14	10	5
3	5	11	4	8	0	0	0,04	85	15	0	0	4	1	3	18	10	4
2	3	9	3	9	0	0	0,55	7	0	0,03	0	6	2	2	16	10	8
2	5	9	2	8	0	0	0	7	0	0	0,31	5	4	4	17	13	6
0,98	5	10	5	7	0	0	0,28	8	2	4	0	6	1	0	15	11	4
2	7	9	2	10	0	0	0	38	37	10	6,37	5	5	0	14	11	9
3	6	10	2	7	0	0	0,04	69	0	0	0	4	0,78	4	16	11	6
2	7	12	3	4	0,03	0,04	0	0	0	2	0	5	0	2	25	11	5
3	6	10	3	6	0,29	0	0,07	10	2	0	0	7	2	13	26	18	5
3	1	8	3	11	0	0	0,33	17	4	2	0	3	0	9	15	11	5
3	5	9	3	6	0	0	0	10	0	3	0	8	0	4	23	17	5
3	6	10	3	7	0	0	0	13	8	0	0	8	4	7	26	16	5

Table 13.2:
Trace element analysis of pressed pellets by panalytical XRF. Results are in ppm.

Handheld XRF sample preparation and analysis

Samples of sherds were prepared by both abrading sections using a silicon carbide grinding paper and by scraping the surface with a stainless steel blade to remove any accretions, surface decoration or slip to ensure the bulk matrix of the ceramic body was exposed. The surface was flattened to simulate the flat smooth samples prepared by fusion and pellet methods and ranged in size from approximately 1 cm^2 to 1.5 cm^2. Samples were cleaned with an air compressor blast prior to analysis. Analysis was conducted using a Bruker handheld Tracer III-SD XRF spectrometer with an Rh tube and a SDD with a resolution of ~ 145 eV. Instrument setting for high Z elements were 40 kV at 20 μA using an Al Ti Cu filter (thickness of 150 μm copper, 25 μm Ti and 300 μm Al) producing valid count rates between 7000 – 9000 cps for between 500 – 600 seconds. Data was analyzed using Bruker SP1XRF software.

Values obtained from the benchtop XRF for trace element analysis from the pressed pellets were used to create an initial calibration using Bruker S1CalProcess based on Compton normalization in conjunction with empirical calibration. The process used was similar to that by Smith (Chapter 2 this volume) and Ferguson (Chapter 12 this volume). The limiting factor was that the data for major light elements that were being calculated from the fused bead samples were not available at the time of calibration so the relevant elements (Al, Si, P, S, K, Ca, and Fe) were not calibrated for in this initial study but will be included in the second round of testing.

The instrument setup for measuring these low Z elements is 15kV and 55 μA under vacuum with no filter. Potentially a 25.4 μm Ti filter could be used, and might be thought of as the right filter to use since it would absorb the L lines of Rh avoiding any peak overlap in that region of the spectrum (i.e. Cl Kα – 2.6 keV). In fact, using no filter provides a better assessment of the lighter elements present. The L lines of Rh enhance the detection of elements with binding energies just below that of Rh Lα (2.69 keV). This enhances the detection of Al, Si, P, and S (1.48 keV, 1.74 keV, 2.0 keV, 2.31 keV respectively; see Figure 13.4 for comparison). This initial testing indicates that when additional data from the benchtop XRF is made available, a similar process for calibration (as described for the trace elements) will be successful.

Determining the limits of quantification using this methodology for our sample set has not yet been undertaken. This is due to the limited number of samples we have to look at and the restricted range of concentrations for each element of interest. As the database continues to grow, a more complete study will be undertaken to provide an accurate assessment of the quantification potential. That being said, the limits of detection achieved using a WDS crystal spectrometer (see above) are not possible with a silicon drift detector (SDD). Although the SDD has a better resolution than a Si-PIN detector (~145eV vs ~200eV respectively), this is approximately 10 times the resolution of the WDS system. There is inevitably some peak overlap that will not be able to be resolved. In addition, WDS systems have a better baseline which improves their detection limit. Previous work on similar Mesoamerican ceramics have shown detection limits as low as 4 ppm (Tagle and Gross, 2010) but a more realistic value for the limit of quantification may be closer to 10 – 14 ppm as found by Speakman *et al.* (2011). This of course varies by element, but provides a rough idea of the potential level of quantification attainable using the handheld XRF. Both of these previous studies used a Tracer III-V) which has a Si-PIN detector and quantification of this level should be considered reasonable for the SDD detector as well. (For more discussion on the determination of the limits of detection and quantification see Thomsen and Schatzlein and Mercuro 2003).

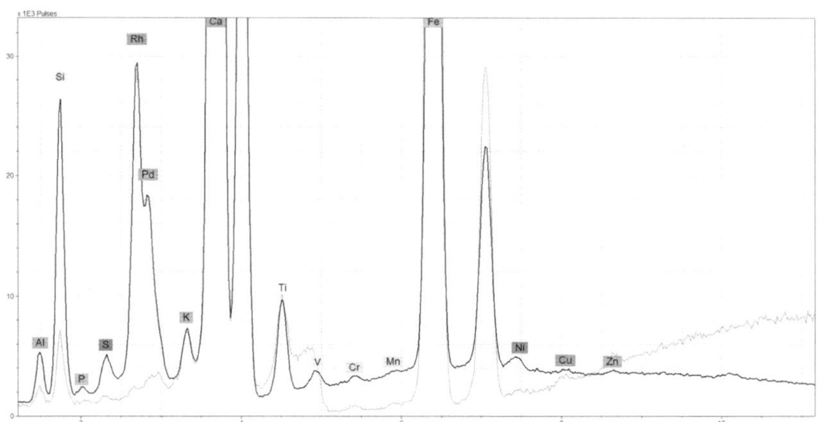

Figure 13.4:
Spectra of light element analysis using a 25.4 μm Ti filter (grey line) versus using no filter (black line) (TRACER III-SD Rh tube 15 kV, 55 μA with vacuum for 180 secs).

The high Z element calibration was able to obtain values for a wide range of other elements outside the low Z ones mentioned previously. The calibration file was created in two parts, one for the identification of the major high Z elements with other elements that have a direct effect on either peak overlap or slope correction (Sr, Y, Zr, Nb, Mo, Cd, Sn and Sb) and the other for the identification of the remaining high Z elements (see Table 13.3). The general matrix of the ceramic appears to be comprised of Ca, Fe, Sr and Zr as these four elements have the strongest $K\alpha_1$ peaks (see Figure 13.5). These matrix elements are included in both calibrations to help account for both elemental peak overlap as well as slope corrections.

All elements found in samples	Elements used for Cal1	Elements used for Cal2
$CaK\alpha_1$	$CaK\alpha_1$	$CaK\alpha_1$
$TiK\alpha_1$	$TiK\alpha_1$	
$V\ K\alpha_1$	$V\ K\alpha_1$	
$CeL\beta_1$	$CeL\beta_1$	
$CrK\alpha_1$	$CrK\alpha_1$	
$NdL\beta_1$	$NdL\beta_1$	
$MnK\alpha_1$	$MnK\alpha_1$	
$FeK\alpha_1$	$FeK\alpha_1$	$FeK\alpha_1$
$CoK\alpha_1$	$CoK\alpha_1$	
$ZnK\alpha_1$	$ZnK\alpha_1$	
$HfL\beta_1$	$HfL\beta_1$	
$GaK\alpha_1$	$GaK\alpha_1$	
$TaL\beta_1$	$TaL\beta_1$	
$W\ L\beta_1$	$W\ L\beta_1$	

All elements found in samples	Elements used for Cal1	Elements used for Cal2
$AsK\alpha_1$	$AsK\alpha_1$	
$BrK\alpha_1$		
$PbL\beta_1$	$PbL\beta_1$	
$ThL\alpha_1$	$ThL\alpha_1$	
$RbK\alpha_1$	$RbK\alpha_1$	
$SrK\alpha_1$	$SrK\alpha_1$	$SrK\alpha_1$
$Y\ K\alpha_1$	$Y\ K\alpha_1$	$Y\ K\alpha_1$
$ZrK\alpha_1$	$ZrK\alpha_1$	$ZrK\alpha_1$
$NbK\alpha_1$	$NbK\alpha_1$	$NbK\alpha_1$
$MoK\alpha_1$	$MoK\alpha_1$	$MoK\alpha_1$
$RhK\alpha_1$	$RhK\alpha_1$	$RhK\alpha_1$
$CdK\alpha_1$		$CdK\alpha_1$
$SnK\alpha_1$		$SnK\alpha_1$
$SbK\alpha_1$		$SbK\alpha_1$

Table 13.3:
Total elements found in the samples and the elements used for each trace element calibration.

Results of calibration

Initial results of calibration look quite promising. The correlation between the benchtop and handheld XRF is surprisingly good considering the small range of values for each element at this point in time. Strontium and Zr show relatively good coefficient of determination values (R^2 values of 0.8574 and 0.8744 respectively). Although these values suggest a relatively good linear fit, a quick look at the plots (Figures 13.6 and 13.7) indicates some discrepancy in datapoints. These outliers will likely improve as we obtain more data-points. For elements with lower concentrations the results are also quite promising. For example, Y, Nb, and Rb have concentrations ranging from ~ 14 – 35 ppm (Y), ~ 6 – 15 ppm (Nb), and ~5 – 57 ppm (Rb). The R^2 values are 0.8814, 0.9447, and 0.09897 respectively. Although we are looking at values that hover around the limit of quantification, it appears that we have a good correlation between the benchtop XRF and the handheld unit.

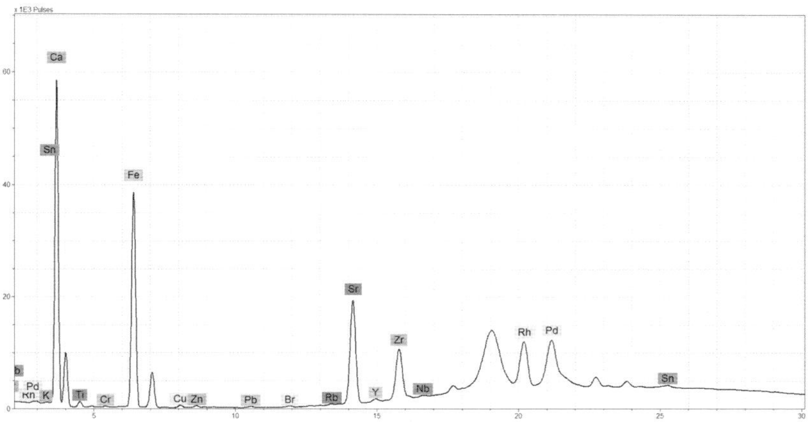

Figure 13.5:
Spectrum of sample SP01 showing the major elements being Ca, Fe, Sr, and Zr (TRACER III-SD Rh tube 40 kV, 20 mA with 279 mm Al, 25.4 mm Ti, 6mm Cu filter collected for 600 secs).

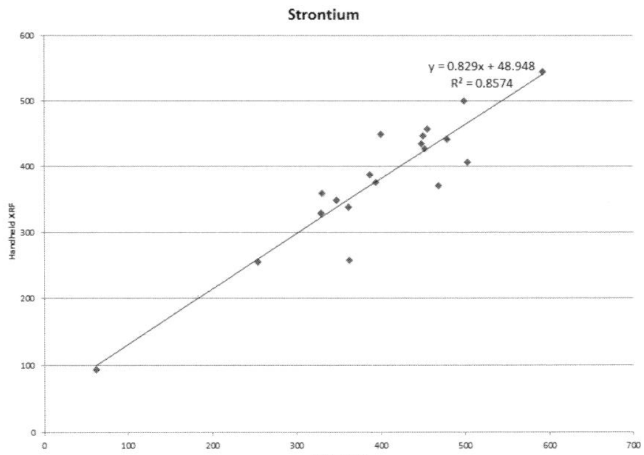

Figure 13.6:
Plot of benchtop XRF and handheld XRF data for Strontium showing an R^2 of 0.8574.

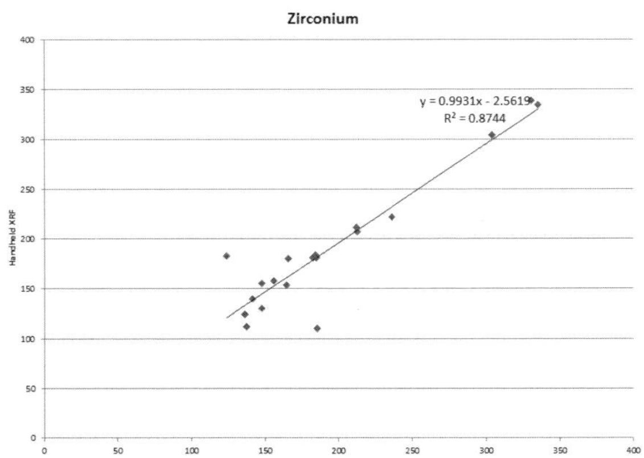

Figure 13.7:
Plot of benchtop XRF and handheld XRF data for zirconium showing an R^2 of 0.8744.

	B	H	B	H	B	H	B	H	B	H	B	H	B	H	B	H	B	H	B	H	B	H	B	H	B	H
Sample	Mn		Zn		Ga		Th		Rb		Sr		Y		Zr		Nb		Mo		Cd		Sn		Sb	
SP1	14	22	26	15	7,0	6,7	10	10	5,3	5,4	456	457	13	15	148	130	5,9	5,9	1,7	1,8	2,3	3,8	9,5	10	2,3	1,8
SP2	11	7,6	30	21	8,3	7,6	10	10	7,1	6,5	450	446	19	20	136	124	6,9	6,7	2,5	1,9	6,2	3,6	9,1	10	2,4	1,5
SP3	11	0	41	100	10	10	12	12	4,8	6,1	449	435	18	18	166	180	8,6	8,9	3,0	2,0	0,6	6,6	7,1	10	1,7	3,2
SP4	17	17	28	17	8,3	6,8	12	11	8,1	8,0	592	544	18	18	185	110	7,2	7,3	2,6	2,1	5,4	8,7	10	7,4	2,0	7,4
SP5	17	13	28	29	6,4	6,1	10	10	8,2	7,0	453	426	20	17	137	112	5,9	6,1	1,7	1,7	6,7	3,4	9,2	10	3,9	3,0
SP6	11	21	41	40	7,7	7,4	10	10	7,2	7,7	330	358	20	23	156	158	7,1	7,7	3,0	2,0	4,7	3,5	7,6	9,5	0,6	1,3
SP8	17	18	27	12	7,8	7,8	9,4	10	6,1	6,4	362	337	15	15	148	155	6,8	6,8	1,8	2,0	4,9	2,4	9,1	10	0,0	1,4
SP9	14	21	31	31	7,0	7,5	10	11	7,9	8,3	394	375	16	16	184	182	7,4	7,4	1,0	2,0	4,2	3,7	10	9,4	3,3	2,5
SP10	0	0	36	41	10	10	10	10	8,0	8,1	255	255	21	20	184	184	8,9	7,7	2,8	1,9	4,8	4,7	11	10	4,0	-0,1
SP12	39	38	33	19	9,5	10	13	12	11	10	387	386	21	20	236	222	9,2	9,0	2,1	2,3	5,2	4,2	9,3	10	2,4	2,5
SP13	13	24	45	46	10	10	11	10	5,6	5,5	363	257	17	15	213	207	9,5	7,9	0,98	2,0	4,7	2,1	10	10	4,8	2,4
SP14	30	29	37	88	10	10	11	12	10	10	348	348	17	17	212	212	9,3	10	2,2	2,0	6,6	7,0	9,3	10	1,6	1,6
SP15	3,1	2,7	40	40	10	10	11	11	7,5	7,1	479	441	19	19	165	154	8,8	8,4	2,6	2,0	6,1	4,0	10	10	1,9	1,7
SP16	11	18	150	150	10	10	11	11	8,5	9,1	504	407	14	14	182	182	8,8	8,7	1,6	2,0	6,9	6,8	12	9,4	3,4	4,2
SP17	57	58	50	49	14	14	18	17	11	11	400	449	36	35	331	339	15	14	3,1	3,0	5,6	5,6	10	10	3,0	3,0
SP18	12	18	29	29	8,2	8,0	11	10	6,3	5,5	469	369	16	17	141	140	6,9	6,8	3,1	2,0	1,1	2,0	8,4	10	3,0	3,6
SP19	80	80	46	46	12	12	17	17	14	13	499	500	32	32	304	304	13	13	2,6	2,8	5,5	5,5	8,9	9,0	2,5	5,2
SP20	74	74	51	51	14	14	16	16	14	14	329	328	28	28	335	335	14	13	3,1	2,8	5,7	6,1	10	10	3,5	3,3
T17B	63	63	52	0	11	13	10	9,4	57	42	62	93	16	23	124	183	8,9	8,0	0,53	1,5	5,6	2,0	11	10	4,2	2,5

Table 13.4:
Comparative data between the benchtop XRF and the Handheld XRF (B = Benchtop XRF, H = Handheld XRF).

Although good correlation exists with some elements there are those that are better correlated than others (Table 13.4). It is possible that some of these elements might be better calibrated when investigating the low Z elements (i.e. Sc, Ti, Ce, V, Nd, Cr, and Mn) and attempts will be made to do some to improve the correlation. During the next phase of this study multiple sample locations will be run on each sample to establish the methodology's precision.

Conclusions

The inherent issues related to XRF analysis of ceramic sherds without major pre-treatment are becoming understood by researchers today. In this particular case, the homogeneity of the ceramic matrix offers a unique opportunity to use handheld XRF for chemical characterization. Where there has been success in calibration of handheld XRFs for other purposes, the intention of collecting data for provenancing purposes is still questionable (Speakman *et al.* 2011) and the issues complicating this matter may never be resolved. At this point of time, for this study, it appears that we have good correlation of data and it is expected that we can only improve as we expand the range of elemental concentrations used for calibration.

The honeymoon between archaeological ceramic specialists and handheld XRF may be ending, but the relationship shows promise. The work described in this chapter is a step toward establishing standard methods for the use of handheld XRF with archaeological ceramics and calibration with benchtop XRF. It is still too early to comment on some of the broader cultural questions about Maya pottery production and exchange discussed earlier, although the results of the calibration work are reason for optimism that handheld XRF can be a valuable tool in this research, which will ultimately require a range of analytical approaches. Research questions that require methods as diverse as type-variety and XRF demand collaboration between specialists with different backgrounds and we hope this chapter is an example of the value of such collaboration.

References

Aimers, James J. 2003. "Abandonment and nonabandonment at Baking Pot, Belize." In *The Archaeology of Settlement Abandonment in Middle America*, edited by Takeshi Inomata and Ronald W. Webb, 149-162. Salt Lake City: University of Utah Press.

———. 2004a. *Cultural Change on a Temporal and Spatial Frontier: Ceramics of the Terminal Classic to Early Postclassic Transition in the Upper Belize River Valley.*

Oxford: BAR International Series No.1325, British Archaeological Reports.

———. 2004b. "The origins and significance of the scroll or slipper support in Maya ceramics." *Mexicon* Vol. XXVI (2):36-42.

———. 2004c. "The Terminal Classic to Postclassic transition in the Belize Valley." In *Archaeology of the Upper Belize Valley: Half a Century of Maya Research*, edited by James F. Garber, 147-177. Gainesville: University of Florida Press.

———. 2007a. "The Curse of the Ware: Using Ceramic Systems in Belize." *Research Reports in Belizean Archaeology* Vol. 4:101-110.

———. 2007b. "What Maya Collapse? Terminal Classic Variation in the Maya Lowlands." *Journal of Archaeological Research* Vol. 15:329-377.

———. 2008. "Snakes on Planes: Sinuous Motifs in the Art of Lamanai." *Research Reports in Belizean Archaeology* Vol. 5:115-123.

———. 2009. "Bring It On: Using Ceramic Systems at Lamanai." *Research Reports in Belizean Archaeology* Vol. 6:245-252.

———. 2010. "You Only Live Twice: The Agency of Ritual Ceramics at Lamanai." *Research Reports in Belizean Archaeology* Vol. 7:119-125.

———. 2012a. *Ancient Maya Pottery: Classification, Analysis, and Interpretation*. Gainesville: University Press of Florida.

———. 2012b. "Problems and Prospects in Maya Ceramic Classification, Analysis, and Interpretation." In *Ancient Maya Pottery: Classification, Analysis, and Interpretation*, edited by James J. Aimers, in press. Gainesville: University Press of Florida.

Aimers, James J., and Elizabeth Graham. 2012a. "Close Encounters at the Mansion: The Columbian Exchange at Structure H12-08, Tipu, Belize." *Research Reports in Belizean Archaeology* Vol. 9:In press.

———. 2012b. "Type-Variety on Trial: Experiments in Classification and Meaning Using Ceramic Assemblages from Lamanai, Belize." In *Ancient Maya Pottery: Classification, Analysis, and Interpretation*, edited by James J. Aimers, in press. Gainesville: University Press of Florida.

Arnold, D. E., H. Neff, and R. L. Bishop. 1991. "Compositional Analysis and "Sources" of Pottery: An Ethnoarcheological Approach." *American Anthropologist* Vol. 93 (1):70-90.

Barone, G., V. Crupi, F. Longo, D. Majolino, P. Mazzoleni, G. Spagnolo, V. Venuti, and E. Aquilia. 2011. "Potentiality of non-destructive XRF analysis for the determination of Corinthian B amphorae provenance." *X-Ray Spectrometry* Vol. 40:333-337.

Bishop, Ronald L., Veletta Canouts, Patricia L. Crown, and Suzanne P. de Atley. 1990. "Sensitivity, Precision, and Accuracy: Their Roles in Ceramic Compositional Data Bases." *American Antiquity* Vol. 55 (3):537-546.

Bishop, Ronald L., Robert L. Rands, and George R. Holley. 1982. "Ceramic Compositional Analysis in Archaeological Perspective." *Advances in Archaeological Method and Theory* Vol. 5:275-330.

Buhrke, V. E., R. Jenkins, and D.K. Smith. 1998. *A Practical Guide for the Preparation of Specimens for X-Ray Fluorescence and X-Ray Diffraction Analysis*. New York: John Wiley & Sons, Inc.

Cecil, Leslie G. 2001a. "Developing Technological Styles of Petén Postclassic Slipped Pottery with Regard to Clay Mineralogy." In *Archaeology and Clays*, edited by I. Druc. Oxford: British Archaeological Reports, International Series 942.

———. 2001b. *Technological Styles of Late Postclassic Pottery from the Central Petén lakes Region, El Petén, Guatemala*. Unpublished Ph. D. dissertation, Department of Anthropology, Department of Anthropology, Southern Illinois University, Carbondale, Carbondale, IL.

———. 2009. "Technological Styles of Slipped Pottery and Kowoj Identity." In *The Kowoj: Identity, Migration, and Geopolitics in Late Postclassic Petén, Guatemala*, edited by Prudence M. Rice and Don S. Rice, 221-237. Boulder: University of Colorado Press.

———. 2012. *Postclassic Ceramic Database* 2012 [cited January 2012]. Available from http://faculty.sfasu.edu/cecillg/Sites.html.

Cecil, Leslie G., and T.W. Pugh. 2004. Kowoj Symbolism and Technology at Late Postclassic Tipuj. Montreal: Paper presented at the 69th Annual Society for American Archaeology Meeting.

Cesareo, R, A Castellano, G Buccolieri, S Quarta, M Marabelli, P Santopadre, M Leole, and A Brunetti. 2004. "Portable equipment for energy dispersive X-ray fluorescence analysis of Giotto's frescoes in the Chapel of the Scrovegni." *Nuclear Instruments and Methods in Physics Research Section B: Beam Interactions with Materials and Atoms* Vol. 213:703-706.

Culbert, T Patrick, and Rober L. Rands. 2007. "Multiple Classifications: An Alternative Approach to the Investigation of Maya Ceramics." *Latin American Antiquity* Vol. 18 (2):181-190.

Culbert, T Patrick, and L. A. Schwalbe. 1987. "X-Ray Fluorescence Survey of Tikal Ceramics." *Journal of Archaeological Science* Vol. 14:635-657.

Day, Peter M., Evangelia Kiriatzi, Alexandra Tsolakidou, and Vassilis Kilikoglou. 1999. "Group Therapy in Crete: A Comparison Between Analyses by NAA and Thin Section Petrography of Early Minoan Pottery." *Journal of Archaeological Science* Vol. 26 (8):1025-1036.

Fermo, Paola, Eleonora Delnevo, Marina Lasagni, Silvia Polla, and Mariette de Vos. 2008. "Application of chemical and chemometric analytical techniques to the study of ancient ceramics from Dougga (Tunisia)." *Microchemical Journal* Vol. 88 (2):150-159.

Forster, N., P. Grave, N. Vickery, and L. Kealhofer. 2011. "Non-destructive analysis using PXRF: methodology and application to archaeological ceramics." *X-Ray Spectrometry* Vol. 40:389-398.

Gifford, J. C. 1976. *Prehistoric Pottery Analysis and the Ceramics of Barton Ramie in the Belize Valley*. Vol. 18, *Memoirs of the Peabody Museum of Archaeology and Ethnology*. Cambridge: Harvard University.

Glascock, Michael D. 1992. "Characterization of Archaeological Ceramics at MURR by Neutron Activation Analysis and Multivariate Statistics." In *Chemical Characterization of Ceramic Pastes in Archaeology*, edited by Hector Neff, 11-26. Madison, WI: Prehistory Press.

Goren, Yuval, Hans Mommsen, and Jörg Klinger. 2011. "Non-Destructive Provenance Study of Cuneiform Tablets Using Portable X-Ray Fluorescence." *Journal of Archaeological Science* Vol. 38 (3):684-696.

Hein, A., A. Tsolakidou, I. Iliopoulos, H. Mommsen, J. Buxeda i Garrigós, G. Montana, and V. Kilikoglou. 2002. "Standardisation of elemental analytical techniques applied to provenance studies of archaeological ceramics: an inter laboratory calibration study." *Analyst* Vol. 127 (4):542-53.

Howie, Linda A. 2005. *Ceramic Production and Consumption in the Maya Lowlands During the Classic to Postclassic Transition: A Technological Study of Ceramics at Lamanai, Belize*. Unpublished Ph.D. thesis, Department of Archaeology, University of Sheffield, Sheffield, U.K.

Ida, Hiroyuki, and Jun Kawai. 2005. "Analysis of wrapped or cased object by a hand-held X-ray fluorescence spectrometer." *Forensic Science International* Vol. 151 (2-3):267-272.

Jenkins, R., R.W. Gould, and D. Gedcke. 1995. *Quantitative X-Ray Spectrometry*. 2nd ed. New York: Marcel Dekker, Inc.

Jenkins, Ron. 1999. *X-ray Fluorescence Spectrometry*. 2 ed, *Chemical analysis*. New York: John Wiley & Sons, Inc.

Jones, Lea D. 1986. *Lowland Maya Pottery: The Place of Petrological Analysis*. Oxford: BAR International Series 288.

———. 1991. "Tempering Trends in Lowland Mayan Pottery." In *Recent Developments in Ceramic Petrology*, edited by Andrew Middleton and Ian Freestone. London: British Museum.

Lohse, Jon C. 2010. "Archaic origins of the Lowland Maya." *Latin American Antiquity* Vol. 21 (3):312-353.

Mannino, Maria Rosaria, and Santino Orecchio. 2011. "Chemical characterization of ancient potteries from Himera and Pestavecchia necropolis (Sicily, Italy) by Inductively Coupled Plasma–Optical Emission Spectrometry (ICP–OES)." *Microchemical Journal* Vol. 97 (2):165-172.

McKillop, Heather. 2005. *In Search of Maya Sea Traders*. College Station: Texas A&M.

Morgenstein, Maury, and Carol A. Redmount. 2005. "Using Portable Energy Dispersive X-Ray Fluorescence (EDXRF) Analysis for On-Site Study of Ceramic Sherds at El Hibeh, Egypt." *Journal of Archaeological Science* Vol. 32 (11):1613-1623.

Neff, H. 2000. "Neutron activation analysis for provenance determination in archaeology." In *Modern Analytical Methods in Art and Archaeology*, edited by E. Ciliberto and G. Spoto, 81-134. New York: Wiley.

Neff, Hector H. 1992. "Introduction." In *Chemical Characterization of Pastes in Archaeology*, edited by Hector H. Neff, 1-10. Madison: Prehistory Press.

Newman, Bonita, and Lawrence Loendorf. 2005. "Portable X-Ray Fluorescence Analysis of Rock Art Pigments." *The Plains Anthropologist* Vol. 50 (195):277-283.

Peacock, D. P. S. 1970. "The scientific analysis of ancient ceramics: A review." *World Archaeology* Vol. 1 (3):375-389.

Piorek, Stanislaw. 1997. "Field-portable X-ray fluorescence spectrometry: Past, present, and future." *Field Analytical Chemistry & Technology* Vol. 1 (6):317-329.

Pohl, Mary D., Kevin O. Pope, John G.Jones, John S. Jacob, Dolores P. Piperno, Susan D. deFrance, David L. Lentz, John A. Gifford, Marie E. Danforth, and J. Kathryn Josserand. 1996. "Early Agriclture in the Maya lowlands." *Latin American Antiquity* Vol. 7:355-372.

Pollard, Mark, Catherine Batt, Ben Stern, and Suzanne M. M. Young. 2007. *Analytical Chemistry in Archaeology*. Cambridge Manuals in Archaeology. Cambridge: Cambridge University Press.

Pool, C. A., and G. J. Bey III. 2007. "Conceptual Issues in Mesoamerican Pottery Economics." In *Pottery Economics in Mesoamerica*, edited by C. A. Pool and G. J. Bey III, 1-38. Tucson: University of Arizona Press.

Rice, Prudence M. 1976. "Rethinking the Ware Concept." *American Antiquity* Vol. 41 (4):538-543.

———. 1977. "Whiteware Pottery Poduction in the Valley of Guatemala: Specialization and Resource Utilization." *Journal of Field Archaeology* Vol. 4 (2):221-233.

———. 1987. *Pottery Analysis: A Sourcebook*. Chicago: University of Chicago Press.

———. 1991. "Specialization, Standardization, and Diversity: A Retrospective." In *The Ceramic Legacy of Anna O. Shepard*, edited by Ronald L. Bishop and Frederick W. Lange, 257-279. Niwot: University Press of Colorado.

———. 1996. "Recent Ceramic Analysis 2: Composition, Production, and Theory." *Journal of Archaeological Research* Vol. 4 (3):165-202.

Rousseau, R.M. 2006. "Corrections for Matrix Effects in X-ray Fluorescence Analysis-A Tutorial." *Spectrochimica Acta-Part B Atomic Spectroscopy* Vol. 61 (7):759-777.

Sanders, William T. 1960. *Prehistoric Ceramics and Settlement Patterns in Quintana Roo, Mexico*. Washington, D.C.: Contributions to American Anthropology and History 12, Carnegie Institution of Washington Publication 606.

Schwalbe, L. A., and T Patrick Culbert. 1988. "Analytical Measures of Variability and Group Differences in X-Ray Flourescence Data." *Journal of Archaeological Science* Vol. 15:669-681.

Shackley, M. Steven. 2010. "Is there Reliability and Validity in Portable X-Ray Flourescence Spectrometry (PXRF)?" *Archaeological Record* Vol. 10 (5).

Smith, Michael E., and Frances Berdan. 2000. "The Postclassic Mesoamerican World System." *Current Anthropology* Vol. 41:283-286.

———. 2003. *The Postclassic Mesoamerican World, The Postclassic Mesoamerican World*. University of Utah Press: Salt Lake City.

Smith, Robert E., G. R. Willey, and J. C. Gifford. 1960. "The Type-Variety Concept as a Basis for the Analysis of Maya Pottery." *American Antiquity* Vol. 25 (3):330-340.

Speakman, Robert J., Nicole C. Little, Darrell Creel, Myles R. Miller, and Javier G. Iñañez. 2011. "Sourcing ceramics with portable XRF spectrometers? A comparison with INAA using Mimbres pottery from the American Southwest." *Journal of Archaeological Science* Vol. 38:3483-3496.

Sullivan, Lauren A., and J.J. Awe. 2012. "Establishing the Cunil Ceramic Complex at Cahal Pech, Belize." In *Ancient Maya Pottery: Classification, Analysis, and Interpretation*, edited by James J. Aimers, in press. Gainesville: University Press of Florida.

Tagle, R., and A. Gross. 2010. ARTAX/TRACER III-V Provenance Studies on Ceramics with Portable (m)XRF Spectrometers. In *Lab Report XRF 437*. Berlin: Bruker Nano GmbH.

Tenorio, Dolores, Melania Jimenez-Reyes, Sonio Rivero-Torres, Fabiola Monroy-Guzman, and Elizabeth Teresita Romero-Guzman. 2010. "Characterization of Archaeological Ceramic from Lagartero, Chiapas, Mexico, by Nuclear and Conventional Techniques." *Journal of the Mexican Chemistry Society* Vol. 54 (1):40-45.

Thomsen, V., D. Schatzlein, and D. Mercuro. 2003. "Limits of Detection in Spectroscopy." *Spectroscopy* Vol. 18 (12):112-114.

Uda, M., S. Sassa, K. Taniguchi, S. Nomura, S. Yoshimura, J. Kondo, N. Iskander, and B. Zaghloul. 2000. "Touch-free in situ investigation of ancient Egyptian pigments." *Naturwissenschaften* Vol. 87 (6):260-263.

Wolf, R.E., and S.A. Wilson. 2007. USGS Fact Sheet 2007-3056. Washington D.C.: United States Geological Survey.

Chapter 14

Glass analysis utilizing handheld X-ray fluorescence

Bruce Kaiser and Aaron Shugar

Introduction

The analysis of glass artifacts, whether they are modern, historical or archaeological, has been a major focus of investigation for cultural heritage researchers. The seminal paper by Sayre and Smith (1961) demonstrated the value in determining the chemical composition of glass to aid in the identification of the manufacturing raw materials. Since then the chemical composition of glass has been used to characterize raw materials use (Freestone 2005), establish base glass composition (Rehren 2000), suggest trade patterns (Kenyon *et al.* 1995; Walton *et al.* 2009), investigate glass colorants (i.e. Sanderson and Hutchings 1987), and help determine firing temperatures used in ancient glass furnaces (Rehren and Shugar 2001; Shugar and Rehren 2002). The analytical techniques that have been used to determine major, minor, and trace element compositions are varied and include scanning electron microscopy (SEM; Wagner *et al.* 2008), inductively coupled plasma spectroscopy (ICP; i.e. Casoli and Mirti, 1992; Dussubieux *et al.* 2009; Dussubieux Robertshaw and Glascock, 2009), neutron activation analysis (NAA; i.e. Franá *et al.* 1987; Kenyon *et al.* 1995; Speakman and Glascock, 2007; Rahman *et al.* 2008), and X-ray fluorescence (XRF; Cox and Gillies, 1986; Shugar 2000).

The use of XRF, which relies on the induced inner atomic electron shell (K,L, M shells) transitions that cause atoms to emit their characteristic low energy X-rays, has been used for glass analysis for over 50 years (Instituut 1957:154). Over the past 5 years there has been rapid advancement in the miniaturization and development of hardware as well as improved computer software that make this analytical technique readily available to cultural heritage researchers. Like all other analytical techniques, XRF has advantages and limitations for the analysis of modern, historical and archaeological glass artifacts. The user's detailed understanding of the physics of the analytical technique is imperative to ensuring proper analysis and determination of accurate compositional data. Without this understanding and having assured that the sample being analyzed meets very strict criteria, the concentration values given by the various instruments' analytical software (calibrations) are often inaccurate. Using concentrations derived in this manner to draw conclusions regarding provenance or technology of manufacture, for example, usually results in erroneous deductions.

Benchtop lab based XRF systems rely on destructive analysis of fragments removed from objects or of material that has to be properly "prepared" to ensure that the samples are homogenous. This typically is done by homogenizing the samples by milling to a uniform particle size and then adding approximately 80% by weight prescribed glass formers such as borax, mixing the batch well and then using a furnace to make fused glass beads (This was being done as early as the late 1950's for XRF analysis; Instituut 1957:154; International Atomic Energy Agency 1960:44). For XRF analysis of glass objects (both lab based and handheld), taking samples and homogenizing them is still the only way to get a "failsafe" accurate elemental analysis of the bulk glass composition. Indeed this is still the very technique required by the quantitative XRF laboratory systems. Unfortunately this process cannot be used on most important cultural heritage objects prepared from glass because its destructive nature is not acceptable for these irreplaceable materials.

With the advent of the most advanced handheld X-ray fluorescence spectrometers that have non-destructive, non-invasive (no sampling required) analytical capabilities, the potential for direct object analysis without causing damage, even the formation of color centers in the glass through radiation damage, becomes practical. However, great care must be taken when applying XRF on any type of unprepared object – surface compositions will be highlighted and non-ideal geometries will compromise data quality. The physics of characteristic X-ray emission from the elements has not changed since it was first discovered in the late 1800s (the beginnings of X-ray fluorescence spectroscopy can be tracked to the finding of high energy photons (X-rays) by Wilhelm Conrad Röntgen in 1895), and is dependent on several factors that can have an exceptional effect on quantification.

The fluorescent X-ray intensity will depend on: the inverse square of the distance from the element/atom to the detector, the matrix density (an exponential factor), elemental X-ray energy emission (an exponential factor), element location in the sample matrix (an exponential factor), any incident beam filtering and the incident beam energy, distribution and angle (all also exponential factors). What this ultimately means is that for any change to any of these parameters in your sample, the *reported* quantitative emission changes very rapidly resulting in large variation in the resulting reported elemental concentrations. The equation that highlights many of these sensitivities is given in the Section on the Depth of Analysis below. The sensitivity to incident beam energy can be understood by reviewing the NIST website dedicated to X-ray attenuation (Hubbell and Seltzer 2012). It also means if an unknown sample is non-uniform (always the case for ancient and historic glass objects, which have surface alkali depletion due to weathering) the automated calibrations built into these instruments will report concentrations that have no real physical basis or meaning. In addition, there is no way any algorithm or 'tuned' calibration can take all these unknown variables into account. Thus, if you do not have good sample uniformity, which is afforded in a prepared sample, quantitative analysis of a non-uniform material, by any XRF system, and the resulting data *must* be wrong. That being said, the raw spectrum contains an exceptional amount of data from which some quantification information can be derived. One should always examine the spectra closely to glean as much as possible from this data, as long as the physics that produced it are well understood.

Non uniform glass – The value of qualitative data analysis

Based on the above discussion on the physics of X-rays interacting with material (glass is focused on here, but the information is relevant to all analyses) and the things one must take into account before accepting any quantitative information, one would think that the usefulness of XRF is quite limited. This is particularly the case with works of art and archaeological materials since a great preponderance of the material to be investigated in these fields is non-uniform as a function of depth and XRF is a surface analysis technique. Actually, this far from the truth.

If one takes into account the physics behind XRF and examines all of the information that can be gathered for the raw photon spectrum (1000 eV to 40000 eV) of a non-uniform material an enormous amount of valuable information can be collected (note that for this discussion the word photon will replace X-ray as this is a more accurate term). Consider all the information available to you when examining an object with 1 eV to 3 eV photons (what is visible to the naked eye in sunlight). You can determine texture, absorption and reflection of light, determine a color, assess its uniformity, and determine transparency (do these photons travel through the object?). The interaction of visible photons and our detection of the beam coming back into our detectors (spectrometers – our eyes), provide a huge amount of information we use in our computer (brain) to sort out many facts about the material we are looking at. Now consider the entire beam of photons from 1000 eV to 40000 eV coming back into the detector in the instrument (our extension of the eye since we cannot see these photons) and the information that can be gained by examining the entire spectrum. Not just the primary fluorescence, but the elastic, inelastic, and coherent scatter as well as the absorption characteristics of the photon beam used to probe the object. All of this information is in the raw spectrum. These features of the spectrum give you information relevant to the sample density, thickness, and average overall atomic composition. In addition to the detailed analysis of the entire fluorescent emission lines of each element detected, you can extract information about the element distribution in the sample.

Very often the algorithms in the automated quantification software provided by manufacturers that provide concentrations will misidentify elements and occasionally not identify elements at all. Visually examining the entire fluorescence spectrum will help identify these algorithm processing errors and allow you to avoid inaccurate conclusions one may draw from those errors. There is thus a huge amount of information for each non uniform glass object (any historic glass object) that can be determined by careful examination of the full raw high energy photon spectra (1000 to 40000 ev) taken from several locations on the artifact and coupled with the information obtained from the visible spectra (1 eV to 3 eV) at those locations. These include acquiring information on the:

1. Relative elemental content of your material either area to area, color to color, or cross object to object,
2. Density of the material,
3. Layered structure of the material,
4. Uniformity of the material,
5. Thickness of the material,
6. Crystalline structure of the material,
7. Any contamination associated with the material (encrustations, diagenesis),
8. And overall sample matrix (silica, potash, soda, lead oxide, Ceramic (Ca))

All of this information can be obtained from careful examination of the raw spectra by judicious application of the physics of photon interaction with matter.

An excellent study using these techniques was performed on mosaic glass tesserea from Herculaneum showing the power of qualitative assessment of the spectra and how one can derive exceptionally valuable data using these principles (Del Frederico *et al.* forthcoming).

Calibration and quantification of glasses

Even with the extensive range of information that qualitative analysis can provide, including the identification of colorants, opacifying agents, and the provenances of some raw materials, there is still a need in glass studies to have quantitative data to make clear inferences about technology of manufacture and provenance. For the last seven years the first author (Kaiser) has been investigating the strengths and weakness of the application of noninvasive XRF to the analysis of a broad array of glass objects. Working with glass reference material provided by various sources he has been developing and studying the best methods and techniques possible to create quantification for glass of all types. The results of this intense study, including what can be determined and what cannot be determined based on the limits of the physics involved are discussed here.

Advantages of energy dispersive XRF (ED-XRF)

Energy dispersive spectrometry refers to the method used to collect the resulting X-rays emitted from a sample once it has been excited by a particular beam of photons typically in the range of 1000 eV to 50,000 eV. Because these photons of this energy typically result from electron energy changes they are referred to as X-rays. The process involves the capture and conversion of individual photons (X-rays) coming from a sample, to a voltage pulse that is proportional to the energy of the X-ray that was captured in the detector. Each voltage pulse amplitude is determined electronically and the number of each is accumulated and displayed as a spectrum over a given period of time set by the experimenter. The relationship between the X-ray energy and the voltage pulse determined and counted is linear. This process is done one X-ray at a time and over a given amount of time a spectrum is produced. The resulting distribution of X-ray energies results in the typical raw energy spectrum one is used to seeing from XRF analysis, having the Y axis convey the number of X-rays of a given energy and the X axis convey the energy range (typically between 0 – 60 keV). The advantages of energy dispersive XRF are many and often lead to researchers overextending the results of the analysis and trusting the numerical value output of the instrument as pre-calibrated by the manufacturer. One should never take such values as correct unless the researcher has very carefully examined the raw spectrum from which they were derived and unless they know their sample meets very strict criteria of sample uniformity, thickness, and matrix type.

The advantages of XRF include not having to physically sample an artifact (as discussed above); the non-destructive/non-invasive analysis of an artifact which insures that the artifact is in exactly the same condition after the analysis as it was before the analysis. The portability of the instrumentation allows analysis to be done in remote areas such as archaeological excavations, research institutions, or museums where the object cannot be removed for traditional laboratory-based analysis; providing 'instant' qualitative analysis; and in some cases if the situation allows (these situations are often misunderstood and quite specific), quantitative elemental analysis.

The limitations of ED-XRF must be very clearly understood prior to reporting any specific results of a given analysis. A clear explanation of each of those limitations, listed below, follows:

- light elements
- surface conditions
- sample uniformity
- measurement depth.

Light elements

Glass can be broken into five main component categories. These are the network formers – they are the basis for glass structural formation, SiO_2 being the most prevalent for us in our consideration of archaeological and historical glasses (although Al_2O_3 is a glass former that is also critical to the stability of ancient and historic glasses); network modifiers – these disrupt the network by reducing the number of Si-O-Si bridges by introducing additional oxygen into the matrix, resulting in a change to the overall Si/O ratio and creating a more open network of non-bridging oxygen ions. Without the network modifiers, a non-crystalline silicate such as fused silica is a glassy or vitreous solid but not a true glass. Modifiers tend to be alkali metal oxides and alkaline earth oxides in nature and have a typical formula of M_2O (i.e. Na_2O, K_2O); network stabilizers act to bridge non-bridging oxygens in the glass matrix offering stability to the overall structure. The concentration of these stabilizers is critical to the stability of ancient and historic glasses – CaO concentrations between 5 wt. % and 10 wt. % are thought to be largely responsible for the high stability of ancient Roman glasses. Network stabilizers have a typical formula of MO (i.e. CaO or MgO); intermediates may be considered network formers, participating in the overall matrix, or they can occupy interstitial sites (i.e. PbO, ZnO and TiO_2); and finally colorants which are typically made of metal oxides or colloidal metals (i.e. copper produces both blue

(Cu^{2+}) and red (Cu^{+} or CuO) (oxidation state dependent), Mn – purple, Fe – green/brown (oxidation state dependent. Fe^{2+} is blue and Fe^{3+} is yellow, a mix of these oxidation states produces green or brown) see Weyl, 1959).

One additional glass component may be added to these five: opacifiers. These are dispersed (undissolved) compounds that create an opaque glass usually through the addition of crystallites that disperse light as it travels through the glass matrix. These inclusions can create serious difficulties when attempting to gain quantitative analysis of glasses (see Shugar 2000 for examples). The main constituents of glasses are all light elements, with only the coloring agents and fluxes such as Pb and Ba typically being higher Z.

The detection and measurement of light elements such as Si, Al, and Na is a particularly difficult challenge when using XRF for glass analysis but highly important since they are the main constituents of all glasses. They typically define the base glass chemistry and are used for basic characterization (for example, historic glasses are characterized as being soda lime, mixed alkali, or potash glasses). In addition, the X-ray energy emitted by these elements is very low and drastically affected in a non-linear fashion by the parameters listed above. In fact, the relative emission of light elements (i.e. Na, Mg, and K) is exponentially lower than the emission of X-rays from heavier elements like Pb and Ba. Therefore, for the same concentration of low Z and high Z elements, the resulting number of X-rays detected from Si and Fe will be drastically lower for Si, resulting in less signal to use to accurately quantify the Si. This is true for all of the lower Z elements from Na to Ca.

The five most common families of historical glasses encountered are listed below (Brill, 1968). All of these classifications are based on the classic defined variation of base glass chemistry as first defined by Sayre and Smith (1961) of either a sodium rich glass or a potash rich glass. Minor variations of these classifications have been found in Asian glasses which tend to be higher in BaO (#4). One will note that there are two main glass compositions in these families, one is SiO_2-rich and the other is a high PbO glass:

1. Na_2O: $CaO:SiO_2$
2. K_2O: $CaO:SiO_2$
3. PbO: (Na_2O/K_2O) :SiO_2
4. PbO: BaO: SiO_2
5. K_2O: SiO_2

Depth of analysis

The attenuation of the emitted K, L, and/or M shell atomic X-rays are of serious concern when looking at materials of variable density. X-rays are attenuated as they pass through matter, meaning that the intensity of the characteristic X-rays being emitted decrease the further they have to travel through a medium. This effect is dependent on three main factors, the thickness of the sample, its density, and its absorption coefficient and it is expressed in the equation (1) below.

$$(1) \qquad I = I_o \exp^{((-\mu/\rho)x)}$$

Where I is the beam intensity with attenuation, I_0 is the beam intensity without attenuation, ρ is the density of the material, μ/ρ is the mass attenuation coefficient for a given material and a given X-ray energy, and ***x*** is the mass thickness which is obtained by multiplying the density ρ by the thickness t (x = ρt) (see Thomsen, *et al.* 2005 for more detail). Since the types of glasses being investigated are variable in their matrix densities and composition (i.e. SiO_2 vs. PbO), the effect these two base glass matrices will have on the attenuation of X-rays will be vastly different. To demonstrate this, one can calculate the way different elements travel through/ out of both SiO_2 base glasses and PbO glasses. Table 14.1 gives the thickness of silica or lead glass required to stop all but 1% of the emitted X-ray energy from escaping the glass from each element that may be found in the glass matrix. This is particularly important to understand, because the only determination one has for the concentration of elements in a samples is based on the number of characteristic X-rays that reach the detector. The equation used to make this determination and calculate the values in Table 14.1 is based on equation (1) and can be written as;

$$(2) \qquad t = -\mathrm{Ln}(I/I_0)/((\mu/\rho) * \rho)$$

Where t is the sample thickness, I is the beam intensity with attenuation, I_0 is the beam intensity without attenuation, μ/ρ is the mass attenuation coefficient for a given material and a given X-ray energy, and ρ is the density of the material that the X-rays are going through.

Since I/I_0 is the ratio, or percent of X-rays that are attenuated, using a value of 0.01 assumes that all but 1.0% of X-rays have been fully attenuated. This gives a value for the $-\mathrm{Ln}(I/I_0)$ term in the equation (2) of 4.61. The mass attenuation coefficient for both SiO_2 and PbO can be found at the NIST X-ray physical reference data site (Hubbell and Seltzer 2011). One can also find the density of these compounds as well as others if you are interested in exploring this effect on other materials (Hubbell and Seltzer 2011).

Element	Emitted energy (keV)	Analysis depth in Si glass (cm)	Analysis depth in Pb glass (cm)
O	0,53	0,000	0,000
Na	1,04	0,001	0,000
Mg	1,2	0,001	0,001
Al	1,47	0,002	0,001
Si	1,74	0,003	0,001
P	2,01	0,001	0,001
Ca	3,69	0,006	0,001
Cr	5,41	0,019	0,003
Fe	6,4	0,03	0,004
Cu	8,01	0,058	0,008
Zn	8,64	0,077	0,009
Pb	10,55	0,113	0,014
Zr	15,78	0,384	0,014

Table 14.1:
A list of elements typically found in glasses and their characteristic X-rays emission depths for both a silicate glass and a lead glass.

Table 14.1 clearly shows that the characteristic X-rays emitted from the light elements that make up the majority of the glass composition can only escape from 7 μm – 64 μm in a silicate glass (Na – Ca) and from 3 μm to 13 μm in a leaded glass. So the resulting information being collected by the detector is not representative of the overall composition of the glass matrix itself. Since the collection of Pb is a silicate glass can escape from depths of 1.13 mm and Ca can only escape from depths of 64 μm, the amount of characteristic lead X-rays collected is far greater than that of Ca. Any additional disturbance to the surface (whether that be corrosion/alkali depletion or non-ideal geometry for example) will significantly alter the measurements further. Since these instruments are run in open air or with a vacuum restricted to within the instrument head and because there are materials covering the window and detector, all the X-rays from oxygen do not even reach the detector and are therefore not capable of being measured. Sodium can only be measured utilizing one handheld system currently available on the market with some aftermarket modification to the instrument, and that system was utilized in this study.

Surface condition

The second feature that can severely affect quantification is the surface condition of the glass. We have established that XRF only measures the major constituents of glass to a depth of approximately 30 μm. If this is the case, any alteration to the surface chemistry will greatly affect the results. Archaeological glasses are notorious for having corrosion crusts, dirt embedded in the surface of the glass as well as heavily leached and alkali depleted surfaces which can extend over 10 – 20 μm into the glass (Sanders and Hench 1973; Clark and Yen-Bower, 1980; Koob 2006 Chapter 2). Even cleaned glasses that have all the visible corrosion removed may still have alkali depletion. General exposure of glass to water in the environment can also alter its surface chemistry with crizzling (a crazed surface with a network of tiny cracks typically caused by network collapse after alkali leaching) and etching affecting the alkali and matrix compositions. In addition, chemicals or surface coating of any kind will alter measurements. If this type of surface is expected to be analyzed it must be noted by the researcher that any quantitative results derived from these types of samples will be inaccurate since X-rays from the various elements come from various depths, and the distribution of the elements in the sample as a function of depth are unknown (due to corrosion/encrustations/alkali depleted layers). Where possible it is highly recommended that the altered surface be removed. The glass should be dry and any dust or dirt must be removed as it is typically composed of similar elements to that of the glass and if not removed, one could be collecting information on the dust composition rather than the glass itself.

Depth of analysis and sample thickness

The third factor that directly affects quantitative results is the thickness and uniformity of the sample. Coated or layered glass such as ancient Hellenistic or Roman mosaic glass cannot be quantitatively analyzed by XRF. There are 3 facts that are immediately clear from the numbers in Table 14.1 above and the Figure 14.1 below. First one cannot use XRF to measure the content of oxygen. Second the detection of Na is extremely difficult as it is a light element and is easily attenuated in air. Third each element is measured to a very *different depth* in the glass, moving from detecting the Al X-rays that come from the surface (~20 μm), to detecting the Zr X-rays from up to ~ 4 mm in Si glass. So any non-uniformity in the glass will greatly alter the total counts of characteristic X-rays detected from each element and consequently the quantification concentrations calculated.

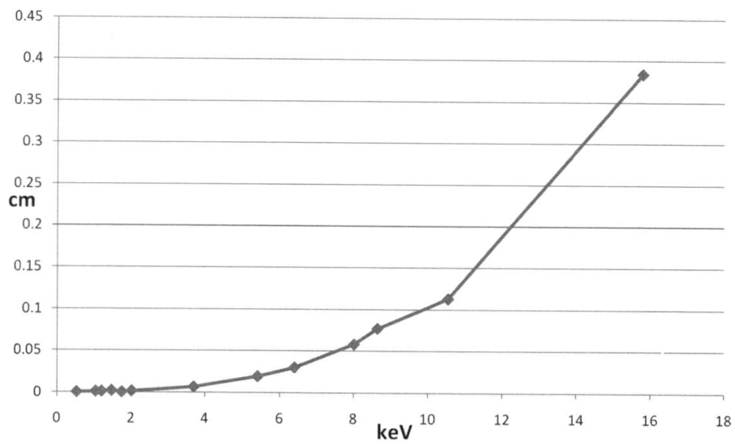

Figure 14.1
Emitted X-ray energy versus depth a given X-ray energy can escape Si glass with a loss of 99% of the X-rays at the surface.

XRF system glass calibration approach

The standard methods of calibration offered by handheld XRF manufacturers are typically based on fundamental parameters and variation of those parameters (see Chapter 1 this volume). These are not a viable option when attempting to obtain accurate quantification from unknown and heterogeneous cultural heritage materials including glasses. For this reason, it is highly recommended that an empirical calibration using reference materials or standards be used for the quantification of ancient and historic glasses (and other materials, see the other chapters in this volume for examples).

By taking into account the issues raised above and adding in a few additional facts regarding the interaction of 1 to 50 keV X-rays in glass, it is possible to create reasonable calibrations for specific glass types and elemental groups if the following rules are followed.

The selected reference material must include all elements of interest in the unknown sample, it must cover the range of elemental concentrations found in the unknowns; include elemental combinations typically found in the unknowns; one must know the concentrations of all elements more accurately than what is expected from the calibration; one must assure that the reference materials are infinitely thick which in the case of the glasses discussed here are at least 4 mm thick (see Table 14.1 and Figure 14.1); one must assure that the reference materials are very uniform, homogenous and composed of similar particles sizes (this is typically not a concern for bulk glass standards), in particular, one must assure the surface of the glass is clean and unaltered (beware of any strong acids used to etch the surface, or bases that might reside on the surface as these will significantly alter the surface structure and chemistry); the reference material must be composed of the same glass matrix as the unknown (Pb glasses and Si glasses are very different matrices and consequently separate calibrations must be done for each group); and the reference materials must cover a similar range of elemental X-ray emission depth *meaning that two separate calibrations are required to cover the full spectrum of elements, one for low Z elements from Na to Fe and the other for high Z elements from Fe to Pb is necessary.*

Once these criteria have been met with the production, or selection of reference materials, the analysis of reference materials and the unknown samples must be run in the exact same way. Following the recommended steps below will offer the best chance at acquiring accurate quantitative data.

Always check the spectrum (raw data) to make sure the backscatter portion of both the unknown and the reference material spectra have the same curvature (see Shugar and Sirois, Chapter 9 this volume). Confirm that you do not have any additional elements in your unknown samples that are not in your reference materials. If you do, you will need to acquire additional reference materials to compensate for them. Ensure that the elemental concentrations found in your unknown samples are not any higher than the concentration established with your reference materials or the resulting data can be skewed (see Ferguson, Chapter 12 this volume).

If you follow these rules, you can overcome the key effects of XRF that hamper accurate quantification. These include the exciting X-ray beam-sample interaction; issues related to the depth of analysis variability from element to element; the problems associated with lack of sample uniformity; additional matrix effects and the associated element to element interaction interferences that can occur in a sample as long as one also takes into account the heterogeneity caused by alkali depletion and/or opacifying agents.

Taking these precautions, several calibrations have been created using reference materials provided by Dr. R. Brill supplemented by NIST standards. Two case studies are now presented showing the power of handheld XRF for the quantification of glass.

Case Study 1: The Na oxide glass calibration (current state of the art)

Compiling the information above and incorporating additional calculations (outside the scope of this chapter), in conjunction with the advent of newer technology in handheld XRF instrumentation, a detailed elemental calibration was created for all the major oxides found in glasses including Na_2O. A unique set of reference materials were assembled by Dr Robert Brill at the Corning Glass Museum over years of research into glass chemistries (Brill, 1999; see Table 14.2 for a list of these reference materials and their oxide compounds). These materials cover the range of glass types listed above and additional reference materials collected by Brill will be made available for further research into calibrations for ancient glasses (pers com Brill 2011). In addition to these reference materials, a subset of glasses that vary in Na_2O concentrations were used for this calibration.

All of the analysis was performed using a Bruker Tracer III SD system with a Rh tube. The Tracer III SD uses a thin window XFlash® silicon drift detector with a resolution of approximately 144 eV. The instrument was run at a tube voltage of 15 kV and a filament current at 24.7 μA with no beam filter inserted. Aftermarket modifications were made by removing the instrument window and running a flow of 0.5 liter per minute of He out through the open window in the nose (helium used was "balloon grade"). This created a slight positive pressure on the system, enveloping the surface of the sample with He, but not with enough force to shift the sample once it was in place. This process removed the air from the head space between the sample, the X-ray tube, and the detector limiting the attenuation effects of air on light elements such as Na in this case. One could argue for the use of the vacuum to do a similar, if not better job at removing the air and creating a suitable environment for the analysis of Na except for two factors. First, using the vacuum requires the use of a nose piece window to maintain the vacuum in the head of the instrument. This window, although very thin, does attenuate Na and reduce the signal that reaches the detector. Second, the vacuum creates a slight depression in the nose piece window allowing for a small air pocket to form. It was for these reasons the He flush system was developed.

Reference	Na oxide	Mg Oxide	Al Oxide	Si Oxide	P Oxide	K Oxide	Sb Oxide	Ca Oxide	Ba Oxide	Ti Oxide	Mn Oxide	Fe Oxide	Co Oxide	Cu Oxide	Pb Oxide
REF A	15,00	2,64	0,90	67,00	0,10	2,90	1,88	5,00	0,52	0,83	1,29	1,10	0,02	1,17	0,07
REF B	16,00	1,01	4,11	60,30	0,92	1,00	0,49	8,50	0,08	0,11	0,30	0,33	0,04	2,63	0,47
BC	18,00	0,00	0,00	71,00	0,00	0,00	0,00	10,00	0,00	0,00	0,00	1,00	0,00	0,00	0,00
AD	18,00	0,00	0,00	72,00	0,00	0,00	0,00	10,00	0,00	0,00	0,00	0,00	0,15	0,00	0,00
DQI	19,00	4,00	2,10	63,30	0,00	2,00	0,00	7,50	0,00	0,00	0,40	0,60	0,15	1,00	0,00
DXS	15,70	1,57	2,72	71,65	0,00	1,08	0,00	6,25	0,00	0,01	0,00	0,33	0,00	0,00	0,00
DFZ	15,00	1,50	2,50	73,50	0,00	1,00	0,00	6,20	0,00	0,00	0,00	0,30	0,00	0,00	0,00
s0	17,00	2,72	6,19	61,87	1,63	3,10	0,00	4,62	0,05	0,32	0,10	2,31	0,00	0,00	0,00
EYC	20,00	0,00	0,00	70,00	0,00	0,00	0,00	9,00	0,00	0,00	1,00	0,00	0,00	0,00	0,00
JQS	14,00	1,40	2,00	66,00	0,00	1,50	0,00	10,00	0,00	0,00	1,00	1,00	0,10	3,00	0,00
s1	17,20	0,84	2,38	67,40	0,17	0,66	0,00	8,11	0,03	0,09	0,86	2,00	0,04	0,14	0,01
s2	17,20	0,42	2,33	70,80	0,14	0,58	0,00	7,99	0,03	0,06	0,03	0,36	0,00	0,00	0,00
s3	18,50	0,41	2,54	68,80	0,12	0,59	0,00	7,66	0,06	0,06	0,48	0,69	0,00	0,00	0,00
s4	16,48	0,51	2,33	69,60	0,16	0,65	0,00	7,49	0,07	0,03	0,53	0,37	0,00	0,01	0,02
s5	19,00	0,45	2,44	67,50	0,06	0,47	0,50	6,82	0,05	0,04	0,01	0,36	0,00	0,01	0,12
s6	15,00	0,55	2,36	68,70	0,17	0,59	0,80	7,71	0,03	0,02	0,66	0,86	0,04	0,10	0,01

Table 14.2:
The concentrations in wt. % of the various reference materials used for calibration.

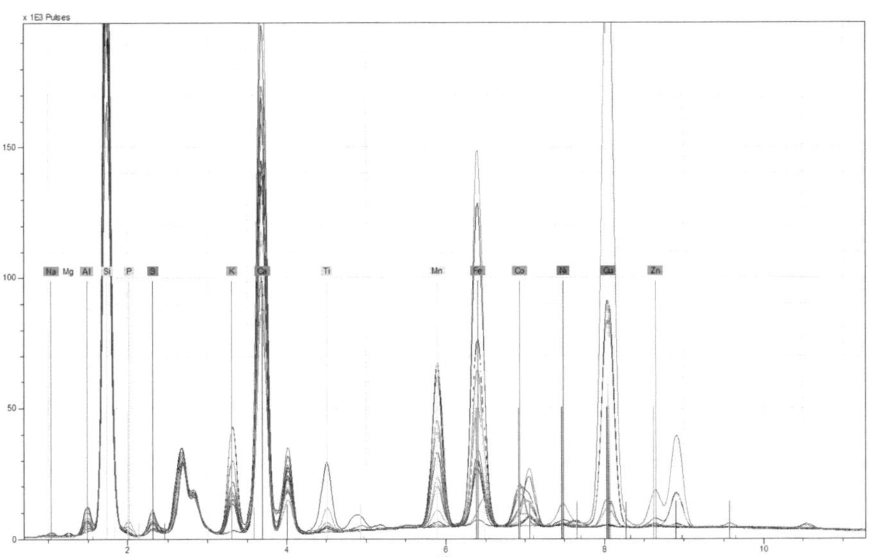

Figure 14.2:
Raw spectra of the reference materials indication the key elements identified (Rh tube 40kV, 25 µA no beam filter).

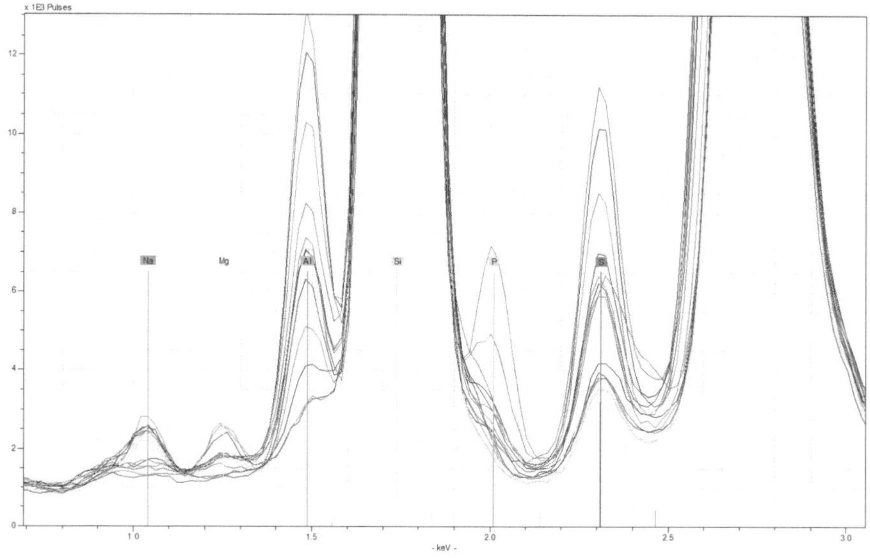

Figure 14.3:
Close up of the raw spectra of the reference materials showing the lighter elements (Rh tube 40kV, 25 μA no beam filter).

All samples were analyzed for 300 sec. The reference material glasses were compiled into the Bruker S1CalProcess software to create the empirical calibration. Table 14.3 contains the result of this unique calibration for the identification of Na using a handheld XRF. This calibration defines the current state of rapidly advancing noninvasive XRF analysis of silicate-based cultural heritage materials. The oxide concentration was used for calibration and is what is reported by the instrument even though it determines pure element concentration. The "val" is the numerical result of the calibration for each compound.

The correlation coefficient determination is excellent for some of these compounds, in particular MgO, Al_2O_3, and Fe_2O_3 (R^2 values: Na_2O – 0.503, MgO – 0.9464, Al_2O_3 – 0.9335, Si_2O – 0.6014, K_2O – 0.8632, CaO – 0.242, and Fe_2O_3 – 0.966). Although some of these R^2 values are significantly low, these initial results are very good considering the number of reference materials available at this point in time and can only be improved as we increase the number of reference materials, and the range of values for each compound used for the calibration.

Reference	Na oxide	Na val	Mg Oxide	Mg val	Al Oxide	Al val	Si Oxide	Si val	K Oxide	K val	Ca Oxide	Ca val	Fe Oxide	Fe val
REF A	15,00	14,56	2,64	2,38	0,90	1,15	67,00	66,48	2,90	2,99	5,00	6,94	1,10	1,09
REF B	16,00	16,27	1,01	1,20	4,11	4,09	60,30	61,21	1,00	0,92	8,50	8,41	0,33	0,43
BC	18,00	18,41	0,00	0,04	0,00	0,10	71,00	70,34	0,00	-0,03	10,00	7,06	1,00	0,78
AD	18,00	18,00	0,00	0,12	0,00	0,12	72,00	67,98	0,00	-0,02	10,00	8,57	0,00	0,07
s0	17,00	16,44	2,72	2,85	6,19	6,37	61,87	66,80	3,10	3,13	4,62	6,49	2,31	2,37
EYC	20,00	18,01	0,00	0,15	0,00	0,21	70,00	70,54	0,00	-0,01	9,00	8,10	0,00	0,01
JQS	14,00	15,27	1,40	1,43	2,00	2,09	66,00	67,57	1,50	1,39	10,00	9,29	1,00	1,15
s1	17,20	15,95	0,84	1,05	2,38	2,27	67,40	68,54	0,66	0,79	8,11	8,43	2,00	1,90
s2	17,20	17,45	0,42	0,58	2,33	3,84	70,80	70,80	0,58	1,24	7,99	7,63	0,36	0,36
s3	18,50	18,71	0,41	0,33	2,54	2,52	68,80	65,68	0,59	0,69	7,66	7,42	0,69	0,36
s4	16,48	17,67	0,51	0,41	2,33	1,80	69,60	70,40	0,65	0,93	7,49	7,45	0,37	0,38
s5	19,00	17,87	0,45	0,78	2,44	2,43	67,50	68,40	0,47	0,58	6,82	9,36	0,36	0,38
s6	15,00	17,66	0,55	0,43	2,36	2,67	68,70	68,79	0,59	1,87	7,71	6,66	0,86	0,87

Table 14.3:
Concentration of derived element oxides using the calibration created.

Case Study 2: An example of glass analysis in the field

At the Greco-Roman site of Karanis in Egypt, XRF analysis was conducted in the field on over 600 glass sherds (Susak Pitzer 2012). As a preliminary test of optimal sample preparation, a small study was conducted which demonstrates the potential use of handheld XRF analysis of glass in the field.

Two key questions drove this research:

1. Does sanding the sherds alter the resulting chemistry, and is it necessary to achieve accurate results?
2. Does it matter how long the samples are run (i.e. if the analysis is done for 180 or 300 seconds)?

A single analysis was performed on the un-sanded sherd for 300 seconds. It was then sanded by hand (using 2000 aluminum oxide grit wet sandpaper to remove approximately 0.05 mm into the surface of the glass) to remove any surface accretions, corrosion products and alkali depletion. It was then subjected to analysis 3 times for both 180 seconds and for 300 seconds. The sample was cleaned with ethanol prior to each XRF analysis.

An initial empirical calibration was created at the Corning Museum of Glass by Drs. Robert Brill and Bruce Kaiser using the same reference materials provided by Dr. Brill. A Bruker Tracer III V system with a Rh tube running at 40 kV and 25 µA with no beam filter was used for this calibration. The calibration was optimized to focus on detecting elements from Mg to Fe. A second instrument, set up with the same beam filter and operating parameters, was used in the field to take several analyses and each analysis was compared to data taken with CMOG Tracer III-V on reference materials to assure the two systems had identical backscatter. When it was found that the response was the same, it was determined that the CMOG glass calibration designed on one instrument could be used with the data taken with the field instrument.

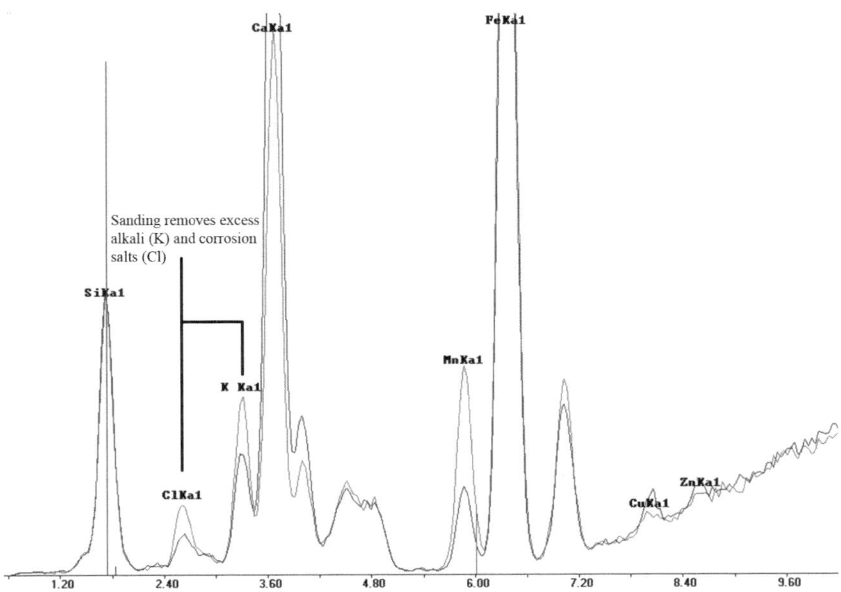

Figure 14.4:
Spectra showing the difference in elemental composition between an uncleaned glass sample and a cleaned glass sample. Instrument setting – Rh tube 15 kV 25 mA, no filter, collection time of 300 seconds.

The resulting spectra are shown in Figure 14.4 and clearly show that surface abrasion/polishing reduced the concentration of Cl, K, Mn and Fe. Interestingly, polishing did not alter the Si intensity greatly which is a key indicator that the corrosion and accretion layers on the glass contained some level of Si in its own right. In addition, the Ca signal more than doubled when the glass was sanded. The sanded analysis still shows some level of Cl, which can be a trace element in glass and have been attributed to natron as the source of sodium (has also been recorded in some cases of Early Bronze Age Egyptian glass – Shugar and Rehren 2002), one can suspect that some of the corrosion materials remain, and that would also indicate some level of alkali depletion remains as well. This is a prime example of the effect that coating can have on the results of analysis. Even very thin coatings (approximately 25.4 μm were removed from the glass) can have a pronounced attenuation effect on light element X-rays from 1 to 7 keV.

The resulting composition derived from the empirical calibration can be seen in Table 14.4. As with the previous study, although the values are reported in element form, they are actually in their appropriate oxide compound (see previous case study). The results show that a misinterpretation of the base chemistry could be made if the glasses had not been cleaned. It also indicates that there has been some clear leaching of K from the glass into the corrosion layers. The slight variation in total wt.% is likely due to the calibration being done on another instrument and suggests that there is about a 1% error built into this process.

With regard to the length of time required to collect data, note that the standard deviation between the three 180 second analysis and the three 300 second analysis shown in Table 14.4 are relatively similar with only Ca showing a marked improvement. For all purposes they are essentially the same, indicating, at least for this experiment, that analysis times of 180 seconds is sufficient to acquire data. Determining your count rate is essential when deciding how long to run your samples, and what level of quantification you desire. The one standard deviation in any random counting system can be estimated by taking the square root of the number of events detected in the photon energy of interest. So if one has 100 events in the photon energy of interest peak then the 1 sigma standard deviation is 10 or about 10%. Thus if you have 10000 events the uncertainly is 100 or just 1%.

An area of larger than 10 mm^2 was cleaned in each case as the spectrometer has a beam spot of 3 by 4mm. The cleaning of the glass shard altered the surface appearance of the glass making it appear more translucent as the encrustation was removed.

The difference in composition between the polished and unpolished glass is substantial. Not cleaning the surfaces and reporting the concentrations of the unpolished glass would have provided inaccurate and misleading results.

This simple example proves how careful one must be when analyzing glass and for that matter any unknown and potentially heterogeneous materials in a quantitative fashion with XRF. Ultimately the value of this experiment is that on site handheld XRF analysis of glass is possible. For researchers at Karanis, the knowledge of the chemical composition of the glass corpus can provide evidence for manufacturing processes and can help distinguish the raw materials employed in the base glass, as well as the colorants, decolorizers, opacifiers, and fluxing agents used. This data will aid in further defining a typology of the glass corpus from Karanis, as well as in identifying any evidence of glass recycling. Consumption patterns and reuse may correlate to fluctuations in the overall economy of the town and/or Roman Empire, as well as to diachronic changes in the Egyptian glass industry. This research comprises the incipient stages of a larger glass study exploring how value relates both to social standing and economic fluctuations in Karanis (Susak Pitzer 2012).

Conclusion

Glass artifacts have the potential to be a very deceiving group of objects. Although they often appear to be of uniform matrix, homogenous and have smooth cleaned surfaces, they rarely do. Whether they have undergone long-term burial corrosion, environmental alterations (i.e. leaching or etching), or purposeful surface etching their chemistry and physical makeup is adversely altered at the surface, making quantitative XRF analysis (a surface analysis technique) difficult at best. Accepting this fact and adjusting experimental protocol to account for it ensures that the use of quantitative analysis is still a powerful means of obtaining useful results. For quantification, many factors have to be taken into account, and with some excellent reference materials it is possible to create accurate results. With aftermarket modification, it is now even possible to quantify sodium. Qualitative analysis is also still a highly useful tool for identifying colorants, opacifying agents, and fluxes in these materials.

The advancement in technology that allows one to create, control and deliver a very precise and spectrally constant and consistent beam of photons of 1000 to 50000 eV coupled with the advancement in detection and processing capabilities of the resulting photons coming back to the detector provides one with a great amount of additional information concerning that object, just as photons from 1 to 3 eV interacting with your sample and detected by your eyes provide you with key information. Elemental make up and distribution, density, uniformity, thickness, molecular matrix, and crystalline structure (apparent in the Bragg scattering present in the raw spectrum) can all be determined by judicious investigation of the entire spectra of 1000 to 50000 eV photons detected from your sample. The fact that all of this is now achievable with a portable handheld device that can be carried anywhere and used on any object (solid, powder, liquid or gas) is a very significant paradigm shift. But with all that capability comes great risk, as those using these instruments must understand the physics behind it. The user must inspect and be able to "read" the raw spectra as there are no algorithms that will do this. Just as the physics of the interaction between visible photons (1 to 3 eV) and your object that provides information that your eyes see can be very complex; no computer algorithm can even begin to sort out what the human brain can!

When using the elemental fluorescence that appear in the spectra of unknown artifacts it is vital that the entire emission spectra including all the K, L and M lines of each element be considered (the analytical software used should give the user all the lines so they can be compared with the actual raw spectra), thus avoiding many of the pitfalls of automated concentration analysis. In understanding the fundamental physical phenomena of a tool as powerful as the handheld 1000 to 50000 eV photon generator/spectrometer a wealth of information can be rapidly learned about glass objects that was not previously possible without invasive and destructive sampling.

Acknowledgements

The authors would like to profoundly thank Dr. Robert Brill for his guidance and his support in this effort. He has graciously provided the data from chemical analysis of glasses he has collected over a lifetime of research. Without his aid, the calibrations created would not have been possible. We would also like to thank Angela Susak Pitzer for her support and the data collected during her research investigation of the glass from Karanis Egypt, and Dr. Eleonora del Federico for her support and for providing the data used to discuss the qualitative advantages of XRF analysis.

References

Brill, R. H. 1968. "The Scientific Investigation of Ancient Glasses." In *The Proceedings of the VIIIth International Congress on Glass*, 47-68.

Casoli, Antonella, and Piero Mirti. 1992. "The analysis of archaeological glass by inductively coupled plasma optical emission spectroscopy." *Fresenius' Journal of Analytical Chemistry* Vol. 344 (3):104-108.

Clark, D. E., and E. Lue Yen-Bower. 1980. "Corrosion of glass surfaces." *Surface Science* Vol. 100 (1):53-70.

Cox, G.A., and K.J.S. Gillies. 1986. "The X-Ray Fluorescence analysis of Medieval Durable Blue Soda Glass fron York Minster." *Archaeometry* Vol. 28 (1):57-68.

Del Federico, E., B. Blümich, G. Chiari, C. Garcia Manzano, A. Haber, C.Kehlet, and W. Zia. Forthcoming. "A multi-analytical, non-invasive investigation of the mosaic and nymphaeum at the house of Neptune and Amphitrite, Herculaneum."

Dussubieux, Laure, Richard Cox, Mark Golitko, and Patrick Ryan Williams. 2009. ICP-MS with adaptable chamber laser: quantitative analysis of glass, metal and ceramic artifacts in the Museum environment. Paper presented at the FACSS 2009, at Louisville, KY.

Dussubieux, Laure, Peter Robertshaw, and Micheal D. Glascock. 2009. "LA-ICP-MS analysis of African glass beads: Laboratory inter-comparison with an emphasis on the impact of corrosion on data interpretation." *International Journal of Mass Spectrometry* Vol. 284 (1-3):152-161.

Frána, J., A. Ma_talka, and N. Venclová. 1987. "Neutron Activation Analysis of some Ancient Glasses from Bohemia." *Archaeometry* Vol. 29 (1):69-89.

Freestone, Ian C. 2005. "The provenance of ancient glass through compositional analysis." In *Materials issues in art and archaeology VII: symposium held November 30-December 3, 2004, Boston, Massachusetts, U.S.A*, edited by Pamela B. Vandiver, Jennifer L. Mass and Alison Murray, 195-208. Warrendale, Pa.: Materials Research Society.

Hubbell, J. H., and S. M. Seltzer. 2011. *Tables of X-Ray Mass Attenuation Coefficients and Mass Energy-Absorption Coefficients from 1 keV to 20 MeV for Elements Z = 1 to 92 and 48 Additional Substances of Dosimetric Interest*. The National Institute of Standards and Technology 2011 [cited December 2011]. Available from http://www.nist.gov/pml/data/xraycoef/index.cfm.

Instituut, Rijksuniversiteit te Utrecht. Geologisch. 1957. *Geologica Ultraiectina*: L. van der Velder.

International Atomic Energy Agency. 1960. *Technical reports series*.

Kenyon, I., R.G.V. Hancock, and S. Aufreiter. 1995. "Neutron Activation analysis of AD 1660 - 1930 European Copper-Coloured Blue Glass Trade Beads from Ontario, Canada." *Archaeometry* Vol. 37 (2):323-337.

Koob, S.P. 2006. *Conservation and Care of Glass Objects*. London: Archetype.

Rahman, S., M. Hamzah, A. Wood, M. Elias, and K. Zakaria. 2008. "INAA of ancient glass beads from Sungai Mas archaeological site, Bujang Valley, Malaysia." *Journal of Radioanalytical and Nuclear Chemistry* Vol. 278 (2):271-276.

Rehren, Th. 2000. "Rationales in Old World base glass compositions." *Journal of Archaeological Science* Vol. 27 (12):1225-1234.

Rehren, Th, and A Shugar. 2001. The composition of Egyptian glass as a function of melting temperature. Paper presented at the XIX International Congress on Glass, at Edinburgh.

Sanders, D. M., and L. L. Hench. 1973. "Mechanisms of Glass Corrosion." *Journal of the American Ceramic Society* Vol. 56 (7):373-377.

Sanderson, D.C.W., and J.B. Hutchings. 1987. "The Origins and Measurement of Colour in Archaeological Glasses." *Glass Technology* Vol. 28:99-195.

Sayre, E.V., and R.W. Smith. 1961. "Compositional Categories of Ancient Glass." *Science* Vol. 133:1824-1826.

Shugar, A. N. 2000. "Byzantine Opaque Red Glass Tesserae from Beit Shean, Israel." *Archaeometry* Vol. 42 (2):375-384.

Shugar, A., and Th Rehren. 2002. "Formation and composition of glass as a function of firing temperature." *Glass Technology* Vol. 43C:145-150.

Speakman, R. J., and M. D. Glascock. 2007. "ACKNOWLEDGING FIFTY YEARS OF NEUTRON ACTIVATION ANALYSIS IN ARCHAEOLOGY." *Archaeometry* Vol. 49 (2):179-183. doi: 10.1111/j.1475-4754.2007.00294.x.

Susak Pitzer, Angela. 2012. Chemical Analysis of Roman Glass from Karanis Using X-Ray Fluorescence Spectrometry. Los Angeles: UCLA.

Thomsen, Volker, Debbie Schatzlein, and David Mercuro. 2005. "Tutorial: Attenuation of X-Rays By Matter." *Spectroscopy* Vol. 20 (9):22-25.

Wagner, Barbara, Anna Nowak, Ewa Bulska, Jerzy Kunicki-Goldfinger, Olivier Schalm, and Koen Janssens. 2008. "Complementary analysis of historical glass by scanning electron microscopy with energy dispersive X-ray spectroscopy and laser ablation inductively coupled plasma mass spectrometry." *Microchimica Acta* Vol. 162 (3):415-424. doi: 10.1007/s00604-007-0835-7.

Walton, M. S., A. Shortland, S. Kirk, and P. Degryse. 2009. "Evidence for the trade of Mesopotamian and Egyptian glass to Mycenaean Greece." *Journal of Archaeological Science* Vol. 36 (7):1496-1503.

Weyl, W.A. 1959. *Coloured Glasses*. London: Dawson's of Pall Mall.

Contributors (alphabetically)

Jim Aimers (State University of New York, Geneseo)
jim_aimers@yahoo.com

Timothy Barrett (University of Iowa),
timothy-barrett@uiowa.edu

Aniko Bezur (The Museum of Fine Arts, Houston),
aniko.bezur@yale.edu

Francesca Casadio (Art Institute of Chicago),
fcasadio@artic.edu

Mary Kate Donais (Saint Anselm College),
MDonais@Anselm.edu

Dori Farthing (State University of New York, Geneseo)
farthing@geneseo.edu

Jeff Furgeson (University of Missouri),
fergusonje@missouri.edu

David George (Saint Anselm College),
DGeorge@Anselm.Edu

Bruce Kaiser (Bruker Elemental),
Bruce.Kaiser@bruker-elemental.net

Art Kaplan (Getty Conservation Institute),
AKaplan@getty.edu

Joseph Lang, (University of Iowa),
joseph-lang@uiowa.edu

Jennifer Mass (Winterthur Museum),
jmass@winterthur.org

Catherine Matsen (Winterthur Museum)
cmatsen@winterthur.org

Chris McGlinchey (Museum of Modern Art),
Chris_McGlinchey@moma.org

Hector Neff (California State University Long Beach),
hneff@csulb.edu

Federico Paredes Umaña (University of Pennsylvania),
afe@sas.upenn.edu

Catherine Schmidt Patterson (Getty Conservation Institute),
CPatterson@getty.edu

Robert Shannon (Bruker-Elemental),
Robert.Shannon@bruker-elemental.net

Aaron Shugar (Buffalo State College),
shugaran@buffalostate.edu

P. Jane Sirois (Canadian Conservation Institute),
Jane_Sirois@pch.gc.ca

Dylan Smith (National Gallery of Art),
DT-Smith@NGA.GOV

Dusan Stulik (Getty Conservation Institute),
DStulik@getty.edu

Karen Trentelman (Getty Conservation Institute),
ktrentelman@getty.edu

Nancy Turner (J. Paul Getty Museum),
nturner@getty.edu

Barbara Voorhies (University of California, Santa Barbara),
voorhies@anth.ucsb.edu

Jennifer Wade (National Science Foundation)
jwad@loc.gov

The Editors

Aaron N. Shugar is an Associate Professor of Art Conservation Science at Buffalo State College, Buffalo, New York.

Jennifer L. Mass is a Senior Scientist in the Scientific Research and Analysis Laboratory Conservation Department at Winterthur Museum, Winterthur, Delaware.

Previously published in the series 'Studies in Archaeological Sciences'

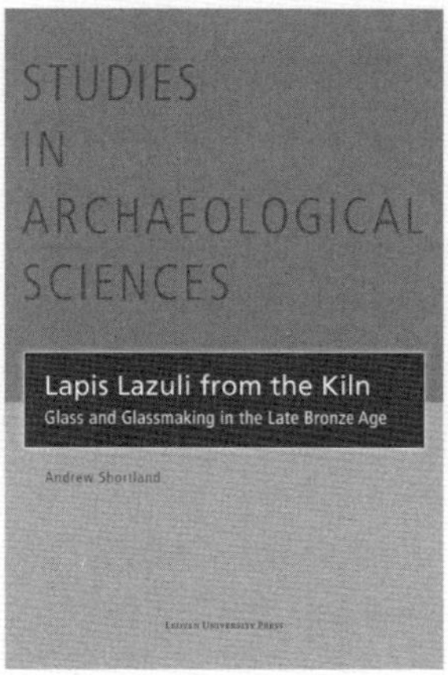

Lapis Lazuli from the Kiln. Glass and Glassmaking in the Late Bronze Age

Andrew Shortland

Lapis Lazuli from the Kiln examines the history of the first glass, from its early sporadic occurrence, through the height of its production in the late second millennium BC, to its disappearance at the end of that millennium. ISBN 978 90 5867 691 7, hardback, 260 p.

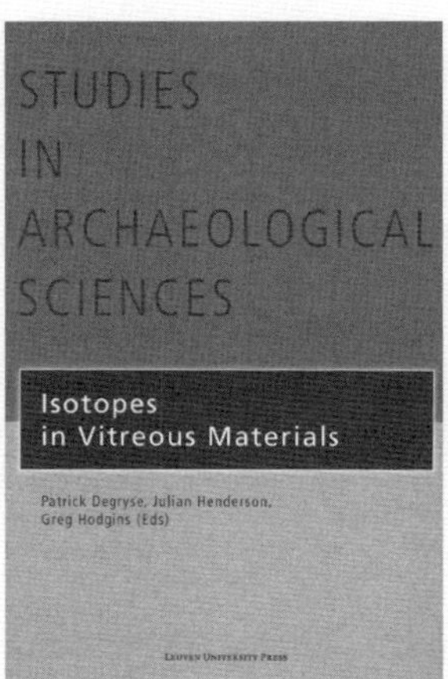

Isotopes in Vitreous Materials

Patrick Degryse, Julian Henderson, Greg Hodgins (eds)

Isotopes in Vitreous Materials includes methodological papers on isotope analysis, innovative applications of several isotope systems to current questions in glass and glaze research, and advances in the knowledge of the economy of vitreous materials. ISBN 978 90 5867 690 0, hardback, 166 p.

For more information visit www.lup.be.

New Online Journal

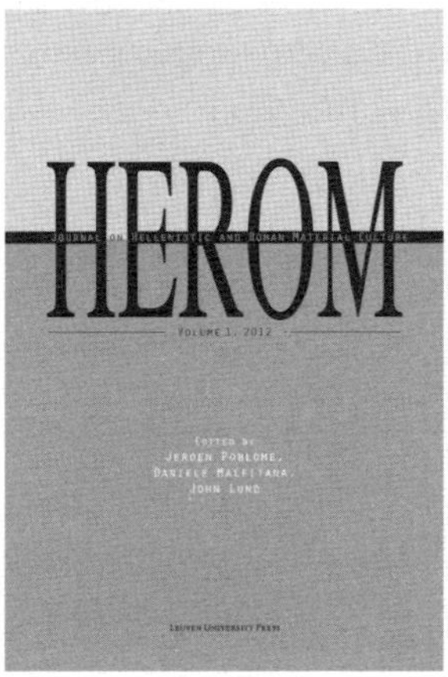

HEROM.
Journal on Hellenistic and Roman Material Culture
Jeroen Poblome, Daniele Malfitana,
John Lund (eds)

Includes new and innovative studies on Hellenistic and Roman material culture in all relevant academic disciplines

HEROM is an annual peer-reviewed online journal presenting innovative contributions to the study of material culture produced, exchanged, and consumed within the spheres of the Hellenistic kingdoms and the Roman world. The journal publishes papers in the full range of the scholarly field and in all relevant academic disciplines within the arts, humanities, social sciences and environmental sciences. *HEROM* creates a bridge between material culture specialists and the wider scientific community, with an interest in how humans interacted with and regarded artefacts from the late 4th century BC to the 7th century AD.

A Scientific Committee with years of expertise
The journal and scientific committee build on their experience of FACTA.
A Journal of Roman Material Culture Studies (2007-2011).

HEROM is published annually.
Online ISSN: 2294-4281
Print ISSN: 2294-4273

Journal available online at www.ingentaconnect.com/content/LUP/HEROM

For ordering information and current subscription rates contact orders@lup.be or visit www.herom.be.

About LEUVEN UNIVERSITY PRESS

Leuven University Press, established in 1971 under the auspices of KU Leuven, is an ambitious academic press of international standing.

Today the press has over a thousand books in print, in a broad range of fields including music, art & theory, text & literature, history & archaeology, philosophy & religion, society, law & economics.

All LUP publications are published with care and attention to detail. Prior to publication, all manuscripts are assessed by an independent editorial board and external specialist readers to ensure the highest academic standards. Combined with a service-oriented and personal approach this peer review process defines our publishing policy.

Leuven University Press is a full member of the Association of American University Presses and a founding member of the Association of European University Presses.

Visit the website www.lup.be for more information about the titles and publishing policy of Leuven University Press.

Contact

Leuven University Press
Minderbroedersstraat 4, box 5602
3000 Leuven
Belgium
T +32 (0) 16 32 53 45
F +32 (0) 16 32 53 52
E info@lup.be
W www.lup.be

Ordering information

Individuals

All Leuven University Press publications are available from your local bookshop or can be ordered directly by mail, fax, or e-mail at orders@lup.be or online at www.lup.be.

Bookshops and libraries

Orders may be sent by mail, fax, or e-mail to orders@lup.be.
Belgian bookshops can also order through boekenbank.be.

Customers in the Netherlands

Bookshops in the Netherlands can order through CB: www.cb-logistics.nl

Customers in France

Bookshops

CID
FMSH-Diffusion
18–20, rue Robert-Schuman
CS 90003
94227 Charenton-le-Pont Cedex, France
Tel +33 (0) 1 53 48 56 30
Fax +33 (0) 1 53 48 20 95
cid@msh-paris.fr

Individual customers

In all bookshops and at Comptoir des presses d'universités, 86, rue Claude Bernard 75005 Paris, France
Tel +33 (0) 1 47 07 83 27
www.lcdpu.fr

Customers in North America and Canada

All English-language publications of Leuven University Press are available through:
Cornell University Press Services
P.O.Box 6525
750 Cascadilla Street
Ithaca, ny 14851-6525, usa
orderbook@cupserv.org
www.cornellpress.cornell.edu

Sales Representation in UK, Ireland, Denmark, Sweden, Norway, Finland, and Greece

University Presses Marketing
The Tobacco Factory
Raleigh Rd. Southville
Bristol, bs3 1tf, United Kingdom
Tel +44 (0) 117 902 0275
Fax +44 (0) 117 902 0294
sales@universitypressesmarketing.co.uk
www.universitypressesmarketing.co.uk

Order fulfilment in UK, Ireland, Denmark, Sweden, Norway, and Finland

NBN International
Airport Business Centre
10 Thornbury Road
Plymouth, pl6 7pp, United Kingdom
Tel +44 (0) 175 220 2301
Fax +44 (0) 175 220 2333
orders@nbninternational.com